Advances in Anatomy, Embryology and Cell Biology

Volume 240

AF798377

Editor-in-Chief

Peter Sutovsky, Division of Animal Sciences and Department of Obstetrics, Gynecology and Women's Health, University of Missouri, Columbia, USA

Series Editors

Z. Kmiec, Department of Histology and Immunology, Medical University of Gdansk, Gdansk, Poland

Ahmed Z. Balboula, Division of Animal Sciences, University of Missouri, Columbia, USA

Maria L. Golson, Division of Endocrinology, Diabetes, Metabolism, Johns Hopkins University, Baltimore, USA

Sven Schumann, Medizinische Hochschule Brandenburg, Institut für Anatomie, Neuruppin, Brandenburg, Germany

Advances in Anatomy, Embryology and Cell Biology publishes critical reviews and state-of-the-art research in the areas of anatomy, developmental and cellular biology. Founded in 1891, this book series has a long standing tradition of publishing focused and condensed information on a given topic with a special emphasis on biomedical and translational aspects.

The series is open to both contributed volumes (each collecting 7 to 15 focused reviews written by leading experts) and single-authored or multi-authored monographs (providing a comprehensive overview of their topic of research).

Advances in Anatomy, Embryology and Cell Biology is indexed in Medline, SCImago, SCOPUS.

Albert Barth · Viv E.A. Perry · Lauren E. Hamilton ·
Peter Sutovsky · Richard Oko

Abnormal Morphology of Bovine Spermatozoa

Second Edition

With 156 Figures

Springer

Albert Barth
Department of Large Animal Clinical Sciences, Western College of Veterinary Medicine
University of Saskatchewan
Saskatoon, SK, Canada

Lauren E. Hamilton
Division of Animal Sciences
University of Missouri
Columbia, MO, USA

Richard Oko
Department of Biomedical and Molecular Sciences
Queen's University
Kingston, ON, Canada

Viv E. A. Perry
Queensland Sperm Morphology Laboratory (QSML)
Goondiwindi, QLD, Australia

Peter Sutovsky
Division of Animal Science and Department of Obstetrics, Gynecology & Women's Health
University of Missouri
Columbia, MO, USA

ISSN 0301-5556 ISSN 2192-7065 (electronic)
Advances in Anatomy, Embryology and Cell Biology
ISBN 978-3-031-70125-2 ISBN 978-3-031-70126-9 (eBook)
https://doi.org/10.1007/978-3-031-70126-9

1st edition: © JOHN WILEY & SONS, INC. 1989
2nd edition: © The Editor(s) (if applicable) and The Author(s), under exclusive license to Springer Nature Switzerland AG 2025, Corrected Publication 2025

This work is subject to copyright. All rights are solely and exclusively licensed by the Publisher, whether the whole or part of the material is concerned, specifically the rights of translation, reprinting, reuse of illustrations, recitation, broadcasting, reproduction on microfilms or in any other physical way, and transmission or information storage and retrieval, electronic adaptation, computer software, or by similar or dissimilar methodology now known or hereafter developed.
The use of general descriptive names, registered names, trademarks, service marks, etc. in this publication does not imply, even in the absence of a specific statement, that such names are exempt from the relevant protective laws and regulations and therefore free for general use.
The publisher, the authors and the editors are safe to assume that the advice and information in this book are believed to be true and accurate at the date of publication. Neither the publisher nor the authors or the editors give a warranty, expressed or implied, with respect to the material contained herein or for any errors or omissions that may have been made. The publisher remains neutral with regard to jurisdictional claims in published maps and institutional affiliations.

This Springer imprint is published by the registered company Springer Nature Switzerland AG
The registered company address is: Gewerbestrasse 11, 6330 Cham, Switzerland

If disposing of this product, please recycle the paper.

Preface

Sperm morphology warrants a strong emphasis in evaluating semen for fertility as it is a trait that is highly correlated to fertility of bulls. The intent of this book is to provide a source of information that is of practical use to veterinarians and laboratory personnel who routinely face questions about, and make judgment decisions upon, the fertility of bulls in natural service following bull breeding soundness examination as well as on the fertility of frozen semen. We have also intentionally designed this book to be useful to the inexperienced undergraduate and yet comprehensive enough to serve as a starting point for graduate students and researchers interested in studying spermatology. A requirement for interpreting the significance of abnormal sperm morphology is a basic understanding of the normal development, ultrastructure and composition of spermatozoa of which a full discussion is included in the first few chapters of this book. In the latter chapters, many of the common bull sperm abnormalities are discussed in terms of origin, incidence, light and electron microscopic appearance, pathogenesis, significance to fertility, mitigation, and possible prognosis. The literature seldom indicates at what level a sperm defect is tolerated in semen without seriously affecting fertility. There is also a conflict of opinion as to whether some of the abnormal morphological forms have any significant effect on fertility at all. This book attempts to clarify many of these ambiguities and brings together a unified classification system for bull sperm abnormalities.

Over 35 years have passed since the first edition of this book and refinements in ultrastructural- and light-microscopic sperm imaging together with advances in proteomics are considered in this new edition, with the result of identifying and localizing many unique sperm proteins. Concomitantly, there have been substantial advances in genomic research enabling scientists to ablate specific genes relevant to male gonad function and analyze their reproductive phenotypes. When applicable, attention is drawn to the consequences (on sperm structure and fertility) of inactivating genes that play key roles in sperm formation and production, especially if their phenotypes appear to match common bull sperm abnormalities. Furthermore, this new edition summarizes advances in the use of sperm quality biomarkers and genetic screening tests in the assessment of bull fertility.

In another chapter of the book dedicated to pre- and post-natal development, attention focuses on early gestation and pre-weaning as two critical periods for sexual development of a bull. Finally, a standardized approach for sperm morphology assessment is proposed.

We are indebted to a large number of veterinarians in practice who have continually provided material and raw data from the 'laboratory of the field'. In addition, we acknowledge a large number of cattle producers and pasture managers who have been generous with their supportive efforts, facilities, and observations. A great deal of what is known about fertility of bulls in natural service and in artificial insemination is due to these people. We would like to express our appreciation to Wei Xu for many of the descriptive drawings found in Chapters 1–3 and her technical support throughout the project.

This current work was supported by the Natural Science and Engineering Research Council of Canada (RGPIN/05305) to Richard Oko, along with the Agriculture and Food Research Initiative Competitive Grant number 2023-67015-39262 from the USDA National Institute of Food and Agriculture, Animal Breeding, Genetics & Genomics Program and seed funding from the College of Agriculture, Food and Natural Resources of the University of Missouri to Peter Sutovsky. The Canadian Association of Animal Breeders provided the foundation for research by Albert Barth. They supported the Development of the Diagnostic Spermatology Laboratory at the Western College of Veterinary Medicine, Saskatoon, SK., along with related support for numerous other research directives.

Saskatoon, Canada Albert Barth
Goondiwindi, Australia Viv E. A. Perry
Columbia, USA Lauren E. Hamilton
Columbia, USA Peter Sutovsky
Kingston, Canada Richard Oko

Contents

Chapter 1
The Ultrastructure and Composition of Bovine Spermatozoa

Contents

Abstract This chapter provides a cytological and compositional evaluation of the various compartments and sub-compartments making up the bull spermatozoon. The intention is to define the sperm head and tail compartments from an ultrastructural perspective and attribute to them their protein constituents gathered from both traditional and modern proteomic approaches. Common to both approaches, the compositional analysis is dependent on the fractionation and isolation of the sperm compartments combined with polyacrylamide gel electrophoresis (PAGE) and Western blotting to detect the identities of the proteins, and immunocytochemistry to confirm their residency. As will be appreciated, the identity of a particular sperm protein together with its residency provide valuable insights not only into its role, but also to the role of the specific sperm compartment it occupies, in development and/or fertilization. Attention is also given in this chapter to the consequences (on sperm structure and fertility) of inactivating genes that play key roles in sperm formation, especially if their phenotypes appear to match common bull sperm abnormalities. The keywords below cover the sperm head and tail compartments addressed in this chapter.

Keywords Bull spermatozoa · Ultrastructure · Protein composition · Sperm head · Nucleus · Perinuclear theca · Acrosome · Acrosome reaction · Sperm tail · Axoneme · Outer dense fibers · Fibrous sheath · Mitochonrial sheath · Gene inactivation · Infertility

© The Author(s), under exclusive license to Springer Nature Switzerland AG 2025

A. Barth et al., *Abnormal Morphology of Bovine Spermatozoa*,
Advances in Anatomy, Embryology and Cell Biology 240,
https://doi.org/10.1007/978-3-031-70126-9_1

1.1 Introduction

An understanding of normal sperm structure and composition is compulsory to begin to comprehend the significance of the various bull sperm abnormalities that develop from genetically- or environmentally-induced disturbances of spermatogenesis and sperm maturation in the epididymis. These sperm abnormalities are analyzed in chapters 4—9 of this book after a foundation in normal structure, protein composition, development and maturation of bull spermatozoa is given in chapters 1—3. As acknowledged in this book's predecessor, 'Abnormal Morphology of Bovine Spermatozoa' (Barth and Oko 1989), many researcher's contributions were compiled to obtain a clearer picture of sperm structure, perhaps none so important to bull sperm morphology as the contribution by Saacke and Almquist (Saacke and Almquist 1964b, a). However, over 30 years have passed since the publication of this book and many refinements in sperm ultrastructure have taken place, which we summarize in this chapter. Concomitantly, with the advent of proteomic research, the molecular identity of many different sperm proteins and their compartmentalizations within the sperm provide valuable insights into their roles during fertilization and may help to explain the pathology of some sperm defects. The intention of this chapter is to define the various sperm head and tail compartments and sub-compartments, attributing to them their protein constituents to the best of our ability. As will be evident in some cases the particular characteristics of the proteins making up a sperm compartment help define its function during spermatogenesis or fertilization. At the same time, there have been huge advances in genomic research enabling scientists to perform genetic 'knock-outs' on specific genes of interest and analyze their phenotypes. When applicable, attention is given to the consequences (on sperm structure and fertility) of 'knocking out' genes that play key roles in sperm formation and production, especially if their phenotypes appear to match common bull sperm abnormalities. Since our overriding theme is on bull spermatology, we will use research data from the bovine species whenever possible, however, when deemed appropriate or helpful, information from studies done in laboratory animals or other domestic species will be included.

1.2 General Features of Bull Spermatozoa

The bull spermatozoon is made up of a spatulate- or paddle-shaped head of approximately 9 μm in length and a long tail of 70 μm. Even at the light microscope level, divisions between some of the regions can be discerned such as the half-moon shaped equatorial segment region, which is positioned at the caudal end of the acrosome (Fig. 1.1). The width of the head is approximately 4 μm at the junction of the acrosome and the postacrosomal region and its thickness is less than 1 μm. The tail is made up of four parts or pieces: the connecting-, mid-, principal- and end-piece (Fig. 1.1). Each piece has its distinguishing charateristics and at their core

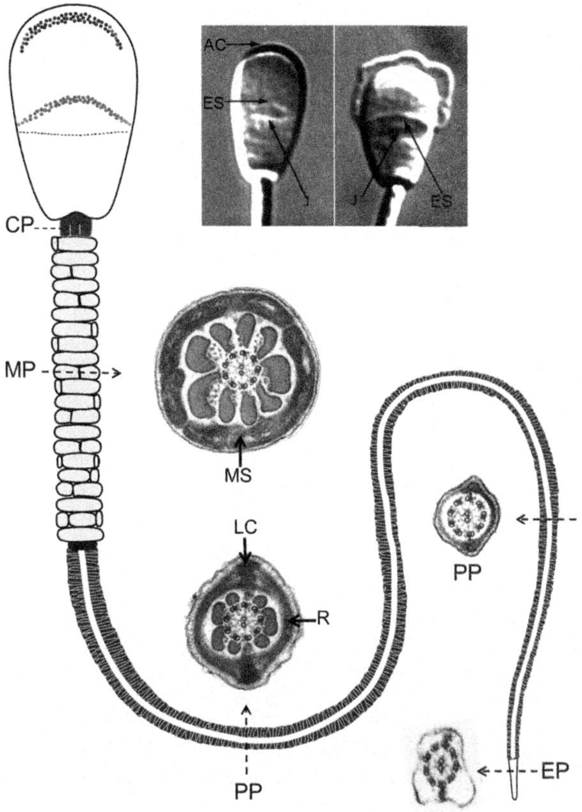

Fig. 1.1 Diagram illustrating the bull spermatozoon with its 9 μm-long, spatulate-shaped head and its 70 μm-long tail. Both the sperm head and tail can be divided up into regions. From a surface perspective seen by differential interference microscopy a junction (J) can be seen dividing the head into an anterior situated acrosome and a posterior situated postacrosomal sheath region. Immediately above this junction is the crescent shaped equatorial segment region whose curved anterior border outlines the beginning of the equatorial segment (ES) and the junction-line its end. The apical crest (AC) is seen in the intact sperm head on the left. In the sperm head on the right, the apical and principal segments of the acrosome begin to disintegrate leaving only the equatorial segment region of the acrosome intact. The four regions of the sperm tail are indicated on the diagram: the connecting piece (CP), joining the head to the tail; the middle piece (MP), showing the transversely orientated mitochondria of the mitochondrial sheath (MS); the principal piece (PP), showing the ribs (R) and longitudinal columns (LC) of the fibrous sheath; and the endpiece (EP) containing only the axoneme, as seen in all cilia. Electron micrographs (EM) seen along the tail of the diagram depict the ultrastructure of the regions as seen in cross-section. These show the central situated axoneme made up of nine microtubular doublets surrounding a central pair of microtubular singlets, the outer dense fibers, the fibrous sheath made up of longitudinal columns and ribs, and mitochondria making up the mitochondrial sheath. The annulus is indicated at the junction of the middle and principal pieces. The plasmalemma, the cytoplasmic membrane surrounding the whole sperm, is seen in the EM cross sections, but not present in the sperm drawing

lies the axoneme, found in all ciliae and flagellae. However, the sperm tail cannot be equated to the latter as it is much more complex, containing additional cytoskeletal elements among which the outer dense fibers and the fibrous sheath are the most prominent. Just as the presence of the fibrous sheath is characteristic of the principal piece, the mitochondrial sheath, made up of helically wound mitochondria surrounding the outer dense fibers, characterizes the midpiece.

1.3 Sperm Head Structure

The basic structural organization of the spermatozoon head is similar among eutherian mammals, with the nucleus and acrosome occupying much of the volume of the sperm head, and the remaining space filled mostly by highly condensed cytosol referred to as the perinuclear theca (Fig. 1.2). The acrosome, which contains receptors and enzymes that allow the sperm to bind and penetrate the outer vestments of the oocyte during fertilization, caps the anterior half of the highly condensed nucleus. It is made up of three segments: the apical-, principal- and equatorial-segment. The former two segments release their contents, in a process called the 'acrosome reaction', just before penetrating the outer vestments of the oocyte, while the equatorial segment (ES) and overlying plasmalemma retain their structural integrity, which insures that membrane fusion with the oocyte first occurs in this region during fertilization. Several reasons are most likely responsible for maintaining the stuctural integrity of the equatorial segment region during the acrosome reaction. First, the inner and outer acrosomal membranes of the ES are in close proximity forming a "zipper-like" structure between each other that is difficult to disrupt (Yu et al. 2009). Secondly, a binding material, refered to as the outer periacrosomal layer, is positioned between the ES and the plasmalemma. This binding material is not found between the acrosome and plasmalemma in the principal and apical segment regions that undergo the acrosome reaction.

The caplike acrosome is derived from the Golgi apparatus during the early steps of spermiogeneis. It is attached to the underlying perinuclear theca (PT) and in the apical segment forms a lateral folding, or ridge, called the apical crest (AC). Within the apical segment of the acrosome the PT protrudes deeply into this structure forming the apical body (AB), which in the past was thought to be a vestige of the perforatorium of murid sperm. This concept has been debunked by research published by the Oko lab (Aul and Oko 2002; Protopapas et al. 2019). In the acrosomal region of the sperm head there are five membranes which need to be considered. Proceeding from the outside to the inside of the sperm head the membranes are plasmalemma, outer acrosomal membrane (OAM), inner acrosomal membrane (IAM) and the nuclear envelope (NE) made up of the outer and inner nuclear membranes (Fig. 1.3).

Sandwiched between the acrosome and the nucleus is a condensed cytosolic protein layer, known as the perinuclear theca (PT), that continues caudally below the acrosome until the basal belt (BB), often termed the posterior ring. The BB results

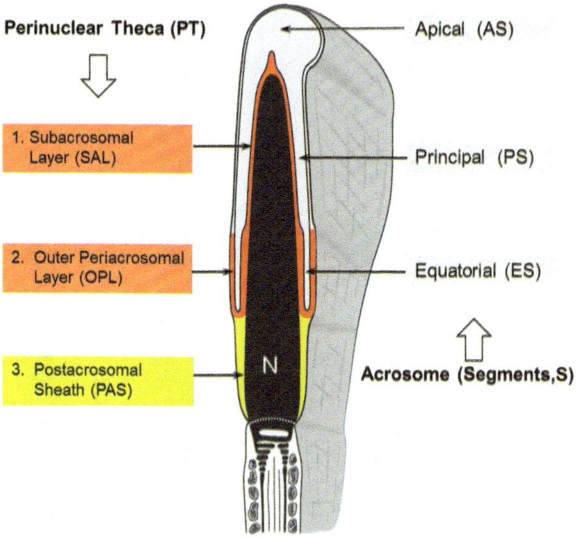

Fig. 1.2 Diagrammatical representation of a mid-sagittal section through the head of a bull spermatozoon in relation to its three-dimensional configuration. The acrosome covers the apical half of the nucleus (N) and is made up of three segments: the apical, principal and equatorial segments. The equatorial segment is the thinned most caudal part of the acrosome and remains intact after the acrosome reaction. Underlying the acrosome is the subacrosomal layer (SAL) of the perinuclear theca (PT) which extends over the outer surface of equatorial segment as the outer periacrosomal layer (OPL). These two layers have the same developmental origin as will be seen in the next chapter and are compositionally similar. The OPL stabilizes the plasmalemma overlying the ES, preserving it for initial fusion with the oolemma of the oocyte during fertilization. Below the acrosome, the SAL is structurally continuous with the postacrosomal sheath (PAS), sandwiched between the plasmalemma and the nucleus. With the exception of a unifying protein, the SAL and PAS are compositionally unique (as the colors indicate) and have different developmental origins as will be seen in the next chapter

from the nuclear envelope and plasmalemma uniting to form a belt-like indentation that separates the cytosolic compartment of the head from the cytoplasm and implantation fossa of the tail. The PT can be divided both stucturally and compositionally into the subacrosomal layer (SAL) and the postacrosomal sheath (PAS) (Fig. 1.2). The SAL not only resides between the IAM and the nuclear envelope but is continuous compositionally with the outer periacrosomal layer (OPL), which is sandwiched between the OAM and the plasmalemma (PM) in the equatorial segment region (ESR) (Figs. 1.2, 1.3 and 1.4)). It is important to make a distinction between the ES and the ESR because many researchers incorrectly use the term ES to refer to the ESR. The ES refers to the caudal segment of the acrosome while the term ESR refers to the entire area within which the ES resides and includes the plasmalemma and OPL overlying the ES and the SAL underlying the ES. The PAS region of the PT begins just below the acrosome and ends at the basal belt (BB) or posterior ring. It lies between the plasmalemma and the nuclear envelope. As

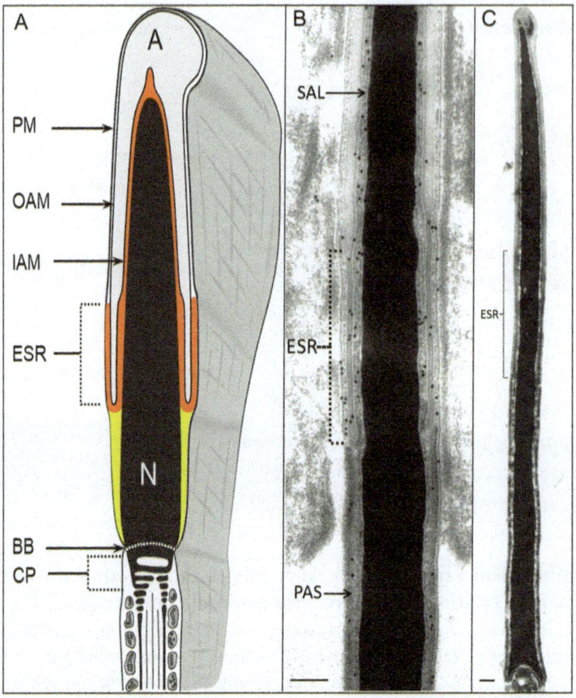

Fig. 1.3 Diagrammatical representation of bull sperm head ultrastructure in (A) is based on ultra-thin sections of bull spermatozoa (B and C) viewed by transmission electron microscopy combined with immuno-gold localization of proteins on these sections. B) A mid-sagittal section, through the equatorial segment region (ESR), immunogold-labeled with an anti-pan perinuclear theca (PT) antibody. The PT proteins as seen by the distribution of the colloidal gold spheres are found throughout the SAL and PAS regions of the PT as well as in the outer peri-acrosomal layer, overlying the equatorial segment of the acrosome. Of all the PT proteins identified so far only calicin has been localized through the entire PT, the rest are either restricted to the SAL (including the OPL) as seen in orange or to the PAS as seen in yellow in the diagrammatical representation. **C)** A mid-sagittal section through the entire head of a bull spermatozoon, immuno-gold labeled with an antibody against histone H2B, that only labels the PAS. However, the intention of this micrograph is to give the viewer a comparison with the diagrammatical sketch in A. Note the toque-shaped apical granule in the apical segment of the acrosome and the relative position of the equatorial segment region in the sperm head. Legends: PM, plasmalemma; OAM, outer acrosomal membrane; IAM, inner acrosomal membrane; ESR, equatorial segment region; BB, basal belt or posterior ring; CP, connecting piece; SAL, subacrosomal layer of the PT; PAS, postacrosomal sheath. Bars = 0.2 μm

will be expained in more detail, most of the sequence-identities of investigated perinuclear theca proteins have been unexpected, revealing novel proteins as well as isoforms of somatic and conventionally nuclear proteins. Their functional characterizations and localizations propose that the perinuclear theca can be regionalized

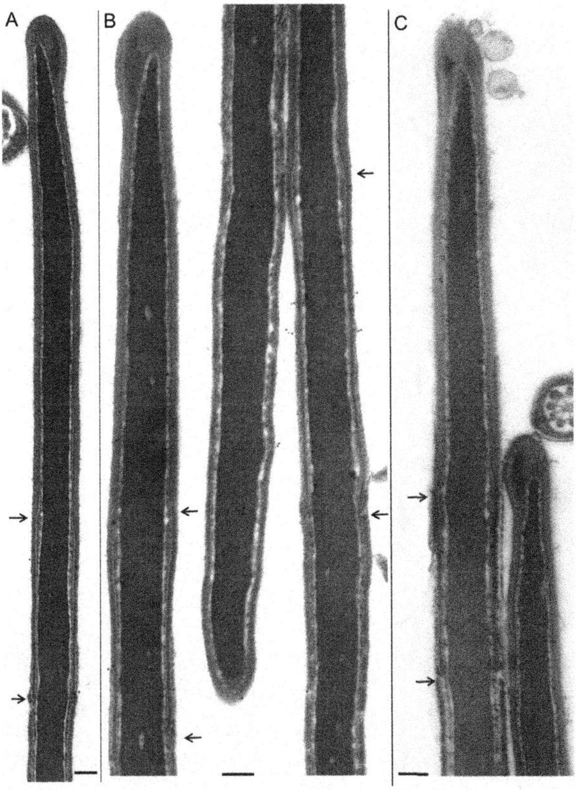

Fig. 1.4 Examples of sagittal sections of the bull sperm head showing various regions and compartments. A-C) Find the apical segment with its toque-shaped granule body, the principal segment, the equatorial segment, the subacrosomal layer, the outer periacrosomal layer, the postacrosomal sheath, and the equatorial segment region between the arrows. **C)** Mid-sagittal sections through portions of the heads of bull spermatozoa, immuno-gold labeled with an antibody against a PT protein that is only found in the subacrosomal layer and outer periacrosomal layer of the perinuclear theca. Bars = 0.2 μm

into two functional categories: the subacrosomal layer, involved in acrosomal assembly during spermiogenesis (Aul and Oko 2002; Mountjoy et al. 2008; Tran et al. 2012), and the postacrosomal sheath, involved in sperm-egg interactions during fertilization such as egg activation and sperm nuclear decondensation and stabilization (Ferrer et al. 2012b; Oko and Sutovsky 2009; Hamilton et al. 2019; Tovich and Oko 2003; Wu et al. 2007a). There is little support for the PT of mature spermatozoa having a traditional cytoskeletal composition.

The inner and outer nuclear membranes continue caudally from the condensed base of the nucleus of the spermatozoa to form a diverticulum, referred to as the

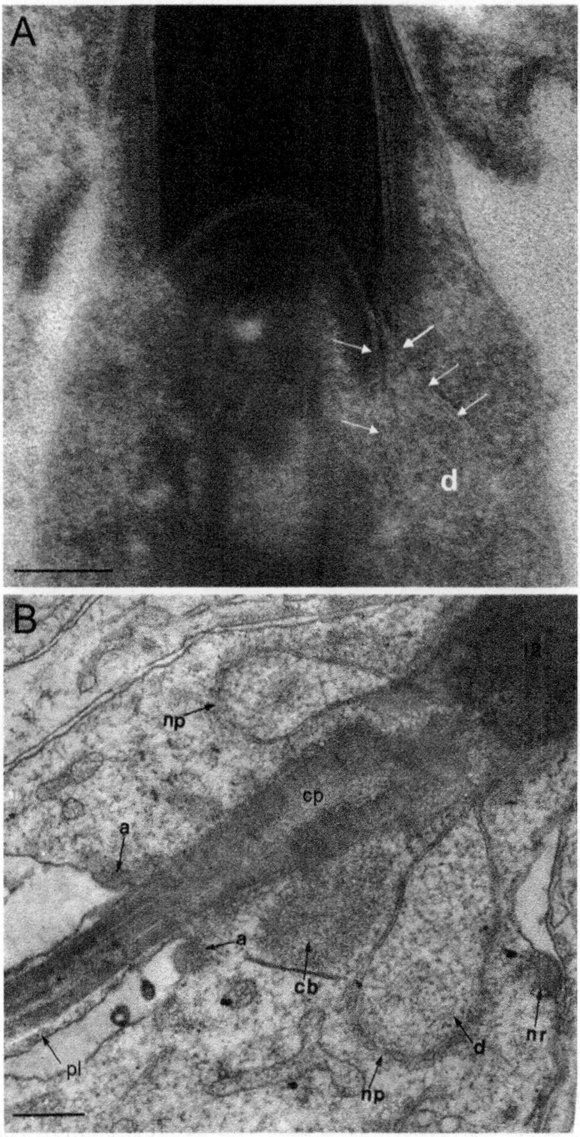

Fig. 1.5 Base of elongated spermatids showing the nuclear envelope (NE) forming a diverticulum termed the redundant nuclear envelope (RNE). A) Spermatid before spermiation showing path of the double membrane NE (arrows) forming the diverticulum (d) of the RNE. Because the thin section is non-osmicated the membranes of the NE appear as white lines with an electron dense material between them. Coming off the diverticulum the NE reflects back onto the base of the nucleus where it bonds with the basal plate of the implantation fossa. **B)** Spermatid nearing the end of nuclear condensation in step 12 of bull spermiogenesis. Here it can be seen that the RNE is studded with nuclear pores. The annulus (a) at the entry of the periaxonemal compartment has not yet migrated distally along the midpiece to expose the outer dense fibers (ODF) to the mitochondria. Legends: cp, connecting piece; cb, chromatoid body; nr, nuclear ring of manchette; pl, plasmalemma. Bars = 0.2 μm

redundant nuclear envelope (RNE) (Fig. 1.5A) On their return to the base of the nucleus they integrate with the basal plate of the implantation fossa (I) that forms the foundation for attachment of the sperm head to the connecting piece of the sperm tail. The RNE is studded with a myriad of nuclear pores and may contain nuclear material that is less electron dense than that of the nucleus proper (Fawcett 1975; Fawcett and Chemes 1979; Mortimer and Thompson 1976; Sutovsky et al. 1999; Tovich et al. 2004; Tran et al. 2012). The nuclear pores migrate down to the base of the spermatid nucleus early in spermiogenesis (Sutovsky et al. 1999; Tovich et al. 2004) and become part of the excess of nuclear envelope (i. e., RNE) that results from the condensation of the spermatid nucleus during the elongation phase of spermiogenesis (Fig. 1.5B). It is important to note that these nuclear pores in the RNE signify that an exchange of molecules between the nucleoplasm of the head and the cytoplasm of the tail could potentially take place.

1.4 Sperm Head Composition

1.4.1 Nucleus

DNA in mammalian sperm nuclei is bound to basic nuclear proteins to form the deoxyribonuclear-protein complex (DNP) called chromatin. The diffuse chromatin characteristic of genetically active germ cells becomes highly condensed and biochemically inert in the mid to late steps of spermiogenesis. Somatic and/or testis specific histones occurring in the spermatogonia and spermatocytes are substituted in these spermatid steps by very basic proteins (i. e., transition proteins followed by protamines) that are synthesized in the cytoplasm during spermiogenesis (Cho et al. 2003, 2001; Meistrich et al. 2003; Oko et al. 1996; Shirley et al. 2004). The mRNA used for this synthesis is produced during the later phases of meiosis and early haploid development. Transition proteins initiate the displacement of histones and the condensation of the spermatid DNA. In turn the transition proteins are displaced by the more basic protamines which complete the condensation process.

Normally the DNA in the chromosomes of somatic cells and developing germ cells is packaged into nucleosomes in which the DNA is wrapped twice around an octamer of histones, every 200 base pairs. The nucleosomes are then further coiled into repeating solenoids, of six nucleosomes per turn, that attach at intervals of about 60 kb to the nuclear matrix at their bases to form solenoid loop domains (Ward 1993). These solenoid loop domains are contained within the nucleus, but when the histones are extracted, they can be visualized outside the nucleus. On the other hand, during spermatid nuclear condensation, histones are removed and eventually replaced by protamines, resulting in sperm nuclei that are approximately 6 times as condensed as somatic cells. Balhorn proposed that protamines are bound to the DNA by lying lengthwise inside the minor groove (Balhorn

1982). He calculated that there were enough arginine groups in the protamines to completely neutralize the negatively charged phosphate groups on the DNA. As a result, the normal electrostatic repulsion between neighboring DNA segments is eliminated and the molecules become tightly packed. The protamine-DNA complex of one strand of DNA fits into the groove of a neighbouring DNA strand so that the DNA strands of the sperm nucleus would be packaged side by side in a linear array. The chromatin is then stabilized by intermolecular and intramolecular covalent disulfide bonds between the protamines. In this manner the whole DNA can be tightly packaged in a small volume. But to insure that packaging of the entire haploid genome could fit into the sperm nucleus, Ward's research uncovered a complementary model refered to as the doughnut-loop model (Ward 1993). In this model, the protamine-bound DNA promotes coiling of the DNA into concentric circles which then collapse together into a doughnut in which the circles of one DNA complex become tightly packed together by Van der Waal's forces. Each doughnut would therefore represent one DNA loop domain attached to the sperm nuclear matrix and the neighboring protamines would form covalent bonds between each other to stabilize the chromatin.

1.4.2 Perinuclear Theca (PT)

The structural integrity and continuity of the PT (i. e., SAL and PAS) are retained after non-ionic detergent extraction of sonicated and isolated sperm heads (Fig. 1.6). Essentially, all that remains is a shell of PT material surrounding the condensed nucleus. This head fraction serves as an ideal starting point for selectively extracting the PT proteins (Ferrer et al. 2012b; Mountjoy et al. 2008; Oko and Maravei 1994; Oko 1995). Sequential extraction steps of this demembranated fraction, first in in 1M NaCl (or KCl), which extracts the PT-anchored histones (Tovich and Oko 2003) and secondly in 100 mM NaOH (Oko and Maravei 1994), which extracts the bulk of PT proteins, allows for their resolution on SDS-PAGE (Fig. 1.6). Antibodies raised against these two extracts only immunolabel the PT suggesting the selectivity of the extraction process (Fig. 1.3). To date, by a variety of proteomic approaches, the sequence identities and localizations of a multitude of PT proteins have been discovered (Aarabi et al. 2014; Aul and Oko 2002; Hamilton et al. 2019; Heid et al. 2002; Herrada and Wolgemuth 1997; Hess et al. 1993; Hess et al. 1995; Hurst et al. 1998; Kann et al. 1991; Leclerc and Goupil 2000; Lécuyer et al. 2000; Mountjoy et al. 2008; Oko and Morales 1994; Tovich and Oko 2003; Tran et al. 2012; von Bülow et al. 1995; von Bülow et al. 1997; Wu et al. 2007a). Antibodies raised against most of these PT proteins label specific protein bands on Western blots of NaOH extracts of the demembranated sperm heads (Fig. 1.7) and localize to either one of the two regions of the PT (i. e., SAL or PAS), or exceptionally to both regions. Although not exhaustive, Table 1.1 lists the types of different PT proteins and their localizations.

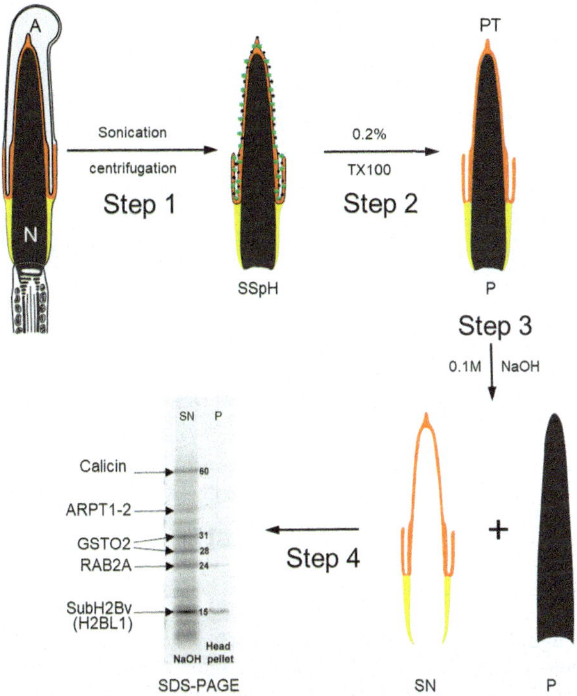

Fig. 1.6 Schematic summary of approach to selectively extract perinuclear theca (PT) proteins from sonicated and isolated sperm heads. After sonication of sperm and separation of heads from tails in a sucrose gradient (Step 1) the sonicated and isolated sperm heads (SSpH) can be incubated with non-ionic detergents (e. g., 0.2% Triton X-100 or NP40) to extract all membranes and membrane-associated proteins (Step 2). Following centrifugation the membrane extracted-sperm head pellet (P), consisting of the condensed nucleus and intact PT (orange and yellow in diagram), can be incubated with an alkaline solution such as 100 mM NaOH (Step 3) to solubilize and extract the PT proteins from the condensed nucleus. After centrifugation the supernatant containing the solubilized PT proteins can be analyzed for protein content by SDS–polyacrylamide gel electrophoresis (SDS-PAGE) and compared to protein content of the pellet, consisting of the sperm nucleus (Step 4). The most prominent proteins of the PT have been identified as can be seen in the SN lane

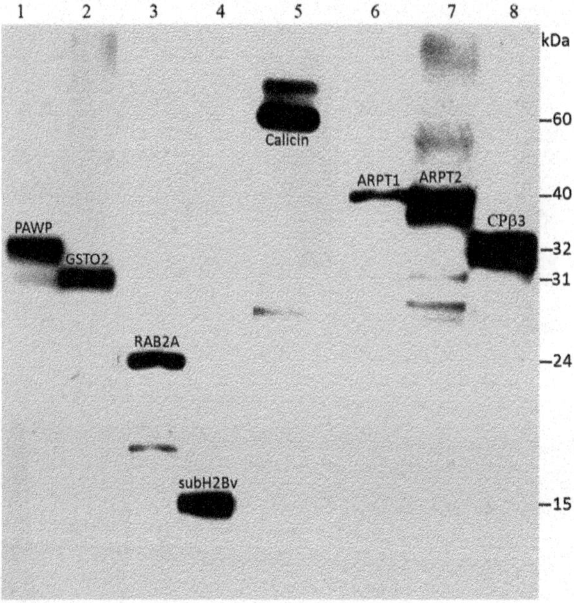

Fig. 1.7 Western blot of NaOH-extracted bull sperm PT proteins labelled with antibodies specific to various well characterized PT proteins. Antibodies: Lane 1: anti-PAWP/WBP2NL, Lane 2: Anti-GSTO2, Lane 3: anti-RAB2, Lane 4: anti-SubH2Bv/H2BL1, Lane 5: anti-Calicin, Lane 6: anti-ARPT1, Lane 7: anti-ARPT2, Lane 8: anti-CPβ3. See Table 1.1 for complementary description

Table 1.1 (A) Perinuclear Theca (PT) Proteins Identified by Dr. Oko's Group.

Identity	kDa	Type of Protein	Location in PT	References
PERF 15	15	Fatty acid Binding Family (only in murids)	SAL*	Oko and Morales 1994, Dev. Biol
SubH2Bv/ H2BL1 (PT15)	15	Histone variant, containing bipartite nuclear localization signal	SAL	Aul and Oko 2002, Dev. Biol. and Tran et al. 2011, Biol. Reprod
RAB2A (PT24)	24	Vesicular Transport	SAL	Mountjoy et al. 2008, Biol. Reprod
KPNA6	60	Karyopherin or Importin alpha	SAL	Tran et al. 2012, Biol. Reprod
KPNB1	75	Karyopherin or Importin beta	SAL	Tran et al. 2012, Biol. Reprod
Calicin (PT60)	60	Basic & Structural	SAL & PAS**	Oko et al. 2001, Andrology In The 21st Century

Identity	kDa	Type of Protein	Location in PT	References
PAWP/ WBP2NL (PT32)	32	WW Domain Binding Protein (Signal transduction)	PAS	Wu et al. 2007a, JBC
GSTO2 Isoform (PT31)	31	Glutathione S-transferase	PAS	Hamilton et al. 2017, 2019, Biol. Reprod
GSTO2 Isoform (PT28)	28	Glutathione S-transferase	PAS	Hamilton et al. 2017, 2019, Biol. Reprod
H3	19	Somatic Core Histones	PAS	Tovich and Oko 2003, JBC
H2B	17.5	Somatic Core Histones	PAS	Tovich and Oko 2003, JBC
H2A	16.5	Somatic Core Histones	PAS	Tovich and Oko 2003, JBC
H4	14	Somatic Core Histones	PAS	Tovich and Oko 2003, JBC

* Subacrosomal Layer
** Postacrosomal Sheath

(b) Perinuclear Theca Proteins Identified by Dr. Franke's Group

Identity	kDa	Type of Protein	Location in PT	References
Cylicin I	74	Basic & Structural	PAS	Hess et al. 1993, JCB
Cylicin II	58	Basic & Structural	PAS	Hess et al. 1995, Exp. Cell Res
Calicin (PT60)	60	Basic & Structural	PAS	vonBulow et al. 1995, Exp. Cell Res
CPα3	31	Actin Capping α3	PAS	vonBulow et al. 1997, Exp. Cell Res
CPβ3	37	Actin Capping β3	PAS	vonBulow et al. 1997, Exp. Cell Res
ARPT1	40	Actin related Protein T1	PAS	Heid et al. 2002, Exp. Cell Res
ARPT2	40	Actin related Protein T2	PAS	Heid et al. 2002, Exp. Cell Res

Figure 1.8 depicts the location of bovine PT proteins whose characterizations have contributed to understanding the roles of the two different regions of the PT. The SAL proteins not only reside beneath the acrosome (between the IAM and NE) but also continue around the outer aspect of equatorial segment of the acrosome (between the OAM and plasmalemma) in the region referred to as the OPL. Ultrastructural immunocytochemistry has shown that antibodies raised against these SAL proteins invariably immunolabel the OPL as well (Fig. 1.9).

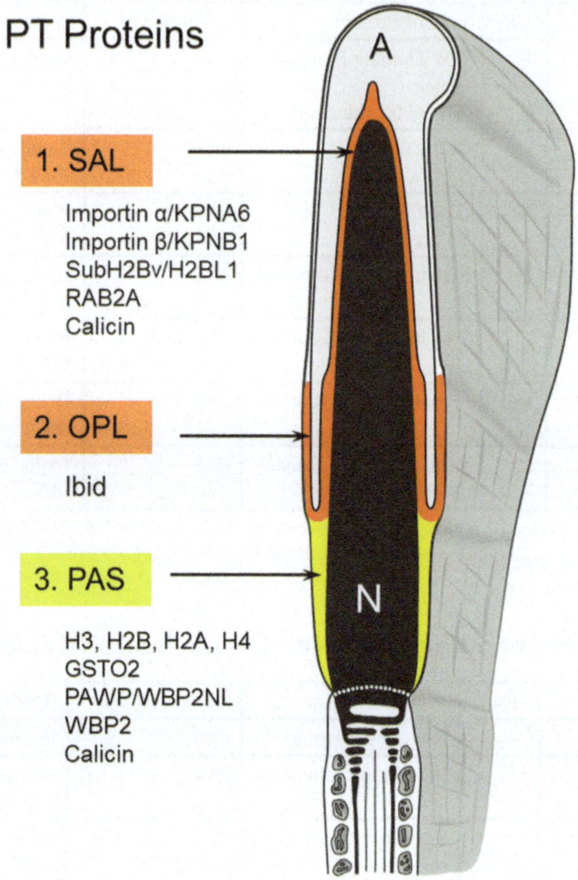

Fig. 1.8 Localization of bovine PT proteins, the characterizations of which have led to an understanding of the different roles the subacrosomal (SAL) and postacrosomal sheath (PAS) regions of the PT play during spermiogenesis and zygotic development, respectively. Of the various proteins listed only calicin, is a constituent of the entire PT and may play a major structural role in unifying the two compartments

Developmental analyses of the SAL proteins have led to the conclusion that these cytosolic proteins get to their peri-acrosomal destination by coating, guiding and attaching the acrosomic vesicle to the spermatid nucleus (Aul and Oko 2002; Oko et al. 2011; Oko and Sutovsky 2009; Tran et al. 2012; Mountjoy et al. 2008). In other words, they are essential for acrosomal-nuclear docking which is a cellular process unique to spermiogenesis and not found in somatic cells. The pathways of assembly giving rise of the different regions of the PT during spermiogenesis are illustrated and discussed in detail in Chap. 2.

The PAS of the PT begins at the end of the equatorial segment of the acrosome and continues caudally between the plasmalemma and nuclear envelope until the

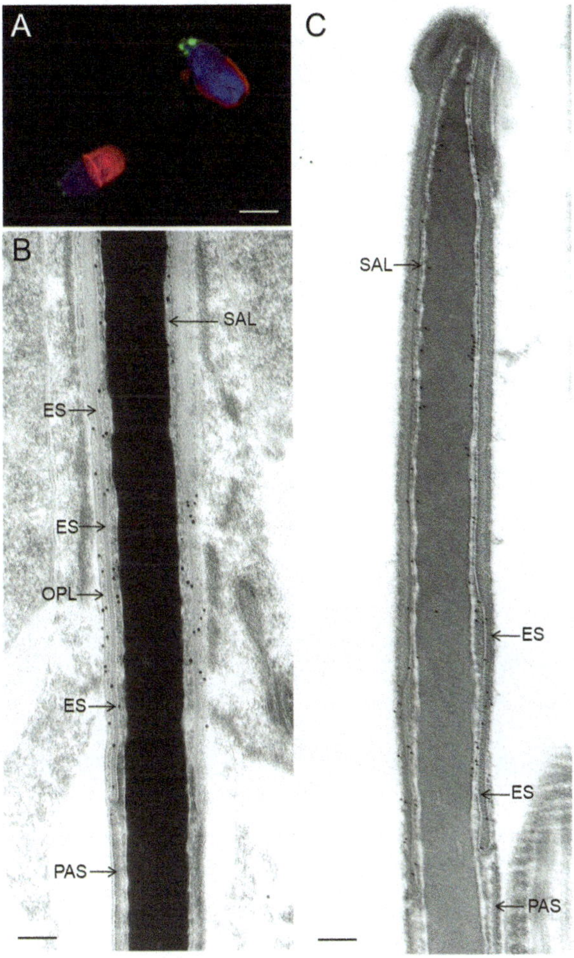

Fig. 1.9 Immunofluorescent- and Immunogold-labelling patterns of bull spermatozoa with antibodies raised against PT proteins residing in the subacrosomal layer (SAL). A) Note that only the apical half of the sperm head is immunoreactive (red) to the anti-SAL antibody (anti-subH2Bv). Blue is the nucleus stained with DAPI and green are nuclear pores labelled with an antibody against nucleoporins. In the lower sperm head the equatorial segment region can be clearly identified. Bar = 5 µm. **B,C)** At the ultrastructural level the immunogold labelling with antibodies raised against two different SAL proteins, in B (SubH2Bv) and C (RAB2A), can clearly be seen in the subacrosomal layer and around the lip of the equatorial segment (ES) in the outer periacrosomal layer (OPL). Note that labelling in the OPL does not continue above the ES and is not found in the postacrosomal sheath (PAS). Bar = 0.2 µm

basal belt or posterior ring (Figs. 1.8 and 1.10). This cytosolic sheath is assembled towards the end of spermiogenesis, coincident with the final steps in the condensation of the caudal half of the nucleus (Tovich et al. 2004; Wu et al. 2007b; Barth and

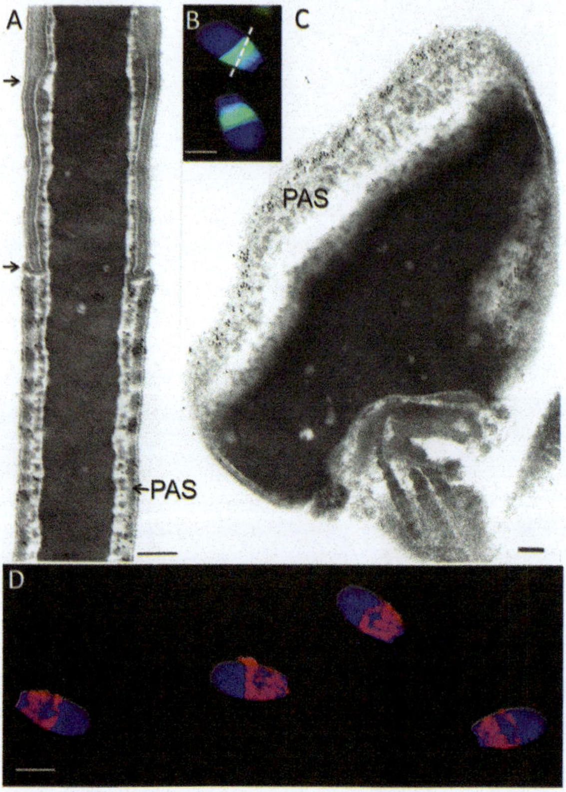

Fig. 1.10 Immunogold- and immunofluorescent-labelling of bull spermatozoa with antibodies raised against PT proteins residing in the postacrosomal sheath (PAS). A) Mid-sagittal section through portion of sperm head showing immunogold labelling with anti-PAWP antibody is restricted to PAS region. No labeling is evident in the subacrosomal layer of the PT. Equatorial segment region is indicated between arrows. Bar = 0.2 μm. **B)** Immunofluorescent labelling (green) with the same antibody covering the caudal half of the sperm head, over the postacrosomal sheath region. The dashed line in the upper sperm head indicates at what level the oblique ultra-section resides in C. Bar = 5 μm. **C)** Oblique section grazing through PAS, immunogold labelled with anti-PAWP antibody. Bar = 0.5 μm. **D)** Immunofluorescent labelling (red) with anti-GSTO2 antibody. Note that labelling is also found in the equatorial segment region above the postacrosomal sheath. The tell-tale sign is the crescent shaped labelling above the postacrosomal sheath region – compare with B. Bar = 5 μm

Oko 1989). In addition to having an obvious structural role, the PAS houses proteins that play essential roles in sperm-egg interactions during fertilization and zygotic development. These interactions include oocyte activation during fertilization (Aarabi et al. 2014; Sutovsky et al. 2003; Wu et al. 2007a), nuclear decondensation during zygotic development (Hamilton et al. 2017, 2019) and chromatin stabilization of the male pronucleus during zygotic development (Hamilton et al. 2021).

1.4.3 The Acrosome and Acrosome Reaction

The acrosome is a secretory vesicle that forms a cap-like structure over the sperm nucleus (Clermont et al. 1994; Clermont and Tang 1985; Tang et al. 1982; Thorne-Tjomsland et al. 1988). It is bound to the nucleus by the perinuclear theca, which intervenes between the nuclear envelope and the inner acrosomal membrane (Oko and Maravei 1994; Oko and Sutovsky 2009). The interior of the acrosome is compartmentalized and contains a variety of proteins including several protease zymogens of which proacrosin/acrosin is best-characterized (Baba et al. 1994; Gerton 2002; Ohmura et al. 1999; Tulsiani et al. 1998; Yamagata et al. 1998; Yu et al. 2009; Toshimori 1998).

Mammalian fertilization requires penetration of the sperm through the zona pellucida (ZP), an extracellular coat of the egg (Wassarman 1999; Yanagimachi 1994). In order for this to occur, either passage through the cumulus oophorus, a follicular cell layer surrounding the ZP, or primary binding with the ZP induces the sperm to undergo the acrosome reaction (Wassarman et al. 2001; Yanagimachi 2011). The acrosome reaction is a form of exocytosis, where multiple fusions occur between the outer acrosomal membrane (OAM) and overlying plasma membrane allowing for the release of acrosomal contents (Fig. 1.11). After completion of the acrosome reaction, most of the inner acrosomal membrane (IAM) is exposed, except in the region of the equatorial segment where the OAM and the plasmalemma retain their integrity and connect to form a continuous membrane (Fig. 1.12). For many years, it was believed that the released acrosomal contents, of which acrosin was the major serine protease, would digest a path for the sperm through the ZP. However, three main experiments stand in the way of this hypothesis. First, the results of an acrosin gene knock-out showed conclusively that mouse sperm do not require acrosin for ZP penetration, although the rate of ZP penetration is affected (Baba et al. 1994). Secondly, it was observed that most fertilizing mouse spermatozoa already undergo the acrosome reaction and release their acrosomal contents in the cumulus oophorus before reaching the ZP (Jin et al. 2011). Thirdly, and most importantly, documentation of a secondary and more adhesive binding step (involving the exposed IAM and the ZP surface), occurring after the acrosome reaction and after the dispersal of acrosomal contents, precludes the possibility that the released contents predigest a path in the ZP (Bleil et al. 1988; Mortillo and Wassarman 1991). Although there are those who argue in favor of motility being sufficient to enable a sperm to push its way through the ZP (Bedford 1998), the calculated force generated by the sperm does not appear strong enough to penetrate the ZP (Green 2002) and trypsin inhibitors, which block sperm-zona penetration, do not block motility (Yamagata et al. 1998).

A consequence of acrosomal exocytosis and the release of acrosomal contents, is the exposure of the IAM (Figs. 1.11, 1.12 and 1.13), allowing it to directly interact with the ZP2 protein on the ZP, a compulsory step for subsequent IAM-directed sperm penetration through the ZP (Huang and Yanagimachi 1985; Wassarman 1988; Yanagimachi 1994). The documentation of this secondary binding step by

Wassarman and colleagues (Bleil et al. 1988; Mortillo and Wassarman 1991) stimulated a search for the protein receptor on the IAM that interacts with the zona. Several acrosomal proteins were proposed, including proacrosin/acrosin (Jones

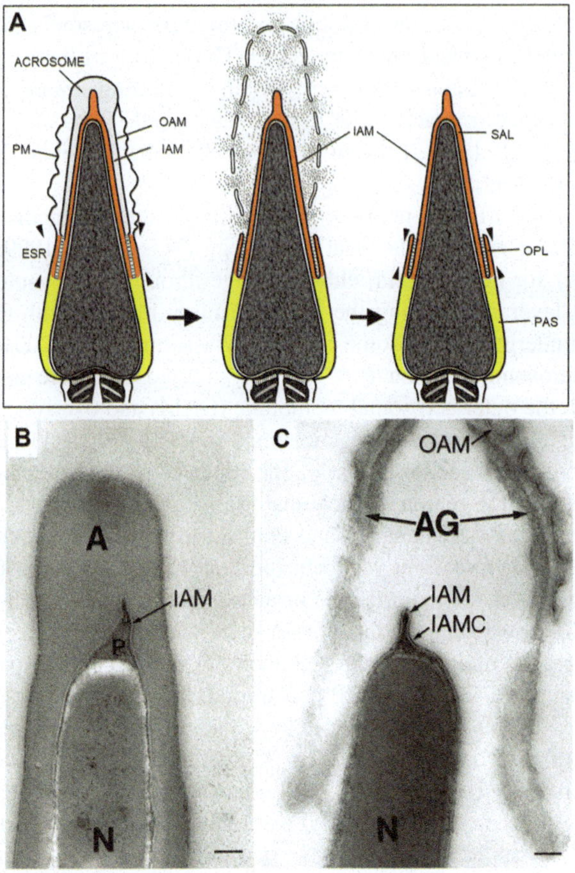

Fig. 1.11 A) Schematic representation of sperm head during the acrosomal reaction leading to the exposure of the inner acrosomal membrane. Spot fusions of the plasma membrane (PM) and the outer acrosomal membrane (OAM) in the apical and principal segments of the acrosome result in hybridization of these membranes to produce multiple vesicles. Vesiculation in turn allows for the release of the acrosomal contents and the exposure of the inner acrosomal membrane (IAM). Note that the equatorial segment region (ESR, between the two arrowheads) remains intact and the PM and OAM unite at the proximal part of the equatorial segment (ES). The yellow region represents the postacrosomal sheath (PAS); the red region is the subacrosomal layer (SAL) on which the IAM rests. **B,C)** Longitudinal sections through the hamster sperm head before (B) and after (C) acrosomal exocytosis (reaction). An electron dense layer, termed the inner acrosomal membrane coat (IAMC), is retained after acrosomal exocytosis. Note that even before exocytosis, the IAMC is visible within the acrosome. The coat contains receptors and proteinases involved in penetration of the oocyte's extracellular matrix. The IAM is seen as a white line in these non-osmicated sections. A, acrosome; P, perinuclear theca; N, nucleus; OAM, outer acrosomal membrane; AG, acrosome ghost. Bars = 0.2 μm

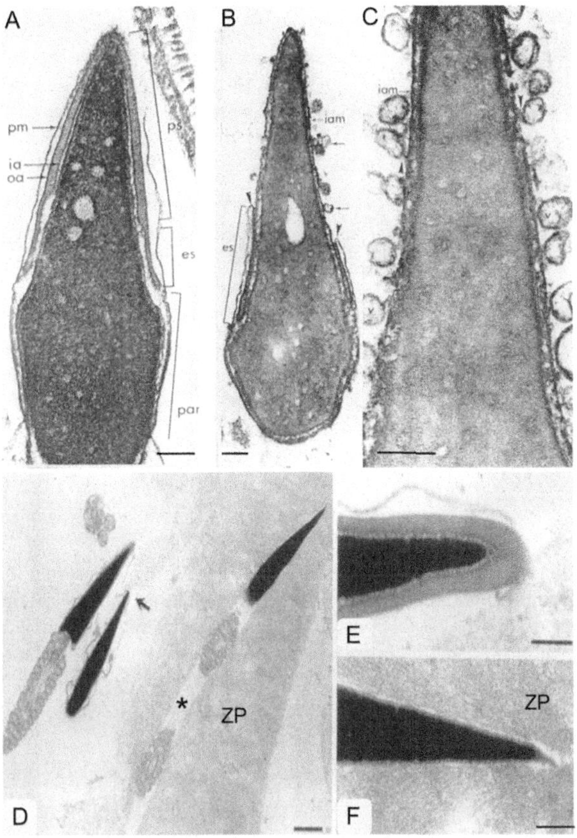

Fig. 1.12 The acrosome reaction and exposure of the IAM is a pre-requisite for penetration of the zona pellucida. A) Electron micrographs of human spermatozoa before acrosome reaction. Bar = 0.5 µm. **B,C)** Low and high magnification of monkey spermatozoa undergoing the acrosome reaction. Note the union of the plasmalemma (pm) and outer acrosomal membrane (oa) at arrowheads at the proximal end of the equatorial segment (es). The plasmalemma overlying the equatorial segment will initiate fusion with the oolemma of the oocyte during fertilization. The hybrid vesicles made up of the outer acrosomal membrane and plasmalemma are conspicuous in C. Bars = 0.2 µm. Source: Micrographs A-C reproduced with permission from (Russell et al. 1979). **D-F)** Monkey spermatozoa undergoing penetration of the zona pellucida (ZP). In D, note the fertilization cone (asterisk) of a spermatozoan penetrating the ZP. In E, tip of sperm contacting the ZP, which will initiate the acrosome reaction. In F, tip of sperm penetrating the ZP, showing that the inner acrosomal membrane forms the leading edge of penetration. Bars = 0.2 µm. Source: Micrographs D-F reproduced with permission from (Sutovsky et al. 1996)

1991; Topfer-Peterson and Henschen 1988; Urch and Patel 1991), PH-20 (Cowan et al. 1991; Hunnicutt et al. 1996; Yudin et al. 1999), SP17 (Kong et al. 1995; Richardson et al. 1994), MC41 (Tanii et al. 2001), SAMP32 (Hao et al. 2002) and SAMP14 (Shetty et al. 2003) but their ZP2-binding ability in situ was not established. By devising a cell fractionation method that allowed for the analysis of

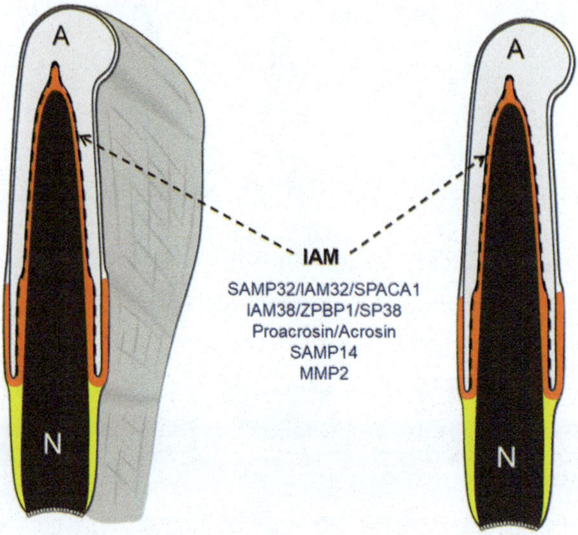

Fig. 1.13 Proteinases and receptors associated with the inner acrosomal membrane (IAM) involved with binding and penetration of the extracellular matrix of the oocyte during fertilization. Dashed line represents position of the IAM in diagrammatical representation of mid-sagittal section of bull sperm head. Diagram on the right is a representation of an ultrathin mid-sagittal section as seen by transmission electron microscopy. A, acrosome; N, nucleus

the protein composition of the IAM directly, Yu et al. (Yu et al. 2006) were able to identify, localize (Fig. 1.14) and characterize such a receptor, termed IAM38/ZPBP1 and reinforce the above concept of a compulsory 'secondary' binding step.

Since sperm-zona penetration only begins after this compulsory 'secondary' binding step, it was hypothesized that the exposed IAM surface, which is the leading edge of the 'fertilization cone' through the ZP (Figs. 1.12 and 1.13), is engaged in lytic activity. Taking advantage of a sperm head fractionation procedure by which detergent extraction can selectively solubilize IAM associated proteins (Fig. 1.15) Ferrer et al. (Ferrer et al. 2012a) was able to identify, localize and characterize proteinases on the IAM that were candidates for enzymatic digestion of the ZP during sperm penetration. They discovered that proacrosin/acrosin and matrix metalloproteinase 2 (MMP2) are IAM-associated enzymes (Figs. 1.16, 1.17). MMP2, also known as gelatinase A, is an MMP that is widely found in somatic tissue and is involved in the breakdown of extracellular matrix material (Werb 1997). Anti-MMP2 antibodies and inhibitors incorporated into in vitro fertilization media significantly decreased fertilization rates leading to the speculation that MMP2 and acrosin cooperate in enzymatic digestion of the ZP during penetration (Ferrer et al. 2012a).

In addition to ZPBP1, acrosin and MMP2, two other important protein constituents of the IAM that were identified as potential contributors to the fertilization process are SAMP32/SPACA1 (Hao et al. 2002) and SAMP14 (Shetty et al. 2003)

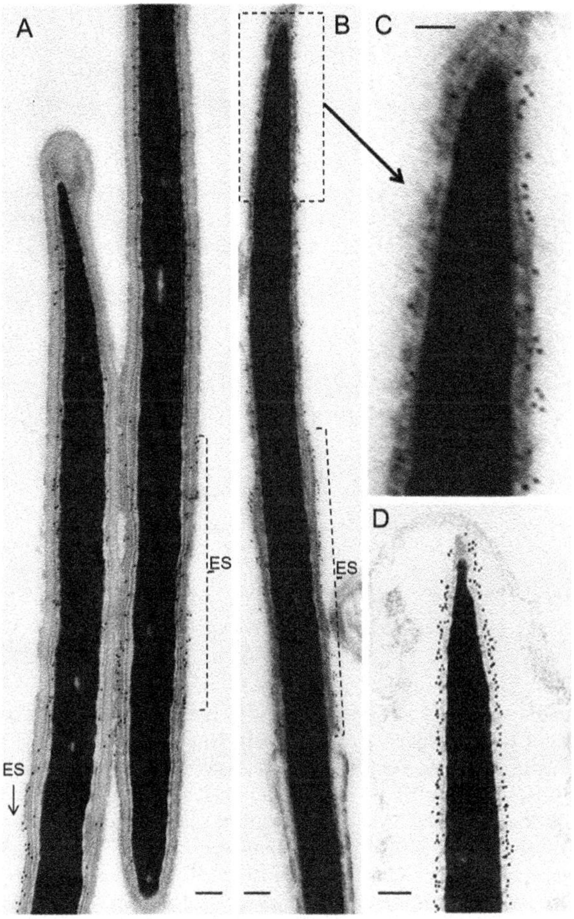

Fig. 1.14 Immunogold-localization of IAM38/ZPBP1 and SPACA1 on sagittal sections through the bull sperm head. A,B) Labelling with anti-IAM38 antibody. Note that the colloidal gold label is associated with the entire inner acrosomal membrane and the outer acrosomal membrane, limited to the equatorial segment (ES). In B, sperm head has undergone the acrosome reaction yet the label remains associated with the IAM and acrosomal membranes in the ES. The area outlined in the dashed box is magnified in the inset **(C)**, directed to by the arrow. **D)** Acrosome reacted sperm head labelled with an anti-SPACA1 antibody. Note that labelling is associated with the IAM. Bars = 0.2 μm

(Fig. 1.13). SAMP14 is a glycosylphosphatidylinositol (GPI)-anchored plasminogen activator receptor whose main role appears to tether plasminogen activator protein, released during the acrosome reaction, which in turn activates plasminogen to plasmin, its active enzyme form (Ferrer et al. 2016). In examination of the in vitro fertilization (IVF) interactions between plasminogen, which is abundant in oviductal fluid, and SAMP14 and MMP2, Ferrer et al. (Ferrer et al. 2016)

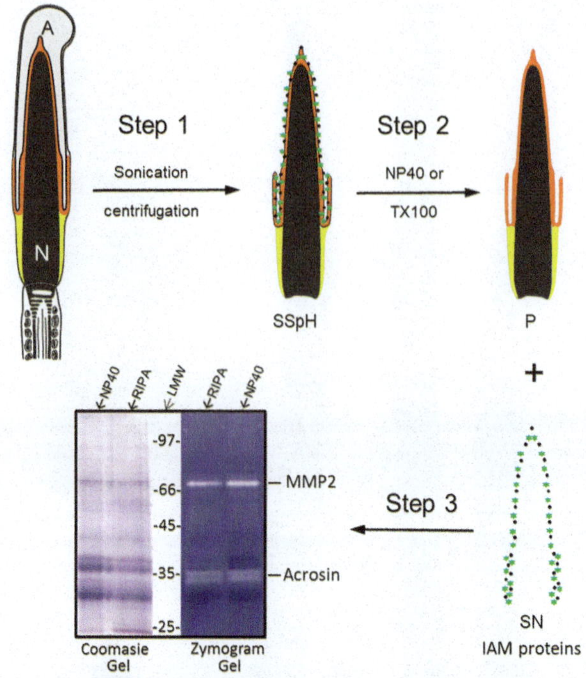

Fig. 1.15 Schematic summary of approach to selectively extract IAM associated proteins from sonicated and isolated sperm heads (SSpH). After sonication of sperm and separation of heads from tails in a sucrose gradient (**Step 1**) the SSpH can be incubated with non-ionic detergents (i. e., 0.1% NP40 or 0.2% Triton X-100) which selectively extract both peripherally attached (black dots) and integral IAM proteins (green stars) from the SSpH (**Step 2**). Centrifugation of this fraction separates IAM proteins into supernatant (SN) and SSpH into pellet (P), which now contains only the sperm nucleus and attached PT. The supernatant is then analyzed on Zymogram gels for active proteinases of which MMP2 and acrosin are most prominent (**Step3**). (Note, prior incubation of SSpH in a high salt solution (e. g., 1 M KCL) before non-ionic detergent (step 2), selectively extracts only peripherally attached IAM proteins (black dots) which make up the inner acrosomal membrane coat, including IAM38, proacrosin and MMP2.)

concluded that several receptor and enzyme interactions on the IAM could be at play leading to extracellular matrix (ECM) penetration of the oocyte by sperm (Fig. 1.18). First, plasmin, resulting from the cleavage of plasminogen, activates MMP2 by cleavage of its pro-segment allowing MMP2 to effect extracellular matrix (ECM) penetration (i. e., cumulus and zona pellucida). Secondly, plasmin effects ECM penetration directly. Thirdly, plasmin activates other serine proteases to their active form, such as proacrosin to acrosin, enabling them to partake in ECM digestion as well. This study also raises the possibility that the female reproductive tract by supplying plasminogen to the fertilization process can in fact compensate for deficiencies or defects in acrosomal associated enzymes as long as the plasminogen activator complex remains intact on the IAM.

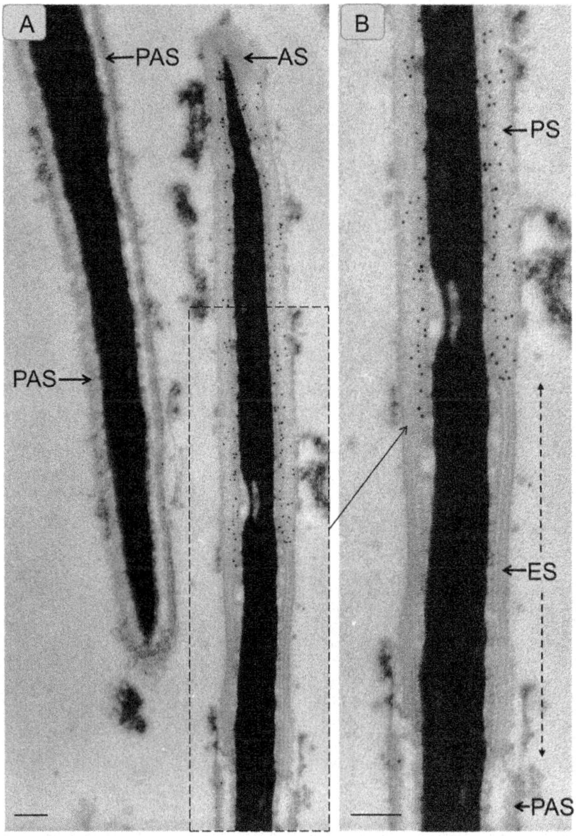

Fig. 1.16 Immunogold localization of proacrosin/acrosin in sagittal sections of bull ejaculated sperm. A) Immunogold labeling with anti-bull acrosin antibody is confined to the inner acrosomal membrane of the principal (PS) and apical (AS) segments of the acrosome. Little or no labeling is evident in the equatorial segment (ES) of the acrosome nor in the postacrosomal sheath (PAS) region. **B**) Outlined area with dashed line in A, shown at higher magnification in B. Bars = 0.2 um

Figure 1.19 summarizes the final distribution of known IAM associated proteins (i. e., receptors, enzymes and integral membrane proteins). Their positions to some extent reflect their function. For example, proacrosin and MMP2 are associated only with the IAM spanning the principal and apical segments, the region of the acrosome undergoing the acrosome reaction. It is this region where the IAM becomes exposed forming the leading lytic edge of sperm-ZP penetration. SPACA1/SAMP32 the only integral membrane protein shown most likely forms an ionic bond with ZPBP1, which is not only involved in zona-binding during fertilization (Yu et al. 2006), but also in compaction of the acrosome, including the equatorial segment, during spermiogenesis (Lin et al. 2007; Yu et al. 2009).

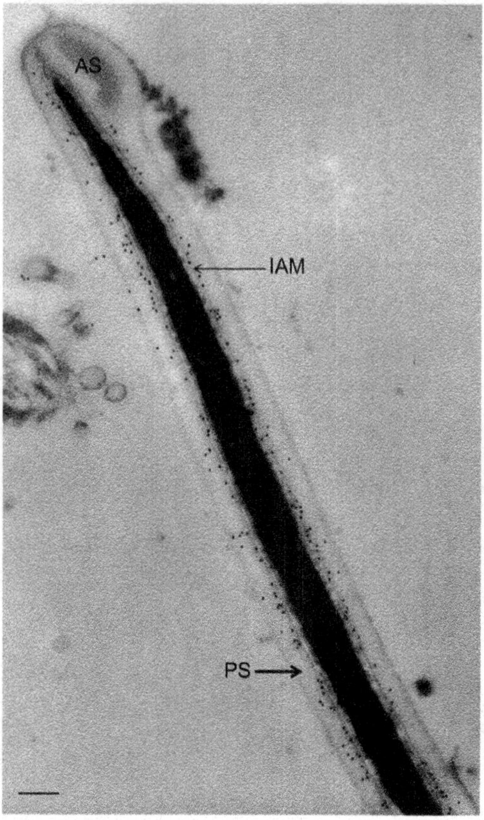

Fig. 1.17 Immunogold localization of MMP2 in sagittal section of bull ejaculated sperm. Like proacrosin, MMP2 is confined to the inner acrosomal membrane of the principal (PS) and apical (AS) segments of the acrosome. Bar = 0.2 μm

SAMP14 is the plasminogen activator receptor, which forms the complex with plasminogen released by the oviduct to effect ECM penetration.

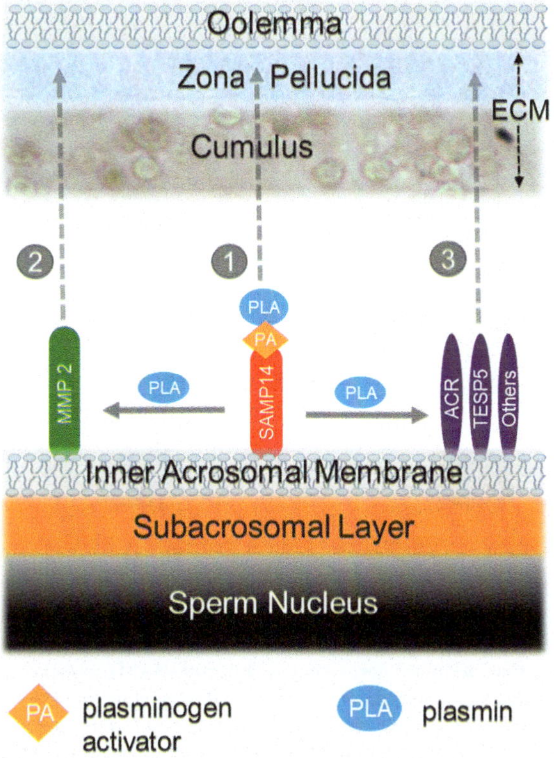

Fig. 1.18 Proposed receptor and enzyme interactions on the inner acrosomal membrane (IAM) leading to extracellular matrix (ECM) penetration of the oocyte by sperm. SAMP14, a plasminogen activator receptor on the IAM, binds plasminogen activator (PA) released during the acrosome reaction which in turn binds plasminogen, released by the oviduct and oocyte, and converts it into the active serine protease, plasmin (PLA). Plasmin can: (1) effect ECM penetration alone, (2) act on MMP2 and (3) other IAM associated enzymes (e. g., ACR (acrosin), TESP5 (PRSS21)) converting them into their active enzymatic forms, which in turn can bolster ECM penetration

1.5 Tail Structure

Understanding the normal structure and development of the sperm tail is critical for interpreting tail anomalies. The mammalian sperm tail is made up of many structurally and functionally dependent components that are assembled at different periods in the process of spermiogenesis. Although there are variations in size and shape of components between species, the general structural organization of the tail is similar. The tail of the sperm can be divided into four discrete anatomical divisions: the connecting piece (CP, neck), the midpiece (MP), the principal piece (PP), and the endpiece (EP), all surrounded by a common membrane (Fig. 1.1). Common to all parts of the tail is the centrally located axoneme composed of nine evenly spaced microtubule doublets surrounding a central pair of microtubule

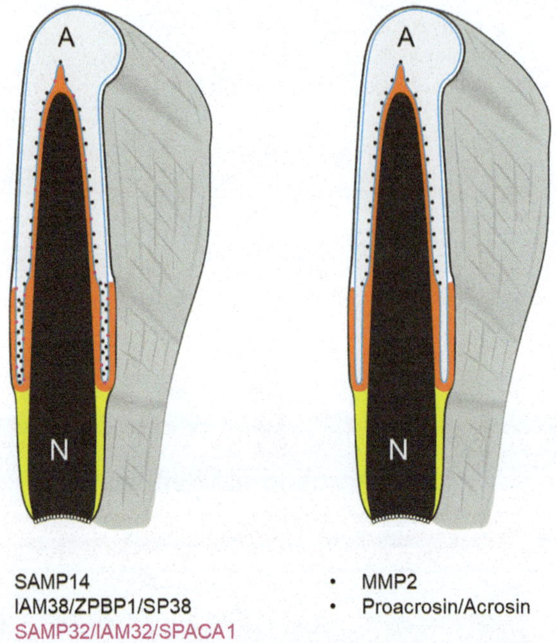

Fig. 1.19 Distribution of inner acrosomal membrane (IAM) associated proteins in bull spermatozoa. Black and pink dots show peripheral and integral protein association, respectively, with the acrosomal membrane. Note that MMP2 and proacrosin, which are peripherally bound to the inner aspect of the IAM, are restricted to the apical and principal segments of the acrosome. SAMP14 and IAM38, which are also peripherally bound to the inner aspect of the IAM, extend into the equatorial segment and are also bound to the outer acrosomal membrane in this most caudal segment of the acrosome. SAMP32, which is an integral protein of the acrosomal membrane, follows the same course as SAMP14 and IAM38

singlets (Fig. 1.20). The nine microtubule doublets are surrounded by and linked to nine corresponding outer dense fibers (ODFs) in the midpiece (Fig. 1.21) and seven in the PP (Fig. 1.22). There is no doubt that the ODFs provide structural stability to the sperm tail, but because of their direct link to the microtubule doublets they probably enhance motility by elastic recoil. It should be emphasized that there are structural linkages between outer microtubular doublets, between each outer doublet and central pair of singlets, and between each outer doublet and corresponding ODF (Oko and Clermont 1990; Fawcett 1981). Together the axoneme and the ODFs make up the axial fiber bundle. There are two structural collars or sheaths that surround the axial fiber bundle and distinguish the midpiece from the principal piece. The midpiece is surrounded by a compact and helically arranged mitochondrial sheath (Fig. 1.21), whereas the principal piece is surrounded by a rib-like fibrous sheath (Fig. 1.22). The annulus (Jensen's ring), a proteineous ring, separates the two sheaths (Fig. 1.23).

The longitudinal columns of the fibrous sheath (FS) replace corresponding ODFs, numbers 3 and 8 along the entire PP, while the ribs of the fibrous sheath (FS) bridge the opposing longitudinal columns on either side (Fig. 1.22).Thus the

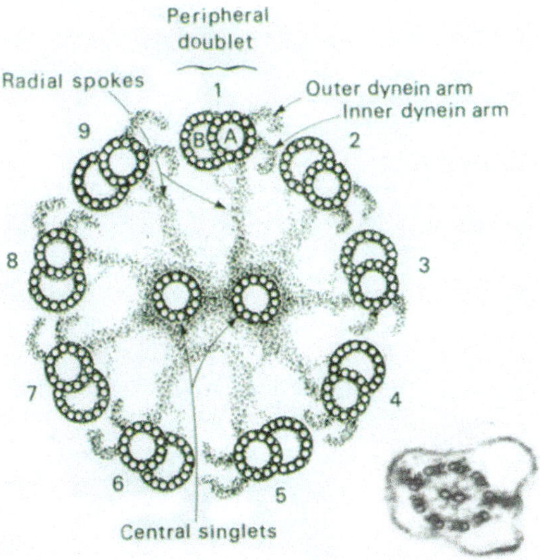

Fig. 1.20 Diagrammatic representation of a cross-section through the axoneme. An electron micrograph of a transverse section of the axoneme through the endpiece of the sperm tail is shown for comparison in the bottom right hand corner. Note that microtubular doublets 3 and 8 form an attachment with the plasmalemma. Bar = 0.2 μm

PP contains seven ODFs as compared to nine in the MP and these ODFs gradually decrease in size distally and even disappear at the end of the PP so that all that remains is the FS surrounding the axoneme (Fig. 1.22). It is important to note that there is no space between the ribs of the FS as they are linked by a thin covering that acts as a barrier, preventing ions and molecules from entering into the axial fiber bundle region. It has been suggested that the position of the longitudinal columns in the plane of the central pair of microtubules restrict lateral bending of the tail, while the periodic spacing of the ribs permit bending of the tail in a plane perpendicular to that of the central singlets and longitudinal columns. It should also be noted that the position of the longitudinal columns creates two unequal or asymetric tail compartments, the smaller made up of microtubular doublets 1, 2 and 9 and corresponding ODFs and the larger one made up of doublets 4, 5, 6 and 7 and corresponding ODFs. This partly explains why sperm motility would be expected to be asymetric. Of the two compartments the larger one is more susceptible to disruption and consequential eviction of ODFs and microtubule doublets due to various tail defects that will be explained in the following chapters. Therefore, from a completely structural point of view the FS to a large degree determines the plane of tail motion and the flexibility and shape of the sperm tail beat. The endpiece is the most distal region of the tail and begins where the fibrous sheath ends. Initially, the 9+2 axoneme arrangement remains intact in the endpiece, while more distally, the doublet tubules gradually dissociate and terminate, the central tubules being the last to do so.

At the proximal end of the tail the outer dense fibers are firmly attached to nine striated columns, which form the striated collar of the connecting piece

Fig. 1.21 Diagrammatic tridimensional representation of the arrangements of components making up the midpiece of the sperm tail. The outer dense fibers (ODF) are numbered according to their corresponding microtubular doublets to which they are attached. ODFs number 1, 5 and 6 are characteristically larger than the others. The mitochondrial sheath is helically arranged around the ODFs. In this diagram the facing mitochondria are cut open longitudinally to reveal the complexity of their intracristal and matricial (dark) spaces that undergo conformational changes dictated by state of respiration. An electron micrograph of a transverse section of the midpiece is shown for comparison in the bottom right hand corner. Note that satellite fibrils of unknown significance are found between the ODFs and although not clearly shown in this preparation of the midpiece section, a fibrous material termed the submitochondrial reticulum binds the mitochondria to the ODFs. Of course, the plasmalemma surrounds the mitochondrial sheath, which is not shown in the diagrammatic representation

(Fig. 1.24). The axoneme terminates at this juncture and is not attached proximally. The connecting piece (CP) is a relatively short segment composed of dense material that surrounds two cavities and shares prominent protein constituents with the ODFs (Figs. 1.23, 1.24). It consists of two distinct regions, the capitulum surrounding the proximal cavity and the striated collar (or columns) surrounding the distal cavity. The striated collar is formed by nine columns which are firmly attached to each other. During development the cavities or vaults of the CP contain the proximal and distal microtubular triplets of the centrioles, the latter directing the assembly of the axoneme of the tail. The convex structure of the proximal part of the capitulum is adapted to fit the shallow concave depression or implantation fossa at the base of the sperm nucleus. Within the implantation fossa the two membranes of the nuclear envelope are fortified with proteins to provide a

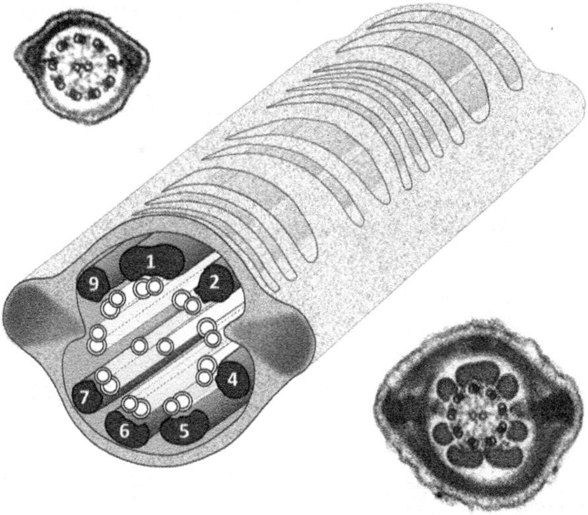

Fig. 1.22 Diagrammatic tridimensional representation of the arrangements of components making up the principal piece of the sperm tail. In the principal piece, ODFs #3 and #8 are replaced by the longitudinal columns of the fibrous sheath (FS), which attach to the corresponding numbered microtubular doublets. The longitudinal columns running opposite of each other are bridged on either side by the ribs of the FS. Slung between the ribs is a thin layer of FS material, which on this diagram is made to look transparent in order to see the ODFs running beneath. Electron micrographs (EM) of transverse sections of the proximal and distal parts of the principal piece are shown for comparison in the bottom right hand corner and top left corner, respectively. In the distal part of the principal piece the ODFs gradually disappear. The EM sections are surrounded by the plasmalemma which is not shown in the diagrammatic representation

sturdy base for implantation of the tail, which is referred to as the basal plate (not to be confused with the basal belt). The capitulum is separated from the basal plate by a narrow gap which is filled by bridging filaments that covalently attach the two structures together insuring that the sperm tail is firmly anchored to the sperm head. Interestingly, if using sonication to break the sperm head–tail junction, separation occurs between the two protein studded nuclear membranes of the basal plate rather than in the gap filled by the bridging filaments.

1.6 Sperm Tail Composition

1.6.1 Axoneme

The microtuble is made up of longitudinally and linearly aranged pairs of alternating α- and β- tubulin subunits forming protofilaments (Tilney et al. 1973). The α and β tubulins are related proteins of approximately 55 kDa (Bryan and Wilson 1971; Stephens 1970). Each microtubular doublet consists of a complete

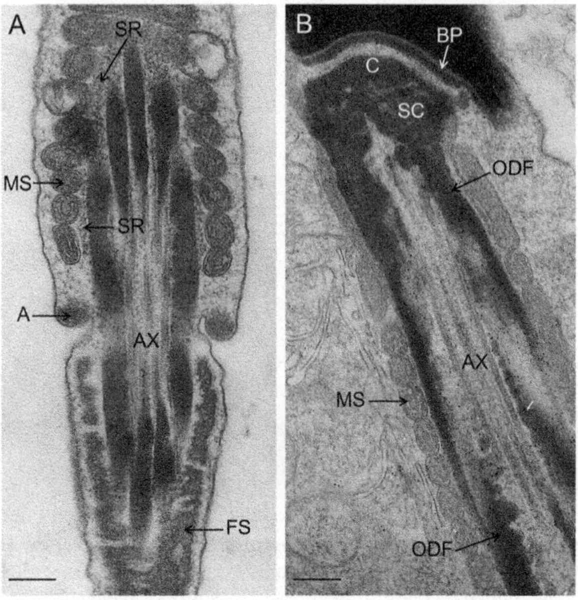

Fig. 1.23 Electron micrographs of sections through the middle-principal piece junction (A) and through the connecting-middle piece junction (B) of the sperm tail, immunogold labelled with a pan anti-ODF antibody. Note that the gold labelling is over the ODFs and the striated collar (SC), capitulum (C) and basal plate (BP) of the connecting piece. Antibodies specific to prominent protein members of the ODFs have all been shown to label the connecting piece, confirming that the connecting piece and ODF share a similar composition. The proteinaceous ring termed the annulus (A) divides the midpiece from the principal piece. Axoneme (AX); Fibrous sheath (FS); Mitochondrial sheath (MS); Submitochondrial reticulum (SR). Bars = 0.2 μm

A-microtubule, which in cross section consists of 13 protofilaments and an attached incomplete B-microtubule, which in cross section consists of 11 protofilaments (Fig. 1.20). Microtubule diversity and degrees of stabilization within different organisms and tissues, including the testis, is maintained by differential genetic expression of the multiple alpha- and beta-tubulin polypeptides (Cleveland and Sullivan 1985; Distel et al. 1984; Kemphues et al. 1982) and by post-translational modifications of alpha-tubulin, i. e., acetylation and detyrosination (Hermo et al. 1991).

Other proteins of the axoneme are Mg-stimulated dynein ATPases which are associated with both the inner and outer dynein arms (Gibbons 1981). The arms extend from subfiber-A of one microtubule doublet towards subfiber-B of the adjacent microtubule doublet (Fig. 1.20) and are responsible for generating ATP-dependent sliding movements between doublet microtubules (Sale and Satir 1977; Satir et al. 2014). The axial spacings of dynein arms and radial spoke along the axoneme are given by Linck (Linck 1979) and Goodenough and Heuser (Goodenough and Heuser 1985b). The inner and outer dynein arms have

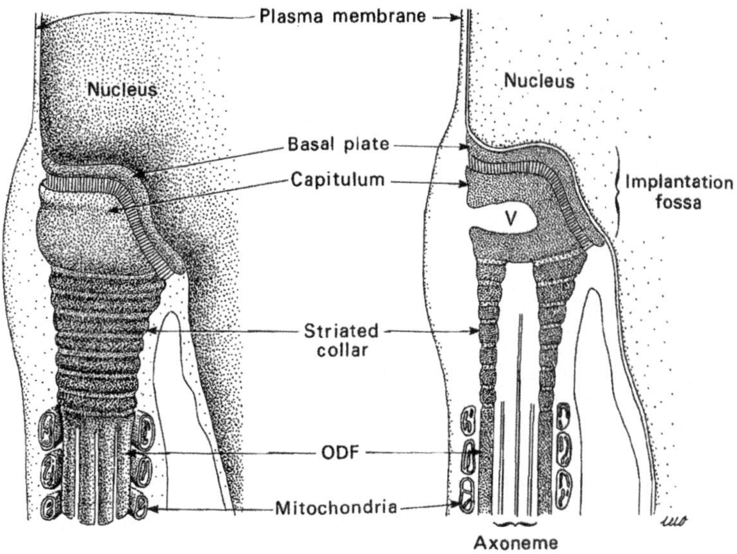

Fig. 1.24 Diagram showing the general structure of the rat connecting piece and its connections with the implantation fossa of the sperm head and the ODFs. Drawing conceptualized by Dr. Yves Clermont. On the left a tridimensional view is shown, while on the right a sagittal section is shown. The connecting piece is composed of the striated collar, the capitulum and the basal plate, the latter two components joined by bridging fibers. The basal plate is firmly attached to the modified nuclear envelope of the nucleus, while the striated collar is structurally continuous with the ODFs. The capitulum and striated collar house vaults (V), which developmentally contained the proximal and distal centrioles, respectively, the distal centriole giving rise to the axoneme. In contrast to the rat, the bull spermatozoon retains its proximal centriole and the angle and shape of the CP-nuclear implantation is different

similarities and differences in composition and function. Of the two arms, the outer arm is best characterized among species because it is readily isolated. High resolution structural studies on outer arm dyneins isolated from Tetrahymena cilia, Chlamydomonas flagella, sea urchin and bull sperm flagellae show that the basic structure of the dynein molecules consists of two or three globular heads (species dependent) joined together by strands to a common stem (Goodenough and Heuser 1985a; Sale et al. 1985; Johnson and Wall 1983; Marchese-Ragona et al. 1987). It is the globular heads of the dynein molecule which are associated with the ATPase activity.

Revealing insights into the mode of assembly and function of related axonemal structures (dynein arms, radial spokes) have come from comparing the protein composition and motility of flagellar-defective mutants with wild-type flagellae of various species (Goodenough and Heuser 1985a; Luck 1984). Mutations have been identified that result in deletions of outer dynein arms, inner dynein arms, radial spokes (which bridge the peripheral doublets to the central singlets, Fig. 1.20), and central tubules. Mutants lacking outer arms are capable of normal motility albeit at a reduced rate (Kamiya and Okamoto 1985; Mitchell and

Rosenbaum 1985), while an inner armless mutant identified is completely paralyzed (Huang et al. 1979). These studies, together with outer arm extraction experiments in sea urchin flagella of Gibbons and Gibbons (Gibbons and Gibbons 1976), suggest that the inner arms are the prime movers of cilia and flagella and that the outer arms form a supplemental system by increasing the speed and force of axonemal bending. Inner arms of Chlamydomonas have five distinct heavy polypeptide chains and a unique complement of intermediate to light chains (Piperno and Luck 1979, 1981). Inner arms are dimorphic, some are two headed while others are three headed, and unlike the 24-nm period of the outer arm, they are spaced at repeating intervals of 40, 32, and 24 along subfiber A of each microtubule doublet so that each triad associates with the base of each radial spoke S, while the dyads associate with the bases of radial spokes S2 and S3 (Goodenough and Heuser 1985b).

The shafts of the radial spokes emanate from subfiber A (Fig. 1.20) of the microtubule-doublets, while the heads of the spokes contact the central pair projections which are spaced at 16-nm intervals (Goodenough and Heuser 1985b). The absence of radial spokes in Chlarnydomonas mutants has been correlated with the loss of 17 polypeptides, while the absence of the spoke heads has been correlated with the loss of five of these 17 polypeptides (Huang et al. 1981). Luck et al. have elegantly shown that the failure in radial spoke or spokehead assembly may be due to the deletion of only one of the component polypeptides (Huang et al. 1981; Luck et al. 1977). The interdependence of assembly in component polypeptides was also shown for mutants affecting outer and inner dynein arms and for central pair microtubules (Dutcher et al. 1984; Huang et al. 1979).

1.6.2 Outer Dense Fibers (ODFs) and Connecting Piece

Incubation of isolated sperm tails in SDS-DTT results in rapid solubilization of all tail components except for the connecting piece-outer dense fiber complex (Fig. 1.25b) which can then be purified by centrifugation, solubilized in sample buffer and run on SDS-PAGE (Fig. 1.25c) to analyze its protein composition (Oko 1988; Olson and Sammons 1980). Five prominent as well as several less prominent proteins are revealed by Coomasie-blue staining (a quantitative stain) of which the 27 kDa protein band is dominant. Antibodies raised against the prominent proteins (Fig. 1.25d) predominantly label the ODFs (Figs. 1.25e,f and 1.23a) and connecting piece (Fig. 1.23b) of spermatozoa, however, the 14 kDa protein also immunocross-reacts with the fibrous sheath (Oko 1988). Thus the ODFs appear to be both structurally and compositionally an extension of the outer shell (i. e., capitulum and striated collar) of the connecting piece (CP) and as a result one would expect defects in the composition of the ODFs to lead to abnormalities in the CP such as head–tail detachments. The 20 kDa ODF protein is a shorter homologue of the 27 kDa ODF protein and is not present in bull sperm ODFs (Morales et al. 1994; Schalles et al. 1998).

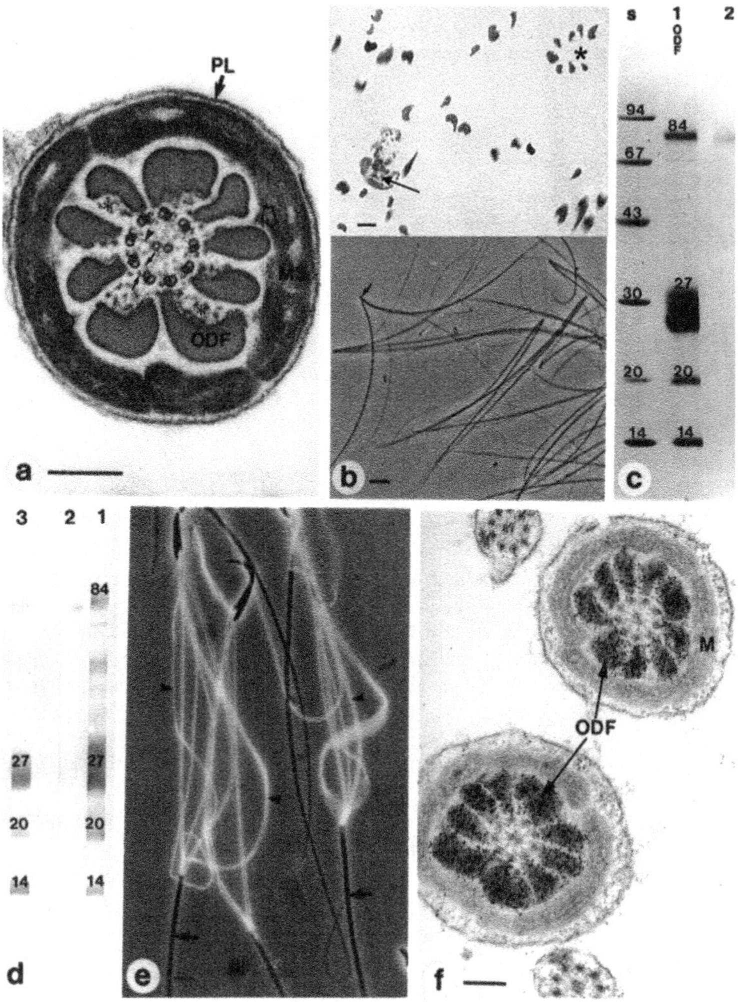

Fig. 1.25 Isolation of outer dense fibers (ODF) of the rat and PAGE, Western and immunocytochemical analyses of ODF protein composition and location. **a)** Electron micrograph of cross section through the mid-piece. PL, plasmalemma; MS, mitochondrial sheath; arrows, dynein arms; arrowhead, radial spoke. Bar = 0.2 μm **b)** Bottom micrograph, phase microscopy of isolated ODFs. Note the connection of the ODFs to the connecting piece (arrow). Bar = 5 μm. Top micrograph, electron microscopy of transverse sections of isolated ODFs and an oblique section through the connecting piece (arrow). Note the rosette (asterisk) formed by the isolated ODFs indicating that they are fettered at the connecting piece. Bar = 0.5 μm. **c)** Coomassie blue dye-stained 8–18% linear gradient SDS polyacrylamide gel of the isolated ODFs showing the prominent protein bands (lane 1). S, are the protein standards in kDa. **d)** Westerns of isolated ODFs labelled with immune serum raised against whole isolated ODFs (lane 1), pre-immune serum control (lane 2) and immune serum affinity purified against the prominent 27 kDa ODF protein (lane 3). Note the immunocrossreactivity of the 27 kDa ODF protein (termed ODF1) with the lower molecular mass ODF proteins. **e)** Regardless of antibody or immune serum preparation against the ODFs, the antibodies specifically label the ODFs (arrowheads) by immunofluorescence on sperm preparations where the mitochondrial sheath is removed to expose the axial fiber bundle. It is clear the mitochondrial sheath's role is not only to produce ATP for motility but to act as a structural conduit for the retention of the ODFs. Arrows, fibrous sheath. Bar = 0.5 μm. **f)** Electron micrograph immunogold labelled with immune serum preparation against the ODFs. Bar = 0.2 μm

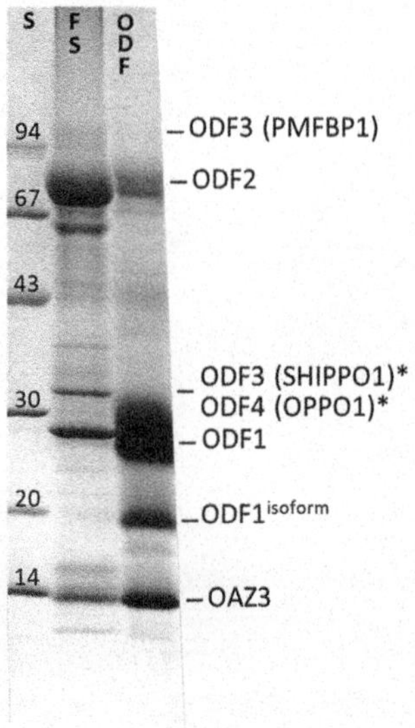

Fig. 1.26 Coomassie blue dye-stained 8–18% linear gradient SDS polyacrylamide gel of the isolated ODF and FS showing the prominent protein members. A list of sequence-identified and characterized ODF proteins, utilizing their official names and juxtaposing them with the protein-banding pattern of isolated ODFs on the polyacrylamide gel is shown. Asterisks signify position of ODF protein on gel based on deduced molecular mass of amino acids in open reading frame of respective gene/mRNA (in this case ODFs were not isolated to locate protein on Westerns in relation to other ODF proteins). S, are the protein standards in kDa

Figure 1.26 provides a list of sequence-identified and characterized ODF proteins, utilizing their official names and juxtaposing them with the protein banding pattern of isolated ODFs on Commassie-blue stained gels. The 27 kDa, ODF1 protein was first to be cDNA cloned and sequence identified (Burfeind and Hoyer-Fender 1991; Morales et al. 1994; van der Hoorn et al. 1990). It has a high proportion of a repetitive motif, Cys-Gly-Pro, at the carboxy-end and a very high content of proline and cysteine, prevalent in ODFs as a whole. At the beginning of epididymal maturation, many of the sulfhydryl (-SH) groups of cysteines in ODFs are weakly complexed with zinc, preventing oxidation of neigboring groups to form the disulfide (-S–S-) derivative cystine. However, as maturation proceeds with transit of sperm through the epididymis zinc is gradually released allowing many of the exposed sulfhydrals to oxidize to -S–S- forming cystines (Calvin et al. 1975). The resulting covalent bonds formed throughout the ODF are the main reason the ODFs are resistant to solubilization in non-ionic detergents unless a reducing agent is added. Importantly, these bonds provide structural continuity throughout the ODF-connecting piece complex. This is a prime example of why the chemical property of sperm obtained from the caput and cauda epididymis can vary substantially.

A yeast two-hybrid method was employed to analyze the interactions of ODF1 in vivo. It was found that ODF1 was able to self interact, through its leucine zipper in its N-terminus and its PCX repeats in its C-terminus (Shao and van der Hoorn 1996). The leucine zipper of ODF1 was also used as bait, in the yeast hybrid methodology, to clone and sequence identify another prominent 84kDa ODF protein, called ODF2 (Shao et al. 1997). ODF2 contains two C-terminal leucine zippers and it was demonstrated that the upstream leucine zipper was required for the critical interaction with the leucine zipper of ODF2, whereas, the other one allows it to self interact. Although both proteins are detected in ODF and connecting piece, ODF2 is present in the cortex and medulla of the ODF, while ODF1 is only detectable in the medulla (Schalles et al. 1998). The thin cortex of the ODFs can be clearly differentiated from the thick medulla in ruthenium red-stained EM X-sections of the sperm tail (Figs. 1.25a, 1.27a). ODF1 was also found to complex with SPAG4, by means of leucine zipper, during spermatid development (Shao et al. 1999). SPAG4, a 49 kDa protein not found in mature sperm, associates with the microtubules of the axoneme during the early and late steps of ODF assembly and appears to form a molecular transport link between the ODF1 and the microtubules. This is in keeping with the observation that ODF1, and not ODF2, forms the surface of ODFs that face and link the axoneme (Schalles et al. 1998). Other proteins that may aid in ODF assembly and found to be associated with ODF1 by leucine zipper interactions during spermatid tail development are SPAG5 (Fitzgerald et al. 2006b) and KLC3 (kinesin light chain 3) (Zhang et al. 2004, 2012).

Unlike the Odf1 gene which appears testis specific, the Odf2 gene expresses different ODF2 protein variants that have different functions in somatic and male germ cells. The somatic version, ODF2/Connexin, contains a centriolar localization motif (Hüber et al. 2008) and a different C-terminus required for formation of cilia (Ishikawa, Kubo et al. 2005), bipolar spindle formation (Soung et al. 2006), and regulation of mitotic progression (Soung et al. 2009). Utilizing a $Odf2^{+/-}$ ES cell line, XL169, with a β-geo cassette integrated between exons 4 and 5 of Odf2, Tarnasky et al. (Tarnasky et al. 2010) showed that high percentage, chimaeric male mice produced from this cell line were infertile. From an ultrastructural perspective, elongated spermatids displayed reduced numbers of ODFs in the sperm midpiece and principal piece, as well as thinning of the ODFs in various regions of the tail. Because of the weakness caused by the missing and thinned ODFs, secondary effects occurred concomittant with the acquistion of motility in the cauda epididymis. These were bending, coiling and breaking of the tail, leading to membrane fusions of bent or broken regions of the tail as well as extrusion of ODFs and microtubular doublets from the axial fiber bundle. The results were in contrast to an earlier study, where the β-geo cassette was integrated into exon 9 of Odf2 (Salmon et al. 2006). In that study it was concluded that one copy of Odf2 sustains normal development and male fertility, but a homozygous interruption of the Odf2 gene is incompatible with early development. More recently it was found that ODF2 is indispensable for the CP (neck)-midpiece connection (Ito et al. 2019). In

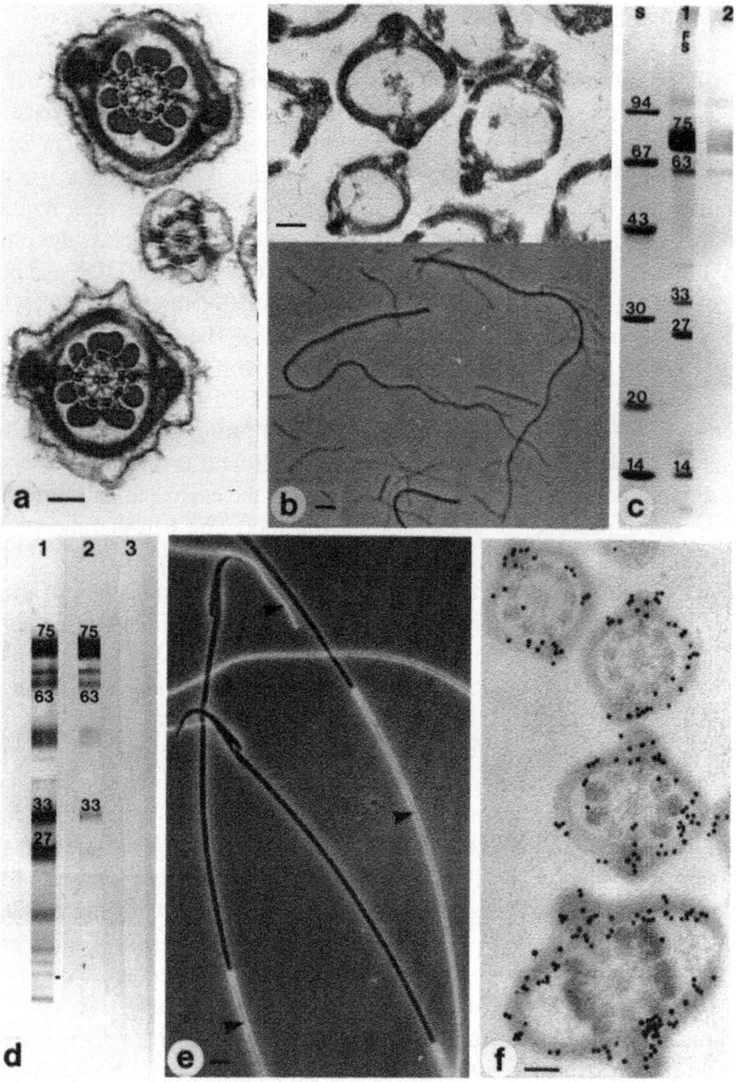

Fig. 1.27 Isolation of fibrous sheath (FS) of the rat and PAGE, Western and immunocytochemical analyses of ODF protein composition and location. a) Electron micrograph showing cross sections through two principal pieces, exposing the FS, and one end-piece. The longitudinal columns have suspensions connecting to microtubular doublet numbers 3 and 8. Bar = 0.2 μm. **b)** Bottom micrograph, phase microscopy of isolated FS. Bar = 5 μm. Top micrograph, electron microscopy of transverse sections of isolated FS. Note some of the suspensions that connected the longitudinal columns to the microtubular doublets (#3 and 8) are retained on the isolated FS. Bar = 0.2 μm. **c)** Coomassie blue dye-stained 8–18% linear gradient SDS polyacrylamide gel of the isolated ODFs showing the prominent protein bands (lane 1). S, are the protein standards in kDa. **d)** Westerns of isolated FS labelled with immune serum raised against whole isolated FS (lane 1), immune serum affinity purified against the prominent 75 kDa ODF protein (lane 2) and pre-immune serum control (lane 3). Note the immunocrossreactivity of the 27 kDa ODF protein (termed AKAP4) with some of the lower molecular mass ODF protein bands. **e)** Regardless of antibody or immune serum preparation against the FS, the antibodies specifically label the FS (arrowheads) by immunofluorescence. **f)** Electron micrograph showing transverse sections through the principal piece immunogold labelled with immune serum preparation against the FS. Bar = 0.2 μm

this study, 129 ES-derived Odf2 chimeric males with deletion of exons 6 and 7 of Odf2 were used to establish heterozygous Odf2$^{+/-}$ male mice that were infertile. This haploinsufficiency led to sperm heads that retained their necks and conversely tails without necks. Thus mutations in different regions of the Odf2 gene can give rise to different phenotypes affecting the ODF-connecting piece complex such as a bending and breakage of the tail or decapitation ocurring between the connecting piece and axial fiber bundle. As will be seen deficiencies in many of the ODF proteins can lead to sperm decapitations, albeit at different levels of the connecting piece. This is not surprising as the main constituents of the connecting piece or neck are the ODF proteins.

In this context, homozygous Odf1-knockout (KO) mice lacking the promoter region and exon 1 were shown to be infertile and one of the major phenotypes reported was head-neck separation, although it was not pointed out at what level of the connecting piece this separation occurred (Yang et al. 2012). Equally as important, this ODF1 deficiency caused a reduction in ODF size and density with many ODF cross-sections showing a hollow appearance, most likely leading to excessive coiling of the tail observed in the epididymis. Two other major phenotypes reported were the loosening of the microtubular doublet- and mitochondrial sheath-ODF attachments, leading to misplacement of ODFs and derangement of mitochondria, respectively. The weakness in microtubular doublet-ODF binding may be due to the inability of SPAG4 to bind ODF1$^{-/-}$ during ODF assembly (Shao et al. 1999), while the weakness in mitochondrial sheath-ODF association may be due to the inability of KLC3 to bind ODF1$^{-/-}$ during mitochondrial sheath assembly (Zhang et al. 2012). The primary ODF1$^{-/-}$-defects, incurred during the assembly of the tail in the testis, appeared to lead to a number of secondary defects that were concomittant with the acquistion of motility in the epididymis, reminiscent of the transition of primary to secondary defects reported in Odf2$^{+/-}$ Chimeric mice (Tarnasky et al. 2010).

The next most prominent ODF protein after ODF1 and ODF 2 is OAZ3 (Fig. 1.26). The ornithine decarboxylase antizyme gene (Oaz3), when first discovered to be expressed exclusively in spermatids, was proposed to encode a 22 kDa antizyme protein involved in polyamine synthesis similar to the 22 kDa OAZ1 and OAZ2 proteins found in somatic cells (Tosaka et al. 2000). Homozygous Oaz3 knock-out male mice were found to be infertile due to the separation of sperm heads and tails occuring between the basal plate and capitulum (Tokuhiro et al. 2009). No gross structural abnormality was observed with the separated heads and tails. Surprisingly, the knock-out of Oaz3 did not result in alteration of polyamine levels in testis or sperm supporting the possibility that OAZ3 has a function other than as an antizyme. Previous to this knock-out it was suggested that both the prominent 14 kDa proteins of the ODF and FS are at least partially encoded by Oaz3 mRNA explaining their tendancy to immuno-crossreact (Fitzgerald et al. 2006a). This was substantiated by a follow up investigation in which an antibody to a peptide within the 1st open reading frame (ORF) of Oaz3 was shown to recognize a sperm protein of the same molecular density (12 kDa) as the recombinant protein encoding the 1st ORF of Oaz3 (Ruan et al. 2011). Mass spectrometry of the immunoreactive sperm protein only recognized sequences of the 1st ORF of OAZ. Furthermore, by immuno-electron microscopy this antibody detected the ODFs, FS and the connecting piece, but not the basal plate. It was also shown by these investigators that translation of Oaz3 starts from a CUG codon, upstream of the first AUG codon from which translation of Oaz1 and 2 begin, and that N-terminal sequences of the prominent 14 kDa ODF and FS proteins are identical to that of OAZ3. In addition, through a yeast hybrid approach, they identified myosin phosphatase targeting subunit 3 (MYPT3) as a binding partner of OAZ3 and showed that OAZ3-MYPT3 modulates the activity of phosphatases PP1β and PP1γ2 (Ruan et al. 2011). The role of OAZ3 in regulating phosphatase activity may therefore provide clues as to how a deficiency of OAZ3 contributes to the separation of sperm heads and tails since it does not appear to play a structural role. Although it has not been definitively proven that OAZ3 (the first ORF) is the sole component of the prominent 14 kDa ODF and FS proteins, it certainly dominates the composition of these two proteins providing the impetus to name them OAZ3.

Another ODF protein of interest is polyamine-modulated factor 1-binding protein (PMFBP1) which was first cloned and sequence identified by Petersen et al. (Petersen et al. 2002) and submitted to GenBank in 1998. At the time of its discovery it was given the name of ODF3 by the authors but is also designated as PMFBP1 in GenBank. It is a coiled-coil protein of ~ 110 kDa with little resemblance in amino acid composition to ODF1 and 2. However, it contains three leucine zipper regions, characteristic of ODF proteins and likely used in ODF assembly. A more recent study that generated Pmfbp1 knockout mice using CRISPR/Cas9 technology found that disruption of this gene resulted in male infertility caused by the production of acephalic heads (Zhu et al. 2018). This study failed to acknowledge the previous characterization of this gene by Petersen et al. who localized its product throughout the ODF. By immuno-fluorescence Zhu et al. localized PMFBP1 to the connecting piece, which they renamed the head–tail coupling apparatus. Oversight by these investigators to recognize, by this technique, that PMFBP1 is also found throughout the ODF was most likely due to the fact that both the mitochondrial sheath and FS form an impenetrable barrier and must be removed for antibodies to access the ODFs. However, the remarkable phenotype of this gene knockout was that the separation ot head and tail occurred between the nuclear envelope and the basal plate during spermiogenesis. Some other defects arising in Pmfbp1-null spermatozoa, found in the epididymis, were misarranged mitochondria, ODF and microtubular doublets, reminiscent of some of the effects caused by deficiencies in ODF1. A take home message is that sperm head–tail separations can be caused by deficiencies in almost any of the ODF proteins as they are constituents of the connecting piece, although the level of separation in the connecting piece varies.

Two other ODF proteins that have been cloned and sequence identified are SHIPPO1(Egydio de Carvalho et al. 2002) and OPPO1 (Nakamura et al. 2002). They are both in the 30 kDa range and as seen on Coomassie-blue stained PAGE (Fig. 1.26) are less prominent members of ODF as far as quantity is concerned, but not necessarily in importance. SHIPPO1 has been given the official name of ODF3 by GenBank which conflicts with the designation of ODF3 that Petersen et al. (Petersen et al. 2002) first gave PMFBP1 (https://www.ncbi.nlm.nih.gov/nuccore/AF092090.2) in 1998, much earlier than the submission of SHIPPO1. OPPO1 is designated as ODF4 in GenBank. SHIPPO1 is highly hydrophobic and contains 6 PRO-GLY-PRO repeats. OPPO1 has a 22-amino acid sequence common to the flagella basal body rod protein of Salmonella and is speculated to be involved with flagella basal body formation in mammals. Consequences of deficiences in SHIPPO1 and OPPO1 have yet to be reported.

1.6.3 Fibrous Sheath (FS)

Incubation of isolated and demembranated sperm tails in Urea-DTT results in solubilization of all tail components except for the fibrous sheath (Fig. 1.27b) which can then be purified by centrifugation, solubilized in sample buffer and run on SDS-PAGE (Fig. 1.27c) to analyze its protein composition (Oko 1988; Olson et al. 1976; Krisfalusi et al. 2006). Many distinct polypeptide bands are revealed of which the 75, 27, and 14 kDa bands are most prominent. Antibodies raised against the prominent proteins (Fig. 1.27d) predominantly label the FS (Fig. 1.27e-f) of spermatozoa, however, the 14 kDa protein immunocross-reacts with the ODF (Oko 1988) and has been identified as OAZ3 (Ruan et al. 2011). Figure 1.28 provides a list of sequence-identified and characterized FS associated proteins, utilizing their official names and juxtaposing them with the protein banding pattern of isolated FS on Commassie-blue stained gels. As will be seen the characterization of these proteins reveals that besides playing an important organizational and stuctural role in the sperm tail, the FS sheath functions as a scaffold for proteins involved in signaling pathways that regulate important sperm maturation processes such as capacitation, for enzymes in the glycolytic pathway that provide ATP for motility, and for proteins involved in redox regulation.

The AKAP (A-kinase anchoring protein) family of proteins are cAMP-dependent protein kinase A (PKA) anchoring proteins. PKA plays a critical role in sperm capacitation and motility that are required for fertilization (Burton and McKnight 2007). Its activity is often refined by attachment to AKAPs, which ensure that PKA is in a specific subcellular domain close to its substrates (Scott et al. 2013). There are at least 25 members of the AKAP family, with overall very little sequence similarity between them except for the PKA binding sites. Three of them, AKAP4, AKAP3 and TAKAP82, are constituents of the FS with AKAP4 being the most abundant FS protein. AKAP4, present in both ribs and columns of the FS, was the first FS protein to be cDNA cloned and identified (Carrera et al. 1994; Fulcher et al., 1995a; Carrera et al. 1996; El-Alfy et al. 1999). Utilizing a yeast two-hybrid screen of a mouse testis cDNA library to first identify proteins that interact with AKAP4, followed by two-hybrid assays to localize the binding domains between the proteins, two PKA binding domains were identified

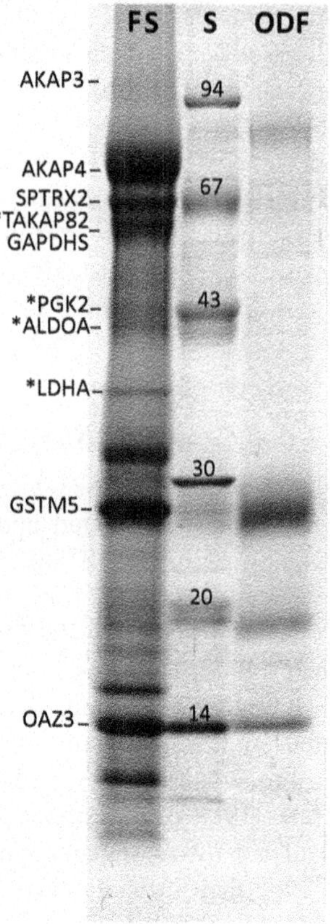

Fig. 1.28 Coomassie blue dye-stained 8–18% linear gradient SDS polyacrylamide gel of the isolated FS and ODF showing the prominent protein members. A list of sequence-identified and characterized FS proteins, utilizing their official names and juxtaposing them with the protein banding pattern of isolated FS on the polyacrylamide gel is shown. Asterisks signify position of FS protein on gel based on deduced molecular mass of amino acids in open reading frame of respective gene/mRNA (in this case FS was not isolated to locate protein on Westerns in relation to other FS proteins). S, are the protein standards in kDa

on AKAP4. One had specificity for binding RIα or RIIα subunits of PKA while the other just bound Riα (Miki and Eddy 1998, 1999). Targeted disruption of the mouse Akap4 resulted in sperm immotility and infertility (Miki et al. 2002). Although the precursors (anlagen) for the assembly of the FS appeared to form, the definitive structure of the FS (i. e., ribs and columns) did not develop. As a consequence proteins normally associated with the FS, some that are essential for motility, were abscent or reduced and the principal piece was shortened. This phenotype is not surprising considering that AKAP4 makes up ~ 40% of the FS structure and binds to another AKAP, AKAP3, as well as two other FS interacting proteins, FSIP1 and FSIP2 (Brown et al. 2003). Surprisingly, AKAP4 does not self associate and since it has only one AKAP3 binding site, its assembly most likely is dependent on linker proteins that have not been determined, although the FSIP proteins are possible candidates. An important observation by Miki et al. (Miki et al. 2002) concerning the phenotype of the disrupted Akap4 gene was that, unlike the stunted assembly of the FS, other components (i. e., axoneme, ODF and mitochondrial sheath) of the sperm tail appeared to develop normally. This could be a tell tale sign of a FS origin in the evaluation of sperm tail defects.

Two other testis- and FS-specific AKAPs, TAKAP80 (Mei et al. 1997) and AKAP3 (Vijayaraghavan et al. 1999; Horowitz et al. 1984; Lin et al. 1995) were discovered on the basis of their interaction with the RII subunit of PKA. AKAP3 undergoes capacitation-induced tyrosine-phosphorylation (Mandal et al. 1999) just as AKAP4 does (Carrera et al. 1996). Unlike AKAP4 which resides in the ribs and columns of the FS, AKAP3 has only been localized to the ribs (Mandal et al. 1999). In addition to binding PKA, yeast hybrid screening also showed that AKAP3 selectively binds ropporin, a protein which interacts with the RHO signalling pathway, and an AKAP-associated protein (ASP), both of which have strong sequence similarity with N terminus of RIIα that is involved in binding AKAP3 (Carr et al. 2001; Fujita et al. 2000). Electronmicroscopy showed that ropporin is localized to the surfaces of the FS but mainly on the inner surface and that its binding partner, rhophilin, which has a binding site for RHO, appears to reside mainly on the surface of the ODF in close apposition to the inner surface of the FS (Fujita et al. 2000). Akap3 gene null mice, that were generated using CRISPR/Cas methodology and compared to Akap4 null mice for phenotypic effects, are infertile and their sperm immotile, even in capacitation inducing media (Xu et al. 2020). A striking difference between Akap3 and Akap4 null spermatozoa is that in $Akap3^{-/-}$ the longitudinal columns appear to assemble normally, opposite microtubule doublets 3 and 8, but the ribs of the FS fail to develop, while in $Akap4^{-/y}$ both the columns and ribs are undeveloped. These results indicate that AKAP3 plays an important role in organizing the development of the ribs, both from a structural and signalling perspective, whereas AKAP4 affects the formation of the entire FS, no doubt providing essential building blocks for its assembly and also serving as a scaffold for anchoring signal transduction pathway components. Considering that AKAP3 only resides in the ribs and not the columns of the FS, the $Akap3^{-/-}$ phenotype is not surprising. Both Akap3 and Akap4 null spermatozoa have principal pieces with reduced length, the shortening of the sperm tail in $Akap4^{-/y}$ being more severe, approaching half the length of a normal tail. On transit through the epididymis these null spermatozoa can acquire bent, coiled and split tails leading to misplacements of microtubular doublets and ODFs from the axial fiber bundle in different regions of the principal piece. In addition to structural impairments in the FS, the absence of AKAP3 and AKAP4 causes misregulated expression of PKA subunits and PKA activity, likely contributing to decreased protein phosphorylation within these sperm and hence impairment in motility (Miki et al. 2002; Xu et al. 2020). As will be discussed, a contributing factor of sperm immotility in the absence of either of these two AKAPs is the failure of glycolytic enzymes to associate with the FS.

A transformation in thinking about where most of the energy required for sperm motility is generated from, arose from gene targeting Glyceraldehyde 3-phosphate dehydrogenase-S (GAPDHS) and phosphoglycerate kinase 2 (PGK2), two isozymes tethered to the FS (Miki et al. 2004; Danshina et al. 2010). Both of these sperm-specific, glycolytic enzymes in mice were found to be essential for sperm ATP production, progressive motility and male fertility, although there are subtle differences in phenotype between mice lacking GAPDHS and PGK2

(Huang et al. 2017). GAPDHS and PGK2 are positioned at key transition points between the ATP-consuming and ATP-generating steps in the sperm glycolytic pathway. Since pyruvate produced in the glycolytic pathway is a major substrate for mitochondrial respiration a defect in sperm glycolysis could alter sperm mitochondrial ATP production. To ensure that the significantly reduced sperm motiliy and ATP levels in GAPDS$^{-/-}$ and PGK2$^{-/-}$ sperm were not due to a lack of substrate provided to the tricarboxylic acid cycle, pyruvate was added to the sperm medium with similar results. Furthermore, WT and GAPDS$^{-/-}$ and PGK2$^{-/-}$ sperm assayed in this medium all had similar oxygen consumption levels providing evidence that the low ATP levels observed in the targeted sperm were not due to mitochondrial deficiency. Taken together, these results indicate that glycolysis rather than oxidative phosphorylation, generates most of the energy required for sperm motility and fertility in the mouse. In a comprehensive review of the research on how energy is generated by mammalian spermatozoa for motlity, Storey (Storey 2008) concluded that glycolysis (sugar is friend) appears to be the dominant pathway in mouse sperm while oxidative phosphorylation (oxygen is friend) is the predominant pathway in bull sperm.

Unlike GAPDHS, PGK2 is not firmly bound to the fibrous sheath (Krisfalusi et al. 2006). Regardless there were no distinguishable ultrastuctural features between wild-type, GAPDS$^{-/-}$ and PGK2$^{-/-}$ spermatozoa, signifiying that at least these glycolytic enzymes are not essential for FS assembly (Danshina et al. 2010; Miki et al. 2004). However, disruption of the gene encoding enolase 4 (ENO4), a sperm-specific glycolytic enzyme which is firmly bound to the FS, disorganizes the normal FS assembly process (Nakamura et al. 2013). A characteristic ultrastuctural feature of elongated spermatids in the testis of the Eno4 disrupted gene was the presence of one or three longitudinal columns in elongating spermatids and a disorganization of the ribs and axial fiber compartments within the PP. Because the longitudinal columns form before the ribs (Irons and Clermont 1982; Clermont et al. 1990), Nakamura et al. (Nakamura et al. 2013) suggested that ENOs is most likely involved in the initial part of FS assembly and is not required for integration of AKAP4 into the FS. Just as with Gapds$^{-/-}$ and Pgk2$^{-/-}$ mice, male mice homozygous for the gene trap allele, Eno4$^{Gt/Gt}$ were infertile, ATP levels were substantially reduced and sperm motility was impaired. In addition to GAPDS and ENO4, several other glycolytic enzymes remain firmly bound to the FS after urea-DTT extraction, including ALDOA (the A isoform of aldolase 1), LDHA (lactate dehydrogenase A) and pyruvate kinase (Krisfalusi et al. 2006). On the other hand, like PGK2, HK1S (spermatogenic cell-specific hexokinase 1 variant) and LDHC (lactate dehydrogenase C) are more readily solubilized from the principal piece by detergents (Nakamura et al. 2010; Krisfalusi et al. 2006; Odet et al. 2008). From the studies above it is clear that almost every step in the sperm glycolytic pathway is catalyzed by male germline-specific isozymes arising from distinct genes or alternatively spliced transcripts. Most likely, the localization of these glycolytic enzymes to the FS or in its near vicinity secures ATP production in close proximity to the dynein ATPases that are distributed along the axoneme in the PP.

Finally, the FS has the potential to regulate spermatid or sperm related oxidation–reduction (redox) reactions during development, capacitation and fertilization as three redox-proteins are incorporated into its structure during spermiogenesis. These are GSTM5 (GSTM3 in human), belonging to an mu class of glutathione S-transferases and SPTRX1 and 2, thioredoxin family members. GSTM5 was cloned, identified and developmentally characterized by Fulcher et al. (Fulcher et al. 1995b). The mu class of glutathione S-transferases normally function in the detoxification of electrophilic compounds and products of oxidative stress, by conjugation with glutathione (Listowsky 2005). To date though it is not yet known whether GSTM5 retains this type of functional enzymatic activity in the FS and when in development or sperm transit it would be employed. Unrelated to redox activity, it was shown that GSTM5 tethers, muscle-type phosphofructokinase (PFKM) which in turn tethers a cell-specific variant isozyme of hexokinase type 1 (HK1S), bringing these glycolytic enzymes in closer association with the dynein ATPases (Nakamura et al. 2010).

Utilizing an in silico search for human/mouse expressed sequence tags (EST) with the conserved active site CGPC, a group of thioredoxins (SPTRX1, SPTRX2, and SPTRX3) with a tissue-specific expression in spermatids and mature spermatozoa were identified, localized and characterized (Miranda-Vizuete et al. 2004). SPTRX1 and SPTRX2, which are localized to the FS, are composed of thioredoxin domains with additional non-thioredoxin domains. SPTRX1's additional domain is novel, while SPTRX2's additional domain is composed of three nucleoside diphosphate (NDP) kinase domains. SPTRX1's incorporation into the longitudinal columns of the FS occurs earlier in spermatid development than the incorporation of SPTRX2, although the latter also locates to the ribs, which begin to form after the assembly of the columns is initiated (Miranda-Vizuete et al. 2003; Yu et al. 2002). The incorporation of SPTRX2 into the FS is also concomitant with the gradual disappearance of SPTRX1 from the longitudinal columns. This replacement of one thioredoxin for another in the FS is intriguing, indicating a specific developmental/maturational function for each of these thioredoxins. This particular expression pattern supports the possibility that SPTRX1 could be a part of a nucleation center for the formation of the longitudinal columns and ribs of the FS, while SPTRX2 may be necessary for post-testicular events such as epididymal sperm maturation, hyperactivation/capacitation and fertilization. Although thioredoxins are traditionally regarded to be general protein-disulfide reductases, the fact that recombinant SPTRX1 can act as both a reducing and oxidizing agent suggests that during spermiogenesis it may be able to facilitate the formation of disulfide bonds among individual FS proteins and destabilize misfolded or improperly incorporated sperm tail proteins. It is reasonable to assume that SPTRX-2 might play a crucial role in disulfide bonding within the FS as well, especially during sperm maturation in the epididymis. However, in both cases the reducing or oxidizing activity of these two thioredoxins in situ has not yet been demonstrated. Phosphorylation is also a major regulatory mechanism in spermatozoa underlying important processes during the acquisition of fertilizing capability by the spermatozoon. Thus the well conserved NDP kinase domains in SPTRX2 could play a critical role as a phosphate donor in these processes, although the phosphorylation activity of this protein has not yet been demonstrated.

1.6.4 Mitochondrial Sheath Arrangement and Composition

During the course of spermatogenesis the mitochondrium gradually undergoes a number of modifications and volume changes which finally provide the spermatozoon with mitochondrial features of unusual ultrastructure (Fig. 1.29A). The prominent changes as summarized by Andre (Andre 1962) are: "(1) the appearance of a new, clear substance, named pseudomatrix, which dilates the intermembranous spaces of the cristae (intracristal spaces); (2) an increase in opacity of the matricial spaces (intercristal spaces), with a concomitant decrease of their volume, which produces a reversal of contrast with respect to normal ultrastructure; (3) an expulsion of the pseudomatrix; (4) a lengthening and coiling of these modified mitochondria around the flagellum to form the mitochondrial sheath (MS)". As the mitochondria elongate and coil around the axial fiber bundle of the elongated spermatid to form the MS they condense and release excess membrane and matrix into the extracellular mileau by a process not unlike exocytosis (Fig. 1.30) (Oko 1984). At this end-point of MS formation the mitochondria are attached to the ODFs by a filamentous, granular substance characterized by Olson et al. (Olson and Winfrey 1990) and coined 'the submitochondrial reticulum' (Fig. 1.30A). Purportedly, the cresent shape of sperm mitochondria is partially determined by a mitochondrial capsule, which was isolated from bull and rat sperm and shown to contain a prominent protein, which retains selenium (Calvin et al. 1981; Pallini 1979). In a landmark study this protein was identified as the phospholipid hydroperoxide glutathione peroxidase (GPx4/PHGPx), a well known selenium retaining enzyme whose redox activity is undetectable once covalently bonded within the capsule (Ursini et al. 1999). It is estimated 50% of the capsule consists of GPx4, which most likely cross-links with itself and another capsular protein termed sperm mitochondria cysteine-rich protein (SMCP) (Cataldo et al. 1996; Flohé 2007). By pepsin-extraction, which leaves the mitochondrial capsule intact and detaches the mitochonrium from the axial fiber bundle, Hrudka (Hrudka 1978) was able to isolate a high yield of morphologically intact mitochondria from bull sperm (Fig. 1.29B) and study their fine structure and conformational states. He had previously found that in bull sperm, about 90 crescentic mitochondria are coupled end by end in parallel helices, four or more in the proximal, three in the central, and two in the distal part of the mitochondrial sheath; each mitochondrion takes about three-quarters of each of a total of 68 gires.

Selenoprotein P (SEPP1) is an extracellular glycoprotein whose main role is to deliver selenium to peripheral tissues including the testis (Hill et al. 2003). It was found that when the normal supply of selenium to the testis was depleted in

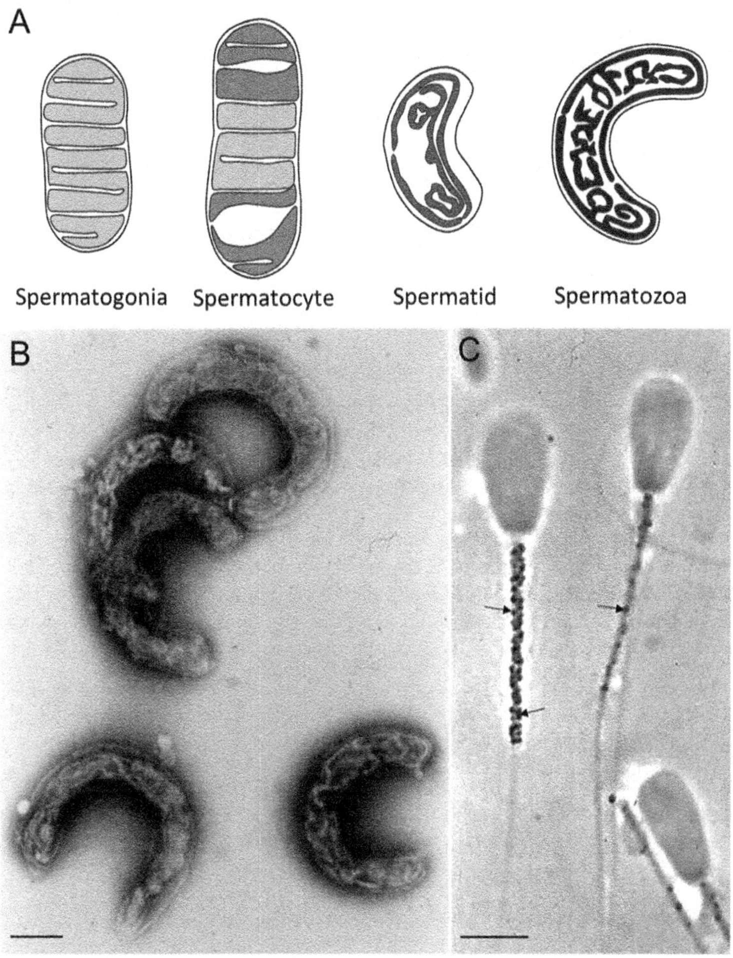

Fig. 1.29 **A) Conformational changes that the mitochondrium goes through in spermatogenesis to provide the spermatozoon with its final mitochondrial characteristic.** The intracristal spaces are white while the matricial spaces (intercristal spaces) of the mitochondria are different shades of gray to black in this diagrammatical rendering of Andre's analysis (Andre 1962). **B) Electron micrograph of PTA stained, isolated mitochondria from the mitochondrial sheath of bull spermatozoa.** The undulating contours of the narrowed intercristal spaces are seen. Micrograph donated from Dr. F Hrudka's collection. Methodology found in Hrudka (Hrudka 1978). Bar = 0.2 μm. **C) Differential measurement of citric acid enzyme activity between bull sperm.** In this case sperm was incubated in a Tris–HCl buffer containing nitro blue tetrazolium (NBT) and Na-succinate as substrate to test the activity of succinate dehydrogenase. Reduction of NBT results in a blue residual deposit in the mitochondria (arrows) as seen under light microscopy. Bar = 5 μm. For methodology refer to Hrudka (Hrudka 1979)

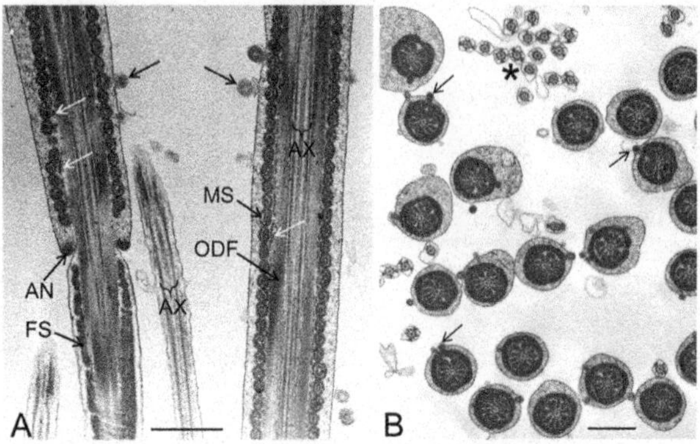

Fig. 1.30 Sperm tails in the last step of spermiogenesis releasing excess mitochondrial-membranes and -matrix in the form of whorls. A) Electron micrograph of longitudinal section of tails, primarily through the mid-piece, showing the release of membranous whorls (buds) from the mitochondria (black arrows), an exocytosis-like process that most likely ensures compaction of the mitochondrial sheath (MS). Note the presence of granular substance (white arrows) interspersed between the MS and the axial fiber bundle, which attaches the mitochondria to the ODF. AN, annulus; AX, axoneme; FS, fibrous sheath; ODF, outer dense fiber. **B)** Cross sections of tails through the mid-piece showing the release of mitochondrial whorls. Note the congregation of axonemes (asterisk) in the lumen originating from round spermatids that have barely begun to assemble the accessory fibers of the sperm tail (i. e., FS and ODFs). Bars = 0.2 μm

Sepp1$^{-/-}$ knockout mice, they became infertile showing a sequelae of sperm-developmental defects in the epididymis originating from a defective assembly of the mitochondrial sheath (MS) in the testis (Olson et al. 2005). The intial lesion in the testis is a truncated MS containing some tell tale mitochondrial deformities. No doubt the weakness in the sheath predisposes the sperm to break in this region during epididymal transit leading to a hairpin configuration of the tail and extrusion of outer dense fiber-microtubular doublets, especially numbers 4–7. Selenium-deficient wild-type males display a comparable temporal development of sperm tail defects. However, unlike the selenium-deficient wild-type males, Sepp1$^{-/-}$ males are resistant to selenium supplementation and do not recover fertility as do the former, suggesting that the supply of selenium to the testis is dependent on SEPP1 transport. The characteristic mitochondrial sheath defect induced by selenium deficiency is probably caused by reduced transport of selenium to GPx4, the prominent mitochondrial capsule protein, as gene knockout of mitochondrial Gpx4$^{-/-}$ reveals almost identical phenotypes (Schneider et al. 2009; Imai et al. 2009). Likewise, a gene knockout encoding ApoER2, a Sertoli cell receptor which binds SEPP1 and uptakes selenium from it for distribution to the spermatids, has a similar phenotype (Olson et al. 2007).

Reminiscent of the characteristic mitochondrial sheath defect induced by selenium deficiency is the resultant segmental aplasia of the mitochondrial sheath and sequelae of tail defects induced by gossypol treatment (Oko and Hrudka 1982, 1984; Oko 1984). Gossypol is a phenolic compound which inhibits mitochondrial

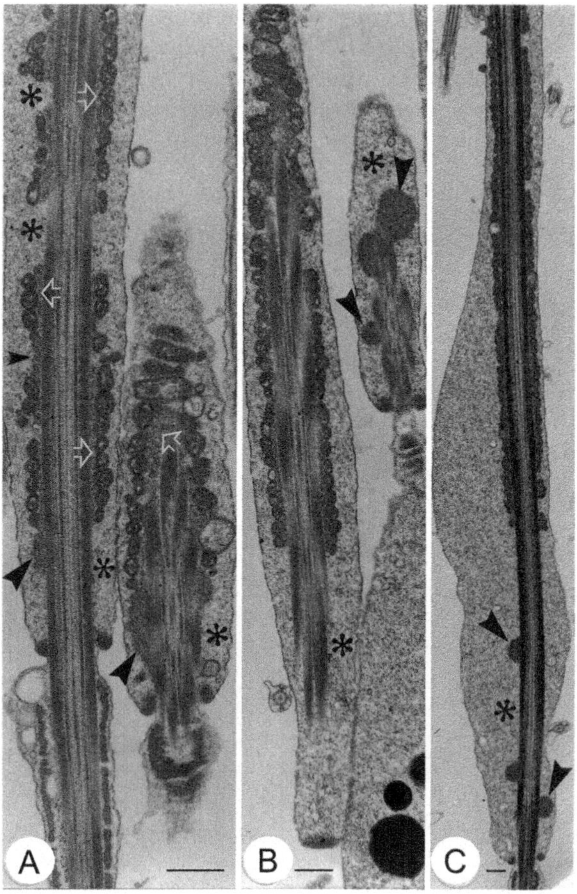

Fig. 1.31 Segmental aplasia and truncation of the mitochondrial sheath caused by inhibition of mitochondrial oxidative phosphorylation. A-C) Longitudinal sections through the mid-piece of the sperm tail at the level of the middle-principal piece junction showing the development of segmental aplasia in the last two steps of spermiogenesis. Due to the swelling and deterioration of mitochondria, there is a shortage of mitochondria to form the complete mitochondrial sheath (MS). As a result, before the last step of spermiogenesis, aplastic regions (gaps) are found throughout the MS, as demarcated by the asterisks in A. However, in the last step of spermiogenesis, the unaffected MS segments, separated by the aplastic segments, reconnect along the sheath with the consequence of truncating the MS and leaving a sizeable gap in the MS (asterisks) directly above the annulus as shown in B and C. Note that a granular substance, normally found between the MS and the axial fiber bundle (white arrows in A), aggregates in the aplastic regions of the MS into hemispherical bodies along the ODFs (arrowheads in A-C). Bars = 0.2 μm

oxidative phosphorylation and leads to a progressive degeneration of individual mitochondria during their assembly of the mitochondrial sheath, resulting in its truncation. The length of the gap (aplasic segment) between the annulus and truncated part of the mitochondrial sheath is dependent on the the dose and duration of treatment and ultimately on how many mitochondria are affected (Fig. 1.31). The weakness in the sheath then predisposes a sequelae of secondary tail defects as the sperm pass through the epididymis, which are identical to those found in selenium deficient sperm (Figs. 1.32, 1.33). They begin with a protrusion or eruption of axial fibers (usually # 4–7) at the level of the gap (Fig. 1.32), a process

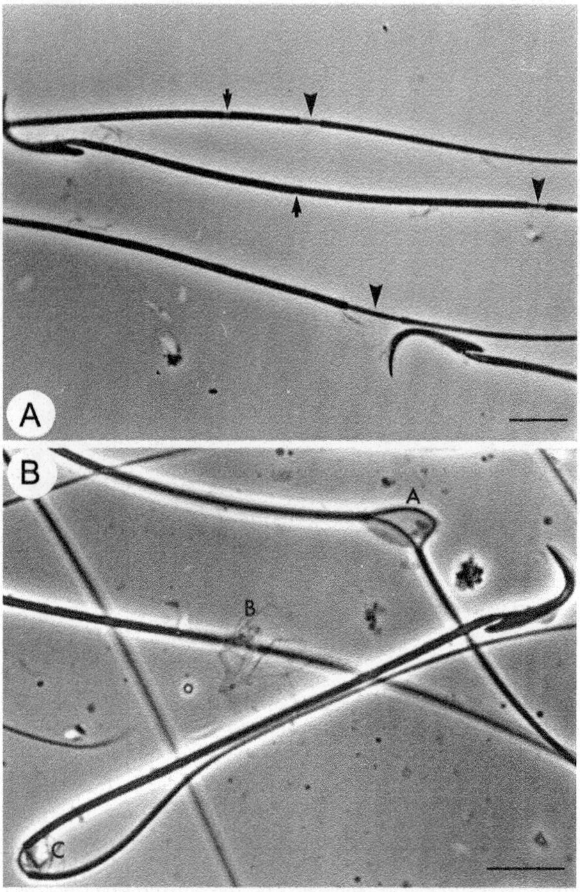

Fig. 1.32 Phase contrast micrographs of gossypol affected rat sperm extracted from the caput (A) and cauda (B) epididymis. A) The caput extracted sperm show the characteristic gap (arrowheads) directly above the annulus caused by the truncation of the mitochondrial sheath (MS). Smaller lesions (arrows) in the MS can also be seen. With dosage increase of gossypol the gaps become longer and the MS more truncated. **B)** Spermatozoa extracted from the cauda epididymis display various secondary lesions, no doubt as a result of gaining the capacity for motility combined with the weakness in the MS. Some of these secondary lesions include bulging and splitting of the axial fiber bundle at the level of the gap (A), fraying and protrusion of axial fibers at the level of the gap (B) and bending of the tail at the level of the gap and reflection of the principal piece over the midpiece (C). This classic hairpin configuration of the tail is common in sperm tail defects affecting the MS and ODFs. Bars = 5 μm

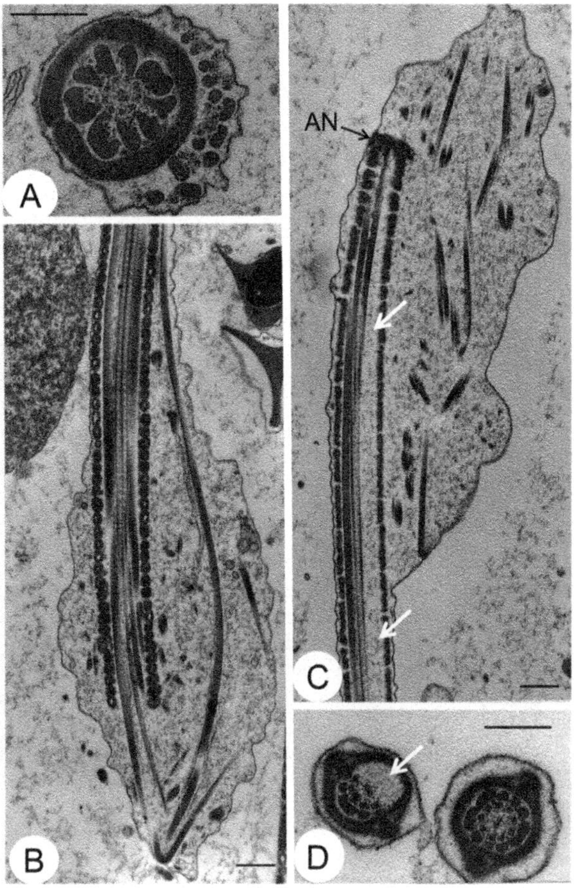

Fig. 1.33 Electron micrographs of cross- and longitudinal-sections through the tail of gossypol-affected spermatozoa, extracted from the cauda epididymis of the rat, displaying dislocated axial bundle fibers (i. e., ODF and microtubular doublets). A) Cross section of midpiece showing dislocated ODFs with their corresponding microtubular doublets between the plasmalemma and mitochondrial sheath. **B)** As seen in longitudinal sections of the midpiece, these displaced axial fibers arise at the level of the gap in the mitochondrial sheath and **C,D)** are usually retracted from the larger compartment of the principal piece (white arrows), normally containing outer dense fiber-microtubular doublets, numbers 4 to 7. AN, annulus. Bars = 0.2 μm

inevitably associated with the retraction of fibers from segments of the tail below the gap (Fig. 1.33). The dislocated fibers often insert between the plasmalemma and the mitochondrial sheath or burst the plasmalemma. Other sequelae are bending and breaking of the tail at the level of the gap, leading to a hairpin configuration of the tail, and ocassionally decapitation. Breakage of the tail invariably leads to either the detachment of the principal piece or its reflection and fusion with the midpiece. The development of primary to secondary lesions is diagrammatically

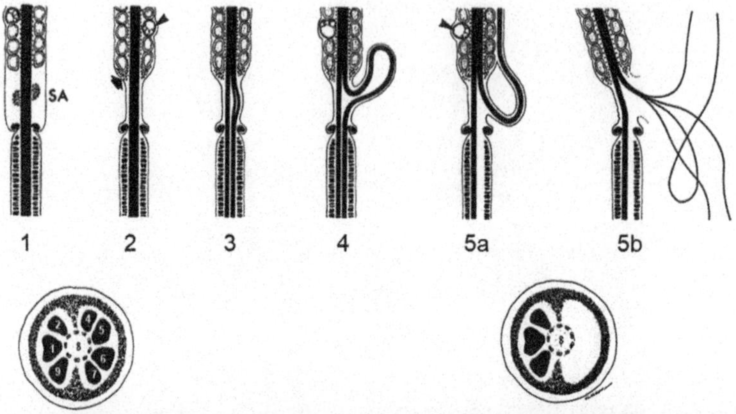

Fig. 1.34 Diagrammatic representation of the developing sperm tail lesion, segmental aplasia (SA) of the mitochondrial sheath, induced by gossypol. 1,2: Spermatozoa from the testis and caput epididymis, respectively. The primary lesion of the midpiece (SA) is characterized by an aplastic segment (gap) of the mitochondrial sheath, found directly above the annulus. In this region a granular substance, normally found interspersed between the mitochondria and ODFs, aggregates to form granular bodies on the axial fiber bundle in testicular sperm. In caput sperm (2) the granular substance or bodies often appear to form a makeshift annulus at the terminal end of the truncated mitochondrial sheath. Additionally, the gap region is reduced in width and degeneration of mitochondria (arrowheads) continue. **3,4:** Spermatozoa from the caput and corpus epididymis display split axial bundle fibers at the level of the gap and below. Axial fibers begin to erupt and bulge at the level of the gap. **5a,b:** Spermatozoa from the cauda epididymis with dislocated fibers #4 to #7, emanating from the principal piece. The ectopic fibers (ODFs and corresponding microtubular doublets) are usually insinuated between the plasmalemma and the mitochondrial sheath (a) or freed and frayed (b). Importantly tails numbered from 3 to 5 are all prone to bend at the level of the gap, with principal piece often reflecting over the midpiece in the classic hairpin configuration or occasionally principal piece separating from midpiece

represented in (Fig. 1.34). The progressive disintegration of the motor apparatus affects the pattern of movement and thus it is not surprising that the treated rats lose their fertility completely. The take home message here is that mitochondrial- and fibrous-sheaths not only provide the ATP required for the motor apparatus but form a supportive yet flexible tubular structure along the length of the tail with the purpose of holding in the motile axial fiber bundle. If these sheaths are breached it is a natural tendency for the axial fibers (microtubular doublets attached to corresponding ODFs) to erupt and dislocate from the aplastic segment and the tail to bend or break in this region.

Mutations or deletion of genes that affect spermatid-mitochondrial dynamics (fusion vs fission) can cause a similar defective sperm phenotype described above. Examples of proteins encoded by such genes, that if deficient lead to segmental aplasia of the mitochondrial sheath, include:

PA-PLA1(Phosphatidic acid-preferring phospholipase A1), which in the seminiferous epithelium is exclusively expressed in round and elongating spermatids and possibly regulates mitochondrial dynamics by hydrolyzing mitochondrial surface phosphatidic acid (PA) (Baba et al. 2014). It is proposed that production of PA by mitochondriol phospholipase D (MitoPLD) facilitates Mitofusin-mediated mitochondrial fusion, whereas the consumption of PA and the production of diacylglycerol by Lipin 1 promote mitochondrial fission (Choi et al. 2006; Huang et al. 2011). *GYKL1(Glycerol kinase-like 1) and GK2 (glycerol kinase 2)*, which exclusively localize to mitochondria in spermatids and spermatoza and may interact with Pld6 (MitoPLD) and induce Pld6 and phosphatidic acid-dependent mitochondrial clustering in cells (Chen et al. 2017). *SPATA19/SPERGEN-1 (spermatogenic cell-specific gene-1)*, localizes to the surface of mitochondria in the mitochondrial sheath, has an N-terminal mitochondrial targeting signal and has the property to induce mitochondrial aggregation in cultured cells (Doiguchi et al. 2002). The spermatid/sperm morphological phenotype of the knockout of this gene was not fully characterized but of what was shown, the sperm were immotile, the mitochondrial sheath appeared disorganized and the mitochondria of the sheath were of irregular appearance (Mi et al. 2015). However, no bending or breaking of the affected sperm tail in the epididymis was reported which most likely was due a severe reduction in mitochondrial respiratory chain activity and ATP and hence immotility. This completely immotile phenotype, with no bending or breaking of the sperm tail during epididymal transit, was also seen with higher dosages of gossypol treatment as uncoupling of oxidative phosphorylation affected all the mitochondria on the sheath resulting in their swelling and deterioration (Fig. 1.35).

Golgi-associated PDZ- and coiled-coil motif-containing protein (GOPC) localizes to the trans-Golgi region in round spermatids (Yao et al. 2002). The primary morphological defect in GOPC-deficient spermatids is the failure of proacosomic vesicles to fuse to form the definitive acrosome, which then appears to trigger further sperm head malformations, making it a good model to study globozoospermia (Ito et al. 2004). The sperm tail, however, seems unaffected until the spermatozoa reach the middle caput epididymis where the tail coils around the sperm nucleus. It appears that a defective posterior ring (basal belt) allows the tail to move up between the plasmalemma and nuclear envelope and coil (Suzuki-Toyota et al. 2007). Surprisingly, a characterisitic of this coiling is that the mitochondria gather at the proximal part of the midpiece, forming a ‘stratified mitochondrial sheath’ leaving the ODFs of the distal midpiece denuded of mitochondria. A similar midpiece disorganization was reported in NECTIN-2/PVRL2 deficient mouse spermatozoa (Bouchard et al. 2000). What prediposes the mitochondria to detach from the axial fiber bundle and then stratify is unknown but based on immunolocalization studies in Gopc$^{-/-}$ spermatozoa, Suzuki-Toyota et al. speculate that a deficiency in mitochondrial capsule protein, GPX4, allows for the detachment while an abundance in SPATA19 allows for mitochondrial adhesion into a stratified mass (Suzuki-Toyota et al. 2007). The take home message here is that even though a protein such as GOPC is not a constituent of mitochondria, but rather of the Golgi apparatus, its disruption can indirectly affect the physiological makeup of other organelles, in this case the sperm mitochondia.

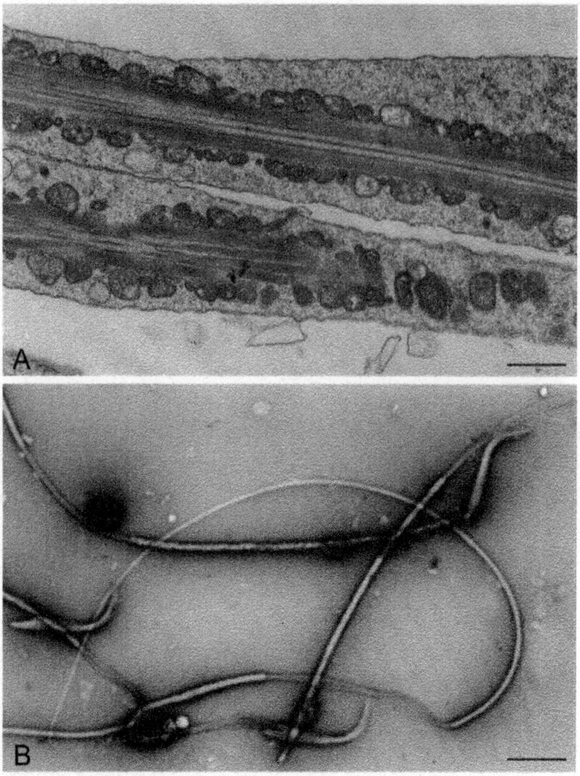

Fig. 1.35 High dosages of gossypol result in swelling and deterioration of most mitochondria in the rat mitochondrial sheath (MS) and consequentially a less motile phenotype. **A)** Electron micrograph of longitudinal section of two midpieces at the end of spermiogenesis afflicted with extensive mitochondrial swelling and degeneration. Bars = 0.2 μm. **B)** Phase micrograph of immotile cauda sperm with very large gaps in the MS and split axial fibers. Secondary lesions such as dislocations of axial fibers (see Figs. 1.32, 1.33) are not as readily found because of sperm immotility. Bars = 5 μm

Compliance with Ethical Standards

All applicable international, national and institutional guidelines for the care and use of animals were followed. This current work was supported by the Natural Science and Engineering Research Council of Canada (Grant number RGPIN/05305) to Richard Oko at Queen's University, along with the Agriculture and Food Research Initiative Competitive Grant number 2023–67015-39262 from the USDA National Institute of Food and Agriculture, Animal Breeding, Genetics & Genomics Program and seed funding from the College of Agriculture, Food and Natural Resources of the University of Missouri to Peter Sutovsky. Work for this book was also supported by a 2021–2022 Lalor Foundation Postdoctoral Research Fellowship to Lauren Hamilton. The research support to Viv Perry was funded by Australian Research Council (Grant number LP110100649),

English Beef and Lamb Executive a division of the Agriculture and Horticulture Development Board (Grant number RC3815), Queensland Sperm Morphology Laboratories, Ridley AgriProducts Pty Ltd and the Australian cattle industry. The Canadian Association of Animal Breeders provided the foundation for research by Albert Barth. They supported the Development of the Diagnostic Spermatology Laboratory at the Western College of Veterinary Medicine, Saskatoon, SK., along with related support for numerous other research directives. Many grants were also received from Saskatchewan Agriculture, including the Cattle Marketing Deductions Fund, the Horned Cattle Trust Fund, the Cattle Breeders Research Council, the Beef Development Fund and the Agriculture Development Fund. Animal research approval for research or clinical studies, summarized in these chapters were given by respective University Animal Care Committees. The authors declare no conflict of interest as there is no perceived interest in prejudicing the impartiality of the research reported.

References

Aarabi M, Balakier H, Bashar S, Moskovtsev SI, Sutovsky P, Librach CL, Oko R (2014) Sperm-derived WW domain-binding protein, PAWP, elicits calcium oscillations and oocyte activation in humans and mice. Faseb j 28(10):4434–4440. https://doi.org/10.1096/fj.14-256495

Andre J (1962) Contribution to the knowledge of the chondriome. Study of its ultrastructural changes during spermatogenesis. J Ultrastruct Res Suppl 3:1–185

Aul RB, Oko RJ (2002) The major subacrosomal occupant of bull spermatozoa is a novel histone H2B variant associated with the forming acrosome during spermiogenesis. Dev Biol 242(2):376–387

Baba T, Azuma S, Kashiwabara S, Toyoda Y (1994) Sperm from mice carrying a targeted mutation of the acrosin gene can penetrate the oocyte zona pellucida and effect fertilization. J Biol Chem 269(50):31845–31849

Baba T, Kashiwagi Y, Arimitsu N, Kogure T, Edo A, Maruyama T, Nakao K, Nakanishi H, Kinoshita M, Frohman MA, Yamamoto A, Tani K (2014) Phosphatidic acid (PA)-preferring phospholipase A1 regulates mitochondrial dynamics. J Biol Chem 289(16):11497–11511. https://doi.org/10.1074/jbc.M113.531921

Balhorn R (1982) A model for the structure of chromatin in mammalian sperm. J Cell Biol 93(2):298–305. https://doi.org/10.1083/jcb.93.2.298

Barth AD, Oko R (1989) Abnormal morphology of bovine spermatozoa. Iowa State University Press, Ames

Bedford JM (1998) Mammalian fertilization misread? Sperm penetration of the eutherian zona pellucida is unlikely to be a lytic event. Biol Reprod 59(6):1275–1287. https://doi.org/10.1095/biolreprod59.6.1275

Bleil JD, Greve JM, Wassarman PM (1988) Identification of a secondary sperm receptor in the mouse egg zona pellucida: role in maintenance of binding of acrosome-reacted sperm to eggs. Dev Biol 128(2):376–385. https://doi.org/10.1016/0012-1606(88)90299-0

Bouchard MJ, Dong Y, McDermott BM Jr, Lam DH, Brown KR, Shelanski M, Bellvé AR, Racaniello VR (2000) Defects in nuclear and cytoskeletal morphology and mitochondrial localization in spermatozoa of mice lacking nectin-2, a component of cell-cell adherens junctions. Mol Cell Biol 20(8):2865–2873. https://doi.org/10.1128/mcb.20.8.2865-2873.2000

Brown PR, Miki K, Harper DB, Eddy EM (2003) A-kinase anchoring protein 4 binding proteins in the fibrous sheath of the sperm flagellum. Biol Reprod 68(6):2241–2248. https://doi.org/10.1095/biolreprod.102.013466

Bryan J, Wilson L (1971) Are cytoplasmic microtubules heteropolymers? Proc Natl Acad Sci U S A 68(8):1762–1766. https://doi.org/10.1073/pnas.68.8.1762

Burfeind P, Hoyer-Fender S (1991) Sequence and developmental expression of a mRNA encoding a putative protein of rat sperm outer dense fibers. Dev Biol 148(1):195–204. https://doi.org/10.1016/0012-1606(91)90329-2

Burton KA, McKnight GS (2007) PKA, germ cells, and fertility. Physiology (Bethesda) 22:40–46. https://doi.org/10.1152/physiol.00034.2006

Calvin HI, Cooper GW, Wallace E (1981) Evidence that selenium in rat sperm is asociated with a cysteine-rich structural protein of mitochondrial capsules. Gamete Res 4(2):139–142

Calvin HI, Hwang FH, Wohlrab H (1975) Localization of zinc in a dense fiber-connecting piece fraction of rat sperm tails analogous chemically to hair keratin. Biol Reprod 13(2):228–239. https://doi.org/10.1095/biolreprod13.2.228

Carr DW, Fujita A, Stentz CL, Liberty GA, Olson GE, Narumiya S (2001) Identification of sperm-specific proteins that interact with A-kinase anchoring proteins in a manner similar to the type II regulatory subunit of PKA. J Biol Chem 276(20):17332–17338. https://doi.org/10.1074/jbc.M011252200

Carrera A, Gerton GL, Moss SB (1994) The major fibrous sheath polypeptide of mouse sperm: structural and functional similarities to the A-kinase anchoring proteins. Dev Biol 165(1):272–284. https://doi.org/10.1006/dbio.1994.1252

Carrera A, Moos J, Ning XP, Gerton GL, Tesarik J, Kopf GS, Moss SB (1996) Regulation of protein tyrosine phosphorylation in human sperm by a calcium/calmodulin-dependent mechanism: identification of A kinase anchor proteins as major substrates for tyrosine phosphorylation. Dev Biol 180(1):284–296. https://doi.org/10.1006/dbio.1996.0301

Cataldo L, Baig K, Oko R, Mastrangelo MA, Kleene KC (1996) Developmental expression, intracellular localization, and selenium content of the cysteine-rich protein associated with the mitochondrial capsules of mouse sperm. Mol Reprod Dev 45(3):320–331. https://doi.org/10.1002/(sici)1098-2795(199611)45:3<320::Aid-mrd9>3.0.Co;2-u

Chen Y, Liang P, Huang Y, Li M, Zhang X, Ding C, Feng J, Zhang Z, Zhang X, Gao Y, Zhang Q, Cao S, Zheng H, Liu D, Songyang Z, Huang J (2017) Glycerol kinase-like proteins cooperate with Pld6 in regulating sperm mitochondrial sheath formation and male fertility. Cell Discov 3:17030. https://doi.org/10.1038/celldisc.2017.30

Cho C, Jung-Ha H, Willis WD, Goulding EH, Stein P, Xu Z, Schultz RM, Hecht NB, Eddy EM (2003) Protamine 2 deficiency leads to sperm DNA damage and embryo death in mice. Biol Reprod 69(1):211–217. https://doi.org/10.1095/biolreprod.102.015115

Cho C, Willis WD, Goulding EH, Jung-Ha H, Choi YC, Hecht NB, Eddy EM (2001) Haploinsufficiency of protamine-1 or -2 causes infertility in mice. Nat Genet 28(1):82–86. https://doi.org/10.1038/ng0501-82

Choi SY, Huang P, Jenkins GM, Chan DC, Schiller J, Frohman MA (2006) A common lipid links Mfn-mediated mitochondrial fusion and SNARE-regulated exocytosis. Nat Cell Biol 8(11):1255–1262. https://doi.org/10.1038/ncb1487

Clermont Y, Oko R, Hermo L (1990) Immunocytochemical localization of proteins utilized in the formation of outer dense fibers and fibrous sheath in rat spermatids: an electron microscope study. Anat Rec 227(4):447–457. https://doi.org/10.1002/ar.1092270408

Clermont Y, Rambourg A, Hermo L (1994) Connections between the various elements of the cis- and mid-compartments of the Golgi apparatus of early rat spermatids. Anat Rec 240(4):469–480. https://doi.org/10.1002/ar.1092400405

Clermont Y, Tang XM (1985) Glycoprotein synthesis in the Golgi apparatus of spermatids during spermiogenesis of the rat. Anat Rec 213(1):33–43. https://doi.org/10.1002/ar.1092130106

Cleveland DW, Sullivan KF (1985) Molecular biology and genetics of tubulin. Annu Rev Biochem 54:331–365. https://doi.org/10.1146/annurev.bi.54.070185.001555

Cowan AE, Myles DG, Koppel DE (1991) Migration of the guinea pig sperm membrane protein PH-20 from one localized surface domain to another does not occur by a simple diffusion-trapping mechanism. Dev Biol 144(1):189–198

Danshina PV, Geyer CB, Dai Q, Goulding EH, Willis WD, Kitto GB, McCarrey JR, Eddy EM, O'Brien DA (2010) Phosphoglycerate kinase 2 (PGK2) is essential for sperm function and male fertility in mice. Biol Reprod 82(1):136–145. https://doi.org/10.1095/biolreprod.109.079699
Distel RJ, Kleene KC, Hecht NB (1984) Haploid expression of a mouse testis alpha-tubulin gene. Science 224(4644):68–70. https://doi.org/10.1126/science.6701535
Doiguchi M, Mori T, Toshimori K, Shibata Y, Iida H (2002) Spergen-1 might be an adhesive molecule associated with mitochondria in the middle piece of spermatozoa. Dev Biol 252(1):127–137. https://doi.org/10.1006/dbio.2002.0833
Dutcher SK, Huang B, Luck DJ (1984) Genetic dissection of the central pair microtubules of the flagella of Chlamydomonas reinhardtii. J Cell Biol 98(1):229–236. https://doi.org/10.1083/jcb.98.1.229
Egydio de Carvalho C, Tanaka H, Iguchi N, Ventelä S, Nojima H, Nishimune Y (2002) Molecular cloning and characterization of a complementary DNA encoding sperm tail protein SHIPPO 1. Biol Reprod 66(3):785–795. https://doi.org/10.1095/biolreprod66.3.785
El-Alfy M, Moshonas D, Morales CR, Oko R (1999) Molecular cloning and developmental expression of the major fibrous sheath protein (FS 75) of rat sperm. J Androl 20(2):307–318
Fawcett DW (1975) The mammalian spermatozoon. Dev Biol 44(2):394–436. https://doi.org/10.1016/0012-1606(75)90411-x
Fawcett DW (1981) The Cell. W. B. Saunders Company, Philadelphia, PA
Fawcett DW, Chemes HE (1979) Changes in distribution of nuclear pores during differentiation of the male germ cells. Tissue Cell 11(1):147–162. https://doi.org/10.1016/0040-8166(79)90015-6
Ferrer M, Rodriguez H, Zara L, Yu Y, Xu W, Oko R (2012a) MMP2 and acrosin are major proteinases associated with the inner acrosomal membrane and may cooperate in sperm penetration of the zona pellucida during fertilization. Cell Tissue Res 349(3):881–895. https://doi.org/10.1007/s00441-012-1429-1
Ferrer M, Xu W, Oko R (2012b) The composition, protein genesis and significance of the inner acrosomal membrane of eutherian sperm. Cell Tissue Res 349(3):733–748. https://doi.org/10.1007/s00441-012-1433-5
Ferrer MJ, Xu W, Shetty J, Herr J, Oko R (2016) Plasminogen Improves Mouse IVF by Interactions with Inner Acrosomal Membrane-Bound MMP2 and SAMP14. Biol Reprod 94(4):88. https://doi.org/10.1095/biolreprod.115.133496
Fitzgerald C, Sikora C, Lawson V, Dong K, Cheng M, Oko R, van der Hoorn FA (2006a) Mammalian transcription in support of hybrid mRNA and protein synthesis in testis and lung. J Biol Chem 281(50):38172–38180. https://doi.org/10.1074/jbc.M606010200
Fitzgerald CJ, Oko RJ, van der Hoorn FA (2006b) Rat Spag5 associates in somatic cells with endoplasmic reticulum and microtubules but in spermatozoa with outer dense fibers. Mol Reprod Dev 73(1):92–100. https://doi.org/10.1002/mrd.20388
Flohé L (2007) Selenium in mammalian spermiogenesis. Biol Chem 388(10):987–995. https://doi.org/10.1515/bc.2007.112
Fujita A, Nakamura K, Kato T, Watanabe N, Ishizaki T, Kimura K, Mizoguchi A, Narumiya S (2000) Ropporin, a sperm-specific binding protein of rhophilin, that is localized in the fibrous sheath of sperm flagella. J Cell Sci 113(Pt 1):103–112
Fulcher KD, Mori C, Welch JE, O'Brien DA, Klapper DG, Eddy EM (1995a) Characterization of Fsc1 cDNA for a mouse sperm fibrous sheath component. Biol Reprod 52(1):41–49. https://doi.org/10.1095/biolreprod52.1.41
Fulcher KD, Welch JE, Klapper DG, O'Brien DA, Eddy EM (1995b) Identification of a unique mu-class glutathione S-transferase in mouse spermatogenic cells. Mol Reprod Dev 42(4):415–424. https://doi.org/10.1002/mrd.1080420407
Gerton GL (2002) 8 - Function of the Sperm Acrosome. In: Hardy DM (ed) Fertilization. Academic Press, San Diego, pp 265–302. https://doi.org/10.1016/B978-012311629-1/50010-3
Gibbons BH, Gibbons IR (1976) Functional recombination of dynein 1 with demembranated sea urchin sperm partially extracted with KC1. Biochem Biophys Res Commun 73(1):1–6. https://doi.org/10.1016/0006-291x(76)90488-5

Gibbons IR (1981) Cilia and flagella of eukaryotes. J Cell Biol 91(3 Pt 2):107s–124s. https://doi.org/10.1083/jcb.91.3.107s
Goodenough UW, Heuser JE (1985a) Outer and inner dynein arms of cilia and flagella. Cell 41(2):341–342. https://doi.org/10.1016/s0092-8674(85)80003-9
Goodenough UW, Heuser JE (1985b) Substructure of inner dynein arms, radial spokes, and the central pair/projection complex of cilia and flagella. J Cell Biol 100(6):2008–2018. https://doi.org/10.1083/jcb.100.6.2008
Green D (2002) Fertilization biophysics. In: Fertilization. Elsevier, pp 387–399
Hamilton LE, Acteau G, Xu W, Sutovsky P, Oko R (2017) The developmental origin and compartmentalization of glutathione-s-transferase omega 2 isoforms in the perinuclear theca of eutherian spermatozoa. Biol Reprod 97(4):612–621. https://doi.org/10.1093/biolre/iox122
Hamilton LE, Lion M, Aguila L, Suzuki J, Acteau G, Protopapas N, Xu W, Sutovsky P, Baker M, Oko R (2021) Core Histones Are Constituents of the Perinuclear Theca of Murid Spermatozoa: an Assessment of Their Synthesis and Assembly during Spermiogenesis and Function after Gametic Fusion. Int J Mol Sci 22 (15). https://doi.org/10.3390/ijms22158119
Hamilton LE, Suzuki J, Aguila L, Meinsohn MC, Smith OE, Protopapas N, Xu W, Sutovsky P, Oko R (2019) Sperm-borne glutathione-S-transferase omega 2 accelerates the nuclear decondensation of spermatozoa during fertilization in mice†. Biol Reprod 101(2):368–376. https://doi.org/10.1093/biolre/ioz082
Hao Z, Wolkowicz MJ, Shetty J, Klotz K, Bolling L, Sen B, Westbrook VA, Coonrod S, Flickinger CJ, Herr JC (2002) SAMP32, a testis-specific, isoantigenic sperm acrosomal membrane-associated protein. Biol Reprod 66(3):735–744. https://doi.org/10.1095/biolreprod66.3.735
Heid H, Figge U, Winter S, Kuhn C, Zimbelmann R, Franke W (2002) Novel actin-related proteins Arp-T1 and Arp-T2 as components of the cytoskeletal calyx of the mammalian sperm head. Exp Cell Res 279(2):177–187. https://doi.org/10.1006/excr.2002.5603
Hermo L, Oko R, Hecht NB (1991) Differential post-translational modifications of microtubules in cells of the seminiferous epithelium of the rat: a light and electron microscope immunocytochemical study. Anat Rec 229(1):31–50. https://doi.org/10.1002/ar.1092290106
Herrada G, Wolgemuth DJ (1997) The mouse transcription factor Stat4 is expressed in haploid male germ cells and is present in the perinuclear theca of spermatozoa. J Cell Sci 110(Pt 14):1543–1553
Hess H, Heid H, Franke WW (1993) Molecular characterization of mammalian cylicin, a basic protein of the sperm head cytoskeleton. J Cell Biol 122(5):1043–1052. https://doi.org/10.1083/jcb.122.5.1043
Hess H, Heid H, Zimbelmann R, Franke WW (1995) The protein complexity of the cytoskeleton of bovine and human sperm heads: the identification and characterization of cylicin II. Exp Cell Res 218(1):174–182. https://doi.org/10.1006/excr.1995.1145
Hill KE, Zhou J, McMahan WJ, Motley AK, Atkins JF, Gesteland RF, Burk RF (2003) Deletion of selenoprotein P alters distribution of selenium in the mouse. J Biol Chem 278(16):13640–13646. https://doi.org/10.1074/jbc.M300755200
Horowitz JA, Toeg H, Orr GA (1984) Characterization and localization of cAMP-dependent protein kinases in rat caudal epididymal sperm. J Biol Chem 259(2):832–838
Hrudka F (1978) A morphological and cytochemical study on isolated sperm mitochondria. J Ultrastruct Res 63(1):1–19. https://doi.org/10.1016/s0022-5320(78)80040-9
Hrudka F (1979) Cytochemistry of oxidoreductases in spermatozoa: the technique revisited. Andrologia 11(5):337–353. https://doi.org/10.1111/j.1439-0272.1979.tb02217.x
Huang B, Piperno G, Luck DJ (1979) Paralyzed flagella mutants of Chlamydomonas reinhardtii. Defective for axonemal doublet microtubule arms. J Biol Chem 254 (8):3091–3099
Huang B, Piperno G, Ramanis Z, Luck DJ (1981) Radial spokes of Chlamydomonas flagella: genetic analysis of assembly and function. J Cell Biol 88(1):80–88. https://doi.org/10.1083/jcb.88.1.80
Huang H, Gao Q, Peng X, Choi SY, Sarma K, Ren H, Morris AJ, Frohman MA (2011) PiRNA-associated germline nuage formation and spermatogenesis require MitoPLD profusogenic

mitochondrial-surface lipid signaling. Dev Cell 20(3):376–387. https://doi.org/10.1016/j.devcel.2011.01.004

Huang TT Jr, Yanagimachi R (1985) Inner acrosomal membrane of mammalian spermatozoa: its properties and possible functions in fertilization. Am J Anat 174(3):249–268. https://doi.org/10.1002/aja.1001740307

Huang Z, Danshina PV, Mohr K, Qu W, Goodson SG, O'Connell TM, O'Brien DA (2017) Sperm function, protein phosphorylation, and metabolism differ in mice lacking successive sperm-specific glycolytic enzymes. Biol Reprod 97(4):586–597. https://doi.org/10.1093/biolre/iox103

Hüber D, Geisler S, Monecke S, Hoyer-Fender S (2008) Molecular dissection of ODF2/Cenexin revealed a short stretch of amino acids necessary for targeting to the centrosome and the primary cilium. Eur J Cell Biol 87(3):137–146. https://doi.org/10.1016/j.ejcb.2007.10.004

Hunnicutt GR, Primakoff P, Myles DG (1996) Sperm surface protein PH-20 is bifunctional: one activity is a hyaluronidase and a second, distinct activity is required in secondary sperm-zona binding. Biol Reprod 55(1):80–86

Hurst S, Howes EA, Coadwell J, Jones R (1998) Expression of a testis-specific putative actin-capping protein associated with the developing acrosome during rat spermiogenesis. Mol Reprod Dev 49(1):81–91. https://doi.org/10.1002/(sici)1098-2795(199801)49:1<81::Aid-mrd9>3.0.Co;2-k

Imai H, Hakkaku N, Iwamoto R, Suzuki J, Suzuki T, Tajima Y, Konishi K, Minami S, Ichinose S, Ishizaka K, Shioda S, Arata S, Nishimura M, Naito S, Nakagawa Y (2009) Depletion of selenoprotein GPx4 in spermatocytes causes male infertility in mice. J Biol Chem 284(47):32522–32532. https://doi.org/10.1074/jbc.M109.016139

Irons MJ, Clermont Y (1982) Kinetics of fibrous sheath formation in the rat spermatid. Am J Anat 165(2):121–130. https://doi.org/10.1002/aja.1001650204

Ito C, Akutsu H, Yao R, Yoshida K, Yamatoya K, Mutoh T, Makino T, Aoyama K, Ishikawa H, Kunimoto K, Tsukita S, Noda T, Kikkawa M, Toshimori K (2019) Odf2 haploinsufficiency causes a new type of decapitated and decaudated spermatozoa, Odf2-DDS, in mice. Sci Rep 9(1):14249. https://doi.org/10.1038/s41598-019-50516-2

Ito C, Suzuki-Toyota F, Maekawa M, Toyama Y, Yao R, Noda T, Toshimori K (2004) Failure to assemble the peri-nuclear structures in GOPC deficient spermatids as found in round-headed spermatozoa. Arch Histol Cytol 67(4):349–360. https://doi.org/10.1679/aohc.67.349

Jin M, Fujiwara E, Kakiuchi Y, Okabe M, Satouh Y, Baba SA, Chiba K, Hirohashi N (2011) Most fertilizing mouse spermatozoa begin their acrosome reaction before contact with the zona pellucida during in vitro fertilization. Proc Natl Acad Sci U S A 108(12):4892–4896. https://doi.org/10.1073/pnas.1018202108

Johnson KA, Wall JS (1983) Structure and molecular weight of the dynein ATPase. J Cell Biol 96(3):669–678. https://doi.org/10.1083/jcb.96.3.669

Jones R (1991) Interaction of zona pellucida glycoproteins, sulphated carbohydrates and synthetic polymers with proacrosin, the putative egg-binding protein from mammalian spermatozoa. Development 111(4):1155–1163

Kamiya R, Okamoto M (1985) A mutant of Chlamydomonas reinhardtii that lacks the flagellar outer dynein arm but can swim. J Cell Sci 74:181–191

Kann ML, Feinberg J, Rainteau D, Dadoune JP, Weinman S, Fouquet JP (1991) Localization of calmodulin in perinuclear structures of spermatids and spermatozoa: a comparison of six mammalian species. Anat Rec 230(4):481–488. https://doi.org/10.1002/ar.1092300407

Kemphues KJ, Kaufman TC, Raff RA, Raff EC (1982) The testis-specific beta-tubulin subunit in Drosophila melanogaster has multiple functions in spermatogenesis. Cell 31(3 Pt 2):655–670. https://doi.org/10.1016/0092-8674(82)90321-x

Kong M, Richardson RT, Widgren EE, O'Rand MG (1995) Sequence and localization of the mouse sperm autoantigenic protein, Sp17. Biol Reprod 53(3):579–590

Krisfalusi M, Miki K, Magyar PL, O'Brien DA (2006) Multiple glycolytic enzymes are tightly bound to the fibrous sheath of mouse spermatozoa. Biol Reprod 75(2):270–278. https://doi.org/10.1095/biolreprod.105.049684

Leclerc P, Goupil S (2000) Distribution and localization of calmodulin-binding proteins in bull spermatozoa. Biol Reprod 62(6):1875–1881. https://doi.org/10.1095/biolreprod62.6.1875

Lécuyer C, Dacheux JL, Hermand E, Mazeman E, Rousseaux J, Rousseaux-Prévost R (2000) Actin-binding properties and colocalization with actin during spermiogenesis of mammalian sperm calicin. Biol Reprod 63(6):1801–1810. https://doi.org/10.1095/biolreprod63.6.1801

Lin RY, Moss SB, Rubin CS (1995) Characterization of S-AKAP84, a novel developmentally regulated A kinase anchor protein of male germ cells. J Biol Chem 270(46):27804–27811. https://doi.org/10.1074/jbc.270.46.27804

Lin YN, Roy A, Yan W, Burns KH, Matzuk MM (2007) Loss of zona pellucida binding proteins in the acrosomal matrix disrupts acrosome biogenesis and sperm morphogenesis. Mol Cell Biol 27(19):6794–6805. https://doi.org/10.1128/mcb.01029-07

Linck RW (1979) Advances in ultrastructure analysis of the sperm flagellar axoneme. The Spermatozoon. Urban and Schwarzenberg, Baltimore

Listowsky I (2005) A subclass of mu glutathione S-transferases selectively expressed in testis and brain. Methods Enzymol 401:278–287. https://doi.org/10.1016/s0076-6879(05)01018-9

Luck D, Piperno G, Ramanis Z, Huang B (1977) Flagellar mutants of Chlamydomonas: studies of radial spoke-defective strains by dikaryon and revertant analysis. Proc Natl Acad Sci U S A 74(8):3456–3460. https://doi.org/10.1073/pnas.74.8.3456

Luck DJ (1984) Genetic and biochemical dissection of the eucaryotic flagellum. J Cell Biol 98(3):789–794. https://doi.org/10.1083/jcb.98.3.789

Mandal A, Naaby-Hansen S, Wolkowicz MJ, Klotz K, Shetty J, Retief JD, Coonrod SA, Kinter M, Sherman N, Cesar F, Flickinger CJ, Herr JC (1999) FSP95, a testis-specific 95-kilodalton fibrous sheath antigen that undergoes tyrosine phosphorylation in capacitated human spermatozoa. Biol Reprod 61(5):1184–1197. https://doi.org/10.1095/biolreprod61.5.1184

Marchese-Ragona SP, Gagnon C, White D, Isles MB, Johnson KA (1987) Structure and mass analysis of 12S and 19S dynein obtained from bull sperm flagella. Cell Motil Cytoskeleton 8(4):368–374. https://doi.org/10.1002/cm.970080409

Mei X, Singh IS, Erlichman J, Orr GA (1997) Cloning and characterization of a testis-specific, developmentally regulated A-kinase-anchoring protein (TAKAP-80) present on the fibrous sheath of rat sperm. Eur J Biochem 246(2):425–432. https://doi.org/10.1111/j.1432-1033.1997.t01-1-00425.x

Meistrich ML, Mohapatra B, Shirley CR, Zhao M (2003) Roles of transition nuclear proteins in spermiogenesis. Chromosoma 111(8):483–488. https://doi.org/10.1007/s00412-002-0227-z

Mi Y, Shi Z, Li J (2015) Spata19 is critical for sperm mitochondrial function and male fertility. Mol Reprod Dev 82(11):907–913. https://doi.org/10.1002/mrd.22536

Miki K, Eddy EM (1998) Identification of tethering domains for protein kinase A type Ialpha regulatory subunits on sperm fibrous sheath protein FSC1. J Biol Chem 273(51):34384–34390. https://doi.org/10.1074/jbc.273.51.34384

Miki K, Eddy EM (1999) Single amino acids determine specificity of binding of protein kinase A regulatory subunits by protein kinase A anchoring proteins. J Biol Chem 274(41):29057–29062. https://doi.org/10.1074/jbc.274.41.29057

Miki K, Qu W, Goulding EH, Willis WD, Bunch DO, Strader LF, Perreault SD, Eddy EM, O'Brien DA (2004) Glyceraldehyde 3-phosphate dehydrogenase-S, a sperm-specific glycolytic enzyme, is required for sperm motility and male fertility. Proc Natl Acad Sci U S A 101(47):16501–16506. https://doi.org/10.1073/pnas.0407708101

Miki K, Willis WD, Brown PR, Goulding EH, Fulcher KD, Eddy EM (2002) Targeted disruption of the Akap4 gene causes defects in sperm flagellum and motility. Dev Biol 248(2):331–342. https://doi.org/10.1006/dbio.2002.0728

Miranda-Vizuete A, Sadek CM, Jiménez A, Krause WJ, Sutovsky P, Oko R (2004) The mammalian testis-specific thioredoxin system. Antioxid Redox Signal 6(1):25–40. https://doi.org/10.1089/152308604771978327

Miranda-Vizuete A, Tsang K, Yu Y, Jiménez A, Pelto-Huikko M, Flickinger CJ, Sutovsky P, Oko R (2003) Cloning and developmental analysis of murid spermatid-specific

thioredoxin-2 (SPTRX-2), a novel sperm fibrous sheath protein and autoantigen. J Biol Chem 278(45):44874–44885. https://doi.org/10.1074/jbc.M305475200
Mitchell DR, Rosenbaum JL (1985) A motile Chlamydomonas flagellar mutant that lacks outer dynein arms. J Cell Biol 100(4):1228–1234. https://doi.org/10.1083/jcb.100.4.1228
Morales CR, Oko R, Clermont Y (1994) Molecular cloning and developmental expression of an mRNA encoding the 27 kDa outer dense fiber protein of rat spermatozoa. Mol Reprod Dev 37(2):229–240. https://doi.org/10.1002/mrd.1080370215
Mortillo S, Wassarman PM (1991) Differential binding of gold-labeled zona pellucida glycoproteins mZP2 and mZP3 to mouse sperm membrane compartments. Development 113(1):141–149
Mortimer D, Thompson TE (1976) Nuclear pores in the spermatozoon of the rat. Experientia 32(1):104–106. https://doi.org/10.1007/bf01932648
Mountjoy JR, Xu W, McLeod D, Hyndman D, Oko R (2008) RAB2A: a major subacrosomal protein of bovine spermatozoa implicated in acrosomal biogenesis. Biol Reprod 79(2):223–232. https://doi.org/10.1095/biolreprod.107.065060
Nakamura N, Dai Q, Williams J, Goulding EH, Willis WD, Brown PR, Eddy EM (2013) Disruption of a spermatogenic cell-specific mouse enolase 4 (eno4) gene causes sperm structural defects and male infertility. Biol Reprod 88(4):90. https://doi.org/10.1095/biolreprod.112.107128
Nakamura N, Mori C, Eddy EM (2010) Molecular complex of three testis-specific isozymes associated with the mouse sperm fibrous sheath: hexokinase 1, phosphofructokinase M, and glutathione S-transferase mu class 5. Biol Reprod 82(3):504–515. https://doi.org/10.1095/biolreprod.109.080580
Nakamura Y, Tanaka H, Koga M, Miyagawa Y, Iguchi N, Egydio de Carvalho C, Yomogida K, Nozaki M, Nojima H, Matsumiya K, Okuyama A, Nishimune Y (2002) Molecular cloning and characterization of oppo 1: a haploid germ cell-specific complementary DNA encoding sperm tail protein. Biol Reprod 67(1):1–7. https://doi.org/10.1095/biolreprod67.1.1
Odet F, Duan C, Willis WD, Goulding EH, Kung A, Eddy EM, Goldberg E (2008) Expression of the gene for mouse lactate dehydrogenase C (Ldhc) is required for male fertility. Biol Reprod 79(1):26–34. https://doi.org/10.1095/biolreprod.108.068353
Ohmura K, Kohno N, Kobayashi Y, Yamagata K, Sato S, Kashiwabara S, Baba T (1999) A homologue of pancreatic trypsin is localized in the acrosome of mammalian sperm and is released during acrosome reaction. J Biol Chem 274(41):29426–29432. https://doi.org/10.1074/jbc.274.41.29426
Oko R (1984) The effects and possible mechanisms of gossypol action on sperm organelles, spermatogenesis and accessory sex glands of the adult rat. University of Saskatchewan,
Oko R (1988) Comparative analysis of proteins from the fibrous sheath and outer dense fibers of rat spermatozoa. Biol Reprod 39(1):169–182. https://doi.org/10.1095/biolreprod39.1.169
Oko R, Clermont Y (1990) Mammalian Spermatozoa: stucture and Assembly of the Tail. Controls of Sperm Motility: biological and Clinical Aspects, 1st Edition edn. CRC Press, Boca Raton, Florida
Oko R, Donald A, Xu W, van der Spoel AC (2011) Fusion failure of dense-cored proacrosomal vesicles in an inducible mouse model of male infertility. Cell Tissue Res 346(1):119–134. https://doi.org/10.1007/s00441-011-1248-9
Oko R, Hrudka F (1982) Segmental aplasia of the mitochondrial sheath and sequelae induced by gossypol in rat spermatozoa. Biol Reprod 26(1):183–195. https://doi.org/10.1095/biolreprod26.1.183
Oko R, Hrudka F (1984) Gossypol-induced early and delayed effects in the seminiferous epithelium of the adult rat. Contracept Deliv Syst 5:22
Oko R, Maravei D (1994) Protein composition of the perinuclear theca of bull spermatozoa. Biol Reprod 50(5):1000–1014. https://doi.org/10.1095/biolreprod50.5.1000
Oko R, Morales CR (1994) A novel testicular protein, with sequence similarities to a family of lipid binding proteins, is a major component of the rat sperm perinuclear theca. Dev Biol 166(1):235–245. https://doi.org/10.1006/dbio.1994.1310

Oko R, Sutovsky P (2009) Biogenesis of sperm perinuclear theca and its role in sperm functional competence and fertilization. J Reprod Immunol 83(1–2):2–7. https://doi.org/10.1016/j.jri.2009.05.008

Oko RJ (1995) Developmental expression and possible role of perinuclear theca proteins in mammalian spermatozoa. Reprod Fertil Dev 7(4):777–797. https://doi.org/10.1071/rd9950777

Oko RJ, Jando V, Wagner CL, Kistler WS, Hermo LS (1996) Chromatin reorganization in rat spermatids during the disappearance of testis-specific histone, H1t, and the appearance of transition proteins TP1 and TP2. Biol Reprod 54(5):1141–1157. https://doi.org/10.1095/biolreprod54.5.1141

Olson GE, Hamilton DW, Fawcett DW (1976) Isolation and characterization of the fibrous sheath of rat epididymal spermatozoa. Biol Reprod 14(5):517–530. https://doi.org/10.1095/biolreprod14.5.517

Olson GE, Sammons DW (1980) Structural chemistry of outer dense fibers of rat sperm. Biol Reprod 22(2):319–332. https://doi.org/10.1095/biolreprod22.2.319

Olson GE, Winfrey VP (1990) Mitochondria-cytoskeleton interactions in the sperm midpiece. J Struct Biol 103(1):13–22. https://doi.org/10.1016/1047-8477(90)90081-m

Olson GE, Winfrey VP, Nagdas SK, Hill KE, Burk RF (2005) Selenoprotein P is required for mouse sperm development. Biol Reprod 73(1):201–211. https://doi.org/10.1095/biolreprod.105.040360

Olson GE, Winfrey VP, Nagdas SK, Hill KE, Burk RF (2007) Apolipoprotein E receptor-2 (ApoER2) mediates selenium uptake from selenoprotein P by the mouse testis. J Biol Chem 282(16):12290–12297. https://doi.org/10.1074/jbc.M611403200

PalliniE. VB (1979) Bull sperm selenium is bound to a structural protein of mitochondria. J Submicrosc Cyto 11:6

Petersen C, Aumüller G, Bahrami M, Hoyer-Fender S (2002) Molecular cloning of Odf3 encoding a novel coiled-coil protein of sperm tail outer dense fibers. Mol Reprod Dev 61(1):102–112. https://doi.org/10.1002/mrd.1136

Piperno G, Luck DJ (1979) Axonemal adenosine triphosphatases from flagella of Chlamydomonas reinhardtii. Purification of two dyneins. J Biol Chem 254 (8):3084–3090

Piperno G, Luck DJ (1981) Inner arm dyneins from flagella of Chlamydomonas reinhardtii. Cell 27(2 Pt 1):331–340. https://doi.org/10.1016/0092-8674(81)90416-5

Protopapas N, Hamilton LE, Warkentin R, Xu W, Sutovsky P, Oko R (2019) The perforatorium and postacrosomal sheath of rat spermatozoa share common developmental origins and protein constituentsdagger. Biol Reprod 100(6):1461–1472. https://doi.org/10.1093/biolre/ioz052

Richardson RT, Yamasaki N, Michael G (1994) Sequence of a rabbit sperm zona pellucida binding protein and localization during the acrosome reaction. Dev Biol 165(2):688–701

Ruan Y, Cheng M, Ou Y, Oko R, van der Hoorn FA (2011) Ornithine decarboxylase antizyme Oaz3 modulates protein phosphatase activity. J Biol Chem 286(33):29417–29427. https://doi.org/10.1074/jbc.M111.274647

Russell L, Peterson RN, Freund M (1979) Morphologic characteristics of the chemically induced acrosome reaction in human spermatozoa. Fertil Steril 32(1):87–92. https://doi.org/10.1016/s0015-0282(16)44121-x

Saacke RG, Almquist JO (1964a) ULTRASTRUCTURE OF BOVINE SPERMATOZOA. I. THE HEAD OF NORMAL, EJACULATED SPERM 1,2,3. Am J Anat 115:143–161. https://doi.org/10.1002/aja.1001150109

Saacke RG, Almquist JO (1964b) ULTRASTRUCTURE OF BOVINE SPERMATOZOA. II. THE NECK AND TAIL OF NORMAL, EJACULATED SPERM 1,2,3. Am J Anat 115:163–183. https://doi.org/10.1002/aja.1001150110

Sale WS, Goodenough UW, Heuser JE (1985) The substructure of isolated and in situ outer dynein arms of sea urchin sperm flagella. J Cell Biol 101(4):1400–1412. https://doi.org/10.1083/jcb.101.4.1400

Sale WS, Satir P (1977) Direction of active sliding of microtubules in Tetrahymena cilia. Proc Natl Acad Sci U S A 74(5):2045–2049. https://doi.org/10.1073/pnas.74.5.2045

Salmon NA, Reijo Pera RA, Xu EY (2006) A gene trap knockout of the abundant sperm tail protein, outer dense fiber 2, results in preimplantation lethality. Genesis 44(11):515–522. https://doi.org/10.1002/dvg.20241

Satir P, Heuser T, Sale WS (2014) A Structural Basis for How Motile Cilia Beat. Bioscience 64(12):1073–1083. https://doi.org/10.1093/biosci/biu180

Schalles U, Shao X, van der Hoorn FA, Oko R (1998) Developmental expression of the 84-kDa ODF sperm protein: localization to both the cortex and medulla of outer dense fibers and to the connecting piece. Dev Biol 199(2):250–260. https://doi.org/10.1006/dbio.1998.8931

Schneider M, Förster H, Boersma A, Seiler A, Wehnes H, Sinowatz F, Neumüller C, Deutsch MJ, Walch A, Hrabé de Angelis M, Wurst W, Ursini F, Roveri A, Maleszewski M, Maiorino M, Conrad M (2009) Mitochondrial glutathione peroxidase 4 disruption causes male infertility. Faseb j 23(9):3233–3242. https://doi.org/10.1096/fj.09-132795

Scott JD, Dessauer CW, Taskén K (2013) Creating order from chaos: cellular regulation by kinase anchoring. Annu Rev Pharmacol Toxicol 53:187–210. https://doi.org/10.1146/annurev-pharmtox-011112-140204

Shao X, Tarnasky HA, Lee JP, Oko R, van der Hoorn FA (1999) Spag4, a novel sperm protein, binds outer dense-fiber protein Odf1 and localizes to microtubules of manchette and axoneme. Dev Biol 211(1):109–123. https://doi.org/10.1006/dbio.1999.9297

Shao X, Tarnasky HA, Schalles U, Oko R, van der Hoorn FA (1997) Interactional cloning of the 84-kDa major outer dense fiber protein Odf84. Leucine zippers mediate associations of Odf84 and Odf27. J Biol Chem 272 (10):6105–6113. https://doi.org/10.1074/jbc.272.10.6105

Shao X, van der Hoorn FA (1996) Self-interaction of the major 27-kilodalton outer dense fiber protein is in part mediated by a leucine zipper domain in the rat. Biol Reprod 55(6):1343–1350. https://doi.org/10.1095/biolreprod55.6.1343

Shetty J, Wolkowicz MJ, Digilio LC, Klotz KL, Jayes FL, Diekman AB, Westbrook VA, Farris EM, Hao Z, Coonrod SA, Flickinger CJ, Herr JC (2003) SAMP14, a novel, acrosomal membrane-associated, glycosylphosphatidylinositol-anchored member of the Ly-6/urokinase-type plasminogen activator receptor superfamily with a role in sperm-egg interaction. J Biol Chem 278(33):30506–30515. https://doi.org/10.1074/jbc.M301713200

Shirley CR, Hayashi S, Mounsey S, Yanagimachi R, Meistrich ML (2004) Abnormalities and reduced reproductive potential of sperm from Tnp1- and Tnp2-null double mutant mice. Biol Reprod 71(4):1220–1229. https://doi.org/10.1095/biolreprod.104.029363

Soung NK, Kang YH, Kim K, Kamijo K, Yoon H, Seong YS, Kuo YL, Miki T, Kim SR, Kuriyama R, Giam CZ, Ahn CH, Lee KS (2006) Requirement of hCenexin for proper mitotic functions of polo-like kinase 1 at the centrosomes. Mol Cell Biol 26(22):8316–8335. https://doi.org/10.1128/mcb.00671-06

Soung NK, Park JE, Yu LR, Lee KH, Lee JM, Bang JK, Veenstra TD, Rhee K, Lee KS (2009) Plk1-dependent and -independent roles of an ODF2 splice variant, hCenexin1, at the centrosome of somatic cells. Dev Cell 16(4):539–550. https://doi.org/10.1016/j.devcel.2009.02.004

Stephens RE (1970) Thermal fractionation of outer fiber doublet microtubules into A- and B-subfiber components. A- and B-Tubulin. J Mol Biol 47(3):353–363. https://doi.org/10.1016/0022-2836(70)90307-4

Storey BT (2008) Mammalian sperm metabolism: oxygen and sugar, friend and foe. Int J Dev Biol 52(5–6):427–437. https://doi.org/10.1387/ijdb.072522bs

Sutovsky P, Hewitson L, Simerly CR, Tengowski MW, Navara CS, Haavisto A, Schatten G (1996) Intracytoplasmic sperm injection for Rhesus monkey fertilization results in unusual chromatin, cytoskeletal, and membrane events, but eventually leads to pronuclear development and sperm aster assembly. Hum Reprod 11(8):1703–1712. https://doi.org/10.1093/oxfordjournals.humrep.a019473

Sutovsky P, Manandhar G, Wu A, Oko R (2003) Interactions of sperm perinuclear theca with the oocyte: implications for oocyte activation, anti-polyspermy defense, and assisted reproduction. Microsc Res Tech 61(4):362–378. https://doi.org/10.1002/jemt.10350

Sutovsky P, Ramalho-Santos J, Moreno RD, Oko R, Hewitson L, Schatten G (1999) On-stage selection of single round spermatids using a vital, mitochondrion-specific fluorescent probe

MitoTracker(TM) and high resolution differential interference contrast microscopy. Hum Reprod 14(9):2301–2312. https://doi.org/10.1093/humrep/14.9.2301

Suzuki-Toyota F, Ito C, Toyama Y, Maekawa M, Yao R, Noda T, Iida H, Toshimori K (2007) Factors maintaining normal sperm tail structure during epididymal maturation studied in Gopc-/- mice. Biol Reprod 77(1):71–82. https://doi.org/10.1095/biolreprod.106.058735

Swaminathan J, Baxter EM, Corces VG (2005) The role of histone H2Av variant replacement and histone H4 acetylation in the establishment of Drosophila heterochromatin. Genes Dev 19(1):65–76. https://doi.org/10.1101/gad.1259105

Tang XM, Lalli MF, Clermont Y (1982) A cytochemical study of the Golgi apparatus of the spermatid during spermiogenesis in the rat. Am J Anat 163(4):283–294. https://doi.org/10.1002/aja.1001630402

Tanii I, Oh-oka T, Yoshinaga K, Toshimori K (2001) A mouse acrosomal cortical matrix protein, MC41, has ZP2-binding activity and forms a complex with a 75-kDa serine protease. Dev Biol 238(2):332–341

Tanphaichitr N, Sobhon P, Taluppeth N, Chalermisarachai P (1978) Basic nuclear proteins in testicular cells and ejaculated spermatozoa in man. Exp Cell Res 117(2):347–356. https://doi.org/10.1016/0014-4827(78)90148-9

Tarnasky H, Cheng M, Ou Y, Thundathil JC, Oko R, van der Hoorn FA (2010) Gene trap mutation of murine outer dense fiber protein-2 gene can result in sperm tail abnormalities in mice with high percentage chimaerism. BMC Dev Biol 10:67. https://doi.org/10.1186/1471-213x-10-67

Thorne-Tjomsland G, Clermont Y, Hermo L (1988) Contribution of the Golgi apparatus components to the formation of the acrosomic system and chromatoid body in rat spermatids. Anat Rec 221(2):591–598. https://doi.org/10.1002/ar.1092210205

Tilney LG, Bryan J, Bush DJ, Fujiwara K, Mooseker MS, Murphy DB, Snyder DH (1973) Microtubules: evidence for 13 protofilaments. J Cell Biol 59(2 Pt 1):267–275. https://doi.org/10.1083/jcb.59.2.267

Tokuhiro K, Isotani A, Yokota S, Yano Y, Oshio S, Hirose M, Wada M, Fujita K, Ogawa Y, Okabe M, Nishimune Y, Tanaka H (2009) OAZ-t/OAZ3 is essential for rigid connection of sperm tails to heads in mouse. PLoS Genet 5(11):e1000712. https://doi.org/10.1371/journal.pgen.1000712

Topfer-Peterson E, Henschen A (1988) Zona pellucida-binding and fucose-binding of boar sperm acrosin is not correlated with proteolytic activity. Bio Chem Hoppe-Seyler 369(1):69–76

Tosaka Y, Tanaka H, Yano Y, Masai K, Nozaki M, Yomogida K, Otani S, Nojima H, Nishimune Y (2000) Identification and characterization of testis specific ornithine decarboxylase antizyme (OAZ-t) gene: expression in haploid germ cells and polyamine-induced frameshifting. Genes Cells 5(4):265–276. https://doi.org/10.1046/j.1365-2443.2000.00324.x

Toshimori K (1998) Maturation of mammalian spermatozoa: modifications of the acrosome and plasma membrane leading to fertilization. Cell Tissue Res 293(2):177–187. https://doi.org/10.1007/s004410051110

Tovich PR, Oko RJ (2003) Somatic histones are components of the perinuclear theca in bovine spermatozoa. J Biol Chem 278(34):32431–32438. https://doi.org/10.1074/jbc.M303786200

Tovich PR, Sutovsky P, Oko RJ (2004) Novel aspect of perinuclear theca assembly revealed by immunolocalization of non-nuclear somatic histones during bovine spermiogenesis. Biol Reprod 71(4):1182–1194. https://doi.org/10.1095/biolreprod.104.030445

Tran MH, Aul RB, Xu W, van der Hoorn FA, Oko R (2012) Involvement of classical bipartite/karyopherin nuclear import pathway components in acrosomal trafficking and assembly during bovine and murid spermiogenesis. Biol Reprod 86(3):84. https://doi.org/10.1095/biolreprod.111.096842

Tulsiani DR, Abou-Haila A, Loeser CR, Pereira BM (1998) The biological and functional significance of the sperm acrosome and acrosomal enzymes in mammalian fertilization. Exp Cell Res 240(2):151–164. https://doi.org/10.1006/excr.1998.3943

Urch U, Patel H (1991) The interaction of boar sperm proacrosin with its natural substrate, the zona pellucida, and with polysulfated polysaccharides. Development 111(4):1165–1172

Ursini F, Heim S, Kiess M, Maiorino M, Roveri A, Wissing J, Flohé L (1999) Dual function of the selenoprotein PHGPx during sperm maturation. Science 285(5432):1393–1396. https://doi.org/10.1126/science.285.5432.1393

van der Hoorn FA, Tarnasky HA, Nordeen SK (1990) A new rat gene RT7 is specifically expressed during spermatogenesis. Dev Biol 142(1):147–154. https://doi.org/10.1016/0012-1606(90)90158-f

Vijayaraghavan S, Liberty GA, Mohan J, Winfrey VP, Olson GE, Carr DW (1999) Isolation and molecular characterization of AKAP110, a novel, sperm-specific protein kinase A-anchoring protein. Mol Endocrinol 13(5):705–717. https://doi.org/10.1210/mend.13.5.0278

von Bülow M, Heid H, Hess H, Franke WW (1995) Molecular nature of calicin, a major basic protein of the mammalian sperm head cytoskeleton. Exp Cell Res 219(2):407–413. https://doi.org/10.1006/excr.1995.1246

von Bülow M, Rackwitz HR, Zimbelmann R, Franke WW (1997) CP beta3, a novel isoform of an actin-binding protein, is a component of the cytoskeletal calyx of the mammalian sperm head. Exp Cell Res 233(1):216–224. https://doi.org/10.1006/excr.1997.3564

Ward WS (1993) Deoxyribonucleic acid loop domain tertiary structure in mammalian spermatozoa. Biol Reprod 48(6):1193–1201. https://doi.org/10.1095/biolreprod48.6.1193

Wassarman PM (1988) Zona pellucida glycoproteins. Annu Rev Biochem 57:415–442. https://doi.org/10.1146/annurev.bi.57.070188.002215

Wassarman PM (1999) Mammalian fertilization: molecular aspects of gamete adhesion, exocytosis, and fusion. Cell 96(2):175–183. https://doi.org/10.1016/s0092-8674(00)80558-9

Wassarman PM, Jovine L, Litscher ES (2001) A profile of fertilization in mammals. Nat Cell Biol 3(2):E59-64. https://doi.org/10.1038/35055178

Werb Z (1997) ECM and cell surface proteolysis: regulating cellular ecology. Cell 91(4):439–442. https://doi.org/10.1016/s0092-8674(00)80429-8

Wu AT, Sutovsky P, Manandhar G, Xu W, Katayama M, Day BN, Park KW, Yi YJ, Xi YW, Prather RS, Oko R (2007a) PAWP, a sperm-specific WW domain-binding protein, promotes meiotic resumption and pronuclear development during fertilization. J Biol Chem 282(16):12164–12175. https://doi.org/10.1074/jbc.M609132200

Wu AT, Sutovsky P, Xu W, van der Spoel AC, Platt FM, Oko R (2007b) The postacrosomal assembly of sperm head protein, PAWP, is independent of acrosome formation and dependent on microtubular manchette transport. Dev Biol 312(2):471–483. https://doi.org/10.1016/j.ydbio.2007.08.051

Xu K, Yang L, Zhang L, Qi H (2020) Lack of AKAP3 disrupts integrity of the subcellular structure and proteome of mouse sperm and causes male sterility. Development 147 (2). https://doi.org/10.1242/dev.181057

Yamagata K, Murayama K, Kohno N, Kashiwabara S, Baba T (1998) P-Aminobenzamidine-sensitive acrosomal protease(s) other than acrosin serve the sperm penetration of the egg zona pellucida in mouse. Zygote 6(4):311–319. https://doi.org/10.1017/s0967199498000264

Yanagimachi R (1994) The physiology of reproduction. Mammalian fertilization:189–317

Yanagimachi R (2011) Mammalian sperm acrosome reaction: where does it begin before fertilization? Biol Reprod 85(1):4–5. https://doi.org/10.1095/biolreprod.111.092601

Yang K, Meinhardt A, Zhang B, Grzmil P, Adham IM, Hoyer-Fender S (2012) The small heat shock protein ODF1/HSPB10 is essential for tight linkage of sperm head to tail and male fertility in mice. Mol Cell Biol 32(1):216–225. https://doi.org/10.1128/mcb.06158-11

Yao R, Ito C, Natsume Y, Sugitani Y, Yamanaka H, Kuretake S, Yanagida K, Sato A, Toshimori K, Noda T (2002) Lack of acrosome formation in mice lacking a Golgi protein, GOPC. Proc Natl Acad Sci U S A 99(17):11211–11216. https://doi.org/10.1073/pnas.162027899

Yu Y, Oko R, Miranda-Vizuete A (2002) Developmental expression of spermatid-specific thioredoxin-1 protein: transient association to the longitudinal columns of the fibrous sheath during sperm tail formation. Biol Reprod 67(5):1546–1554. https://doi.org/10.1095/biolreprod.102.004838

Yu Y, Vanhorne J, Oko R (2009) The origin and assembly of a zona pellucida binding protein, IAM38, during spermiogenesis. Microsc Res Tech 72(8):558–565. https://doi.org/10.1002/jemt.20696

Yu Y, Xu W, Yi YJ, Sutovsky P, Oko R (2006) The extracellular protein coat of the inner acrosomal membrane is involved in zona pellucida binding and penetration during fertilization: characterization of its most prominent polypeptide (IAM38). Dev Biol 290(1):32–43. https://doi.org/10.1016/j.ydbio.2005.11.003

Yudin AI, Vandevoort CA, Li MW, Overstreet JW (1999) PH-20 but not acrosin is involved in sperm penetration of the macaque zona pellucida. Molecular Reproduction and Development: Incorporating Gamete Research 53(3):350–362

Zhang Y, Oko R, van der Hoorn FA (2004) Rat kinesin light chain 3 associates with spermatid mitochondria. Dev Biol 275(1):23–33. https://doi.org/10.1016/j.ydbio.2004.07.014

Zhang Y, Ou Y, Cheng M, Saadi HS, Thundathil JC, van der Hoorn FA (2012) KLC3 is involved in sperm tail midpiece formation and sperm function. Dev Biol 366(2):101–110. https://doi.org/10.1016/j.ydbio.2012.04.026

Zhu F, Liu C, Wang F, Yang X, Zhang J, Wu H, Zhang Z, He X, Zhang Z, Zhou P, Wei Z, Shang Y, Wang L, Zhang R, Ouyang YC, Sun QY, Cao Y, Li W (2018) Mutations in PMFBP1 Cause Acephalic Spermatozoa Syndrome. Am J Hum Genet 103(2):188–199. https://doi.org/10.1016/j.ajhg.2018.06.010

Chapter 2
Bovine Spermatogenesis

Contents

Abstract The intent of this chapter is to provide a morphological foundation in the normal cellular process of bovine gamete development so that abnormalities occurring are recognizable. The knowledge gained here is essential to begin to understand the significance of many of the common bull sperm abnormalities encountered in the clinics. Spermatogenesis is divided into three phases (i. e., Mitosis, Meiosis and Spermiogenesis) all happening in the seminiferous epithelium. The 'Cycle of the Bovine Seminiferous Epithelium' is explained in relation to these phases. Information is provided as to how to identify the stages of the bovine cycle and the steps of spermiogenesis at the histological and ultrastructural levels in preparation to recognize where and when in the cycle a spermatid abnormality arises. Spermiogenesis, the last phase of spermatogenesis, is the most revealing phase to recognize gamete abnormalities as this is where spermatid head and tail differentiation take place and spermatid compartments materialize. The formation of the nucleus, acrosome, manchette, perinuclear theca, axoneme, outer

© The Author(s), under exclusive license to Springer Nature Switzerland AG 2025

A. Barth et al., *Abnormal Morphology of Bovine Spermatozoa*,
Advances in Anatomy, Embryology and Cell Biology 240,
https://doi.org/10.1007/978-3-031-70126-9_2

dense fibers, fibrous sheath, connecting piece and mitochondrial sheath occur during this phase and are evaluated. The origins and assembly of a number of essential proteins compartmentalizing the sperm head and tail as well as defects arising during spermiogenesis are reviewed.

Keywords Bull · Spermatogenesis · Spermiogenesis · Cycle of the Bovine Seminiferous Epithelium · Sperm nucleus · Acrosome · Manchette · Perinuclear theca · Axoneme · Outer dense fibers · Fibrous sheath · Connecting piece · Mitochondrial sheath

2.1 Introduction

Spermatogenesis is a long process of cellular differentiation in which a spermatogonial stem cell proceeds through several mitotic divisions, meiosis, and numerous cytological transformations leading to the generation of mature elongated spermatids. Sperm cells are formed in the seminiferous tubules of the testis. Figure 2.1 represents a cross section through part of the seminiferous epithelium of a tubule in which the germ cells develop. The stem cells or spermatogonia are situated in the basal compartment or layer at the base of a tubule. They undergo several mitotic divisions to form spermatocytes, which are located a layer higher. The spermatocytes then go through the long process of reductional or meiotic divisions to form the next layer of haploid cells called spermatids. The spermatids, which are originally round cells, undergo a complex series of transformations to form a final layer of elongated spermatids with a head and tail and are then released into the lumen of the seminiferous tubules. This final cellular transformation is referred to as spermiogenesis. Thus, differentiation of the germ cells into their final form is always associated with their upward mobility in the seminiferous epithelium. The Sertoli cell, whose cytoplasm encompasses the germ cells, assists in the upward mobility and development of these cells by providing physical, nutritional, and hormonal requirements.

2.2 Sertoli Cell and Blood Testis Barrier

The testis is divided into an avascular seminiferous epithelium that makes up approximately 90% of the volume and a vascularized interstitial compartment composed of Leydig cells, blood and lymphatic vessels, macrophages, and connective tissue. The latter compartment produces the interstitial fluid that bathes the Leydig cells, vessels, and outside of the seminiferous tubules. It is via this fluid that the hormones and nutrients are transported from the vessels to the energy-demanding seminiferous epithelium. The control of substances that enter the germinal epithelium are regulated by the blood-testis barrier of which there are two, the myoid cell layer that surrounds the basal lamina of the seminiferous tubules and

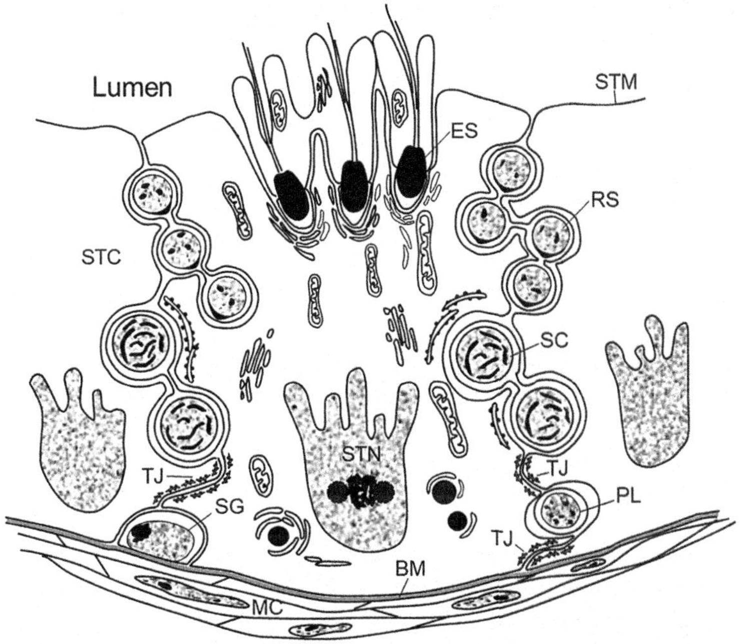

Fig. 2.1 Diagrammatic illustration to show the relationship of germinal cells to the Sertoli cell in the seminiferous epithelium: spermatogonia (SG), preleptotene spermatocytes (PL) spermatocytes (SC), round spermatids (RS), elongated spermatids (ES), Sertoli-Sertoli cell tight junctions (TJ), basal lamina or basement membrane (BM), myoid cells (MC), Sertoli cell membrane (STM), Sertoli cell nucleus (STN). The developing germ cells are enveloped by the Sertoli cell. The basal compartment containing the spermatogonia and preleptotene spermatocytes, is separated from the adluminal compartment, containing more advanced stages of germ cell development, by specialized Sertoli cell junctions. These occluding junctions (tight junctions) are the principal component of the blood-testis barrier. Note that the preleptotene spermatocyte, at the onset of meiosis, is undergoing translocation to enter the adluminal compartment, ensuring that the meiotic and post-meiotic phases of spermatogenesis are totally under Sertoli cell regulation. The tight junction immediately above the preleptotene spermaotcyte will degenerate while the one forming below will become permanent

the Sertoli-Sertoli cell tight junctions that divide the seminiferous epithelium into basal and adluminal compartments (Fig. 2.1) (Setchell 1980; Waites and Gladwell 1982). By far the most important of these barriers is that of the Sertoli cell tight junctions. The basal compartment is composed of mostly spermatogonia, while the adluminal compartment is composed of spermatocytes undergoing meiosis and spermatids undergoing cellular transformations. The existence of the tight junctions restricts the passage of many factors (e. g., electrolytes, nutrients, hormones, antibodies), establishing a unique and favorable environment in the adluminal compartment for meiosis and the development of the haploid spermatids.

Because many substances including androgens cannot penetrate further than the tight junctions between Sertoli cells, the Sertoli cell becomes instrumental in regulating the composition of the seminiferous fluid that it secretes into the adluminal compartment. In fact, the uptake of androgens regulates the transcription of genes in the Sertoli cell rather than in the germ cells of the adluminal compartment. Thus, the androgens indirectly via the Sertoli cell influence the differentiation of germ cell in the adluminal compartment. That the Sertoli cells provide physical support for the germ cells is obvious from their topographic relationship. The Sertoli cell shows both configurational relationships and specialized contacts with the developing germ cells (Russell 1980; Russell and Griswold 1993; Vogl et al. 2008). The configurational relationship and contacts suggest that the Sertoli cell coordinates four to five different germ cell generations in one tubule so that they all evolve synchronously to form the next stage. They also suggest facilitation of germ cell movement upward as each layer of cells matures. Russell (Russell 1977b) has shown that the Sertoli cell plays an active role in segregating maturing germ cells from the basal to the adluminal compartment without disrupting the Sertoli cell-Sertoli cell tight junction barrier. This is accomplished by the development of a new set of tight junctions immediately below the preleptolene (primary) spermatocyte in stage VIII, so that for a period of time the spermatocyte has barriers both above and below it (Fig. 2.1). Eventually the tight junction above undergoes dissolution allowing the spermatocyte to enter the adluminal compartment. The translocation of these germ cells at the onset of meiosis suggests that this process is involved in initiating meiosis.

Besides the tight junctions between Sertoli cells, other specialized contacts between Sertoli and germ cells exist. These include desmosome-like contacts (Russell 1977a), gap junctions (Russell 1980), ectoplasmic junctional specializations (Vogl et al. 1991; Clermont et al. 1980; Ross 1976; Russell 1977c), tubulobulbar complexes (Russell and Clermont 1976; Vogl et al. 2013), and Sertoli cell processes (Morales and Clermont 1982). They serve to ensure that the germ cells are transported in an orderly fashion, to transport substances between the adjoining cells, to hold the elongated spermatids within deep recesses of the Sertoli cells, to eliminate excess spermatid cytoplasm, and to constitute a pathway for the delivery of material from the Sertoli cells into the spermatids respectively. Besides physical or mechanical support for the control of germ cell differentiation, evidence also suggests that the Sertoli cell mediates hormonal effects on spermatogenesis and even that its secretory activity is cyclic and stage-related (Parvinen 1982; Sharpe 1983; Steinberger et al. 1979). One of the most convincing experiments in favor of the hypothesis that the hormone-dependent requirements for germ cell differentiation are mediated through the Sertoli cell is derived from the experiments of Lyon et al. (1975). They found that germ cells deprived of androgen receptors in chimaeric mice could develop into fertile spermatozoa in the presence of normal Sertoli cells. Likewise, evidence implying that the androgen-dependent requirements are stage-related was demonstrated elegantly by Russell and Clermont (Russell and Clermont 1977). They showed that gonadotropin deprivation in rats leads to a discriminative degeneration of germ cells in the adluminal compartment of the seminiferous epithelium in stages VII-VIII. The fact that three germ cell generations were affected simultaneously suggested that the effect was at the Sertoli cell level. They also found that this effect could be prevented by treatment with LH, implying an androgen dependency. In accordance with this observation the highest concentrations of endogenous testosterone (Parvinen and Ruokonen 1982) and maximal ABP secretion (Ritzen et al. 1982), which is a Sertoli cell function, are normally found at this point of the cycle. In summary, Sertoli cells serve many functions that are critical for the normal differentiation of the germinal epithelium, particularly of the germ cells in the meiotic and haploid (post-meiotic) phases of development. They include physical and nutritive support, hormonal regulation, luminal and basal secretions, endocytic (receptor and fluid mediated) and phagocytic activity and formation and upkeep of the tight junctions to keep the blood-testis barrier intact (Skinner and Griswold 2004; Russell and Griswold 1993).

2.3 The Cycle of the Seminiferous Epithelium

There are 13 distinctive cell types in the process of bovine spermatogenesis: A0, A1, A2, A3, intermediate (In), B1 and B2 spermatogonia; preleptotene (PL), leptotene (L), zygotene (Z), and pachytene (P) primary spermatocytes; secondary (II) spermatocytes; and spermatids that develop in chronological order (Fig. 2.2). The spermatids can be divided further into 14 distinctive steps represented by Arabic numerals (1–14). If you were to cut a transverse section through any seminiferous tubule of the bovine testis, you would find that the various germ cell types are not associated at random within the tubule but rather in one of the 12 possible cellular associations or stages represented by Roman numerals (I- XII) in Fig. 2.2. Tubules in stages I-VI are always made up of five distinct germ cell generations, whereas stages VII—XII always contain four generations (Fig. 2.2). When you proceed from the outside to the inside of any given segment of a seminiferous tubule, each generation of cells becomes successively older, or more mature and also more numerous because of the consecutive mitotic and meiotic divisions encountered in the process of spermatogenesis. Over time each cellular association, or stage (I-XII), in any given segment of a seminiferous tubule develops synchronously to form the next consecutive higher stage, that is, stage I will always develop synchronously into stage II, stage II to stage III, stage III to IV, and so on until stage XII is reached. At the end of stage XII, each layer of cells, making up one distinct generation of germ cells in this stage, is shifted up one layer in the seminiferous epithelium to form stage I again; this cycle is repeated continually. By this orderly and cyclic process, called the cycle of the seminiferous epithelium, mature elongated spermatids are continuously formed and released into the lumen of the seminiferous tubule in stage VI.

The duration of the cycle of the seminiferous epithelium was calculated to be 13.5 days in the bull (Hochereau et al. 1964; Hochereau-de Reviers 1970). The method for this determination is to inject tritiated thymidine which is selectively incorporated only into spermatogonia and spermatocytes that are synthesizing DNA in preparation for mitosis or meiosis. Quantitative analysis of round tubular

Stages of the Bovine Cycle

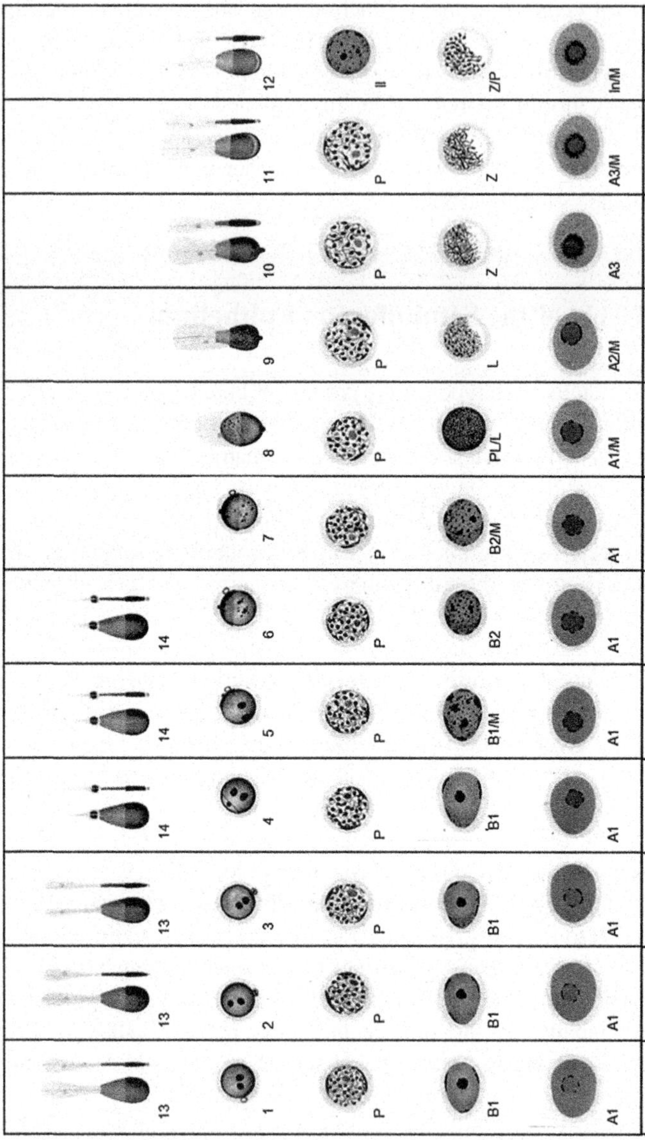

(S)	I	II	III	IV	V	VI	VII	VIII	IX	X	XI	XII
(F)	6.5%	4.5%	5.9%	7.0%	2.5%	4.7%	14.9%	17.8%	11.6%	6.6%	17.4%	0.6%
(D)	0.88	0.61	0.80	0.94	0.34	0.63	1.96	2.40	1.57	0.89	2.35	0.08

Fig. 2.2 Stages of the cycle of the bovine seminiferous epithelium. The diagram is from Berndtson and Desjardins' original article (Berndston and Desjardins 1974) but modified to incorporate Hochereau's mitotic and labeling indexes (Hochereau-de Reviers 1970; Hochereau 1967). Columns, represented by Roman numerals I-XII, depict cellular associations at each of the 12 stages of the cycle of the seminiferous epithelium in the bovine testis as defined by Berndtson and Desjardins. The types of germ cells observed in developmental sequence are: A spermatogonia (A1-3); intermediate spermatogonia (In); B spermatogonia (B1, B2); preleptotene primary spermatocytes (PL); leptotene primary spermatocytes (L); zygotene primary spermatocytes (Z); pachytene primary spermatocytes (P); and secondary spermatocytes (II). Fourteen steps of spermiogenesis are illustrated by spermatids (numbered 1–14) in the upper two rows. As nuclear condensation progresses two profiles (a face view and lateral view) of elongated spermatids become evident in steps 10–14 of spermiogenesis. D, F, and S refer to days, frequency, and stages respectively, as derived by Berndtson and Desjardins. The positioning of A1, A2 and A3 spermatogonia within the stages of the cycle is derived from Hochereau's mitotic and labeling indexes and is in agreement with our light and electron microscope observation. M refers to mitosis of the different developing cell categories occurring at different stages and is approximated from Hochereau's mitotic index. The second mitosis occurring after A2 is an equivalent one giving rise to A3 and renewing stem cells or A1 spermatogonia. Thus, the basal compartment from stages X to XII also accommodates A1 spermatogonia, not shown in these stages in this diagram. The In and B spermatogonia (B1, B2) positions in the stages of the cycle of the bovine seminiferous epithelium are also derived from Hochereau's mitotic and labeling indexes and are in agreement with our light and electron microscope observations. See Barth and Oko (Barth and Oko 1989) for comparison of stage designations in the cycle of the seminiferous epithelium between Berndston and Desjardins' approach and Hochereau's approach

cross sections containing labeled cells, following a controlled lapse of time after injection, is used to determine the stage of cycle to which the most advanced labeled cells have progressed. For example, if 13.5 days were allowed to elapse after the PL spermatocytes at stage VIII (Fig. 2.2) have incorporated the label, the label will show up in the same stage, but in the P spermatocytes located in a layer higher in the seminiferous epithelium. Thus, the disappearance of a particular stage until its reappearance again in any given segment of the seminiferous epithelium constitutes one cycle of the seminiferous epithelium and lasts 13.5 days in the bovine species. Similarly, by allowing 39 days to pass from injection of tritiated thymidine to castration (not quite three cycles of the seminiferous epithelium), the label will show up in the mature elongated spermatids that are being released into the tubular lumen at stage VI (Fig. 2.2). By allowing 50 days to elapse, the label will show up in the ejaculated sperm, which means it takes approximately 11 days for the released spermatids to traverse the epididymis. Earlier intravenous-labeled phosphorus (P^{32}) studies (Orgebin 1961) are in close agreement with these calculations. Once the length of the bovine seminiferous cycle was established, the frequency of occurrence in percent (F, Fig. 2.2) or the days (D, Fig. 2.2) that each stage lasts was easily determined. For example, if out of 1000 random round tubular cross sections examined it is found that 149 belonged to stage VII, then that stage would occupy 14.9% of the 13.5-day cycle, or 1.96 days. For a determination of the frequency of occurrence of the stages in the 12-stage cycle of bovine seminiferous epithelium, refer to Berndtson and Desjardins (1974).

2.4 Three Phases of Spermatogenesis (Salient Features)

Mitosis. This phase is involved with proliferation, renewal and differentiation of Spermatogonia. Types of spermatogonia in this phase: A1-A3; In, Intermediate; B1and B2.
Meiosis. This phase involves Primary (I) and Secondary (II) Spermatocytes. (I) Steps involved in the prophase of the first meiotic division of primary Spermatocytes: Preleptotene (Pl), Leptotene (L), Zygotene (Z), early Pachytene (EP), mid-Pachytene (MP) and late Pachytene (LP). (II) Secondary spermatocytes undergo the second meiotic division that yield haploid spermatids.
Spermiogenesis. Spermatids metamorphose into Spermatozoa (Steps 1 – 14).

2.5 Spermatogenesis and the Cycle of the Bovine Seminiferous Epithelium (Salient Features)

There are thirteen distinctive germ cell types in the process of Bovine spermatogenesis.

The spermatids can be divided further into 14 distinct cell types or **steps**.

If one were to cut a transverse section through any seminiferous tubule of the bovine testis, one would find that the various germ cell types are not associated at random, but rather in one of twelve possible cellular associations or **stages** represented by Roman numerals (Fig. 2.2).

Tubules in stages I—VI are always made up of five distinct cell generations, whereas, stages VII—XII contain four generations (one group of cells at one given stage of development is equal to one generation of germinal cells).

When one proceeds from the outside to the inside of any given segment of a seminiferous tubule, each generation of cells becomes successively older, or more mature and also more numerous because of consecutive mitotic and meiotic divisions encountered in the process of spermatogenesis.

Over time each cellular association or stage in any given segment of a seminiferous tubule develops synchronously to form the next consecutive higher stage, i. e., stage I will form stage II, stage II to stage III and so on until stage XII is reached. At the end of stage XII, each layer of cells making up a distinct generation of germ cells, differentiates and is shifted up one layer in the seminiferous epithelium to form stage I again; this cycle is repeated continually. In bulls this cycle (of the seminiferous epithelium) takes 13.5 days to complete.

Open intercellular bridges interconnect the germ cell in each generation, which arises from one spermatogonial stem cell, or unit, until the end of spermatogenesis.

Because the cycle is 13.5 days, stem cells are expected to enter spermatogenesis at regular 13.5-day intervals for renewal of germ cell generations and maintaining the 13.5-day cycle.

Thus, the disappearance of a particular stage until its reappearance again in any given segment of the seminiferous tubule constitutes one cycle of the seminiferous epithelium. If one were to follow a particular germ cell generation (e. g., PL in stage VIII) through one cycle of the seminiferous epithelium it would end up in the same stage but one layer higher in the seminiferous epithelium (in this case as a P in stage VIII) (See Fig. 2.2). By this orderly and cyclic process, mature elongated spermatids are continuously formed and released into the lumen of the seminiferous tubule in stage VI.

The duration of spermatogenesis is fixed and constant, i. e., 62 days. The duration of each stage and step is also fixed.

Because spermatogenesis is a long and complex process of differentiation, it is sensitive to various toxic factors both chemical and physical. However, so far no toxic factor has been able to change the duration of spermatogenesis.

2.6 The Wave of the Seminiferous Epithelium

The cycle of the seminiferous epithelium refers to the synchronous evolution of one stage to the next **over time** in any one particular segment of the seminiferous tubule. The wave, on the other hand, refers to successive stages arranged in a descending order along the tubule beginning at the rete testis (Perey et al. 1961). The stage segment next to the rete can be any one of the 14 in the rat (e. g., stage VI most proximal in Fig. 2.3). When stage XIV is reached, the order repeats itself (stage XIV descending to I). Since the tubules are horseshoe shaped, with both ends entering into the rete, the two descending orders meet somewhere close to the center of the tubule (Perey et al. 1961). The descending order of stage segments can be interrupted by modulations where, instead of descending, the order ascends for a short distance and then resumes its descending order (Fig. 3, gray colored region). The wave, as defined by Percy and Leblond (Perey et al. 1961), also occurs in the bull (Hochereau 1963).

2.6.1 Important Differences between Cycle and Wave

Cycle	Wave
Duration of stage is constant	Length of segment is variable
Transition between stages is gradual	Boundaries of segments are sharp
Sequence of stages is irreversible	Series of segments may be altered
Duration of cycle is constant	Length of wave is variable

The old maxim, "the wave is in space what the cycle is in time," cannot be substantiated. However, it is important to realize that, even though the length of the

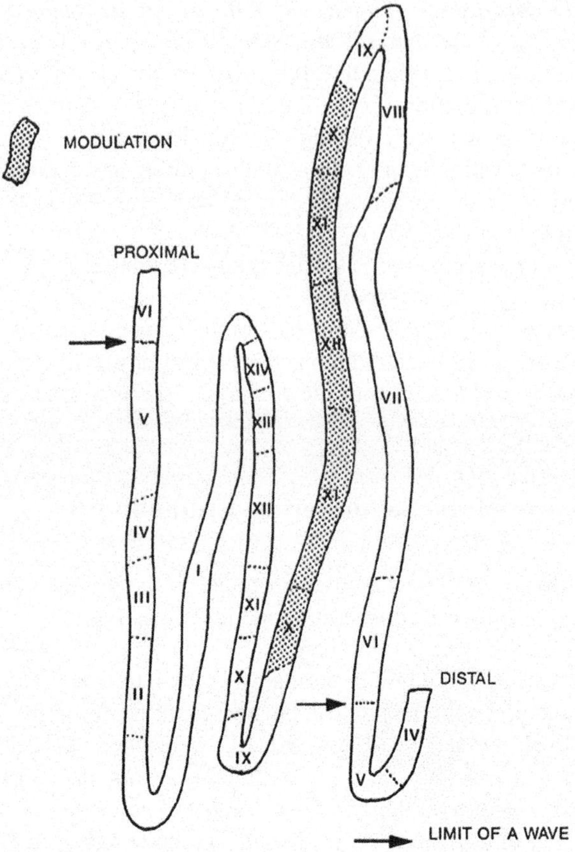

Fig. 2.3 Wave of the seminiferous epithelium. Portion of a tubule showing the extent or limits of a wave. Starting from the proximal end, the usual descending order of stage segments is observed as far as the shaded area where a modulation of the wave occurs. At this point the stage segment order ascends and then descends, finally reestablishing the original descending order. Thus, the portion of tubule between the arrows constitutes a wave with a modulation. Redrawn from Perey et al. (1961)

segments is variable and not proportional to the duration of the stages of the cycle, the average length of the segments in the testis is proportional to duration of the cycle stages.

2.7 Spermatogonial Renewal

To provide continuity in the cycle of the seminiferous epithelium, a renewal of spermatogonia must be assured at some point in the mitotic divisions. Hochereau-de Reviers (Hochereau 1967; Hochereau-de Reviers 1970) succeeded in locating the mitotic division at which bovine spermatogonial renewal occurred by studying the labeling index (thymidine tracing) in conjunction with the mitotic index (colchicine treatment) of developing spermatogonia. The author showed that three successive mitotic generations of A type spermatogonia (A1, A1, A3) were followed by consecutive mitotic divisions to form intermediate (In), B1, and B2 spermatogonia (Fig. 2.2). The renewing stem cells were found to originate after the mitosis of the A2 type, which was an equivalent mitosis resulting in A1, and A3 spermatogonia. Thus, beginning from stage X (Fig. 2.2) there are two populations of A spermatogonia in the basal layer of the seminiferous tubule, with A3 committed to forming spermatids and with A1, renewing the spermatogonial line in stage I. If the onset of bovine spermatogenesis is accepted as the time of formation of A3 spermatogonia, its duration is approximately 48 days; however, if the beginning of spermatogenesis is associated with the renewal of A1 spermatogonia, its duration is approximately 62 days. Hochereau-de Reviers (Hochereau 1967) also demonstrated the existence of pre-type A spermatogonia (A0), which are not immediately involved in the cyclic production of A1 spermatogonia and are considered reserve spermatogonia.

2.8 Development of Cell Types during Spermatogenesis

As noted, there are 13 distinctive cell types involved in the process of bovine spermatogenesis, not including spermatid transformations (Fig. 2.2). Type A spermatogonia are the germ cells that arise from the gonocytes and rest on the basal lamina. There are four distinctive types of A spermatogonia in the bovine, A0, A1, A2, and A3. The A0 spermatogonia give rise by mitosis to either A1 or A0 spermatogonia, which are the renewing and reserve stem cells, respectively (Hochereau 1967; Hochereau-de Reviers 1970). By determining the average number of germ cell nuclei in at least 100 round tubular cross sections of the 12 stages of the seminiferous epithelium cycle, Berndtson and Desjardins (Berndston and Desjardins 1974) concluded that there are three mitotic divisions of the A type spermatogonia, terminating with the intermediate spermatogonia. This is in agreement with the Hochereau's results obtained by measurements of the labeling index and mitotic index (Hochereau 1967; Hochereau-de Reviers 1970). Only subtle morphological differences distinguish spermatogonial cells; cytological differentiation is derived from differential staining characteristics of the nuclei and is dependent on fixation procedures, which precipitate the chromatin to various degrees (Clermont 1972). Generally, renewing type A spermatogonia have ovoid nuclei, a fine textured (dustlike) nucleoplasm and a single nucleolus surrounded by small clumps of chromatin (Berndston and Desjardins 1974). It is important to note, that right from the first mitotic division, all mitotic and meiotic divisions during spermatogenesis are incomplete resulting in daughter cells arising from each division joined by intracellular bridges. So theoretically, one clone beginning from an A1 spermatogonium and going through all the divisions would result in 256 spermatids joined by 255 intercellular bridges. However, germ cell degeneration at specific developmental phases of spermatogenesis normally occurs, significantly reducing the number of sperm expected from one stem cell. It is estimated that there is about a 33% loss of germ cells during A to In spermatogonial differentiation and another similar loss during B spermatogonial differentiation (Berndston and Desjardins 1974; Hochereau-de Reviers 1970).

Intermediate bovine spermatogonia are mitotically derived from A3 spermatogonia. They are distinguishable from A spermatogonia by their decreased size, decreased nuclear volume, heavily stained chromatin adhering to the nuclear membrane; and a smaller, more irregular heterogeneous nucleolus (Berndston and Desjardins 1974; Clermont 1972; Courot et al. 1970).Type B1 bovine spermatogonia are mitotically derived from the intermediate type and divide again to form B2 spermatogonia. Both types are hard to distinguish on a morphological basis and are very similar to intermediate spermatogonia (Clermont 1972); in fact, many authors classify intermediate and B spermatogonia as one group (Courot et al. 1970). Both line the basement membrane and are almost completely surrounded by Sertoli cytoplasm. Berndtson and Desjardins (Berndston and Desjardins 1974) distinguished B2 from B1 spermatogonia by observing that the clumps of chromatin along the nuclear membrane and in the nucleoplasm of B2 spermatogonia were more numerous and smaller. The organization of the cytoplasm in A, intermediate, and B spermatogonia is essentially simple and these cells are relatively poorly equipped with organelles—a Golgi body and a few scattered mitochondria making up their usual complement of organelles (Burgos et al. 1970).

Preleptotene bovine spermatocytes are mitotically derived from B2 spermatogonia and are the last cells produced by mitosis. They appear slightly after the release of bovine spermatozoa into the lumen of the seminiferous tubules and represent a pre-meiotic interphase preceding the prophase of the first meiotic maturation division. Synthesis of DNA occurs in the preleptotene spermatocyte for 23 h (Hochereau 1967), therefore, the term "resting spermatocyte" used to describe preleptotene spermatocytes is inaccurate. The nucleus of the preleptotene spermatocyte resembles that of the B2 spermatogonia in that it is round and dense with granular chromatin (Berndston and Desjardins 1974; Clermont 1972) and the cytoplasm of these cells is poorly developed. As discussed previously the Sertoli cell assists in translocating the preleptotene spermatocyte from the basal to the adluminal compartment most likely initiating the meiotic phase of development.

The next stage in the maturation of the primary mammalian spermatocyte gives rise to a leptotene spermatocyte, which is considered the first or primary stage of the meiotic prophase. The granular chromatin has disappeared and beaded chromatin filaments become visible, marking the beginning of the reorganization of the duplicated chromatin into the diploid number of chromosomes (Berndston and Desjardins 1974; Clermont 1972). ^{3}H-thymidine tracer experiments reveal that DNA synthesis goes on during the beginning of the leptotene stage (Courot et al. 1970). The larger cytoplasm is equipped with a much better selection of membrane-bound organelles, including smooth and rough endoplasmic reticulum, mitochondria, and Golgi membranes (Burgos et al. 1970). The zygotene spermatocyte is the next stage of the meiotic prophase and is characterized by synaptinemal complexes and a sex vesicle, which appear in the nucleus (Berndston and Desjardins 1974; Burgos et al. 1970). The synaptinemal complex consists of three linear dense structures oriented parallel to one another and is considered to represent the pairing element of homologous chromosomes. The sex vesicle is a more condensed mass of chromatin containing synaptinemal cores or unpaired sex chromosomes interspersed between chromatin fibrils (Solari 1964).

The next long stage of the meiotic prophase is referred to as the pachytene stage and begins when the pairing of the homologous chromosomes ceases. The synaptinemal complexes are fully formed and surrounded by large irregular clumps of chromatin; a small nucleolus can be seen amid these clumps (Berndston and Desjardins 1974; Burgos et al. 1970; Clermont 1972; Courot et al. 1970). The pachytene spermatocytes grow in nuclear volume from the early until late pachytene stage. It is throughout pachytene period that exchange or crossover of genes occurs between the homologous pairs of chromosomes by a process known as homologous recombination, adding to the genetic diversity of the species. The pachytene cytoplasm is well endowed with organelles and it is evident that synthesis of membranes has continued since the preleptotene stage (Burgos et al. 1970). The last two stages in the meiotic prophase are the diplotene and diakinesis stages characterized by the formation of tetrads between homologous chromosomes, joined by chiasma and the migration of condensing chromosomes to the periphery of the nucleus, respectively (Burgos et al. 1970; Courot et al. 1970). In regards to the diplotene stage, points of crossing over that occur through the pachytene stage become visible as chiasma (or points of contact) after the synaptonemal complex dissembles and the homologous chromosomes are slightly apart from each other.

Finally, two diploid secondary spermatocytes arise from the first meiotic maturation division, with metaphase I, anaphase I, and telophase I taking place relatively quickly (Clermont 1972). The nuclei of the secondary spermatocytes stain lightly with hematoxylin and contain irregular globules of chromatin dispersed throughout (Berndston and Desjardins 1974). Secondary spermatocytes are easily distinguished from the round young spermatids that contain large regular chromatin bodies and are smaller than the pachytene spermatocytes, but larger than round spermatids. Secondary spermatocytes enter the second maturation division, quickly producing four haploid round spermatids.

2.9 Spermiogenesis (Spermatid Metamorphosis)

Spermiogenesis is the process by which round haploid spermatids formed from the second maturation division of meiosis undergo a complex series of cellular transformations to form mature elongated spermatids, with a head and a tail, which are subsequently released from the seminiferous epithelium. Grossly, the process has been divided into four phases (Leblond and Clermont 1952a, b), namely, Golgi, cap, acrosome, and maturation (Fig. 2.4). In the bull, the Golgi phase is the time from the formation of the round spermatid (step 1) until the formation of a spherical acrosomic vesicle, with a centrally located granule, is associated with the nuclear membrane (step 3). Spermatids with a growing acrosomic head cap are in the cap phase (steps 4–7). The period in which the nucleus and acrosome undergo the most dramatic morphological transformation (consisting of condensation, elongation, and shaping) is referred to as the acrosome phase (steps 8–12) but more accurately can be termed the elongation phase as the acrosome does not grow but only conforms to the shape of the elongating nucleus. The maturation phase (steps 13–14), during which spermatids complete their differentiation, is extremely complex and includes the final formation of the sperm tail and connecting piece, where the bulk of ODF and FS proteins are laid down, and final condensation and shaping of the nucleus. Because a thorough understanding of the cyto-morphogenesis of the head and tail is crucial for interpreting sperm anomalies, the following sections of this chapter are devoted to a review of the essential features of spermatid structure and development.

2.10 Identification of the Stages of the Cycle

The basic approach for distinguishing the 12 stages of the bovine seminiferous epithelium cycle (Berndston and Desjardins 1974) (Fig. 2.2) was originally developed on the rat testis by Leblond and Clermont (1952a, b). In that case, 14 stages in the rat seminiferous epithelium were identified using the sequential morphological development of the acrosome and nuclei of the differentiating spermatid. The 14 stages were defined by combining the periodic acid-Schiff (PAS) and hematoxylin staining techniques on histological sections. PAS specifically stains the acrosome system, whereas hematoxylin stains the nuclei. For the most part as seen in Fig. 2.4 the identification of stages I to VII in the 12-stage cycle of the bovine seminiferous epithelium depends upon the sequential changes in acrosome

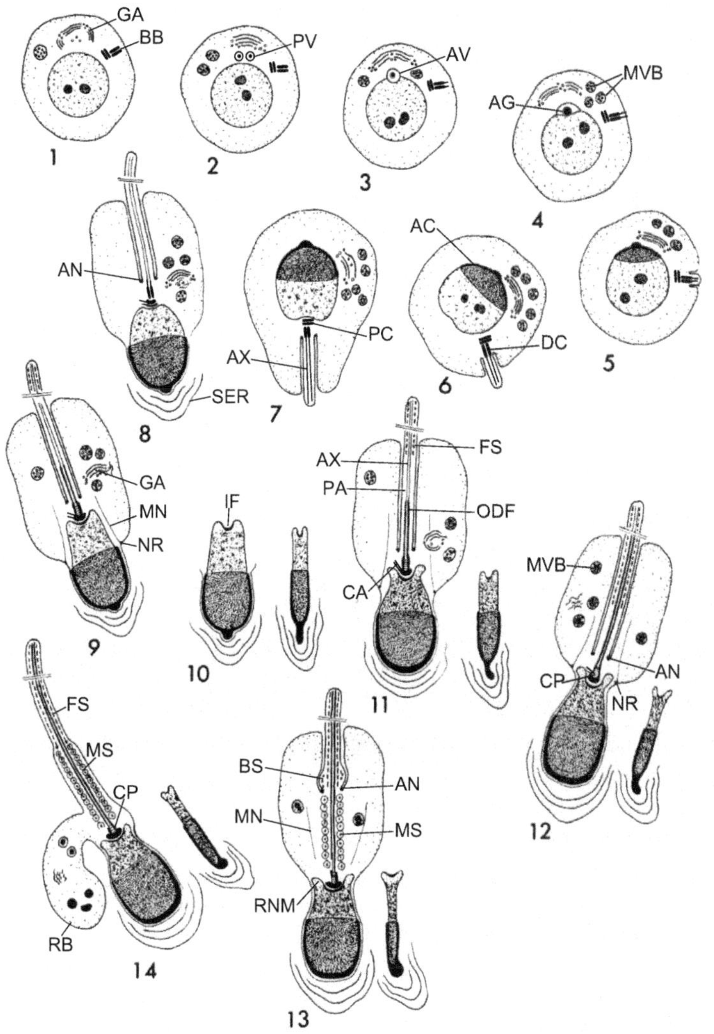

Fig. 2.4 Bovine spermiogenesis. Schematic illustration from light and electron microscope observations of the 14 steps (1–14) involved in bovine sperm head and tail differentiation (see text for more detail): Golgi apparatus (GA) (the Golgi apparatus has also been referred to as the Golgi complex or Golgi zone), basal body (BB), proacrosomic vesicle or granule (PV), chromatoid body (CB), acrosomic granule (AG), multivesicular body (MVB), annulus (AN), Sertoli endoplasmic reticulum (SER), proximal centriole (PC), axoneme (AX), acrosome cap (AC), distal centriole (DC), implantation fossa (IF), manchette (MN), nuclear ring (NR), fibrous sheath (FS), outer dense fibers (ODF), periaxonemal compartment (PA), connecting piece (CP), centriolar adjunct (CA), mitochondrial sheath (MS), residual body (RB), redundant nuclear membrane (RNM)

formation in steps 1 to 7 spermatids, respectively. The identification of stages VIII to XII mostly depend upon identifying the sperm head shape changes occurring in steps 8 to 12, respectively, along with some other characteristic changes in germs cells associated with these stages that are described later.

Ortavant (1958) and Amann (1962), on the other hand, used a different method to divide the cycle of the ruminant seminiferous epithelium. They divided the cycle into 8 stages based on the alteration of shape and position of the spermatid nuclei and the meiotic activity of the spermatocytes. Feulgen-hematoxylin staining of testicular tissue was used. Hochereau de Reviers (Hochereau 1967; Hochereau-de Reviers 1970), using the same approach, increased the specificity by subdividing the stages of longer duration and as a result divided the cycle of the bovine seminiferous epithelium into 14 stages. It should be emphasized that the stages of the cycle are really artificial subdivisions of a continuous process; therefore, the decision as to where the divisions are made in the process is arbitrary, but should depend upon easily identifiable morphological features. The division of the cycle into stages is fundamental to qualitatively and quantitatively study the normal and abnormal differentiation of cellular organelles and components. Accurate subdivisions of this process are required to determine the origin of various sperm defects and to follow their progression and incidence.

A comparison between Berndtson and Desjardins's (1974) and Hochereau's (1967) approach to dividing the cycle of the bovine seminiferous epithelium into stages is covered in Chap. 3 in Fig. 3.2 of our predecessor book (Barth and Oko 1989). Comparative details have been discussed in detail by Oko (Oko 1977), whose study confirmed that the distinguishing features used by Berndtson and Desjardins are more precise and reliable indicators of position in the cycle than those used by Hochereau. By alternating serial 1-μm sections for light microscopy with thin sections for electron microscopy, Oko further defined the development of the acrosome in the early stages of spermiogenesis and incorporated additional cellular features, such as the formation of the microtubular manchette, the assembly of the postacrosomal sheath, and the formation of the accessory components of the tail during spermiogenesis, which can be used as additional indicators of position within the cycle. These features are illustrated schematically in Fig. 2.4 and ultrastructurally in Figs. 2.5, 2.6, 2.7, 2.8, 2.9, 2.10, 2.11 and 2.12.

2.11 Sperm Head Differentiation

Essentially, the chronological steps in spermatid head formation are used as the main criteria for dividing the cycle of the bovine seminiferous epithelium into 12 stages (Berndston and Desjardins 1974; Oko 1977). The first 12 steps (1 to 12) are used as criteria to identify each of the 12 stages (I to XII) of the cycle, while step 13 and 14 spermatids are referred to as the second generation of spermatids and occur from stages I to III and stages IV to VI, respectively (see Figs. 2.2, 2.4).

Step 1 spermatids (Fig. 2.5) are derived from the meiotic division of secondary spermatocytes (Fig. 2.11). They are distinguished by the presence of a large circular Golgi apparatus and a multivesicular body (Figs. 2.4, 2.5A,B). The multivesicular body contains uniform membranous particles, 45–50 nm in diameter, that stain more intensely (red) with PAS stain than the Golgi apparatus (Oko 1977). In the latter part of step 1, the Golgi apparatus gives rise to Golgi vesicles containing granular acrosomal material (Fig. 2.5C). These vesicles are referred to as proacrosomic granules as the vesicles often contain a dense granule that is centrally located. The cytoplasm of step 1 spermatids also contains a chromaphilic mass referred to as the chromatoid body (Fawcett et al. 1970; Susi and Clermont 1970), a basal centriole, mitochondria, and endoplasmic reticulum.

In step 2 spermatids the Golgi vesicles formed by the Golgi apparatus (Fig. 2.5D, E) begin to fuse together and implant onto the nuclear membrane (Figs. 2.4, 2.6A,B). Usually, two or more multivesicular bodies are formed in the vicinity of the Golgi apparatus at this stage, and on PAS-stained histological sections they can be misinterpreted as proacrosomic granules because of their intense red color.

Step 3 spermatids are characterized by a large circular Golgi vesicle that is well attached on the surface of the spermatid nucleus (Figs. 2.4, 2.6C,D). This attached Golgi vesicle, referred to as the acrosomic vesicle, results from the cumulative fusions of the proacrosomic vesicles produced earlier. It contains a granular body, referred to as the acrosomic granule, formed by the accumulation of granular material contained in the proacrosomic vesicles. The acrosomic granule stains intensely red by PAS on histological sections and locates itself centrally within the acrosomic vesicle (Fig. 2.6D). With this step the phase of spermiogenesis referred to as the Golgi phase is completed.

The cap phase of spermiogenesis begins with a marked flattening of the nuclear surface beneath the expanding acrosomic vesicle in step 4 spermatids (Figs. 2.4, 2.6E, 2.7A). The acrosomic granule within the acrosomic vesicle has grown in size and closely opposes the inner acrosomal membrane attached to the nuclear envelope. Two or more multivesicular bodies and a chromatoid body are located in the vicinity of the Golgi apparatus. Note that the nuclear lamina, which directly underlies the nuclear envelope, delineates the region of the nuclear envelope where attachment and capping of the acrosomic vesicle takes place (Fig. 2.6E). In an inducible mouse model of male infertility where the acrosome does not form, the nuclear lamina still delineates the region where the capping of the acrosome would have taken place (Oko et al. 2011; Tran et al. 2012). This observation suggests that the nuclear lamina's presence influences or signifies a molecular change in the overlying nuclear envelope that is crucial for acrosomal attachment and spreading over the nuclear surface.

Step 5 spermatids are characterized by the presence of a newly formed acrosomic head cap that covers a quarter to a third of the spermatid nucleus (Figs. 2.4, 2.7B). The acrosomic granule dominates the center of the head cap while the Golgi apparatus hovers over it. Several multivesicular bodies have formed and are in close association with the Golgi apparatus (Fig. 2.7B,C). As in the previous steps, the spermatid nucleus is round and contains two to three distinct chromatin bodies.

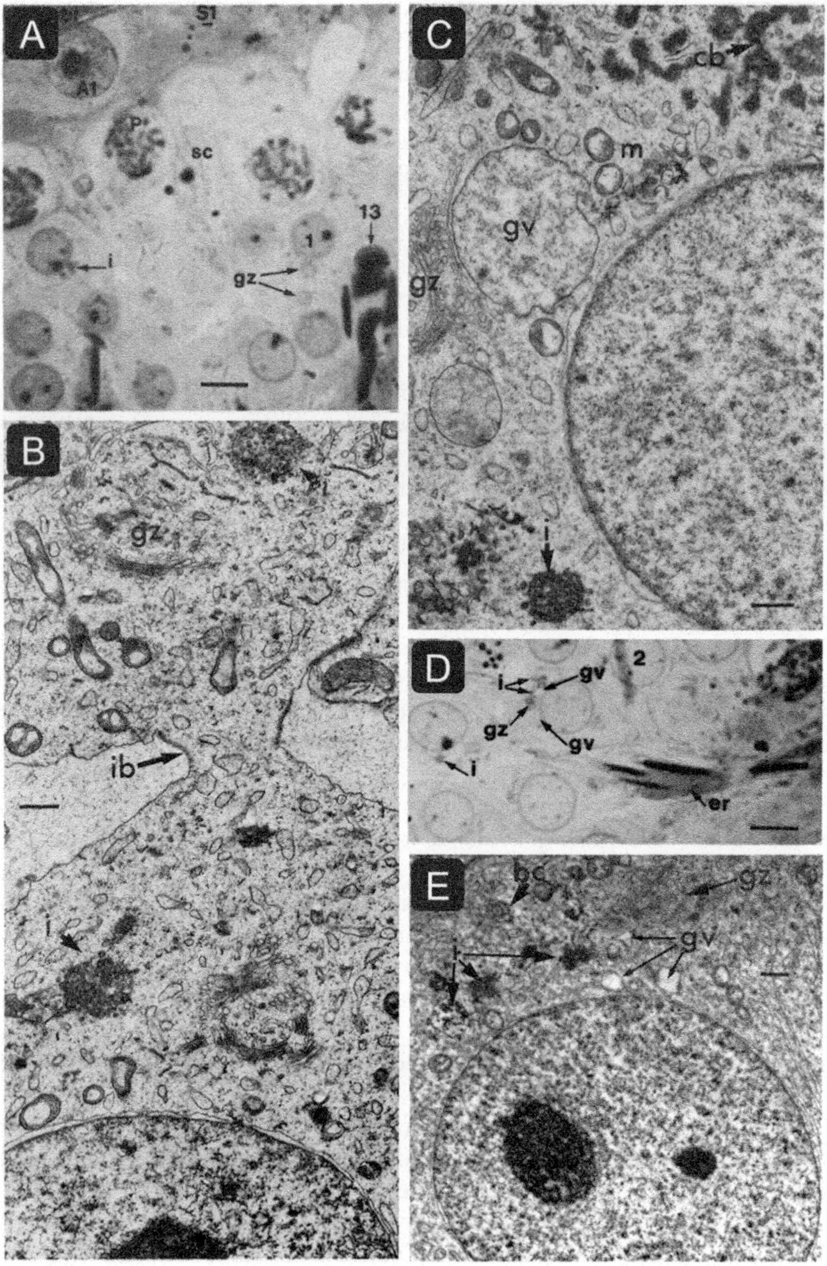
A
A1
sc
13
1
i
gz
B
gz
ib
i
C
cb
m
gv
gz
i
D
i
gv
2
gz
gv
i
er
E
bc
gz
gv
i

◀ **Fig. 2.5 A) Stage I.** Light micrograph (LMG) of histologically stained (PAS-hematoxylin) semi-thin (1μ) epoxy section through part of a bovine seminiferous tubule in stage I of the 12-stage cycle of the seminiferous epithelium. Numbers 1–14 in this and subsequent micrographs indicate the developmental steps of spermatids within the 12-stage cycle. Note the round Golgi apparatus or zone (gz) and multivesicular body or inclusion body (i) in proximity to the nucleus in step 1 spermatids and compare to higher resolution in B. Basal lamina (bl), spermatogonia (A1), pachytene spermatocyte (P), Sertoli cytoplasm (sc). Bar=5 μm. **B) Stage I.** Electron micrograph (EMG) of two early step 1 spermatids joined by an intercellular bridge (ib). Note the round Golgi apparatus (gz) and the multivesicular body (i), both being distinguishing features for step 1 spermatids. Bar=0.5 μm. **C) Stage I**: EMG of a step 1 spermatid in a later stage of development. Note that the Golgi apparatus (gz) has given rise to a Golgi vesicle (gv) or proacrosomic vesicle. The multivesicular body (i), the chromatoid body (cb), and mitochondria (m) are all in the vicinity of the Golgi apparatus. All these organelles play an integral role in spermatid development. Bar = 0.5 μm. **D) Stage II**. LMG of a group of step 2 spermatids. The Golgi apparatus (gz) has given rise to what appears as two or more Golgi vesicles (gv) or proacrosomic vesicles. Two or more PAS-positive multivesicular bodies (i), which stain dark red, are usually located in the vicinity of the Golgi complex in this step. Compare with E. A group of elongated spermatids in step 13 are surrounded by a V-shaped mass of Sertoli endoplasmic reticulum (er) or endoplasmic specialization. Bar= 5 μm. **E) Stage II**. EMG of a step 2 spermatid showing several proacrosomic vesicles (gv) that may be extensions of a larger Golgi vesicle. An extensive Golgi apparatus gz, several multivesicular bodies (i), and an elementary flagellum of the basal centriole (bc) are located in the vicinity of the Golgi vesicle. Bar=0.5 μm

Characteristic of step 6 spermatids is a head cap that extends well over one third of the spermatid nucleus (Figs. 2.4, 2.7D,E). The Golgi apparatus has migrated to the midway mark between the acrosomic granule and the distal end of the head cap and remains in association with several multivesicular bodies (Fig. 2.7E). The basal centriole and chromatoid body also migrate distally toward the base of the nucleus. This step also marks the beginning of disintegration of the chromatin bodies in the spermatid nucleus.

In step 7, which marks the end of the cap phase of spermiogenesis, the head cap is fully extended (Figs. 2.4, 2.7E). The Golgi apparatus and multivesicular bodies are at the distal end of the head cap ready to be extruded with the rest of the organelles into the distal cytoplasm. It is also in this step that the basal centriole implants itself into the basal pole to form the head–tail junction.

In the acrosome phase of development, step 8 spermatids are distinguished by a V-shaped bundle of Sertoli endoplasmic reticulum that surrounds the top half of acrosomic head cap region, an oval-shaped nucleus, and a cytoplasm that is now distal to the head cap and called the cytoplasmic lobe (Figs. 2.4, 2.8A,B). The microtubular manchette is beginning to form around the distal half of the nucleus; the implanted basal centriole, composed of the proximal and distal centriole, is associated with the chromatoid body; and the Golgi apparatus is in the distal cytoplasm. Scattered clumps of chromatin within the spermatid nucleus signify nuclear condensation.

Step 9 spermatids are characterized by a distinctly elongated nucleus that has not yet flattened (Figs. 2.4, 2.8C,D). Nuclear elongation and shaping may be in

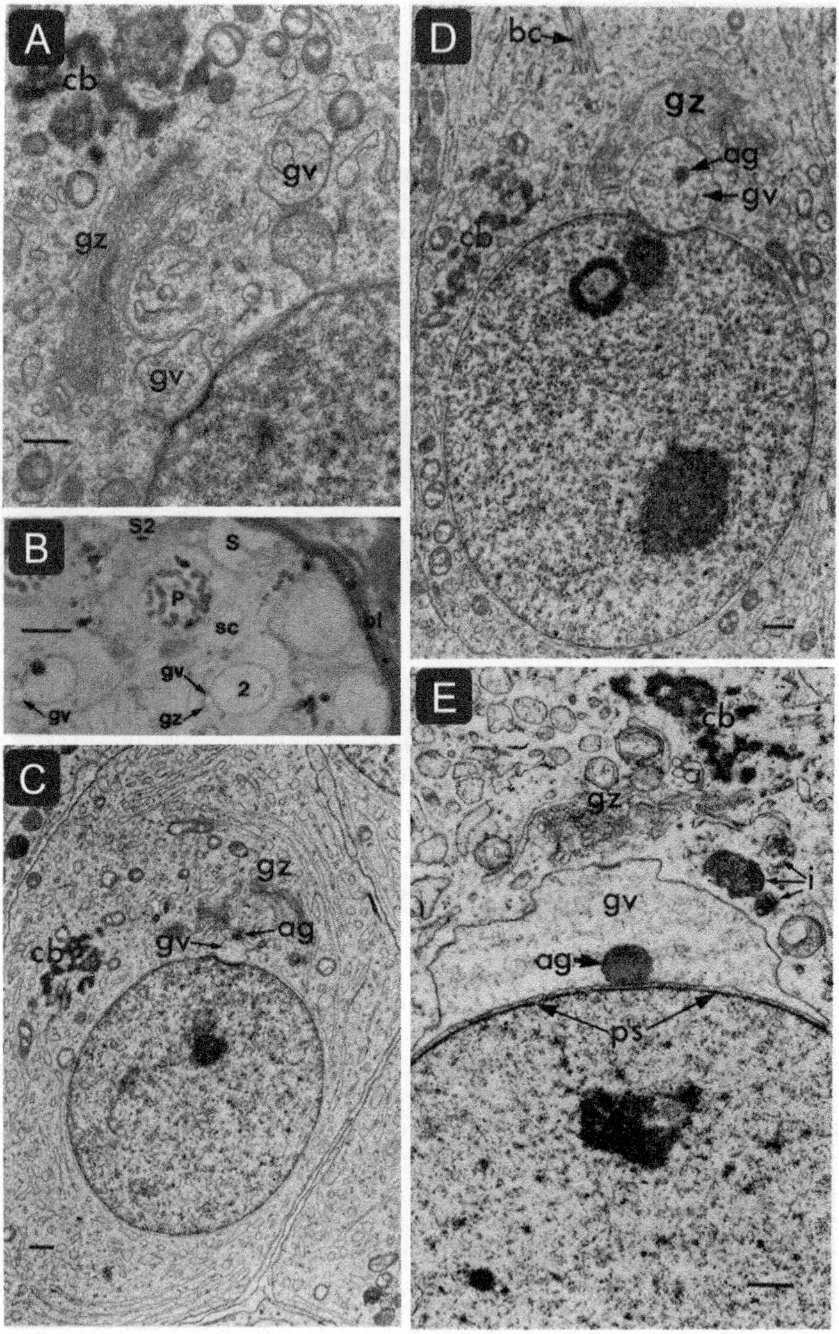
A
cb
gv
gz
gv
B
S2
S
P
sc
gv
gv
gz
2
C
gz
ag
gv
cb
D
bc
gz
ag
gv
cb
E
cb
gz
i
gv
ag
ps

◀ **Fig. 2.6** A) **Stage II.** EMG of a late step 2 spermatid. An extension of the Golgi vesicle (gv) has implanted itself onto the nuclear membrane. At this point after the fusions of smaller proacrosomic vesicles, the resulting larger vesicle is called the acrosomic vesicle. The chromatoid body (cb), which is composed of two distinct components, is situated close to the Golgi apparatus (gz). Bar = 0.5 μm. **B) Stage II.** LMG of a cross section through part of a seminiferous tubule in late stage ll of development. The Golgi vesicle (gv) or acrosomic vesicle is embedded on the spermatid nucleus but no sign of acrosomic granule is evident. Sertoli cell (S), Sertoli cytoplasm (sc), basal lamina (bl), pachytene sperma ~ tocyte (P), Golgi apparatus (gz). Bar = 5 μm. **C) Stage III.** EMG of an early step 3 spermatid. An acrosomic granule (ag) is developing within the implanted acrosomic vesicle (gv). Chromatoid body (cb). Bar = 0.5 μm. **D) Stage III.** EMG of a typical step 3 spermatid, which has a centrally located acrosomic granule (ag) within a well-defined circular acrosomic vesicle (gv). The flagellar extension of the basal centriole (bc) and the chromatoid body (cb) are in proximity to the Golgi apparatus (gz). Bar = 0.5 μm. **E) Stage IV.** Electron micrograph of a step 4 spermatid. In this stage the acrosomic vesicle (gv) begins to extend over the nuclear surface, and a perinuclear substance (ps) or nuclear dense lamina, located immediately beneath the nuclear envelope demarcates the vesicle's extension. This perinuclear substance has underlined the Golgi vesicle since it first attached to the nuclear envelope in stage II. (See A, C and D). The acrosomic granule (ag) has grown in size and dominates a central position within the acrosomic vesicle at its attachment to the nucleus. Several multivesicular bodies (i) and the chromatoid body (cb) surround the vesicle. Bar = 0.5 μm

some way associated with the microtubular manchette. The microtubules making up the manchette are connected to the nuclear ring, which is located just below the end of the acrosomic head cap. Nuclear condensation is more evident than in the previous step. Multivesicular bodies are located in the distal cytoplasm (Figs. 2.8D, 2.9A).

In step 10, nuclear flattening is evident when the spermatids are viewed laterally and condensation of the nucleus continues (Figs. 2.4, 2.9B-D). This step marks the end of the typical appearance of the acrosomic granule and the beginning of the distal migration of the nuclear ring of the manchette (Fig. 2.9C). The postacrosomal sheath begins to be laid down around the distal half of the nucleus, proximal to the migration of the nuclear ring. The multivesicular bodies, which have been located in the distal cytoplasm since step 8 and were distinguishable by their uniform membranous particles, begin to disintegrate and become amorphous (Fig. 2.9D).

Step 11 spermatids are grouped in longitudinally arranged bundles deeply embedded in the Sertoli cytoplasm and surrounded by the Sertoli endoplasmic reticulum (Fig. 2.10A). The spermatid nucleus is almost completely flattened when viewed laterally (Fig. 2.10A, B), and the acrosomic granule of the preceding step is replaced by a long, protruding crescent-shaped apical crest (Fig. 2.10D). The nuclear ring connected to the manchette continues to migrate distally and is located about midway between the end of the head cap and the base of the nucleus (Fig. 2.4, Fig. 2.10B-D). This distance marks the extent of postacrosomal sheath development. The migration of the manchette appears to aid in distal nuclear condensation and to squeeze the excess nuclear membrane, left over by the condensation of the nucleus, into a diverticulum (Fig. 2.10B).

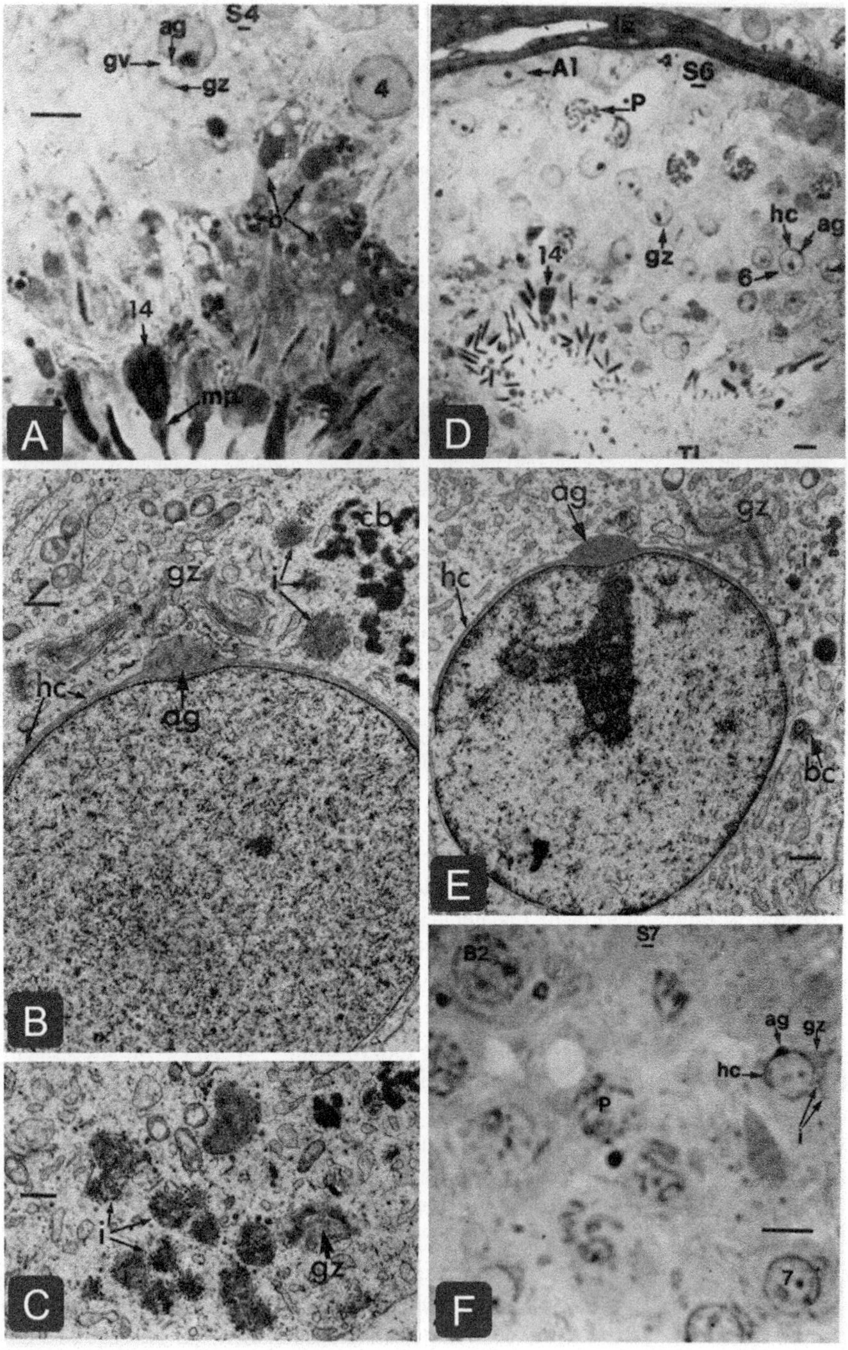
A
ag
S4
gv
gz
4
b
14
mp
D
A1
S6
P
hc
ag
gz
14
6
Tl
B
cb
gz
i
hc
ag
C
i
gz
E
ag
gz
hc
bc
F
S7
B2
ag
gz
hc
P
i
7

◀ **Fig. 2.7 A) Stage IV.** LMG of a cross section through part of a seminiferous tubule in Stage IV of development. Note the flattening of the nuclear surface beneath the acrosomic vesicle (gv) of the step 4 spermatid. The acrosomic granule (ag) occupies a central basal position within the expanding vesicle. Elongated step I4 spermatids have migrated to the lumen of the tubule and are extruding their excess cytoplasm in the form of residual bodies (rb). In some step 14 spermatids the cytoplasmic droplet is evident on the middle piece (mp) of the tail. Bar = 5 μm. **B) Stage V.** EMG of a step 5 spermatid. Note the newly formed acrosomic head cap (hc) and the position of the Golgi apparatus (gz) directly over the acrosomic granule (ag). Several multivesicular bodies (i) and a chromatoid body (cb) appear in the vicinity of the Golgi apparatus. The head cap covers about one quarter of the nuclear surface and is underlined by nuclear dense lamina. Bar = 0.5 pm. **C)** As above, but section cut above the acrosome showing 10 or more multivesicular bodies (i) in the area of the Golgi apparatus (gz). Note the chromatoid body in the right upper corner. Bar = 0.5 μm. **D) Stage VI.** LMG survey section of a tubule in stage VI (S6) of development. The step 14 spermatids are extruded into the tubular lumen in this stage. Refer to E for details of step 6 spermatid structure. Interstitium (IE), spermatogonium (A1), pachytene spermatocyte (P), head cap (hc), acrosomic granule (ag), Golgi apparatus (gz), tubular lumen (TL). Bar = 5 μm. **E) Stage VI.** EMG of a step 6 spermatid characterized by a head cap (hc) that has extended itself over approximately one-third way around the nucleus and a Golgi apparatus (gz) that has migrated distally midway between the acrosomic granule (ag) and the distal end of the head cap. The multivesicular bodies (i) and basal centriole (bc), containing the growing axoneme, migrate distally with the Golgi apparatus. Bar = 0.5 μm. **F) Stage VII.** Light micrograph of a cross section through part of a seminiferous tubule. The Golgi apparatus (gz) and multivesicular bodies (i) are at the distal end of the spermatid head cap (hc), ready to be extruded into the distal cytoplasm. Spermatogonia (B2), pachytene spermatocytes (P). Bar = 5 μm

The main features that distinguish step 12 from step 11 spermatids are the newly completed postacrosomal sheath and the shedding of the nuclear ring and the manchette from the nucleus into the distal cytoplasm (Figs. 2.4, 2.11A,C). However, the most distinguishable feature of stage 12 is the presence of secondary spermatocytes derived from the pachytene spermatocytes by the first reductional division of meiosis (Fig. 2.11A). Step 12 culminates the most dramatic transformation of head condensation and shaping and leads to the maturation phase of spermiogenesis.

Steps 13 and 14 (Figs. 2.4, 2.11B,D, 2.12A,B) are mainly concerned with the final stages of sperm tail differentiation and will be discussed in a following section of this chapter. However, final changes in head shape during the maturation phase concern the transition of the long, protruding crescent-shaped apical crest seen in step 12 into a fully formed apical crest in the early part of step 13 (Fig. 2.11C cf. Fig. 2.11D), the thinning of the caudal part of the acrosome to form the equatorial segment and a reduction in size of the diverticulum during step 14 (Figs. 2.4, 2.12A,B).

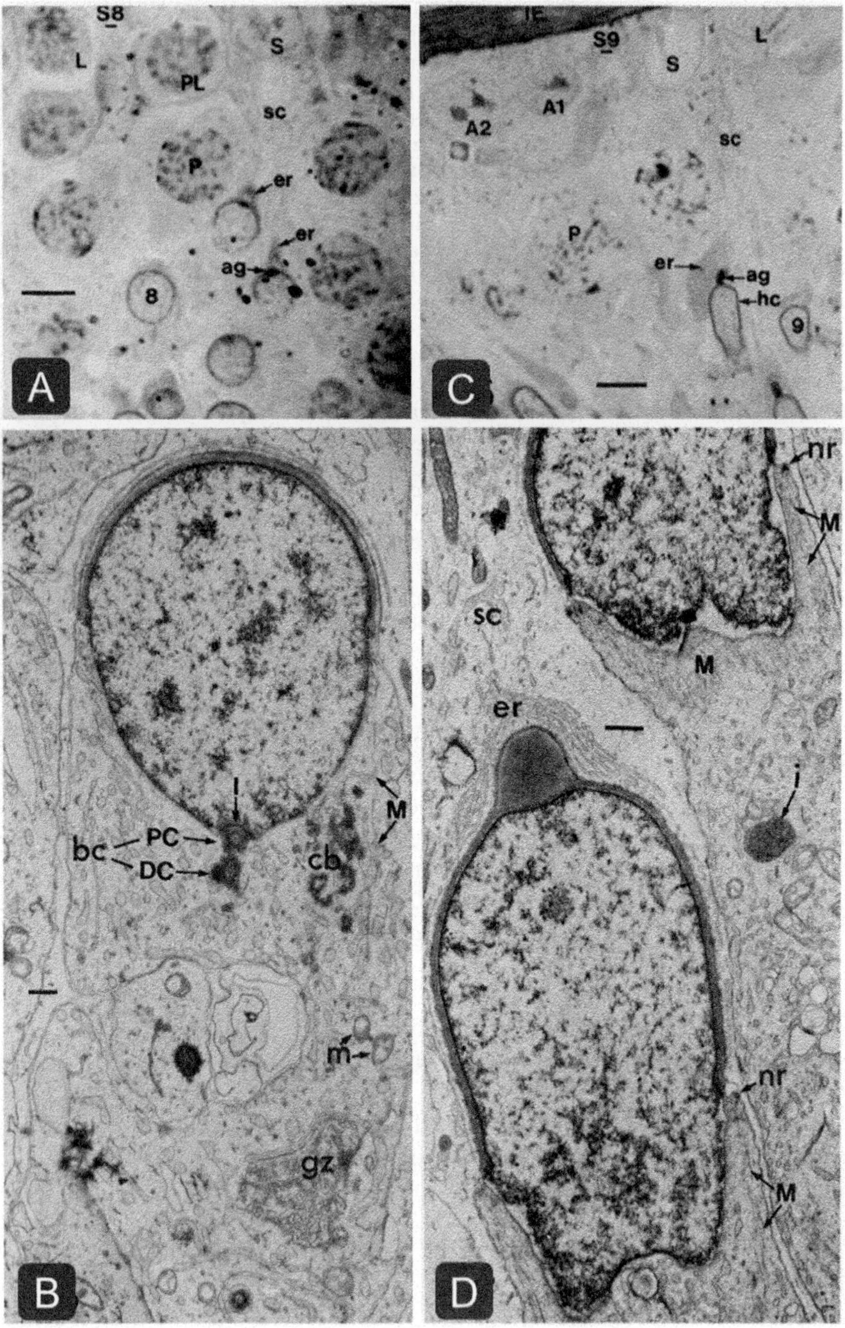
S8
L
PL
S
sc
P
er
er
ag
8
A
IE
S9
S
L
A1
A2
sc
P
er
ag
hc
9
C
I
bc
PC
DC
cb
M
m
gz
B
nr
M
sc
M
er
i
nr
M
D

◀ **Fig. 2.8** **A) Stage VIII.** LMG of part of a cross section through a Stage VIII seminiferous tubule showing leptotene (L), preleptotene (PL), and pachytene spermatocytes (P), step 8 spermatids (8), Sertoli nucleus (S) and Sertoli cytoplasm (sc). Note the V-shaped bundle of Sertoli endoplasmic reticulum (er) or ectoplasmic specializations (Vogl et al. 2000) that surrounds the head cap region of the oval nucleus. Bar = 5 µm. **B) Stage VIII.** EMG of a step 8 spermatid. Note that the microtubular manchette (M) is beginning to form around the distal part of the nucleus in the area destined to become the postacrosomal sheath and that the basal centriole (bc), made up of the proximal centriole (PC) and the distal centriole (DC), is well embedded in the nuclear implantation fossa (I). The chromatoid body (cb) is in proximity to the basal centriole and the Golgi apparatus (gz), and mitochondria (m) are in the distal portion of the cytoplasm. The spermatid cytoplasm is distal to the head cap. Bar = 0.5 µm. **C) Stage IX.** LMG of part of a cross section through a tubule in stage IX of development. Note the nuclear elongation of step 9 spermatids. Interstitium (IE), spermatogonia (A2 and A1), Sertoli nucleus (S), Sertoli cytoplasm (sc), leptotene spermatocyte (L), pachytene spermatocyte (P), Sertoli endoplasmic reticulum (er), acrosomic granule (ag), acrosomic head cap (hc). Bar = 5 µm. **D) Stage IX.** EMG of portions of two step 9 spermatids embedded in the Sertoli cytoplasm (SC). Nuclear elongation and condensation is more prominent than in the previous step. The acrosome is surrounded by Sertoli endoplasmic reticulum (er). The microtubular manchette (M) is attached to the nuclear ring (nr) just below the distal end of the head cap. Note the multivesicular body (i) in the distal cytoplasm, which is enlarged in Fig. 9A. Bar = 0.5 µm

2.12 Sperm Tail Differentiation

Tail differentiation and growth begin early in round spermatid development and continue almost throughout the entire period of spermiogenesis, a process that takes almost 19 days in the bull (Fig. 2.4.). The neckpiece and axoneme are formed by a pair of centrioles (diplosome, basal body, basal centriole), oriented at right angles to each other and composed of nine circularly arranged microtubular triplets. In the course of tail development the distal centriole gives rise to the axoneme and is also associated with the formation of the striated columns, while the proximal centriole is associated with the formation of the capitulum (Fawcett and Phillips 1969; Irons 1983). The outer dense fibers arise from the striated columns and have a similar composition (Schalles et al. 1998). Once these components are formed, the regular structure of the centrioles vanishes. During the Golgi and cap phase of spermiogenesis (steps 1-7 in the bull) the axoneme gradually grows from the end of the distal centriole (Figs. 2.4, 2.5E, 2.7E). By the beginning of the acrosome or elongation phase (steps 8-12 in the bull) the proximal centriole of the basal centriole is situated in the implantation fossa of the nucleus and the axoneme has reached full length (Figs. 2.4, 2.8B). Attachment of the tail to the plasmalemma occurs earlier in the cap phase just below the basal centriole by a ringlike structure called the annulus and results in the formation of a double membrane sleeve around the forming axoneme, which segregates the cytosol of the growing axoneme from the rest of the cell and creates a periaxonemal compartment (Figs. 2.4, 2.10B,C). The only way in for building materials to get into this

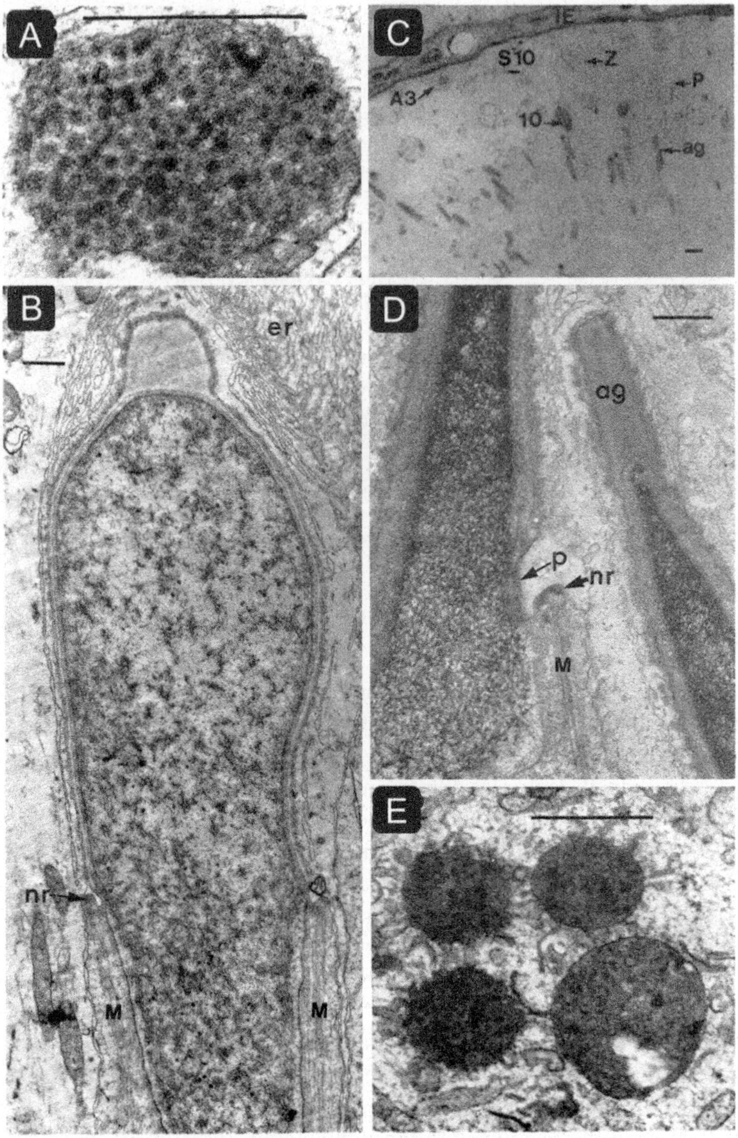

Fig. 2.9 **A) Stage IX.** Enlargement of the multivesicular body seen in Fig. 2.9A. The body contains numerous membrane-bound particles 45–50 nm in size. **B) Stage X.** EMG of a step 10 spermatid. The nuclei of step 10 spermatids continue to elongate and condense and begin to flatten when viewed laterally. See C and D. Endoplasmic reticulum (er), nuclear ring (nr), microtubular manchette (M). Bar = 0.5 μm. **C) Stage X.** LMG of a cross section through part of a seminiferous tubule in stage X (S10) of development. Note the elongation and flattening of spermatids. Zygotene spermatocyte (Z), spermatogonium (A3), interstitium (IE), pachytene spermatocyte (P), acrosomic granule (ag). Bar = 5 μm. **D) Stage X.** EMG of portions of two late step 10 spermatids cut sagittally. The acrosomic granule (ag) begins to elongate and the chromatin is more condensed. Note that the nuclear ring (nr) with the manchette (M) begins its distal migration down the distal half of the nucleus and that the postacrosomal sheath (p) is laid down proximal to the migration of the nuclear ring. Bar = 0.5 μm. **E) Stage X.** EMG of a portion of the distal cytoplasm of a spermatid showing several multivesicular bodies. In this stage the multivesicular bodies begin to lose their membrane-bound particle characteristic and take on an amorphous appearance. Compare with Fig. 9A. Bar = 0.5 μm

periaxonemal compartment is through the annular ring. It is important to note that during the so called 'acrosome phase' the acrosome is already fully formed and only conforms in shape to the elongating and condensing nucleus. Thus, a much more descriptive term for this phase is elongation.

During the elongation phase the rudiments of the capitulum and striated columns differentiate into their final form (Irons 1983). This phase of development is also associated with the initiation of the outer dense fiber (ODF) formation. The nine outer dense fibers develop in a proximal-to-distal direction in association with the microtubule doublets of the axoneme (Figs. 2.4; 2.10B,C). However, a dramatic enlargement of these fibers occurs in the maturation phase (steps 13–14), the final phase of spermiogenesis (Fig. 2.11D). The fibrous sheath (FS), on the other hand, develops in a distal-to-proximal direction (Figs. 2.4, 2.10E) along the axoneme destined to form the principal piece. Its two longitudinal components form slowly, from the early steps of spermiogenesis until the mid-maturation phase, whereas its ribs are formed and assembled during a shorter period, from the mid-acrosome to mid-maturation phase. Assembly of the mitochondrial sheath is initiated just after the annulus descends to form the midpiece-principal piece junction (Figs. 2.4, 2.11D). Descent occurs early in the maturation phase (step 13) and is followed rapidly by a proximal-to-distal distribution of mitochondria around the axial fiber bundle (Figs. 2.4, 2.12A). The mitochondria then condense and become helically arranged, a process that lasts almost to the end of spermiogenesis. Just before spermiation the residual midpiece cytoplasm (residual body) is extruded at the level of the neck-piece (Figs. 2.4, 2.7A, 2.12B). Although the sperm tail and head are fully formed at the time of spermiation, final maturational changes occur in the epididymis.

2.12.1 Assembly of the ODF and FS

Irons and Clermont (Irons and Clermont 1982a, b) investigated the morphogenesis of the ODF and FS of the rat spermatid by relating the precise timing of various steps in their formation to electron microscope radioautographic observations of their incorporation of 3H-amino acids at various intervals after intratesticular injection. In step-8 spermatids, they found that nine very fine fibers of the ODF, termed ODF anlagen, are associated with the corresponding set of microtubule doublets in the most proximal portion of the axoneme (Fig. 2.13). During steps 9 to 14 (equivalent to steps 9 to 12 in the bull), these fibers gradually increase in thickness and in length in a proximal-to-distal direction. In step 16 (equivalent to step 13 in the bull), according to both radioautographic and electron microscope analysis, the ODF undergo a rapid and marked increase in diameter (Fig. 2.13). It is noticeable that this massive increase in size of ODF takes place immediately following the migration of the annulus from its juxtanuclear position to its final position at the distal end of the cytoplasmic lobule that takes place very early in step 16 (step 13 in the bull). Since the narrow periaxonemal cytoplasmic compartment is then eliminated from the proximal segment of the forming tail, one can

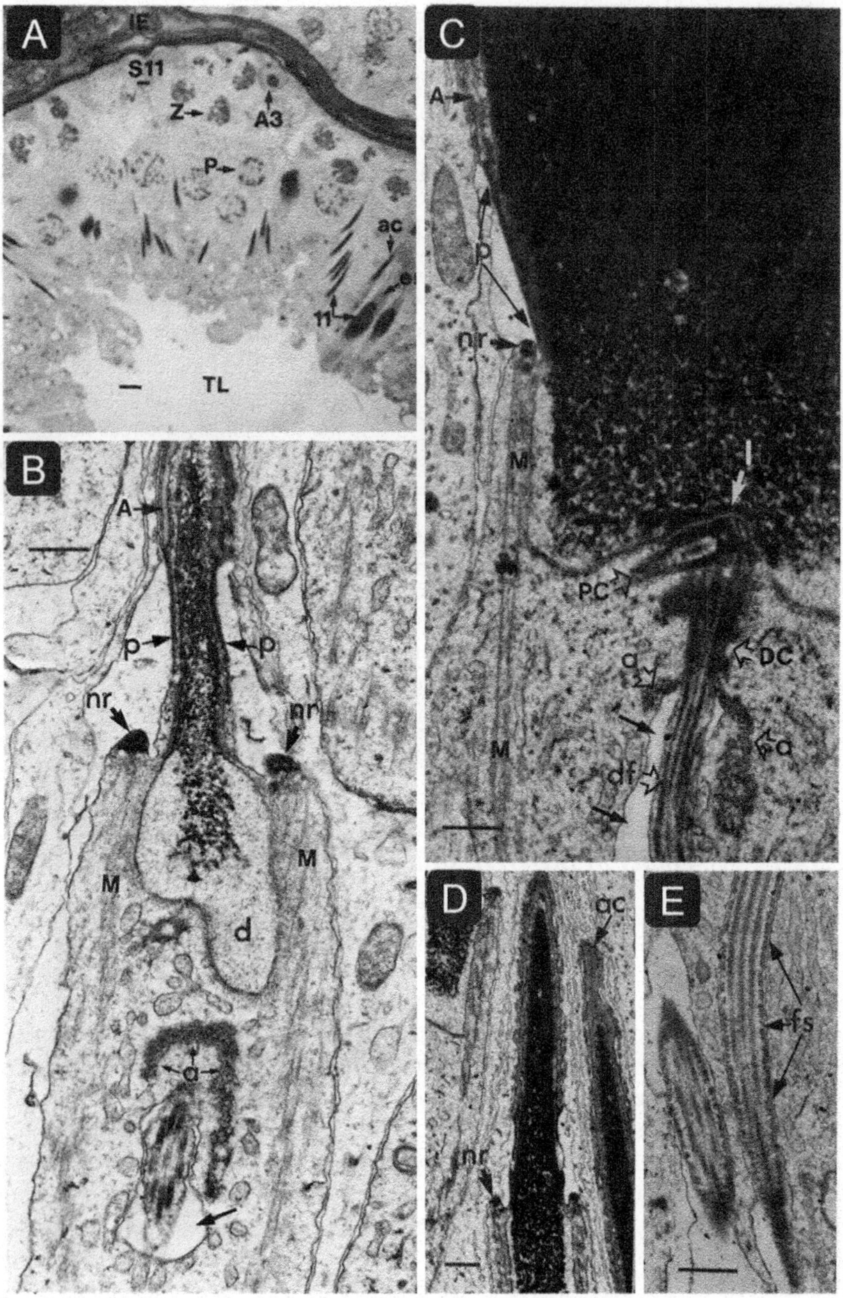
A
IE
S11
Z
A3
P
ac
11
TL
B
A
p
p
nr
nr
M
M
d
a
C
A
p
nr
M
M
I
PC
DC
a
a
df
D
ac
nr
E
fs

◀ **Fig. 2.10** **A) Stage XI.** LMG of a cross section through part of a seminiferous tubule in stage XI of development. Step 11 spermatids are grouped in bundles deep in the Sertoli cytoplasm and are surrounded by Sertoli endoplasmic reticulum (er). The spermatid nucleus is almost completely flattened when viewed sagittally. Interstitium (IE), zygotene spermatocyte (Z), spermatogonia (A3), pachytene spermatocyte (P), apical crest (ac), tubular lumen (TL). Bar=5 μm. **B) Stage XI.** EMG of a distal portion of a step 11 spermatid. Note that the nuclear ring (nr) that connects to the microtubules of the manchette (M) has migrated midway between the distal end of the acrosome (A) and the base of the nucleus. This distance marks the extent of postacrosomal sheath (p) assembly. The manchctte in its distal migration appears to squeeze the excess nuclear membrane into a pouch or diverticulum (d). Bar=0.5 μm. **C) Stage XI.** EMG of part of a frontal section through a step 11 spermatid. Compare with B. Note the implantation fossa (I), the proximal (PC) and distal (DC) centriole and the annulus (a), which is located just below the distal centriole and is attached to the membrane bound cul-de-sac or sleeve (arrows) that surrounds the flagellum. Also note that the outer dense fibers (df) surround the axial fiber bundle in the proximal region of the flagellum. Acrosome (A), postacrosomal sheath (p), nuclear ring (nr), manchette (M). Bar=0.5 μm. **D) Stage XI.** EMG of a portion of a group of step 11 spermatids. The acrosomic granule is replaced by a long protruding crescent-shaped apical crest (ac). Nuclear ring (nr). Bar=0.5 μm. **E) Stage XI.** EMG of a distal portion of the flagellum of a step 11 spermatid. Note that the fibrous sheath (fs) surrounds the axial fiber bundle in this region but that the outer dense fibers as seen in figure C have not yet formed. Bar=0.5 μm

speculate that the precursor proteins have a facilitated access to the ODF anlagen, explaining the rapid growth in ODF size.

In the case of the morphogenesis of the FS, Irons and Clermont (1982b) have shown that anlagen of the longitudinal columns begin to be deposited with the initiation of axoneme assembly and that these miniscule deposits interact with both the plasmalemma and microtubule doublets 3 and 8 in the part of the tail that will become the principal piece. On the other hand, rib anlagen, closely associated with the entire plasmalemma, first appear in the distal end of the principal piece in step 9 (Fig. 2.13). These two forms of anlagen form the basic framework for the assembly of the FS, which begins at step 11 and continues in a distal-to-proximal direction until step 17 (Fig. 2.13). In step 15, according to both radioautographic and electron microscope analysis, the entire existing framework of the FS undergoes a rapid increase in electron density due to the convergence of rib anlagen and deposition of additional protein to form definitive ribs and longitudinal columns. Thus, the assembly of both these cytoskeletal structures is a lengthy multistep process, which extends from steps 8 to 19 in the case of the ODF and from steps 2 to 17 in the case of the FS.

The results of an immunocytochemical investigation on the formation of the FS and ODF verifies the timing of the assembly of these flagellar proteins as presented by Irons and Clermont but in addition, relates the time and site of their synthesis in the cytoplasm of the rat spermatid to their assembly along the axoneme (Clermont et al. 1990; Oko and Clermont 1989). These studies also indicate that the posttranslational expression of protein synthesis is taking place long after the initiation of the condensation of the spermatid's nucleus in step 9, at which time RNA synthesis stops (Monesi et al. 1978; Hecht 1987), implying that transcribed mRNAs coding for the biosynthesis of FS and ODF proteins are first stored and

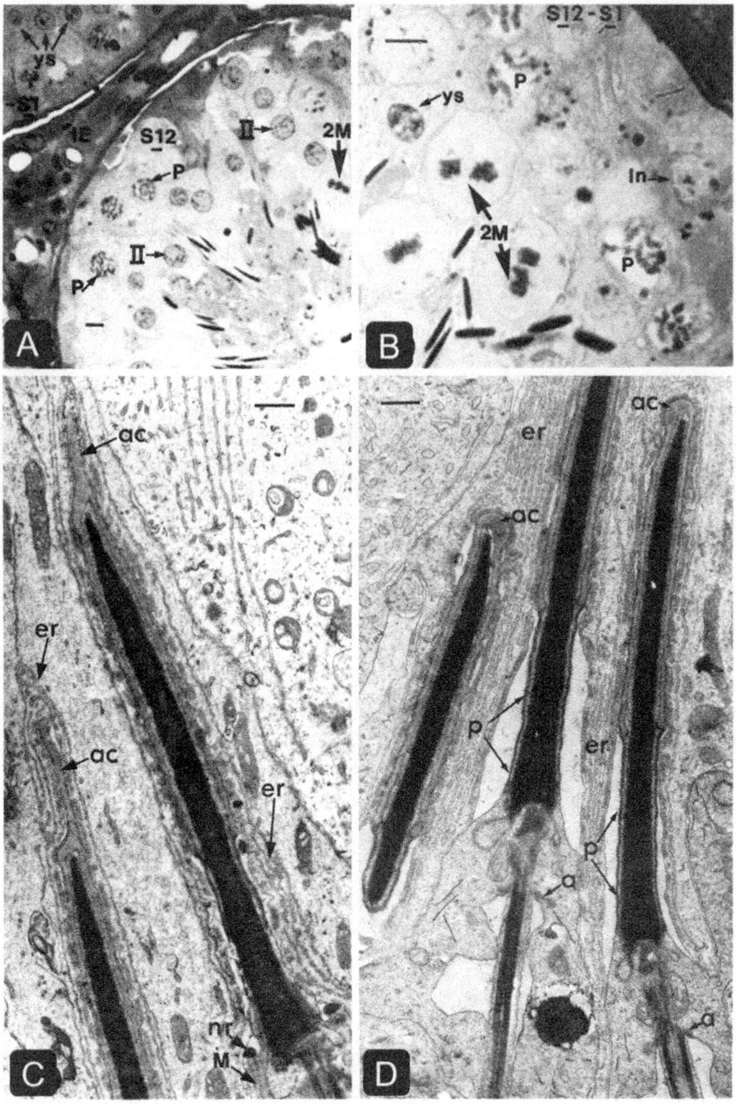

Fig. 2.11 **A) Stage XII.** LMG of a cross section through part of a tubule in stage XII showing secondary spermatocytes (II), which is the most distinguishing characteristic of this stage. Second meiotic division (2M), young spermatids (ys), pachytene spermatocyte (P), interstitium (IE). Bar = 5 μm. **B) Stage XII—I.** LMG of a cross section through part of a tubule showing the transition period between stages XII and I. The second meiotic division (2M) gives rise to young spermatids (ys). Intermediate spermatogonium (In), pachytene spermatocyte (P). Bar = 5 μm. **C) Stage XII.** EMG of a portion of two spermatid heads in step 12. Note that the heads are surrounded by Sertoli endoplasmic reticulum (er) and that the nuclear ring (nr) and manchette (M) have migrated to the base of the head. The apical crest (ac) is a long protruding crescent-like structure in this stage. Bar = 0.5 μm. **D) Stage I.** EMG of a group of step 13 spermatids in stage I. See Fig. 5A for light micrograph representation. The spermatids are grouped in bundles and surrounded by Sertoli endoplasmic reticulum (er). The apical crest (ac) resembles that of a mature sperm, the postacrosomal sheath (p) is fully formed, the annulus (a) is still present below the neck piece, the mitochondria have not yet begun surrounding the midpiece of the flagellum and the manchette has disintegrated. Bar = 0.5 μm

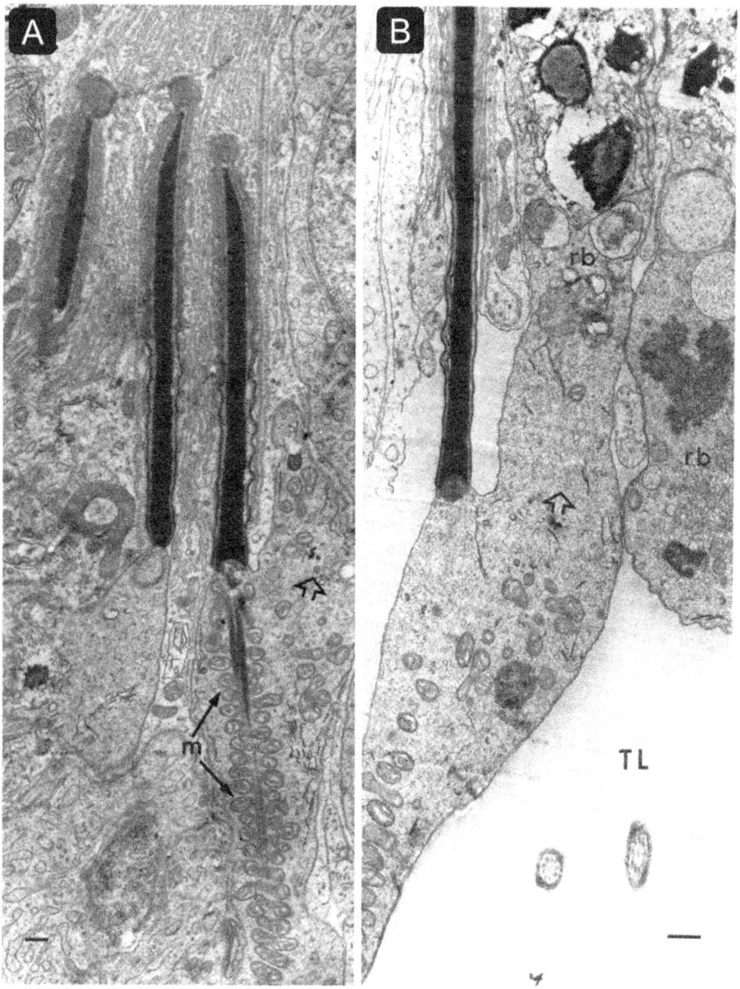

Fig. 2.12 **A) Stage III.** EMG of a group of step 13 spermatids in stage III of development. The annulus has descended already in late stage I, allowing the mitochondria (m) to surround the midpiece of the sperm. Note that the excess spermatid cytoplasm just beginning to be extruded from the flagellum (arrow). Bar = 0.5 μm. **B) Stage IV.** EMG of a distal portion of a step I4 spermatid in stage IV. See Fig. 7A for light micrograph representation. Note that the distal spermatid cytoplasm is being extruded (arrow) in the form of residual bodies (rb) and that the spermatid is positioned in the tubular lumen (TL). Bar = 0.5 μm

then translated later in spermiogenesis. Immunoperoxidase staining of antibodies directed against major ODF and FS polypeptides indicated that the production of these polypeptides in the cytoplasm of rat spermatids begins in step 9, gradually increases in subsequent steps, and reaches a peak in steps 15 to 17 for FS proteins and steps 16 to 18 or ODF proteins (Oko and Clermont 1989).

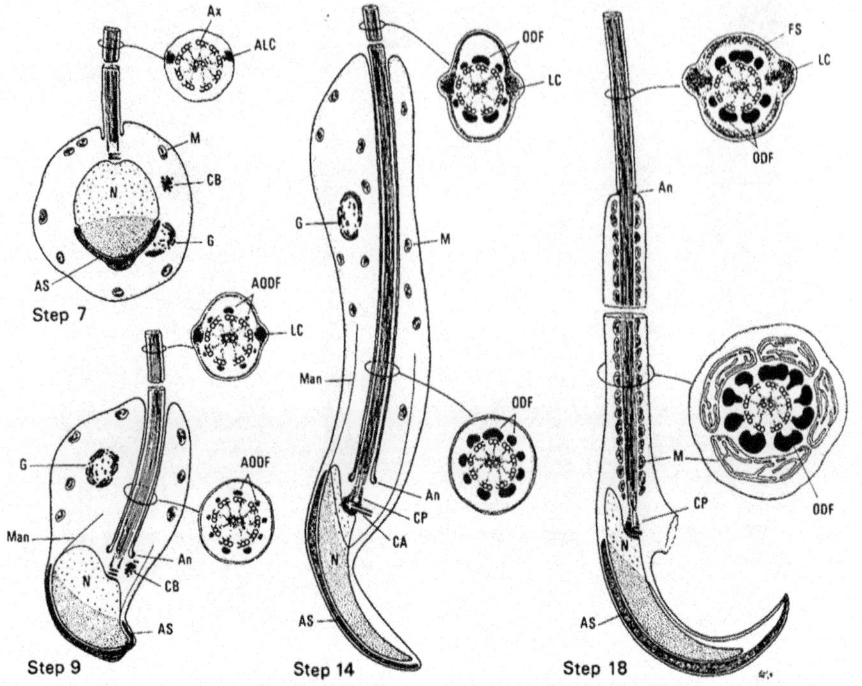

Fig. 2.13 Diagram showing the main ultrastructural changes taking place in the developing tail of the rat spermatid. The drawing was conceptualized by Dr. Yves Clermont. Four steps of rat spermiogenesis are illustrated, i. e., steps 7, 9, 14, and 18 which are equivalent to steps 7, 9, 12 and 14 of bovine spermiogenesis; the assembly of the accessory structures (ODF and FS) of the spermatid tail follow the same sequence in both bull and rat. Cross-sectional outlines of the proximal and distal regions of the tail are shown with the various components, i. e., outer dense fibers (ODF) and their anlagen or precursor material (AODF), and fibrous sheath with its longitudinal column (LC) and its anlagen or precursor material (ALC) and the ribs. The development of the principal piece with its fibrous sheath is illustrated in the four electron microscope cross-sectional profiles above, while the development of the middle piece with the outer dense fibers (ODF) and their anlagen (AODF) are illustrated in the three electron microscope cross-sectional profiles below. Note that until the end of step 14 (step 12 in the bull), the tail structures are enclosed within a periaxonemal compartment delimited in its proximal portion by a double plasma membrane which extends to the annulus (An), located near the connecting piece. Early in step 16 (step 13 in the bull), the annulus slides along the axoneme and concomitantly, the ODF increase in diameter and mitochondria migrate toward the tail. Man, manchette; AS, acrosomic system; N, nucleus; M, mitochondria; CB, chromatoid body; G, Golgi apparatus; CA, centriolar adjunct; CP, connecting piece

The FS and ODF proteins are produced in the cytoplasmic lobule and not in the thin periaxonemal cytoplasmic compartment. However, there is a qualitative difference in their distribution: the FS proteins are diffusely distributed, while the ODF proteins appear to be concentrated in granular masses. The granular masses correspond at the EM level to a particular type of granular body which is especially abundant and selectively immunoreactive to ODF antibodies at the time of peak ODF formation (Clermont et al. 1990). FS and ODF proteins diminish considerably in the cytoplasm of spermatids at the end of spermiogenesis. However, at the completion of spermiogenesis, a large portion of unused FS and ODF protein, detectable by immunocytochemistry, appear to be present in the residual cytoplasm of the spermatids. The superfluous cytoskeletal proteins are then apparently segregated from the cytoplasm into residual bodies or cytoplasmic droplets: the surplus of FS proteins are eliminated via the cytoplasmic droplets, while the majority of ODF proteins are eliminated via the residual bodies (Oko and Clermont 1989). The residual bodies are eventually phagocytosed and degraded by the Sertoli cell, while the contents of the cytoplasmic droplets are eventually endocytosed by clear cells in the rat epididymis.

2.12.2 Formation of the Mitochondrial Sheath

During the first three phases of spermiogenesis, the spherical or ovoid mitochondria are generally located along the plasma membrane of the spermatid (Fig. 2.13). These mitochondria have a peculiar internal organization of their lamellar cristae that has been well analyzed by André (1962). The intracristal spaces are distended by an electron-lucent fluid, while the narrowed intercristal spaces contain an electron-dense mitochondrial matrix (Fig. 1.29 in Chapter 1). Early in spermiogenesis there is a burst in mitochondrial DNA synthesis which may be indicative of a proliferation and increase in number of the mitochondria (Hecht and Liem 1984). During the elongation phase, the mitochondria are prevented from establishing contact with the growing tail by the double layer of plasma membrane, which demarcates the periaxonemal compartment (Figs. 2.4, 2.13). At the onset of the maturation phase in step 15 of the rat (step 13 of bull), the plasma membrane-associated annulus, to which chromatoid body material had been added earlier, suddenly and rapidly slides down the axoneme (Figs. 2.4, 2.13, 2.14). Simultaneously, the mitochondria, which now have free access to the developing tail, migrate toward the ODF and align along its length (Figs. 2.14, 2.15). The mechanisms that are involved in this displacement and alignment remain to be clarified.

An initial scanning electron microscope investigation shows that the disposition and arrangement of mitochondria along the midpiece of the tail is a highly regular process that proceeds in a synchronized stepwise manner (Otani et al. 1988). The process, as described by these investigators in mouse spermatids, occurs in four distinct steps. In a first step (corresponding to step 13 of bull spermiogenesis), mitochondria are arranged in four longitudinal rows along the axial filament and make individual contacts with the ODF. In the second step (corresponding to mid-step 13 of bull spermiogenesis), the longitudinal rows twist dextrally along the axial filament and the spherical mitochondria between rows become staggered. In the third step (corresponding to late-step 13 and early-step14 of bull spermiogenesis), the mitochondria gradually elongate into crescentic rod-like bodies.

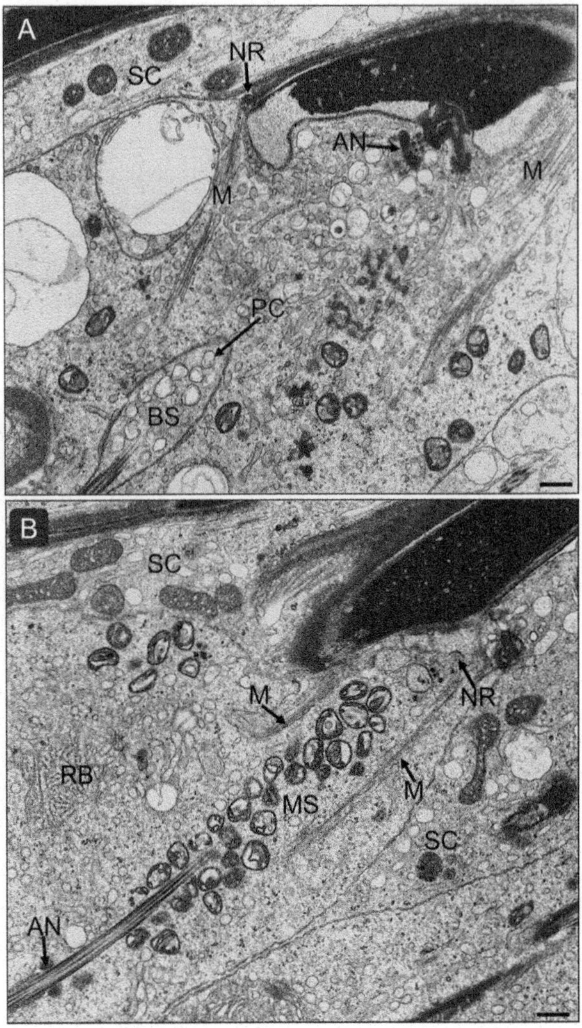

Fig. 2.14 Electron micrographs of elongating rat spermatids before and after descents of the manchette and annulus. A) The microtubular manchette (M) surrounds the caudal half of the nucleus of a step 14 spermatid (step 11 in the bull). The microtubules of the manchette extend from the nuclear ring (NR) just below the acrosome into the distal cytoplasmic lobe. Note the angular nature of the microtubules of the manchette when compared to the bull (Fig. 10C). At this step, the annulus (AN) is located just below the connecting piece. Because the section represents an oblique cut through the longitudinal axis of the spermatid, the periaxonemal compartment (PC) which begins directly below the annulus is invisible until it is revealed distally. The bulbous vesicle-filled swelling (BS) indicates the proximal most portion of fibrous sheath-rib assembly. Bar = 0.5 μm. **B)** The manchette (M) and annulus (AN) have migrated distally to their final destinations in this early step 16 spermatid (step 13 in the bull). The descent of the manchette is associated with a narrowing of its girth about the nucleus. At this point the manchette has served its purpose (i. e., shaping the nucleus and depositing the postacrosomal sheath) and disintegrates. The descent of the annulus eliminates the periaxonemal compartment around the midpiece region of the tail allowing the mitochondria from the cytoplasmic lobe to access the axial fiber bundle and form the mitochondrial sheath around the growing ODFs. The annulus begins its descent once the manchette has fully descended. Sertoli cytoplasm (SC). RB, radial body of smooth endoplasmic reticulum. Bar = 0.5 μm

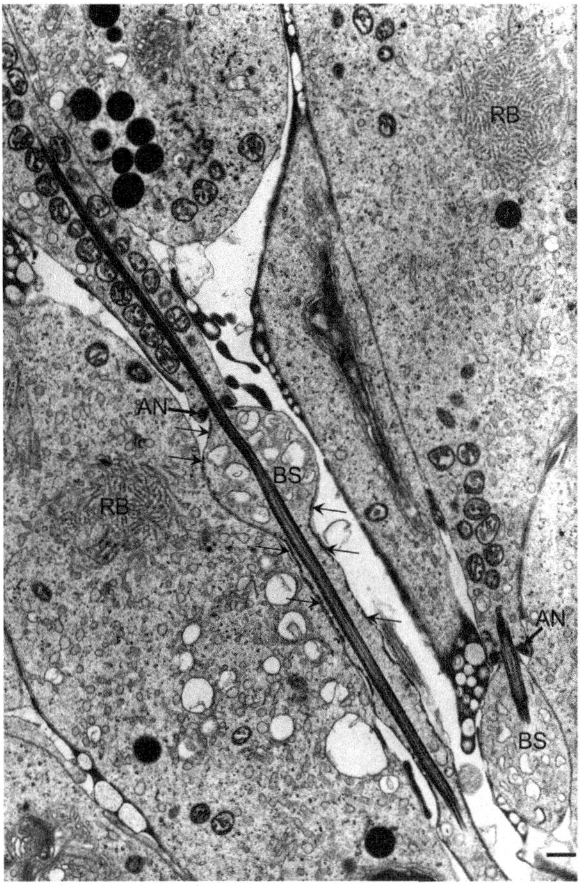

Fig. 2.15 Electron micrograph of the newly formed midpiece—principal piece junction of the spermatid tail just after annular descent. The annulus (AN) marks the division between the midpiece and principal piece of the tail. Directly below the annulus is the bulbous vesicle-filled swelling (BS), which marks the proximal most portion of fibrous sheath-rib assembly. Note that within the BS, the rib material (arrows) is being laid down adjacent to the plasmalemma whereas below the BS it has already been assembled. Above the annulus the mitochondria are surrounding the outer dense fibers of the axial fiber bundle marking the initial process of mitochondrial sheath assembly. Note the abundance of smooth endoplasmic reticulum in the form of a reticular body (RB) in step16 spermatid cytoplasm. Bar = 0.5 µm

Because of their staggered disposition, the mitochondria of opposite rows, and not of neighboring ones, contact one another end-to-end and form a sinistral double helix. At the end of this third step, the end-to-end contacts are observed in every second gyre on four longitudinal lines that surround the core of the axial filament. As suggested by Otani et al. (1988) the elongation of mitochondria in some species whose spermatozoa display a staggered pattern of the mitochondrial sheath (e.g., common-brown bat, marsupial, Chinese hamster) may stop at this point. In

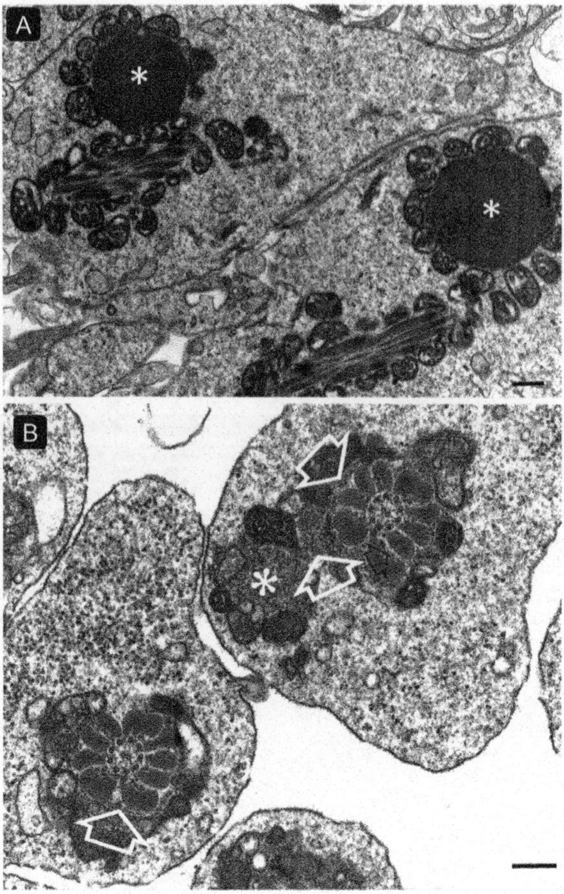

Fig. 2.16 Electron micrographs of rat spermatids, sectioned through midpieces, undergoing mitochondrial sheath formation. A) Step 17 spermatids. Note the presence of a granular body (asterisk) surrounded by superfluous mitochondria that have not been utilized in mitochondrial sheath formation. Bar = 0.5 μm. **B)** Step 18 spermatids. Note that the granulo-filamentous material that resides in the granular body (asterisk) appears interspersed between the mitochondrial sheath and axial fiber bundle (white arrows) and between the outer dense fibers (black arrow). This material was coined the submitochondrial reticulum by Olson et. al. (Olson and Winfrey 1990) and seems to originate from the granular bodies that are abundant throughout these steps of spermiogenesis (Oko 1984). Bar = 0.5 μm

the fourth and final step of development in the mouse (corresponding to mid-step 14 of bull spermiogenesis), the mitochondria further elongate, and as a consequence, the end-on touching of mitochondria in mature spermatids appears every third gyre on now five longitudinal lines which surround the core of the axial filament.

By 3D reconstruction modeling, based on scanning electron microscopy, Ho and Wey (2007) proposed a new model for mitochondrial sheath (MS) formation. In this model each mitochondrion stays where they first attach to the axial fiber bundle and elongate laterally from two directions to reach their final double helical structure without a great distance shift along the outer dense fibers. In the first step, four dextrally arranged spherical mitochondria rows surround the axial fiber bundle. In step 2, mitochondria opposite each other grow towards each other to form a doughnut-shaped structure. In step 3, each pair of doughnut-shaped mitochondria slide past each other in a specific pattern, where one end of each pair moves distally and the other moves proximally, to transform the MS into the sinistral double helical structure. In step 4 the crescent-shaped mitochondria elongate to three-fifths of 1 gyre, becoming rod-shaped. This regular disposition of mitochondria along the midpiece not only indicates that mitochondria are programmed to elongate simultaneously and at the same rate, but also that the distribution and attachment to the ODF is exquisitely controlled by a still unknown mechanism.

The condensation of the mitochondria, which is mainly due to the resorption of the clear fluid present in the initially distended intracristal spaces, as well as the characteristic elongated crescentic shape taken by the organelle appears to be the result of internally programmed molecular process. This view receives support from the fact that superfluous mitochondria, which are not utilized in the formation of the mitochondrial sheath and which remain free in the cytoplasmic lobule, also condense and acquire the typical crescentic shape in synchrony with the mitochondria associated with the ODF. Furthermore, these superfluous, crescentic mitochondria often cluster around large spherical bodies, made up of a granulo-filamentous material (Fig. 2.16A), which is similar to the material seen between the mitochondria and the ODF in the middle piece of mature spermatozoa (Fig. 2.16B) (Oko and Hrudka 1984b; Oko 1984). This granulo-filamentous material or submitochondrial reticulum as coined by Olson et al. (Olson and Winfrey 1990) may serve to bind mitochondria to the ODF.

The morphology of the mitochondria may also be due to mitochondrial capsule proteins that appear to coat the mitochondrial surface (Calvin et al. 1981). The incorporation of selenium into this protein starts during the early phases of spermiogenesis, reaches a peak during the acrosome phase, and ceases early in the maturation phase, as the mitochondria become associated with the ODF (Calvin et al. 1987). A sperm-specific lactate dehydrogenase isozyme also appears in the mitochondria during the early steps of spermiogenesis (Goldberg and Hawtrey 1967; Machado de Domenech et al. 1972). These various observations emphasize the fact that mitochondria undergo complex structural and biochemical modifications before and during the formation of the tail of spermatozoa.

2.13 Origin and Assembly of Proteins Compartmentalizing the Sperm Head

The specialized structure of mammalian spermatozoa reflects a complex process of biochemical transformation taking place during the haploid phase of germ cell development. With the exception of mitochondria none of the standard cytoplasmic organelles (e. g., ER, Golgi, centrioles) are seen in the mature mammalian spermatozoa. Instead, what is observed in the spermatozoon are derivatives of these organelles, such as the acrosome formed by the ER-Golgi system, the transient Golgi-derived saccular elements of the cytoplasmic droplet (Oko et al. 1993), the axoneme derived from one of the two centrioles and a very condensed nucleus. In addition to these transformations, some of the most conspicuous developmental changes are concerned with the elaboration of specialized 'cytoskeletal' head and tail components of which there appear to be no equivalent in any other cell type (Oko 1998). In the sperm head the major cytoskeletal component is the perinuclear theca (PT), which is not composed of traditional cytoskeletal proteins as was originally thought, but rather from a variety of cytosolic and nuclear proteins that form a detergent resistant shell around the nucleus.

2.13.1 Formation of the Subacrosomal Layer of the PT

The formation of the acrosome during the first half of spermiogenesis can be divided into Golgi and cap phases, distinguished by two different phases of secretory activity by the Golgi apparatus (Thorne-Tjomsland et al. 1988; Tang et al. 1982; Clermont and Tang 1985). In the first phase (steps 1–3) the Golgi secretes several small and dense membrane bound secretory granules rich in hydrolytic enzymes. These proacrosomic granules (PA) coalesce to form a single larger, dense cored, spherical acrosomic vesicle (AV), which then associates with the nuclear envelope (NE) via peripheral proteins that surround both the PA and AV and become an integral part of the subacrosomal layer (SAL) of the PT (Fig. 2.17). In the second phase (steps 4–7) the acrosomic vesicle (AV) enlarges by fusing with numerous small carrier vesicles originating from the trans face of the Golgi apparatus. As a result, there is a continual addition of membrane and glycoprotein, allowing the less dense cortex part of the AV to expand over the nucleus and take the shape of a cap. It should be emphasized that the expansion of the acrosomic system is synchronous with the expansion of the underlying SAL and with expansion of the nuclear dense lamina, which in turn underlies the nuclear membrane (Oko et al. 2011).

In spermatozoa the IAM is bound firmly to the nuclear envelope via a proteinaceous layer referred to as the SAL of the PT. The SAL is considered a structural entity, however, developmentally it originates in association with the extravesicular part of the acrosomal membrane (Figs. 2.17, 2.18 and 2.19). SAL proteins (e.g.,

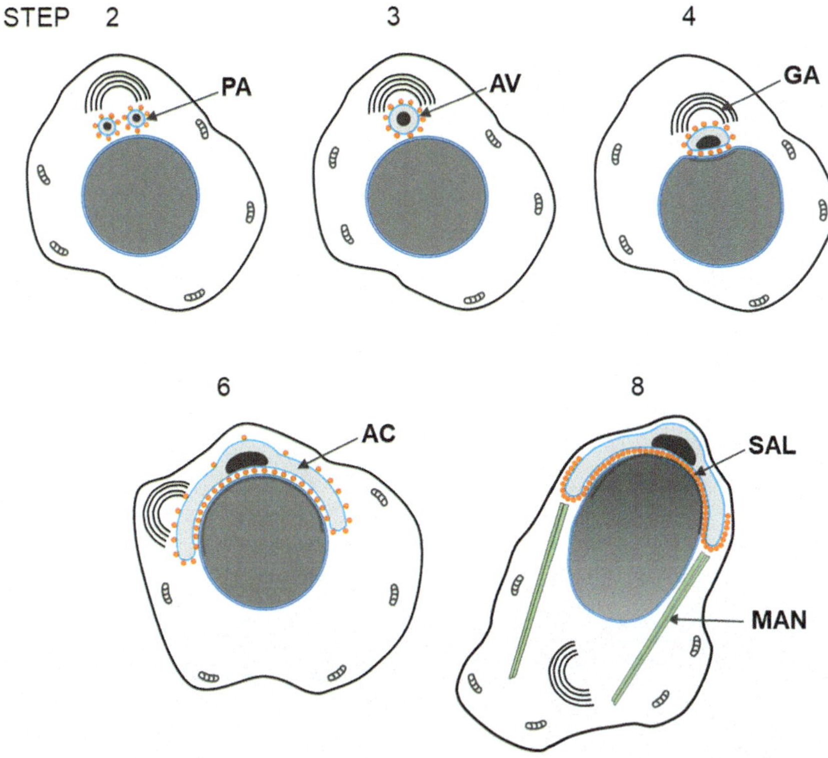

Fig. 2.17 Proteins making up subacrosomal layer (SAL) of the PT of spermatozoa are involved with acrosomal-nuclear docking and capping during spermiogensis. SAL proteins (orange dots) coat the acrosomal membrane of proacrosomic vesicles (PA) and the acrosomic vesicle (AV) during the Golgi phase of spermiogenesis (steps 1–3). As a result, they are the first proteins to intervene between the AV and the nuclear envelope of the round spermatid. During the early cap phase of spermiogenesis (steps 4–5) the SAL proteins continue to attach to the periphery of the expanding acrosome cap (AC) but towards the end of the cap phase (steps 6–7) they begin to diminish from the OAM and finally end up positioned in step 8 spermatids between the IAM and nuclear envelope and between the OAM and plasmalemma, in the region that will become the equatorial segment. So far all of the characterized SAL proteins (i. e., SubH2Bv/H2BL1, KAPα, KAPβ and RAB2A) follow the localization pattern depicted in this figure. Note that the nuclear lamina (represented by the black line underneath the nuclear envelope) outlines and probably regulates acrosomal-expansion over the area of the nucleus. As shown in Fig. 2.25 even when the formation of the acrosome is completely disrupted the nuclear lamina usually outlines the region of the nucleus where the acrosome is missing. GA, Golgi apparatus; MAN, manchette

SubH2Bv, Importin α and β, RAB2A) are initially associated with the outside surface of the acrosomal membrane of proacrosomic and acrosomic vesicles before nuclear attachment of these vesicles (Aul and Oko 2001; Mountjoy et al. 2008; Oko and Maravei 1995; Oko 1998; Oko and Sutovsky 2009) (Figs. 2.18, 2.19).

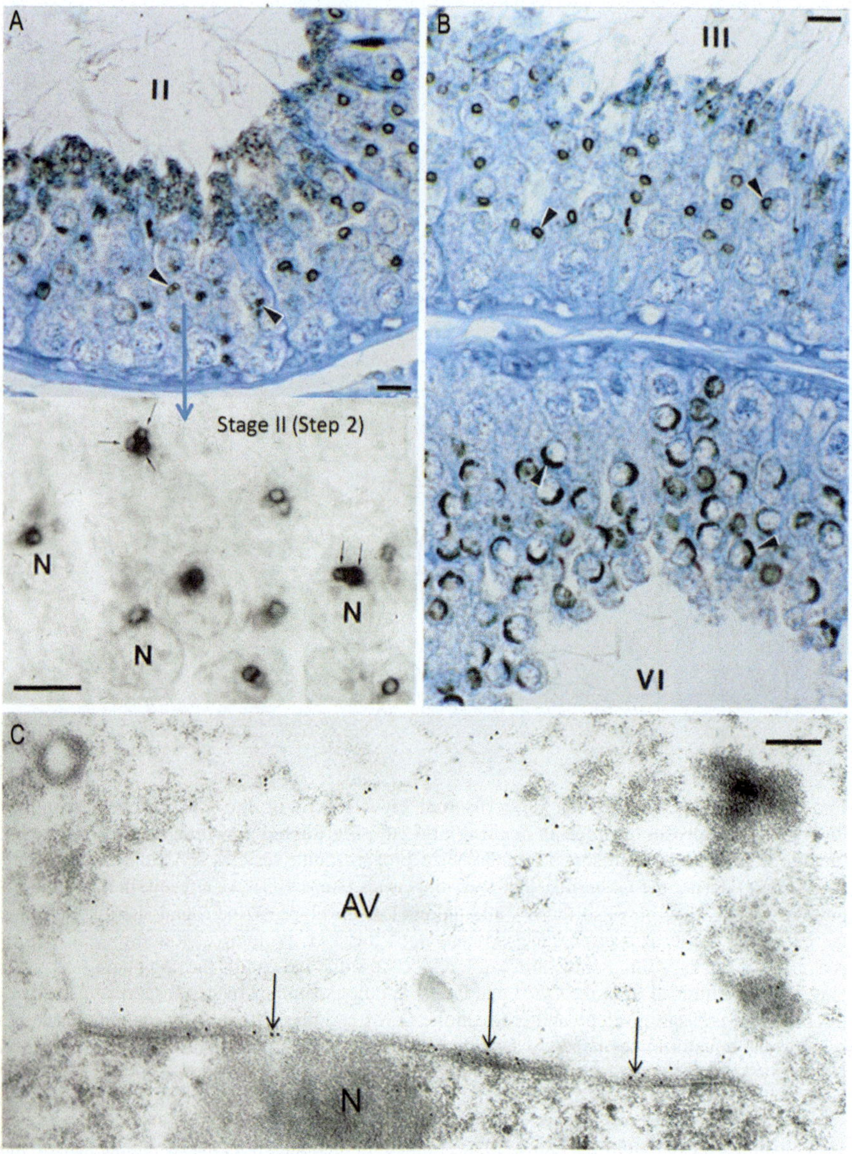

Fig. 2.18 Sections representing stages II, III and VI of the cycle of bovine seminiferous epithelium, immunoperoxidase-stained with affinity-purified anti-SubH2Bv/H2BL1 antibody. Staining with antibodies representing other subacrosomal layer proteins (i. e., KAPα, KAPβ and RAB2A) showed identical results. **A)** In this early step of spermiogenesis (step 2) immunoperoxidase reaction product is seen surrounding individual proacrosomic vesicles as they fuse together (arrowheads or small arrows) and begin to attach to the spermatid nuclei (N). Blue arrow points to higher magnification section, not stained with Commassie blue. Bars = 5 μm. **B)** Stage III showing immunoperoxidase reaction product surrounding the periphery of the forming acrosomic vesicles (arrowheads) that have by this stage attached to the nuclei of the round spermatids. In stage VI, the immunostaining remains associated with the acrosome as it caps (arrowheads) and spreads over the apical portion of the nucleus of the round spermatid (arrows). Bar = 5 μm. **C)** Electron micrograph of a portion of a step 4 spermatid immunogold labeled with anti-SubH2Bv antibody. The labeling (black dots) surrounds the acrosomic vesicle (AV) and is found between the attachment site of the acrosomic vesicle and the nucleus (N), or SAL (arrows). Bar = 0.5 μm

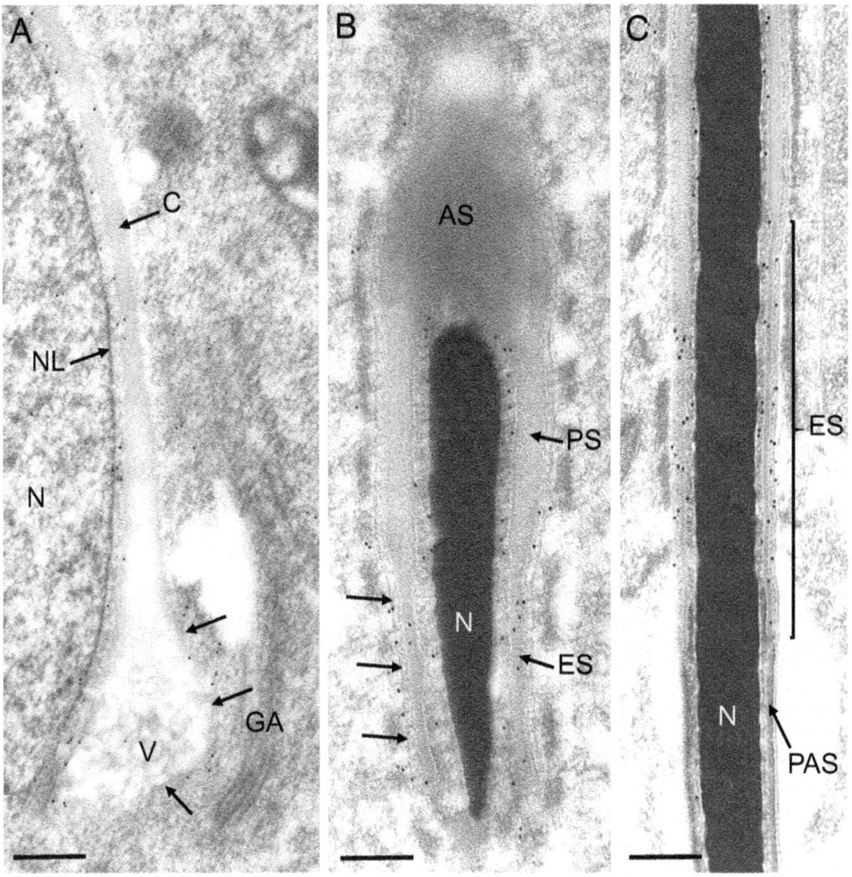

Fig. 2.19 Electron micrographs of sections through the heads of spermatids, immunogold labeled with anti-SAL antibodies. A) At the end of the cap phase, most of the immunolabeling of SAL proteins is between the acrosome cap (C) and the nuclear envelope, which is underlain by the nuclear lamina (NL). Note that the Golgi apparatus (GA), just before it slides off the acrosome, continues to produce small vesicles (V) that are added to the growing acrosome cap and that the immunolabel surrounds the outer border of the newly laid down cap (arrows), between the Golgi apparatus and outer acrosomal membrane. N, nucleus. Bar = 0.5 µm. **B)** In this elongated spermatid, just after the equatorial segment of the acrosome has thinned, note the persistence of the immunolabeling in the subacrosomal layer of the PT and on the outer aspect of the equatorial segment (arrows) in the region termed the outer periacrosomal layer. AS, apical segment of the acrosome; PS, principal segment; ES, equatorial segment. Bar = 0.5 µm. **C)** Elongated spermatid just before spermiation. Note the persistence of the label in the subacrosomal and outer periacrosomal layer. PS, principal segment; ES, equatorial segment; PAS, postacrosomal sheath. Bar = 0.5 µm

Although there may be SAL proteins not yet identified, anti-sera raised to isolated PT, only label the periphery of the acrosomic vesicles in the Golgi and cap phases of spermiogenesis, indicating that SAL proteins have this common origin (Oko and Maravei 1995). Of course, anti-PT sera contains antibodies directed against the postacrosomal sheath (PAS) part of the PT as well, but as will be discussed PAS proteins are not synthesized until the elongation phase of spermiogenesis (Tovich et al. 2004; Wu et al. 2007b). These immunolocalization studies, summarized in Fig. 2.17, imply that the SAL assembles from cytosolic proteins that reside on the periphery of the forming acrosome, which in turn raises the hypothesis that SAL proteins are integral to the mechanism of acrosomal-nuclear docking.

Studies supporting this hypothesis arose from the molecular and developmental characterization of SubH2Bv/H2BL1, a histone H2B variant residing in the subacrosomal compartment of mammalian spermatozoa (Aul and Oko 2001). SubH2Bv was shown to contain a functional bipartite nuclear localization signal (bNLS) (Tran et al. 2012), which, to be targeted to the nucleus, requires interactions with importin-α (KAPα) and -β (KAPβ) (Cingolani et al. 1999; Görlich et al. 1996). Indeed, co-localization of KAPβ, KAPα and SubH2Bv to the periphery of the forming acrosomic vesicle, followed by their co-localization to the SAL at the time of acrosomal-nuclear docking, suggested a trimeric binding interaction between these proteins which was supported by recombinant SubH2Bv affinity pull-down assays on germ cell extracts (Tran et al. 2012). The results of this study provide a compelling argument that these three nuclear homing proteins work in concert to direct the acrosomic vesicle to the nucleus. The nuclear dense lamina's presence undoubtedly influences or signifies a molecular change in the overlying NE that allows for acrosomal attachment via a receptor complex in the NE. Components that make up this complex are just beginning to be identified. SPAG4L-2 is a testis-specific SUN domain transmembrane protein, restricted to the apical nuclear region of round spermatid that faces the AV (Frohnert et al. 2011). Likewise, Dpy19l2 is predominantly spermatid specific and shares a similar localization to SPAG4L-2, both proteins being residents of the inner nuclear membrane (Pierre et al. 2012). Moreover, the Dpy19l2$^{-/-}$ mouse model exhibits a similar sperm phenotype (i. e., acrosome-less and rounded nuclear shape) to that found in human type 1 globozoospermia, caused by the disruption of its ortholog (Pierre et al. 2012). Developmentally in the mouse, the absence of Dpy19l2 destabilizes the nuclear dense lamina and the junction between the nuclear envelope and the SAL, leading to the detachment of the acrosome, which itself is malformed, at the beginning of the elongation phase. A further consequence of the disruption of the nuclear dense lamina is the failure of the manchette to link positionally to the nucleus, leading to the failure of sperm nuclear elongation and postacrosomal sheath assembly. As will be discussed, a defining feature between globozoospermic and 'globozoospermic-like' acrosomeless sperm defects is the absence or presence of the postacrosomal sheath, respectfully.

2.13.2 Formation of the Postacrosomal Sheath of the PT

At the end of the Golgi phase of spermiogenesis the Golgi apparatus disassociates from the acrosome cap region and migrates distally into the cytoplasmic lobe (Fig. 2.17). At this point, the acrosome and the SAL are compositionally complete, but will conform to the shape changes of the condensing and elongating nucleus during the succeeding elongation phase (steps 8–12 in the bull) (Fig. 2.4). Just as the emergence of SAL occurs in concert with acrosome development, postacrosomal sheath (PAS) assembly is concomitant with spermatid elongation. This phase of spermiogenesis is characterized by the caudal manchette, a transient microtubular structure that encircles the caudal-half of the spermatid nucleus and is implicated in spermatid-head shaping (Russell et al. 1991; Meistrich et al. 1990) and "intramanchette transport (IMT)" (Kierszenbaum 2002). At the onset of nuclear elongation, manchette microtubules extend caudally from the nuclear ring to the distal cytoplasmic lobe, creating a girdle-like structure that encapsulates the spermatid head below the acrosome (Figs. 2.17, 2.20). PAS proteins synthesized in the cytoplasmic lobe appear to be stored and transported up the microtubular manchette to the spermatid head. As the manchette migrates distally at the end of spermatid elongation, proteins residing on the manchette are deposited in the space once occupied by the microtubules, thereby forming the PAS of the PT (Barth and Oko 1989; Longo and Cook 1991; Oko and Maravei 1995; Oko and Sutovsky 2009; Tovich et al. 2004; Wu et al. 2007b; Protopapas et al. 2019) (Figs. 2.20, 2.21 and 2.22). In bull, the descent of the manchette occurs between steps 11 and 12 of spermiogenesis (Figs. 2.10, 2.11), after which time the manchette quickly deteriorates in the cytoplasmic lobe. Previous studies on resident PAS proteins have provided evidence that supports this narrative of biogenesis, including WBP2NL/PAWP, the core histones, and GSTO2 (Hamilton et al. 2017, 2021, 2018; Tovich et al. 2004; Wu et al. 2007b). In contrast to the SAL, proteins of the PAS have been functionally implicated in early fertilization events such as oocyte activation and pronuclear formation (Aarabi et al. 2014a, b; Hamilton et al. 2021, 2019; Wu et al. 2007a).

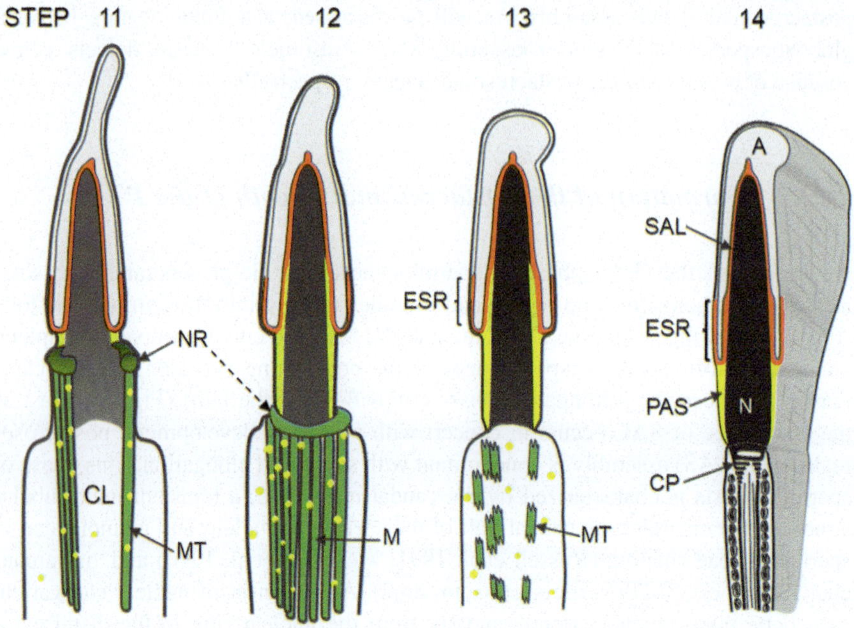

Fig. 2.20 Diagrammatic summary of postacrosomal sheath (PAS) formation in bovine spermiogenesis. Spermatid steps 11–14 are depicted, with step 14 representing a mid-sagital sectional view of a structurally mature spermatid or spermatozoa. The gray to black transition in the nucleus of steps 11 and 12 spermatids represents the final condensation of the nucleus, which is concomitant with the descent of the microtubular manchette (green structure). The microtubules (MT) of the manchette are anchored proximally to the nuclear ring (NR). The manchette serves as a storage site and transportation track for PAS proteins (yellow dots), which are translated from respective mRNAs in the cytoplasmic lobe (CL). Upon manchette descent in steps 11 and 12, manchette associated PAS proteins are translocated to the forming PAS, facilitated by microtubule motor proteins. Initially some of the newly transported PAS proteins share space with the subacrosomal layer (SAL, orange coloured layer) proteins in the equatorial segment region (ESR) but by step 14, with the thinning of the equatorial segment (ES), most PAS proteins become confined to the PAS region (solid yellow) of the PT. Once the manchette has descended in step 13 it disintegrates. Note that in step 11 the NR and MT of the girdle forming the manchette is excised frontally to expose the condensing nucleus. A, acrosome; CP, connecting piece of the sperm tail

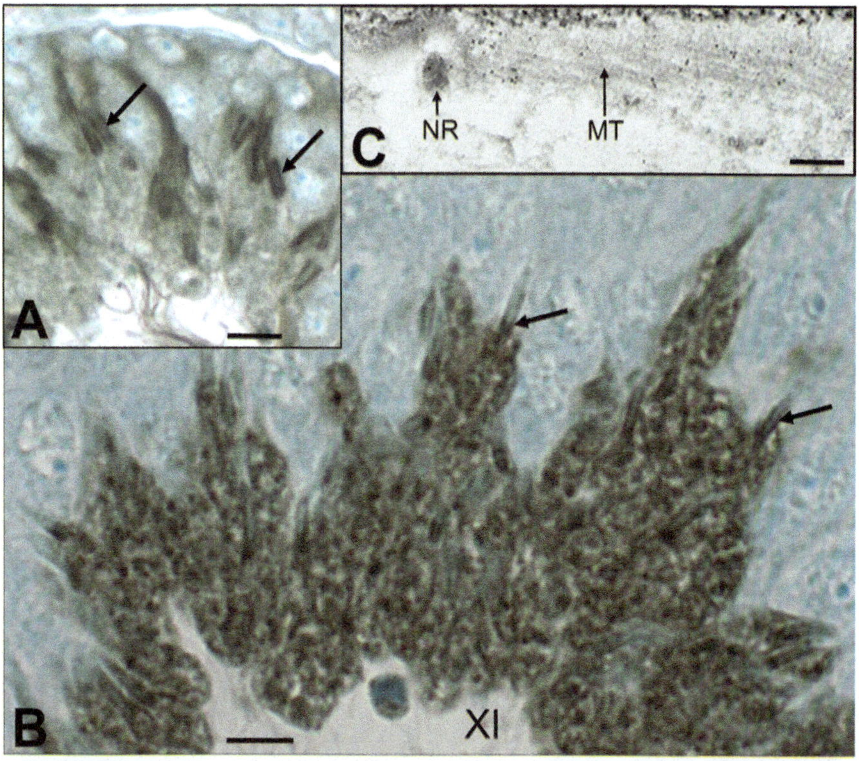

Fig. 2.21 Sections representing stage XI of the cycle of mouse seminiferous epithelium, immunoperoxidase-stained with antibodies against tubulin and PAWP/ WBPL1 showing affinity for the microtubular manchette. A) Light micrograph of manchettes (arrows) of elongated spermatids in step 11 (stage 11) labeled with anti-α-tubulin antibody. Note that the axonemes seen in the lumen of the seminiferous epithelium are also immunoreactive. Bar = 10 µm. **B)** Light micrograph of manchettes (arrows) of elongated spermatids in step 11 labeled with anti-PAWP/WBP2NL antibody. Note that the immunoperoxidase staining of the manchette is almost identical to that found with anti-tubulin antibody. Bar = 10 µm. Labelling with antibodies representing other postacrosomal sheath resident proteins (i. e., H3, H2B, H2A, H4, GSTO2, WBP2, and Calicin) showed identical results. **C)** Electron micrograph of manchette immunogold labeled with anti-H3 antibody. Note the localization of the colloidal gold label to the microtubules (MT) and nuclear ring (NR). Bar = 0.5 µm. Immunogold labelling with antibodies representing other postacrosomal sheath resident proteins showed identical results

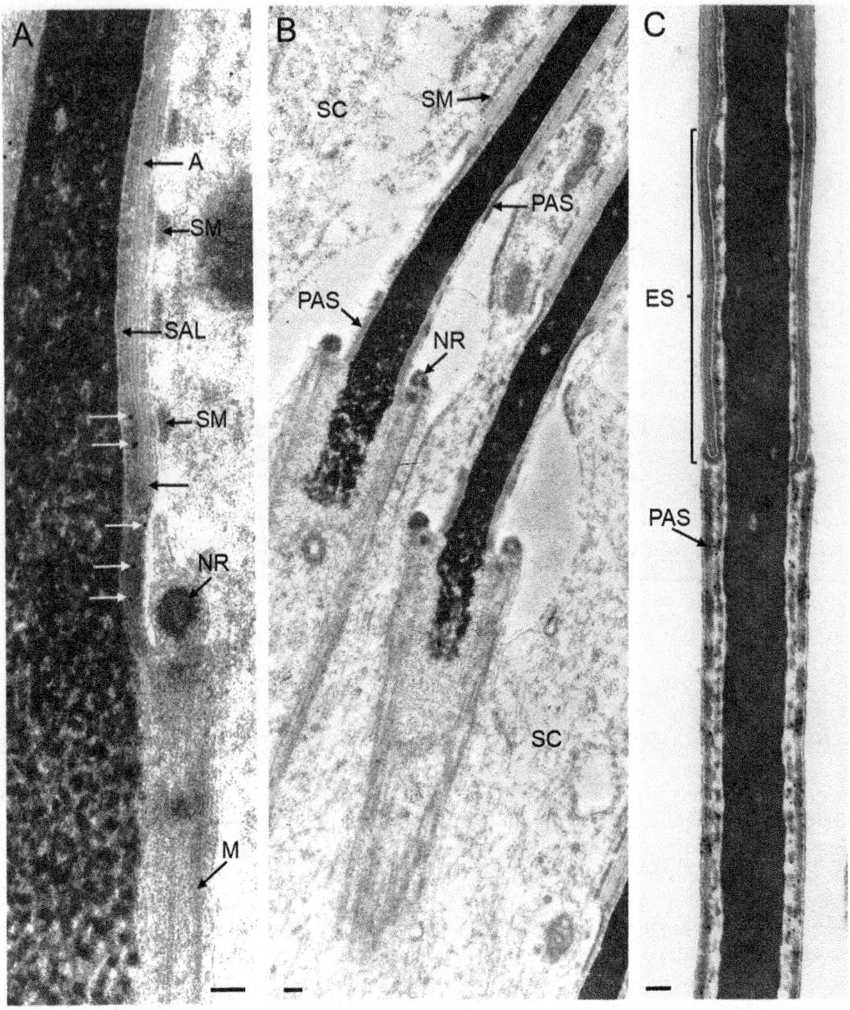

Fig. 2.22 Electron micrographs depicting the assembly of the postacrosomal sheath in the wake of the descending manchette during bovine spermiogenesis. A) A sagittal section through a portion of the head of an early step 11 spermatid, immunogold labeled with anti-PAWP/WBP2NL, showing early deposition of PAWP (arrows) as the manchette (M) begins to descend. The black arrow depicts the caudal end of the acrosome. Note that some of the label is found in the lower subacrosomal region. A, acrosome; NR, nuclear ring of the manchette; SM, Sertoli cell mantle; SAL, subacrosomal layer. Bar = 0.2 μm. **B)** Section through the heads of two elongated spermatids in step 11 showing mid-descent of the manchette. In its wake the postacrosomal sheath (PAS) is laid down. Although difficult to discern at this magnification the nuclear rings are immunogold labeled intensely with anti-PAWP antibody. SC, Sertoli cell. Bar = 0.2 μm. **C)** A sagittal section through the head of a late step 14 spermatid after manchette descent and equatorial segment (ES) thinning. The immuno-gold particles that label PAWP are now exclusively detected over the postacrosomal sheath (PAS). Bar = 0.2 μm

2.13.3 Origin and Assembly of IAM Proteins During Spermiogenesis

The inner acrosomal membrane coat (IAMC) proteins (i. e., IAM38/ZPBP1, proacrosin, MMP2 and SAMP14; see Chap. 1 in Figs. 11, 19) have a similar origin. They are secreted by the Golgi apparatus as part of the electron dense granule within the proacrosomic vesicle (PAV) during the Golgi phase of spermiogenesis (Ferrer et al. 2012a, b, 2016; Oko et al. 2011; Yu et al. 2009) (Fig. 2.23). After coalescence of the PAVs to form the acrosomic vesicle (AV) and its subsequent attachment to the nuclear envelope (NE) of the spermatid, IAMC proteins begin to redistribute from the dense granule to the intravesicular side of the acrosomal membrane during the cap phase of spermiogenesis (Fig. 2.23). By step 6, most of the IAMC proteins have resituated themselves from the acrosomic granule to the acrosomal membrane. Thereafter the proteins are gradually lost from the OAM. However, the timing of the redistribution of IAMC proteins from the acrosomic granule to the acrosomic membrane can vary among species during spermiogenesis. In the bull the redistribution of these proteins takes place during the cap phase so that by the beginning of spermatid elongation (step 8) the majority of IAMC proteins are located along the IAM. At this step, the regions of interaction of the IAMC proteins with the IAM and OAM are dependent on the protein type (Fig. 2.23). IAM38/ZPBP1and SAMP14 are found all along the IAM and in addition, are associated with the OAM in the caudal region of the acrosome that will become the equatorial segment. The thinning of this region of the acrosome occurs late in spermiogenesis as depicted in Fig. 2.20, but this change does not affect the positioning of these two proteins. Essentially, the positioning of IAM38/ZPBP1and SAMP14 in step 8 spermatids (Fig. 2.23, Step 8a) remains the same in spermatozoa (Chap. 1 in Fig. 19). In contrast, Proacrosin and MMP2 end up in step 8 bound solely to the IAM in the apical and principal segments of the acrosome (Fig. 2.23, Step 8b) and retain this position in spermatozoa.

SPACA1/IAM32, on the other hand, appears to be constitutively secreted as an integral part of the acrosomal membrane by the Golgi apparatus during both the Golgi and cap phases of spermiogenesis (Fig. 2.24) (Oko et al. 2011). However, at the end of the cap phase, parallel with the IAMC proteins, it starts to disappear from the OAM and, at the beginning of spermatid elongation, has exactly the same positioning as IAM38 (compare Fig. 2.24 with Fig. 2.23). The coincident localization between these proteins suggests that SPACA1, an integral membrane protein, may be involved in binding IAM38/ZPBP1, a peripheral membrane protein, to the acrosomal membrane during its redistribution from the acrosomic granule and reorganizing it along this membrane during the cap phase. IAM38/ZPBP1 plays a crucial role in acrosome compaction during spermiogenesis (Lin et al. 2007; Yu et al. 2009) and in adhesive binding of acrosome-reacted spermatozoa to the ZP of the oocyte during fertilization (Yu et al. 2006).

Defects associated with the formation of the acrosome and the manchette In humans, mutations in two genes Dpy19l2 (Harbuz et al. 2011; Koscinski

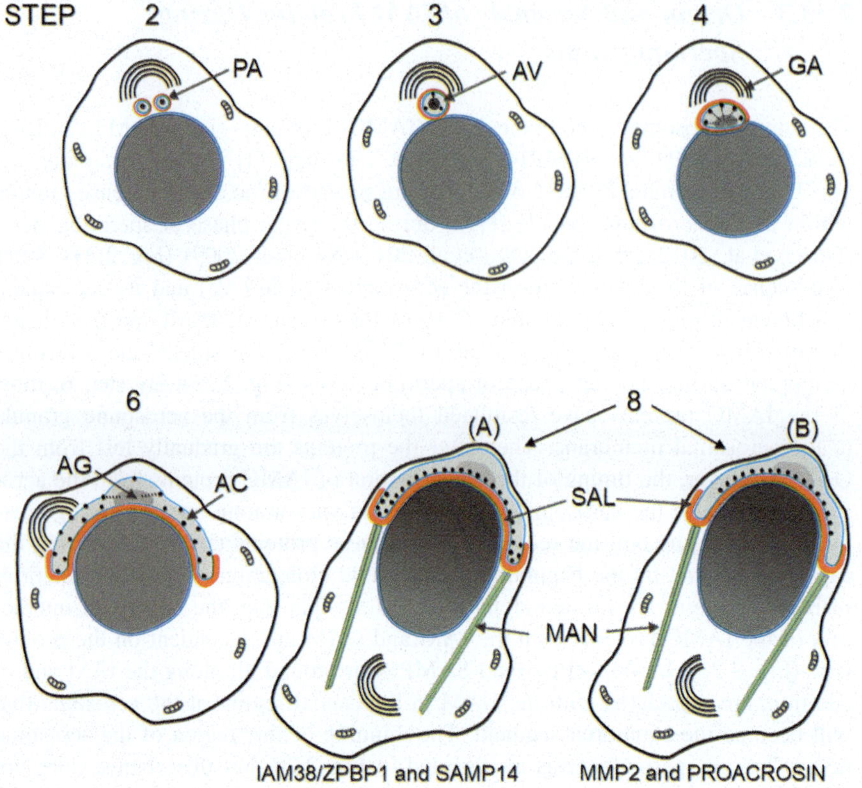

Fig. 2.23 Assembly of inner acrosomal membrane coat (IAMC) proteins (i. e., IAM38/ZPBP1, MMP2 and Proacrosin) during spermiogenesis. The IAMC proteins are secreted by the Golgi apparatus (GA) and form the dense granule within the proacrosomic vesicle (PA) in step 2 of spermiogenesis. After fusion of the proacrosomic vesicles in step 3 to form the acrosomic vesicle (AV), the IAMC proteins (black dots) begin to redistribute from the dense granule to the inside of acrosomal membrane (blue line). As the AV grows and caps the nucleus in steps 4 to 7 the IAMC proteins continue to redistribute from the acrosomic granule (AG) to the acrosomal membrane. However, in steps 7 and 8 there is a gradual diminution of the IAMC proteins (black dots) from the outer acrosomal membrane (OAM), concomitant with a increase of IAMC proteins along the inner acrosomal membrane (IAM). The final location of the IAMC proteins, which is reached in step 8, is dependent on the protein type: IAM38 becomes situated along the entire IAM and along the OAM in the region that is destined to become the equatorial segment (step 8, A), while proacrosin and MMP2 only associate with the IAM in the apical and principal segments of the acrosome (step 8, B). The subacrosmal layer (SAL) of the perinuclear theca (depicted in orange) surrounds the periphery of the acrosomic vesicle at the beginning of acrosomal development and is involved with acrosomal-nuclear docking and capping during spermiogensis (see Fig. 2.16 for explanation). AC, acrosomic cap; MAN, microtubular manchette

et al. 2011) and Spata16 (Dam et al. 2007) were demonstrated to cause globozoospermia. As previously discussed the Dpy19l2$^{-/-}$ mouse model shares a similar sperm phenotype (i. e., acrosome-less and rounded condensed nuclear shape) to that found in its human ortholog associated with type 1 globozoospermia and

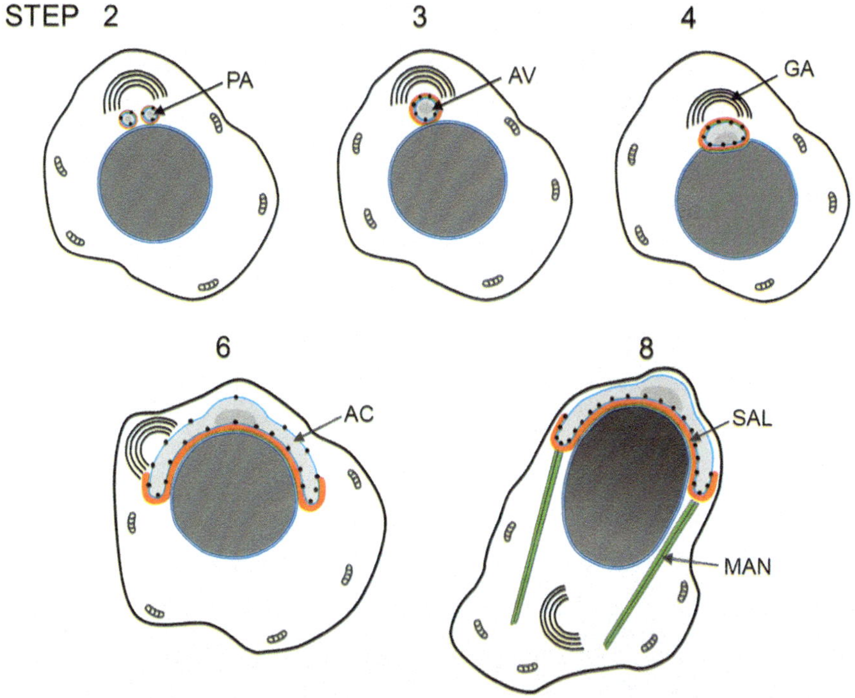

Fig. 2.24 Assembly of integral membrane protein, SPACA1/IAM32, during spermiogenesis. SPACA1 (black dots) is secreted constitutively by the Golgi apparatus (GA) as an integral part of the acrosomic membrane during the Golgi (step 1–3) and cap (step 4–7) phases of spermiogenesis. However, towards the end of the cap phase SPACA1 diminishes from the outer acrosomal membrane (OAM) and resides in its final position in step 8. It is found throughout the inner acrosomal membrane and in the region of the outer acrosomal membrane that will become the equatorial segment. The subacrosmal layer (SAL) of the perinuclear theca (depicted in orange) surrounds the periphery of the acrosomic vesicle from the beginning of acrosomal development and is involved with acrosomal-nuclear docking and capping during spermiogensis (see Fig. 2.16 for explanation). AC, acrosomic cap; AV, acrosomic vesicle; MAN, microtubular manchette; PA, proacrosomic vesicle

has been exploited to trace the pathological development of this infertility syndrome (Pierre et al. 2012). In mice, a myriad of genes are associated with sperm defects that resemble human type 1 globozoospermia, some more than others. Loss-of-function mutations in these genes affect different phases of acrosomal development reflecting different molecular processes in play. For example the blind-sterile recessive mutation (Sotomayor and Handel 1986) as well as inhibition or inactivation of HRB (Kang-Decker et al. 2001), GBA2 (Yildiz et al. 2006; Oko et al. 2011) and AU040320 (Guidi et al. 2018) prevent, either directly or indirectly, fusion of the pro-acrosomal vesicles to form the AV in the Golgi phase (Fig. 2.25). As a consequence the expansion of the acrosomal cap is curtailed (Figs. 2.25, 2.26). However, in Hrb-deficient mice, the resulting acrosomeless round spermatids undergo a further transformation during the elongation phase of

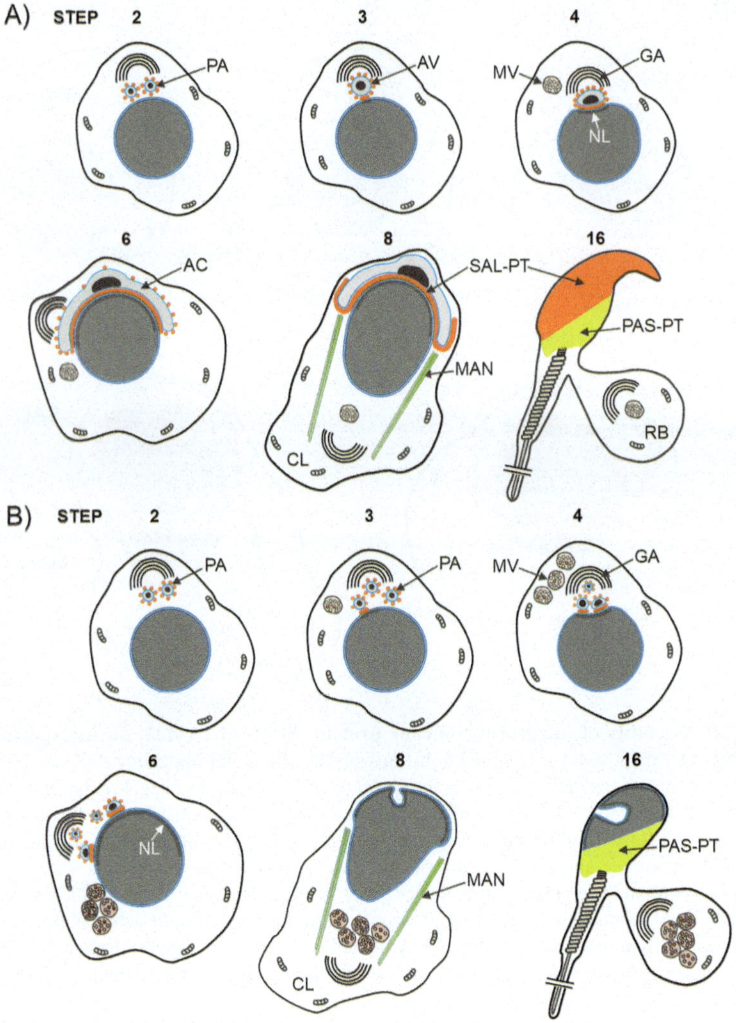

Fig. 2.25 Illustrative comparison of key differences in acrosomal biogenesis in control (A) and GBA2-inhibited (*N*B-DNJ-treated) (B) mice. The AV does not form in GBA2-inhibited step 3 spermatid. Instead proacrosomic vesicles (PA) are continually produced and transiently attach to the nuclear envelope from steps 3 to 7. Even though acrosome growth is inhibited during the cap phase, the nuclear lamina (arrowhead) continues to grow and outlines the area of the nucleus that would have been covered by the acrosome. The subacrosomal layer (illustrated in orange) does not follow the growth of the nuclear lamina in affected spermatids, instead it is only seen where the PA attach to the nuclear envelope over the nuclear lamina. Another outcome of GBA2-inhibition on acrosomal development is the abnormal accumulation of multivesicular bodies (MVB), which are eventually eliminated via the residual body (RB) at spermiation. These MVB accumulate many of the unused proteins (red dots) that make up the subacrosomal layer of the perinuclear theca. The mature affected spermatid is acrosomeless, lacks the SAL and has a stunted, misshapen and vacuolated apical half of the nucleus. The caudal half of the nucleus, however, appears more normal and is surrounded by the postacrosomal sheath (PAS) of the perinuclear theca (coloured in yellow)

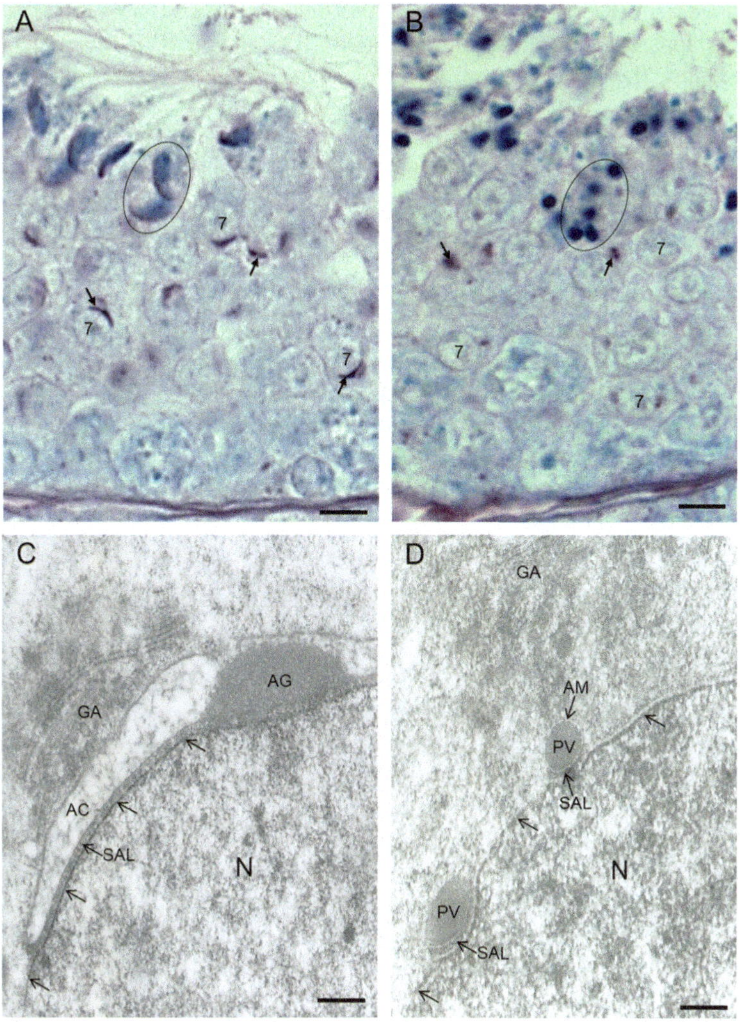

Fig. 2.26 The effect of NB-DNJ (i. e., GBA2 inhibition) on acrosome formation during mouse spermiogenesis. Bouin's fixed mouse testicular sections in stage VII of the 12 stage cycle of the seminiferous epithelium stained with periodic acid Schiff (magenta) to detect the acrosome and methylene blue for general staining of nuclei and cytoplasm. **A)** Normal testicular section showing acrosomal capping (arrows) of step 7 spermatid nuclei (7) and fully developed falciform-shaped acrosome of step 16 spermatids (circled). Bar = 5 µm. **B)** In contrast, NB-DNJ treated mice show no acrosomal capping but rather an accumulation of PAS positive granules (arrows), which are multivesicular bodies, in the spermatid cytoplasm close to the nucleus. Note that the step 16 elongated spermatids (circled) lack an acrosome and nuclei are deformed. Bar = 5 µm. **C,D)** Ultrastructural comparison of normal- and NB-DNJ affected-acrosomal development of step 6 spermatids. In the control (C), note the normal formation of the acrosomal granule (AG) and cap (AC); the Golgi apparatus (GA) is clearly seen contributing to the capping of the acrosome. While in the treated, in place of a growing acrosomal cap, two proacrosomic vesicles (PV) are attached to the nucleus. The acrosomal membrane (AM) outlining the PVs appears white because the because the tissue was not osmicated. Note that intervening between the proacrosomic vesicles and the nucleus (N) is the subacrosomal layer (SAL), which does not form prior to the nuclear attachment of the acrosomic vesicles. The nuclear lamina demarcates the region (arrows) that would normally be covered by the acrosome. The nuclear lamina is the dark line that appears just below the nuclear envelope in the region of acrosomal growth. Bar = 0.2 µm

spermiogenesis (Kierszenbaum et al. 2004; Juneja and van Deursen 2005), which is not as apparent in blind-sterile mice nor GBA- and AU040320-deficient mice. In the latter, the acrosomeless nuclei of spermatids undergo caudal-elongation and -condensation in the presence of a normally positioned manchette (Figs. 2.25, 2.27, 2.28), while the nuclei of Hrb-deficient spermatids remain rounder but condense in the presence of an abnormally positioned and ectopic manchette. This suggests that the disruption of both the acrosome and the manchette are required for a truer Globozoospermic phenotype, in alignment with the conclusions of the Dpy19l2$^{-/-}$ mouse model study by Pierre et al. (2012).

There is considerable evidence for manchette involvement in head shaping, especially of the caudal half of the spermatid nucleus (Cole et al. 1988; Meistrich et al. 1990; Russell et al. 1991). Furthermore, it has been shown that this microtubule tract is responsible for intra-manchette transport of PT proteins for postacrosomal sheath and perforatorial assembly (Protopapas et al. 2019; Tovich et al. 2004; Wu et al. 2007b). Coincidentally, sperm affected with Type 1globozoospermia are devoid of the entire PT, coincident with loss of oocyte-activating ability, while GBA2-inhibited mice (Oko et al. 2011) retain the PAS part of the PT along with their oocyte-activating ability (Aarabi et al. 2014b; Suganuma et al. 2005; Wu et al. 2007b). Missense mutation affecting the Vps54 gene (Paiardi et al. 2011) and inactivation of other mouse genes such as Csnk2a2 (Xu et al. 1999; Escalier et al. 2003) Gopc (Yao et al. 2002), Zpbp1/Zpbp2 (Lin et al. 2007), Pick1 (Xiao et al. 2009), Tmf1 (Lerer-Goldshtein et al. 2010), Hsp90b1 (Audouard and Christians 2011), Smap2 (Funaki et al. 2013), Atg7 (Wang et al. 2014), Mfsd14a (Doran et al. 2016), and GM130 (Han et al. 2017) also cause phenotypes resembling human globozoospermia. However, many of the resulting sperm phenotypes differ from type 1 globozoospermic sperm as their miss-shaped condensed nuclei contain a PAS, indicative of manchette-assembly and retention of at least some oocyte activating ability. These genetic models show that disruption of a variety of different cellular processes involved in acrosome biogenesis can lead to similar 'globozoospermia-like' phenotypes. The proteins encoded by these genes are localized in different cellular compartments such as the endoplasmic reticulum (GBA2, HSP90b1), Golgi apparatus (AU040320, PICK1, GOPC, SMAP2, SPACA1), membrane of the AV (VPS54, HRB), acrosomal matrix (ZPBP1/2, SPACA1), nuclear envelope (DPY19L2, CSNK2a2) or nucleus (CSNK2a2). Most of their molecular functions in acrosomal biogenesis are unknown and though some are localized to different compartments, such as GBA2 to the ER, AU040320 to the Golgi, and HRB to the membrane of the AV, their disruption can have a similar effect, of for example in these cases of preventing fusion of the proacrosomic vesicles (see Fig. 2.25).

Of the gene mutations that directly affect the manchette the most thoroughly characterized from a morphological point of view is the Azh or Hook1 mutation (Cole et al. 1988; Meistrich et al. 1990; Russell et al. 1991). HOOK1 is a protein that normally binds to microtubules and organelles through its N- and C- terminal domains, respectfully but its function in the microtubules of the manchette is unknown. However, the resulting morphological phenotype of its loss-of-function

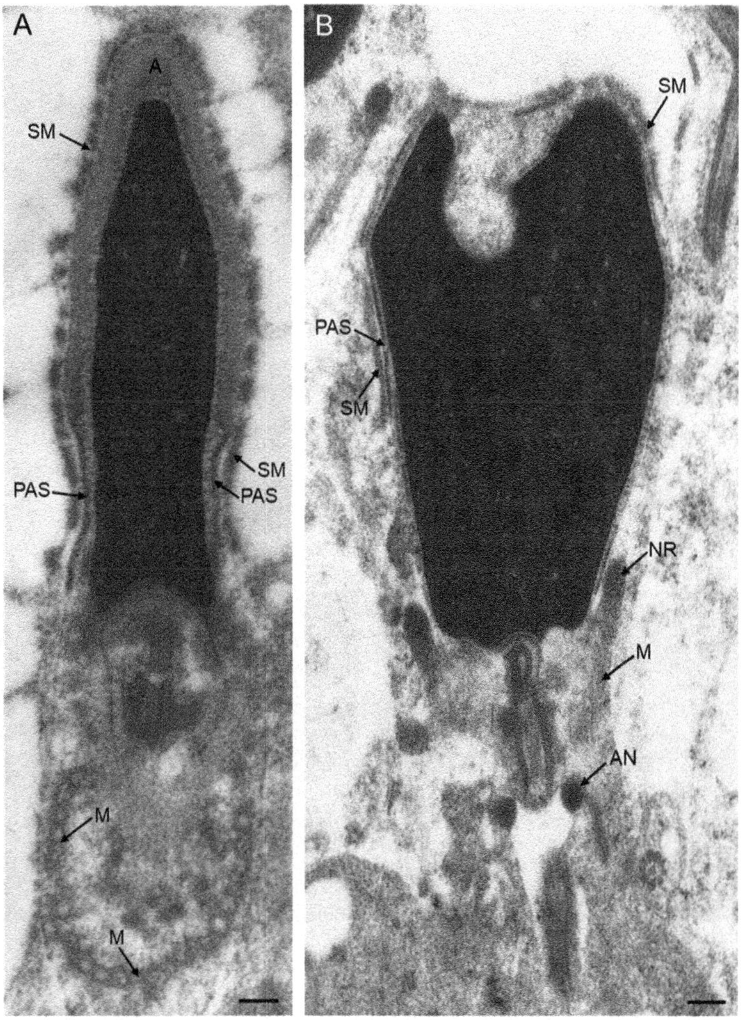

Fig. 2.27 Ultrastructural comparison of development in step 13–14 spermatids in untreated and GBA2-inhibited (NB-DNJ-treated) mice. A) A step14, untreated spermatid with microtubular manchette (M) completely descended, shows a streamlined shape and expected arrangement of acrosome (A) and postacrosomal sheath (PAS). SM, Sertoli mantle. Bar = 0.2 μm. **B)** A step 13 treated spermatid. The acrosomeless nucleus has condensed but is misshaped, remaining paddle shaped, even though the manchette with its nuclear ring (NR) has almost completely descended along the nucleus leaving the PAS in its wake. Note the nuclear vacuole in the acrosomeless region, emanating from spermatid or Sertoli cell cytoplasmic invaginations during round spermatid development. AN, annulus; SM, Sertoli cell-mantle (ectoplasmic specialization). Bar = 0.2 μm

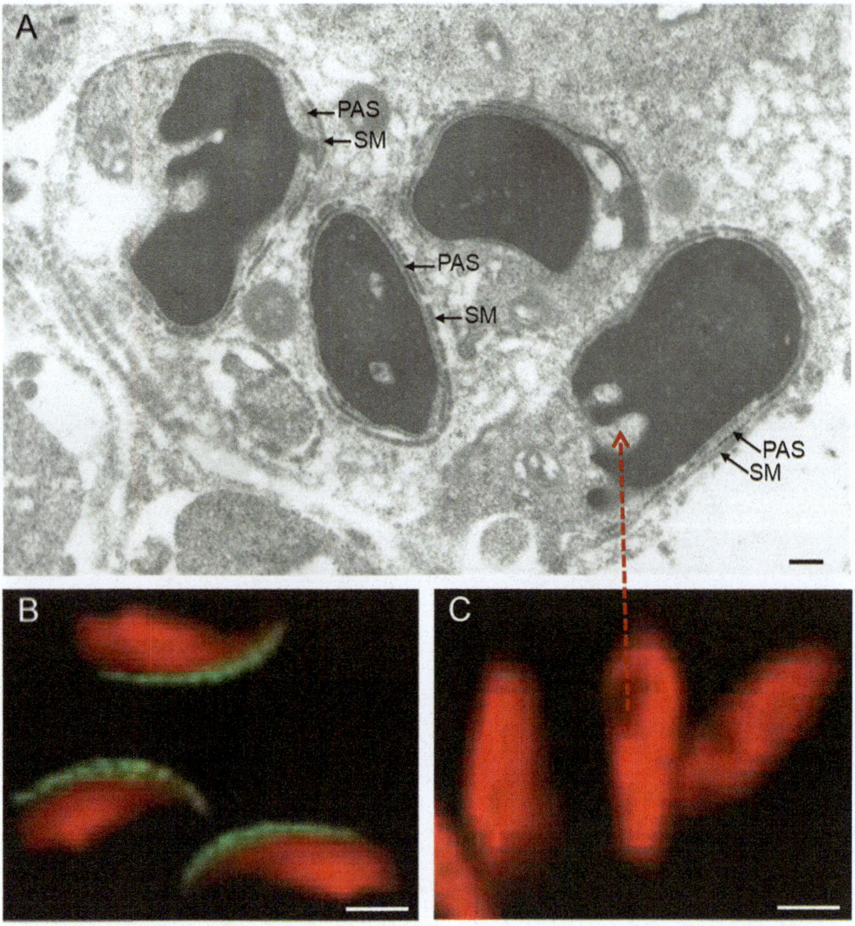

Fig. 2.28 Nuclear deformities in elongated spermatids of GBA2-inhibited (NB-DNJ-treated) mice. A) Oblique sections through step 16 spermatids just before spermiation. The nuclei are acrosome-less and miss-shapened with characteristic vacuolations located apically, below where the acrosome would normally reside. Although the postacrosomal sheath (PAS) has been laid down in the wake of the descending manchette, it is discontinuous in places. Normally at this late step of development, the PAS would be a continuous structure. SM, Sertoli mantle. Scale bar = 0.2 µm. **B, C)** Comparison of head shape between spermatozoa in control (**B**) and GBA2-inhibited (NB-DNJ-treated) (**C**) mice at the light microscope level. Images were obtained by fluorescence microscopy. The apical and principal segments of the acrosome were detected with an anti-acrosomal monoclonal antibody (green) and nuclei were stained with propidium iodide (red). Evident is the lack of acrosome-staining in the NB-DNJ-affected spermatozoa mice, as well as the morphological heterogeneity of the sperm nuclei in comparison with its uniformity in control spermatozoa. Note the vacuolations within the affected nuclei and compare with electron microscope image (red dashed arrow). Scale bar = 5 µm

is a mouse spermatozoa exhibiting a 'club-like' head shape rather than the normal falciform shape. The thickened apical part of the nucleus (the club) covered by the misshapen acrosome can vary in dimensions and shape but the caudal region below the acrosome (the handle) most often has a symmetric conical shape that reflects the tight conical miss-arrangement of the microtubules of the manchette during the elongation phase of spermiogenesis. Ectopically positioned microtubules of the manchette, in Azh mutant mice (Meistrich et al. 1990) as well as in mice injected with other microtubular disruptors (Russell et al. 1991), also have a tendency to direct invaginations or exvaginations of the spermatid nucleus creating microtubular containing nuclear membrane bound vacuoles. Mutations of Azi1 (Hall et al. 2013) and Rim-bp3 (Zhou et al. 2009) produce a similar spermatid developmental phenotype to that described above for the Hook1 mutation. The respective investigators suspect that AZI1 is involved in microtubular trafficking and that RIM-BP3, which can complex with HOOK1, is an adapter in the process of vesicle fusion and release. However, regarding the proposed functions of RIM-BP3 and HOOK1 there is no morphological evidence provided in the literature of vesicle or organelle transport along the microtubules of the manchette. CLIP-170 is another microtubular-associated protein whose function appears to change in association with the manchette in this case from an expected microtubule plus-end-tracking protein to an undefined structural role. Its inactivation in mice leads to an irregular arrangement of microtubules of the manchette and consequently an abnormally formed spermatid nuclear head shape (Akhmanova et al. 2005). Finally, consistent with the role for E-MAP-115 in stabilizing microtubules, microtubule associations in the inactivated E-MAP-115 mutant mouse are morphologically abnormal in the manchette of spermatids, leading to deformities in the spermatid nuclei (Komada et al. 2000), not unlike seen in the manchette mutants described above.

Importantly, in a developmental analysis of mouse mutants known to affect sperm fertility, including most of those affecting acrosome and manchette formation mentioned above, ***it was concluded that mutations that affect sperm head shape always affect motility*** (de Boer et al. 2015). In contrast, when the prime defect is in sperm tail formation, head morphogenesis in these mutants is mostly normal (de Boer et al. 2015). The authors state, 'the striking dependency of motility on proper sperm head development is not easy to understand, but likely is of evolutionary benefit'.

2.14 Endocrine Control of Spermatogenesis

It is established that the endocrine control of spermatogenesis is dependent on follicle-stimulating hormone (FSH), luteinizing hormone (LH), and testosterone, but the degree of involvement of each factor varies with the experimental data and in different species (Steinberger 1971).The most precise documentation of hormonal control of spermatogenesis is in the murid (Berndtson et al. 1974; Means et al. 1976; Oko and Hrudka 1984a; Parvinen 1982; Parvinen and Ruokonen 1982;

Robaire et al. 1979; Russell and Clermont 1977). In contrast, little information is available for the bull and ram (Courtens and Courot 1980; Courot and Ortavant 1981; Curtis and Amann 1981). However, since the process of spermatogenesis is similar in mammalian species, its hormonal control should be similar; it is on this assumption that the following discussion is based. Furthermore, much of the information compiled is on experiments that block gonadotropins (LH, FSH) and androgens by hypophysectomy, steroid treatment, or other means concomitant with hormone replacement therapy. Experiments that employed the transillumination-assisted microdissection technique have helped immeasurably in the understanding of hormone regulation (Parvinen and Ruokonen 1982). By this procedure individual stage segments of the seminiferous tubules can be isolated and analyzed for hormone and molecular content. Hormone binding and end product studies have established that FSH's target is the Sertoli cell (Means et al. 1976), LH's target is the Leydig cell (Catt et al. 1972, 1974; De Kretser et al. 1969, 1971; Sharpe et al. 1990), and testosterone's target is the Sertoli cell (Sanborn et al. 1977; Mulder et al. 1974; Hansson et al. 1976, 1974; French et al. 1974) and possibly to a little extent the germ cell (Wright and Frankel 1980).

Testicular development and establishment of mammalian spermatogenesis are dependent on FSH, LH, and androgens. The primary importance of FSH is in initiating pubertal development of the testis. In adulthood, FSH appears to have more of an optimizing function on the process of spermatogenesis. On the other hand, LH via androgens is important in the initiation of pubertal development and also in maintaining spermatogenesis during adulthood. In the perinatal rat it has been demonstrated that, several days before parturition, FSH stimulates Sertoli cell proliferation of the testes (Orth 1984). In immature hypophysectomized rats supplemented with FSH, this hormone stimulates seminiferous tubule growth, increases the Leydig LH receptor number and thereby the capacity to secrete testosterone, produces androgen-binding protein (ABP), controls the production of seminiferous tubule fluid (STF), and decreases degeneration of spermatogonia (Chen et al. 1976; Jegou et al. 1982; Jégou et al. 1983; Means et al. 1976; Odell et al. 1973; Sharpe 1983). These events are concurrent with the initiation phase of spermatogenesis by which type A spermatogonia, through several mitotic divisions, form primary spermatocytes. Interestingly, this phase of spermatogenesis is at least qualitatively if not quantitatively gonadotropin-independent in the rat; blocking gonadotropin secretion by estrogen had no effect on the development of this phase (Chowdhury and Steinberger 1975). Furthermore, testicular cultures showed that spermatogenesis can proceed up to pachytene (primary) spermatocytes, but not beyond, in both the newborn and adult rat in the absence of gonadotropins or androgens (Steinberger and Steinberger 1965, 1966). The rate of the cycle becomes established in this first phase in most mammals that have been studied (Gondos 1977). The wave of the seminiferous epithelium is also believed to be established in this phase (Perey et al. 1961). The beginning of the meiotic phase of spermatogenesis (the development of early pachytene spermatocytes to spermatids) occurs 20 days after birth in the rat (Chowdhury and Steinberger 1975; Clermont and Perey 1957; de Jong and Sharpe 1977) and after 28 weeks in the bull (Curtis and Amann 1981). This second phase of spermatogenesis is androgen-dependent in the immature as well as the adult rat (Berndtson et al. 1974; Chowdhury and Steinberger 1975; Steinberger 1971) and ram (Courot and Ortavant 1981) (41). The meiotic phase of spermatogenesis coincides with the development of the blood-testis barrier, cessation of Sertoli cell division, formation of a tubular lumen, and a switch in hormone dependence from FSH to testosterone. Testosterone now replaces FSH functions by maintaining ABP of the Sertoli cell (Hansson et al. 1976; Weddington et al. 1975) and controlling STF production (Jégou et al. 1983).

Evidence for the independence of FSH in the adult rat was raised by Desjardins (1978). He demonstrated that spermatogenesis can be rescued and maintained by testosterone in hypophysectomized rats if endogenous FSH levels were high when testosterone supplementation began. However, if testosterone supplementation was delayed until several weeks after hypophysectomy when FSH levels became low, rescue of spermatogenesis was dependent on both testosterone and FSH compensation. Once the process was re-established, testosterone administration alone was sufficient for maintenance. Berndtson et al. (1974) and Oko and Hrudka (1984a) also showed that the spermatogenesis can be maintained both qualitatively and quantitatively in the adult rat implanted with polydimethylsiloxane capsules containing androgens. Analysis of stage-specific segments of the seminiferous tubule reveal that stages VII and VIII of the cycle of the seminiferous epithelium in the rat (equivalent to stages VI and VII in the bull) have the highest concentrations of endogenous testosterone (Parvinen and Ruokonen 1982) and maximal secretion of ABP (Ritzen et al. 1982). Russell and Clermont (1977) found that regression of the rat seminiferous epithelium after hypophysectomy was initially due to the discriminative degeneration of mid-pachytene and step 7–8 and 19 spermatids in these same stages. The fact that LH supplementation significantly reduced the number of degenerating cells implied a stage-specific androgen dependency. Furthermore, Russell et al. (1981) and Oko and Hrudka (1984a) demonstrated that steroidal or other chemical agents that cause an androgen deficiency and/or block androgens at their active sites produced an identical morphological pattern of response in the testis as hypophysectomy did (Figs. 2.29, 2.30 and 2.31). Considering that the degenerating cells are all located in the adluminal compartment of stages VI-VIII suggests that the androgen-dependent requirements of these susceptible cell types are controlled by the Sertoli cell that surrounds them.

With the advent of the transillumination microdissection technique by which stage-specific segments were analyzed, it became clear that stage-related variation in Sertoli secretory activity exist and that the actions of FSH and testosterone on the Sertoli cell vary according to the stage of the cycle in the rat (Lacroix et al. 1981; Parvinen 1982; Parvinen and Ruokonen 1982; Vihko et al. 1984). Whether or not similar stage specificities exist in the bull or ram, especially with respect to androgen deficiency, remains to be shown. Although it is documented that reproductive hormone deficiencies cause testicular degeneration, there is a scarcity of information on the types of sperm defects caused by such insults. Most of the information obtained is of a secondary nature. For example, injecting

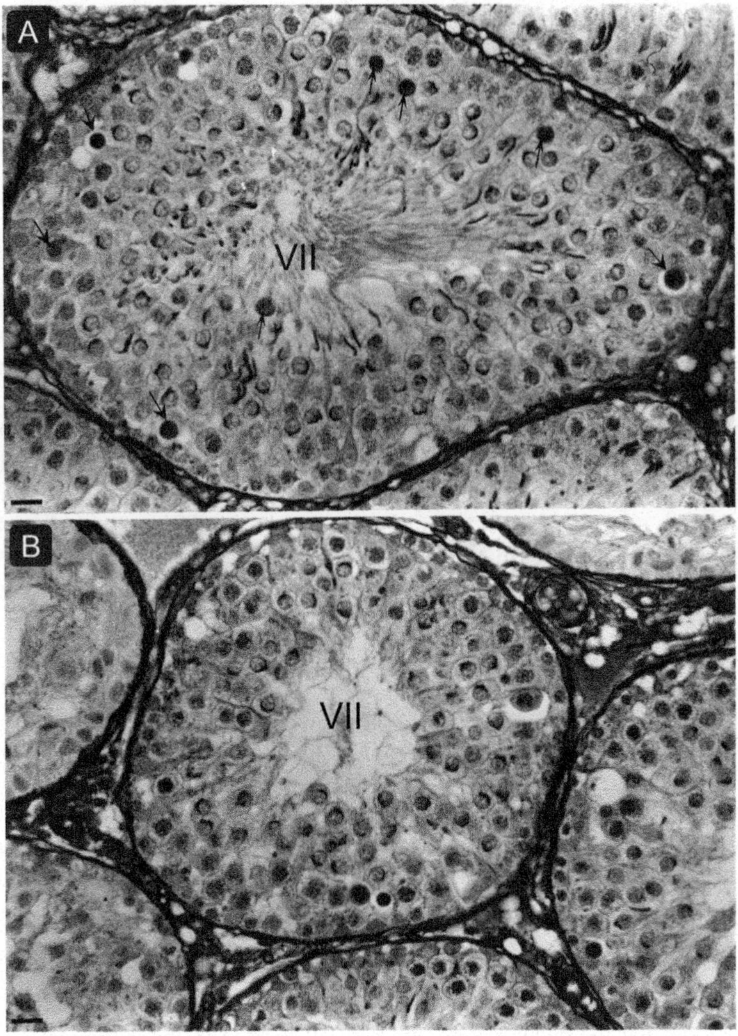

Fig. 2.29 Testicular anti-androgenic response. **A)** Light micrograph of a rat seminiferous tubule at stage Vll of the cycle of the seminiferous epithelium showing degeneration of mid-pachytene spermatocytes (small arrows) and step 7 round spermatids (large arrows) and the vacuoles left after cytolysis. This stage-specific discriminative degeneration of germ cells is typical soon after hypophysectomy and estrogen treatment or exposure to any other chemical agents that cause an androgen deficiency and/ or block androgens at their active sites. Bar = 10 µm. **B)** Rat seminiferous tubules showing different degrees of atrophy after longer anti-androgenic exposure. Note the lack of elongated spermatids in stage VII; these spermatids may have been prematurely sluffed off into the tubular lumen. Stage-specific degeneration depletes the germ cell population and leads to severe tubular atrophy as seen in the seminiferous tubules on the left. Bar = 10 µm

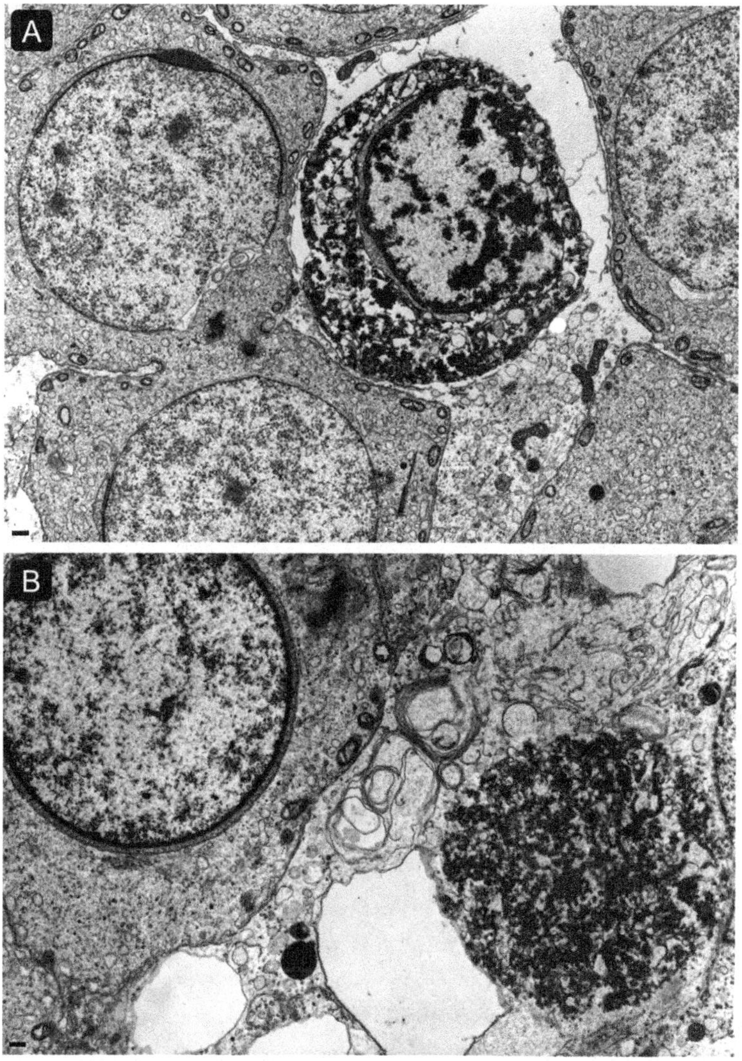

Fig. 2.30 Electron micrographs of stage-specific degeneration in stage VII caused by androgen deficiency. **A)** Degeneration of step 7 spermatid surrounded by unaffected round spermatids. Bar = 0.5 μm. **B)** Degeneration of mid-pachytene spermatid next to a normal looking step 7 spermatid. Bar = 0.5 μm

dexamethasone into bulls can induce the development of nuclear vacuolation in sperm (Oko 1977; Coulter et al. 1976). The question here is whether the vacuolation is due to a direct or indirect effect of dexamethasone. Support for an indirect hormonal effect comes from the studies of Thibier and Rolland (Thibier and Rolland 1976), who showed that dexamethasone strongly suppresses LH

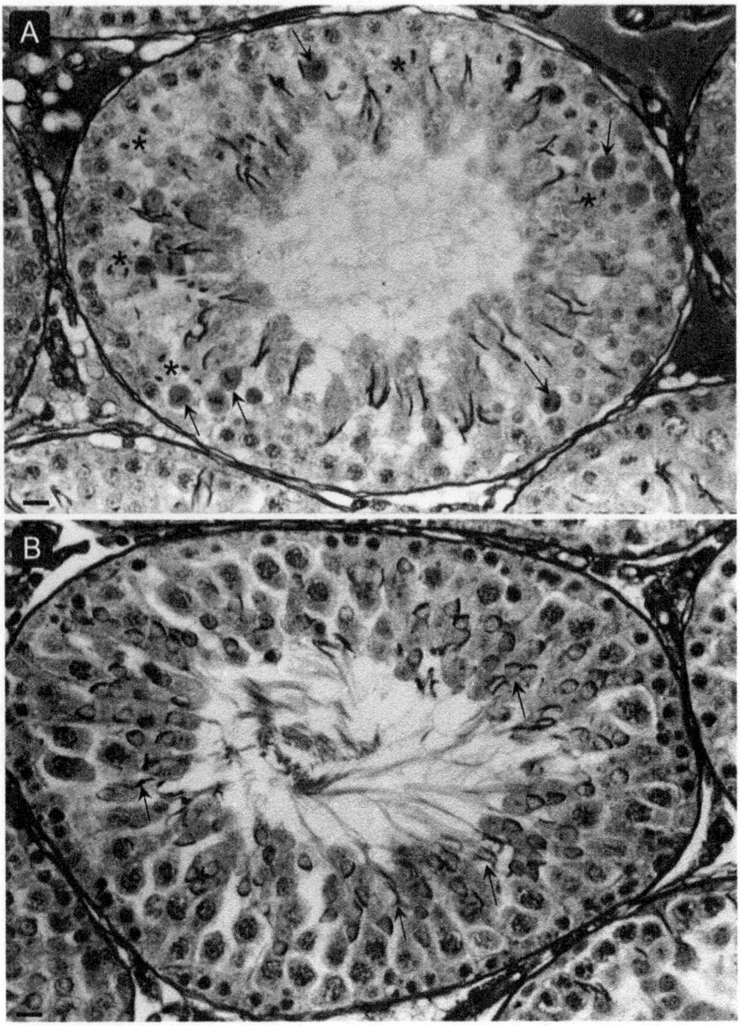

Fig. 2.31 Light micrographs of stage specific effects of androgen deficiency on testicular tissue. A) Degenerating spermatocytes (arrows) prevalent during the second meiotic division in stage XIV of the rat cycle. Note the increased PAS reaction of affected meiotic cells compared to unaffected meiotic cells (asterisk). This degeneration becomes evident after prolonged androgen deficiency. Bar = 10 μm. **B)** Retention of elongated spermatids (arrows) at stage IX of the cycle of the seminiferous epithelium of the rat. Normally by mid-stage VIII all the elongated spermatids should have been released into the tubular lumen. These spermatids will eventually be phagocytosed by the Sertoli cells. This phenomenon is common after prolonged androgen deficiency. Bar = 10 μm

and testosterone in bulls. Ultimately, resolution of this particular query as well as others will depend on preplanned hormone deficiency studies. From years of experience both Lagerlof (Lagerlof 1951) and Knudsen (Knudsen 1954) suggested that the degree of testicular degeneration, caused by a disturbance such as environmental change is dependent on the endocrine constitution of the bull in question. If this statement is true, one can surmise that certain disturbances (i. e., environmental, toxic, physiological) combined with a poor endocrine constitution exacerbate a variety of sperm defects. However, this hypothesis would be very difficult to prove or disprove.

To date we are aware of only one hormone deficiency study that examined the morphogenesis of spermatids in ruminants. This study utilized hypophysectomized and testosterone-supplemented rams and concluded that this treatment combination induced acrosome abnormalities in 20-day, but not in 40- day, treated animals (Courtens and Courot 1980). The timing of acrosome abnormalities in the 20-day treated rams, considered in relation to the duration of spermatogenesis, led us to suggest that acrosome development is under the control of endocrine-dependent cellular events occurring before the beginning of spermiogenesis, possibly involving the Sertoli cell and primary spermatocytes. The only reservation to this hypothesis is that they did not include an experimental group that would have been examined immediately or several days after hypophysectomy. Because the ejaculates were not examined, it is difficult to ascertain whether any of the defective spermatids made it that far. Interestingly, the rams hypophysectomized without supplementation developed spermatids with both abnormal nuclei and acrosomes that eventually became pyknotic and were phagocytosed by Sertoli cells. Furthermore, numerous degenerative meiotic divisions were seen. After 40 days only spermatogonia and primary spermatocytes were encountered. This particular case resembles the results obtained in rodent hormone deficiency studies. It appears that a mechanism exists within the seminiferous epithelium that prevents abnormal germ cells from leaving the testis. However, the question that remains is, what degree of spermatid abnormality (acrosomal, nuclear, or flagellar) is allowed to spermiate and make it as far as the ejaculate.

2.15 Transcriptional and Translational Control of Spermiogenesis

Most of the proteins present in specialized components of spermatozoa are synthesized during spermiogenesis implying a process of haploid regulated transcription and translation. During mammalian spermatogenesis two notable bursts of RNA synthesis occur, one in mid-pachytene spermatocytes and the other between steps 1 to 7 of spermiogenesis during round spermatid development (Monesi, 1978). The consequences of the round spermatid-RNA activity is to contribute to the synthesis of proteins incorporated in the components of the differentiating spermatids (e. g., glycoproteins of the acrosome and cytosolic proteins of the cytoskeletal-like components of the head and tail). Since gene transcription terminates at the beginning of the nuclear elongation in spermiogenesis (step 8), the proteins required for nuclear condensation (e. g., transition proteins and protamines), formation of cytoskeletal-like components of the tail (e. g., ODF and FS), and adhesion of mitochondria to the ODF must be translated from stored mRNAs. This delay in the translation of mRNA until needed is a process referred to as 'post-transcriptional control of gene expression' during spermiogenesis.

Various mechanisms of translational control during spermiogenesis for different mRNA species are proposed. For example, in the case of the protamines, it was shown in transgenic mice that cis-elements of the 3' untranslated region (3'UTR) of protamine-1 mRNA act to delay the translation of transcripts during spermiogenesis (Braun et al. 1989). Northwestern blots provided evidence that an 18 kDa RNA binding protein specifically binds with conserved Y and H elements in 3'UTR of protamine-2 to suppress its translation (Kwon and Hecht 1993). Since the mRNAs of protamine-1 and transition proteins also contain Y and H elements in their 3'UTRs it was proposed that they may be translationally regulated under similar mechanisms. Yang et al. (2005, 2007) found that the DNA/RNA-binding protein Y-box protein 2 (MSY2) plays a role in storing and regulating mRNA during spermiogenesis. Mice lacking Msy2 were infertile, with elongating spermatids unable to develop and presumably phagocytosed by Sertoli cells as there was no evidence of spermatozoa in the epididymis. In addition, a redistribution of mRNAs from stored ribonucleoprotein particles to polysomes was found as well as a marked decrease of many postmeiotic germ cell mRNAs, including the mRNAs encoding the transition proteins and protamines. RNA affinity chromatography and mass spectrometry provided specific evidence that MSY2 interacts with the 3' terminus of the 3'UTR and a Y-box recognition sequence of protamine-1 and sperm-mitochondria cysteine-rich (Smcp) mRNAs (Cullinane et al. 2015). In the follow up Msy2 knockout, protamine-1 and Smcp mRNAs were recruited onto polysomes for translation and were expressed prematurely in round spermatids. Another protein involved in suppressing translation by binding to the 3'UTRs of specific mRNAs during spermiogenesis is testes brain RNA-binding protein (TR-RBP) (Braun 1998). Morales et al. (2002) went on to clarify TB-RBP's role in mRNA transport and regulation of translation. They found that TB-RBP bound to the Y and H elements of Akap4 mRNA (AKAP4 is a prominent protein of the fibrous sheath) and that over 80% of Akap4 and protamine-2 mRNAs were complexed with TR-RBP and Ter ATPase. While complexed with TR-RBP there was no translation of Akap4 and protamine-2 mRNAs allowing them to conclude that the TB-RBP-mRNA complex functions to suppress translation in spermiogenesis.

As for the transcriptional control of spermiogenesis, transcriptional activator CREM (cyclic AMP-responsive element modulator) is the master switch for post-meiotic gene expression, activating a very wide range of haploid germ cell specific genes. CREM mutant mice arrest in round spermatid development and display a dramatic decline in postmeiotic gene expression (Nantel and Sassone-Corsi 1996; Blendy et al. 1996). Some of the post-meiotic proteins whose genes contain cyclic AMP responsive elements on their promoters, which are responsive to CREM, are proacrosin, protamines, transition proteins, calspermin, ODF proteins, SMCP, ACE, RT7, CYP51, mGPDH and Cyclin A (Krausz and Sassone-Corsi 2005). Many of these proteins are essential for the structuring of the spermatid and thus it is not surprising that CREM mutant mice arrest early in round spermatid development.

Compliance with Ethical Standards
All applicable international, national and institutional guidelines for the care and use of animals were followed. This current work was supported by the Natural Science and Engineering Research Council of Canada (Grant number RGPIN/05305) to Richard Oko at Queen's University, along with the Agriculture and Food Research Initiative Competitive Grant number 2023-67015-39262 from the USDA National Institute of Food and Agriculture, Animal Breeding, Genetics & Genomics Program and seed funding from the College of Agriculture, Food and Natural Resources of the University of Missouri to Peter Sutovsky. Work for this book was also supported by a 2021-2022 Lalor Foundation Postdoctoral Research Fellowship to Lauren Hamilton. The research support to Viv Perry was funded by Australian Research Council (Grant number LP110100649), English Beef and Lamb Executive a division of the Agriculture and Horticulture Development Board (Grant number RC3815), Queensland Sperm Morphology Laboratories, Ridley AgriProducts Pty Ltd and the Australian cattle industry. The Canadian Association of Animal Breeders provided the foundation for research by Albert Barth. They supported the Development of the Diagnostic Spermatology Laboratory at the Western College of Veterinary Medicine, Saskatoon, SK., along with related support for numerous other research directives. Many grants were also received from Saskatchewan Agriculture, including the Cattle Marketing Deductions Fund, the Horned Cattle Trust Fund, the Cattle Breeders Research Council, the Beef Development Fund and the Agriculture Development Fund. Animal research approval for research or clinical studies summarized in these chapters were given by respective University Animal Care Committees. The authors declare no conflict of interest as there is no perceived interest in prejudicing the impartiality of the research reported.

References

Aarabi M, Balakier H, Bashar S, Moskovtsev SI, Sutovsky P, Librach CL, Oko R (2014a) Sperm-derived WW domain-binding protein, PAWP, elicits calcium oscillations and oocyte activation in humans and mice. Faseb j 28(10):4434–4440. https://doi.org/10.1096/fj.14-256495

Aarabi M, Balakier H, Bashar S, Moskovtsev SI, Sutovsky P, Librach CL, Oko R (2014b) Sperm content of postacrosomal WW binding protein is related to fertilization outcomes in patients undergoing assisted reproductive technology. Fertil Steril 102(2):440–447. https://doi.org/10.1016/j.fertnstert.2014.05.003

Akhmanova A, Mausset-Bonnefont AL, van Cappellen W, Keijzer N, Hoogenraad CC, Stepanova T, Drabek K, van der Wees J, Mommaas M, Onderwater J, van der Meulen H, Tanenbaum ME, Medema RH, Hoogerbrugge J, Vreeburg J, Uringa EJ, Grootegoed JA, Grosveld F, Galjart N (2005) The microtubule plus-end-tracking protein CLIP-170 associates with the spermatid manchette and is essential for spermatogenesis. Genes Dev 19(20):2501–2515. https://doi.org/10.1101/gad.344505

Amann RP (1962) Reproductive capacity of dairy bulls. IV. Spermatogenesis and testicular germ cell degeneration. Am J Anat 110:69–78. https://doi.org/10.1002/aja.1001100107

Andre J (1962) Contribution to the knowledge of the chondriome. Study of its ultrastructural changes during spermatogenesis. J Ultrastruct Res Suppl 3:1–185

Audouard C, Christians E (2011) Hsp90β1 knockout targeted to male germline: a mouse model for globozoospermia. Fertil Steril 95 (4):1475–1477.e1471–1474. https://doi.org/10.1016/j.fertnstert.2010.12.006

Aul RB, Oko RJ (2001) The major subacrosomal occupant of bull spermatozoa is a novel histone H2B variant associated with the forming acrosome during spermiogenesis. Dev Biol 239(2):376–387. https://doi.org/10.1006/dbio.2001.0427

Barth AD, Oko R (1989) Abnormal morphology of bovine spermatozoa. Abnormal morphology of bovine spermatozoa

Berndston WE, Desjardins C (1974) The cycle of the seminiferous epithelium and spermatogenesis in the bovine testis. Am J Anat 140(2):167–179. https://doi.org/10.1002/aja.1001400204

Berndtson WE, Desjardins C, Ewing LL (1974) Inhibition and maintenance of spermatogenesis in rats implanted with polydimethylsiloxane capsules containing various androgens. J Endocrinol 62(1):125–135. https://doi.org/10.1677/joe.0.0620125

Blendy JA, Kaestner KH, Weinbauer GF, Nieschlag E, Schütz G (1996) Severe impairment of spermatogenesis in mice lacking the CREM gene. Nature 380(6570):162–165. https://doi.org/10.1038/380162a0

de Boer P, de Vries M, Ramos L (2015) A mutation study of sperm head shape and motility in the mouse: lessons for the clinic. Andrology 3(2):174–202. https://doi.org/10.1111/andr.300

Braun RE (1998) Post-transcriptional control of gene expression during spermatogenesis. Semin Cell Dev Biol 9(4):483–489. https://doi.org/10.1006/scdb.1998.0226

Braun RE, Peschon JJ, Behringer RR, Brinster RL, Palmiter RD (1989) Protamine 3'-untranslated sequences regulate temporal translational control and subcellular localization of growth hormone in spermatids of transgenic mice. Genes Dev 3(6):793–802. https://doi.org/10.1101/gad.3.6.793

Burgos MH, Vitale-Calpe R, Apoki A (1970) Fine sructure of the testis and its functional significance. In: Johnson AD, Gomes WR, L. VN (eds) The Testis, vol 1. Academic Press, New York, pp 551–649

Calvin HI, Cooper GW, Wallace E (1981) Evidence that selenium in rat sperm is associated with a cysteine-rich structural protein of mitochondrial capsules. Gamete Res 4(2):139–142

Calvin HI, Grosshans K, Musicant-Shikora SR, Turner SI (1987) A developmental study of rat sperm and testis selenoproteins. J Reprod Fertil 81(1):1–11. https://doi.org/10.1530/jrf.0.0810001

Catt KJ, Tsuruhara T, Dufau ML (1972) Gonadotrophin binding sites of the rat testis. Biochim Biophys Acta 279(1):194–201. https://doi.org/10.1016/0304-4165(72)90254-1

Catt KJ, Tsuruhara T, Mendelson C, Ketelslegers JM, Dufau ML (1974) Gonadotroping binding and activation of the interstitial cells of the testis. Curr Top Mol Endocrinol 1:1–30. https://doi.org/10.1007/978-1-4684-2595-6_1

Chen YI, Payne AH, Kelch RP (1976) FSH stimulation of Leydig cell function in the hypophysectomized immature rat. Proc Soc Exp Biol Med 153(3):473–475. https://doi.org/10.3181/00379727-153-39571

Chowdhury AK, Steinberger E (1975) Effect of 5alpha reduced androgens on sex accessory organs, initiation and maintenance of spermatogenesis in the rat. Biol Reprod 12(5):609–617. https://doi.org/10.1095/biolreprod12.5.609

Cingolani G, Petosa C, Weis K, Müller CW (1999) Structure of importin-beta bound to the IBB domain of importin-alpha. Nature 399(6733):221–229. https://doi.org/10.1038/20367

Clermont Y (1972) Kinetics of spermatogenesis in mammals: seminiferous epithelium cycle and spermatogonial renewal. Physiol Rev 52(1):198–236. https://doi.org/10.1152/physrev.1972.52.1.198

Clermont Y, McCoshen J, Hermo L (1980) Evolution of the endoplasmic reticulum in the Sertoli cell cytoplasm encapsulating the heads of late spermatids in the rat. Anat Rec 196(1):83–99. https://doi.org/10.1002/ar.1091960109

Clermont Y, Oko R, Hermo L (1990) Immunocytochemical localization of proteins utilized in the formation of outer dense fibers and fibrous sheath in rat spermatids: an electron microscope study. Anat Rec 227(4):447–457. https://doi.org/10.1002/ar.1092270408

Clermont Y, Perey B (1957) Quantitative study of the cell population of the seminiferous tubules in immature rats. Am J Anat 100(2):241–267. https://doi.org/10.1002/aja.1001000205

Clermont Y, Tang XM (1985) Glycoprotein synthesis in the Golgi apparatus of spermatids during spermiogenesis of the rat. Anat Rec 213(1):33–43. https://doi.org/10.1002/ar.1092130106

Cole A, Meistrich ML, Cherry LM, Trostle-Weige PK (1988) Nuclear and manchette development in spermatids of normal and azh/azh mutant mice. Biol Reprod 38(2):385–401. https://doi.org/10.1095/biolreprod38.2.385

Coulter GH, Oko R, Costerton JW (1976) Incidence and ultrastructure of "Crater" defect of bovine spermatozoa. Theriogenology 9:165–173

Courot M, Hochereau-de Reviers MT, Ortravant R (1970) Spermatogenesis. In: Johnson AD, R. GW, L. VN (eds) The Testis, vol 1. Academic Press, New York, pp 339–432

Courot M, Ortavant R (1981) Endocrine control of spermatogenesis in the ram. J Reprod Fertil Suppl 30:47–60

Courtens JL, Courot M (1980) Acrosomal and nuclear morphogenesis in ram spermatids: an experimental study of hypophysectomized and testosterone-supplemented animals. Anat Rec 197(2):143–152. https://doi.org/10.1002/ar.1091970204

Cullinane DL, Chowdhury TA, Kleene KC (2015) Mechanisms of translational repression of the Smcp mRNA in round spermatids. Reproduction 149(1):43–54. https://doi.org/10.1530/rep-14-0394

Curtis SK, Amann RP (1981) Testicular development and establishment of spermatogenesis in Holstein bulls. J Anim Sci 53(6):1645–1657. https://doi.org/10.2527/jas1982.5361645x

Dam AH, Koscinski I, Kremer JA, Moutou C, Jaeger AS, Oudakker AR, Tournaye H, Charlet N, Lagier-Tourenne C, van Bokhoven H, Viville S (2007) Homozygous mutation in SPATA16 is associated with male infertility in human globozoospermia. Am J Hum Genet 81(4):813–820. https://doi.org/10.1086/521314

de Jong FH, Sharpe RM (1977) The onset and establishment of spermatogenesis in rats in relation to gonadotrophin and testosterone levels. J Endocrinol 75(2):197–207. https://doi.org/10.1677/joe.0.0750197

De Kretser DM, Catt KJ, Burger HG, Smith GC (1969) Radioautographic studies on the localization of 125I-labelled human luteinizing and growth hormone in immature male rats. J Endocrinol 43(1):105–111. https://doi.org/10.1677/joe.0.0430105

De Kretser DM, Catt KJ, Paulsen CA (1971) Studies on the in vitro testicular binding of iodinated luteinizing hormone in rats. Endocrinology 88(2):332–337. https://doi.org/10.1210/endo-88-2-332

Desjardins C (1978) Endocrine regulation of reproductive development and function in the male. J Anim Sci (Suppl) 2:56–79

Doran J, Walters C, Kyle V, Wooding P, Hammett-Burke R, Colledge WH (2016) Mfsd14a (Hiat1) gene disruption causes globozoospermia and infertility in male mice. Reproduction 152(1):91–99. https://doi.org/10.1530/rep-15-0557

Eliasson R, Mossberg B, Camner P, Afzelius BA (1977) The immotile-cilia syndrome. A congenital ciliary abnormality as an etiologic factor in chronic airway infections and male sterility. N Engl J Med 297 (1):1–6. https://doi.org/10.1056/nejm197707072970101

Escalier D, Silvius D, Xu X (2003) Spermatogenesis of mice lacking CK2alpha': failure of germ cell survival and characteristic modifications of the spermatid nucleus. Mol Reprod Dev 66(2):190–201. https://doi.org/10.1002/mrd.10346

Fawcett DW, Eddy EM, Phillips DM (1970) Observations on the fine structure and relationships of the chromatoid body in mammalian spermatogenesis. Biol Reprod 2(1):129–153. https://doi.org/10.1095/biolreprod2.1.129

Fawcett DW, Phillips DM (1969) The fine structure and development of the neck region of the mammalian spermatozoon. Anat Rec 165(2):153–164. https://doi.org/10.1002/ar.1091650204

Ferrer M, Rodriguez H, Zara L, Yu Y, Xu W, Oko R (2012a) MMP2 and acrosin are major proteinases associated with the inner acrosomal membrane and may cooperate in sperm penetration of the zona pellucida during fertilization. Cell Tissue Res 349(3):881–895. https://doi.org/10.1007/s00441-012-1429-1

Ferrer M, Xu W, Oko R (2012b) The composition, protein genesis and significance of the inner acrosomal membrane of eutherian sperm. Cell Tissue Res 349(3):733–748. https://doi.org/10.1007/s00441-012-1433-5

Ferrer MJ, Xu W, Shetty J, Herr J, Oko R (2016) Plasminogen Improves Mouse IVF by Interactions with Inner Acrosomal Membrane-Bound MMP2 and SAMP14. Biol Reprod 94(4):88. https://doi.org/10.1095/biolreprod.115.133496

French FS, McLean WS, Smith AA, Tindall DJ, Weddington SC, Petrusz P, Nayfeh SN, Ritzén EM, Hansson V, Trystad O (1974) Androgen transport and receptor mechanisms in testis and epididymis. Nature 250(465):387–391. https://doi.org/10.1038/250387a0

Frohnert C, Schweizer S, Hoyer-Fender S (2011) SPAG4L/SPAG4L-2 are testis-specific SUN domain proteins restricted to the apical nuclear envelope of round spermatids facing the acrosome. Mol Hum Reprod 17(4):207–218. https://doi.org/10.1093/molehr/gaq099

Funaki T, Kon S, Tanabe K, Natsume W, Sato S, Shimizu T, Yoshida N, Wong WF, Ogura A, Ogawa T, Inoue K, Ogonuki N, Miki H, Mochida K, Endoh K, Yomogida K, Fukumoto M, Horai R, Iwakura Y, Ito C, Toshimori K, Watanabe T, Satake M (2013) The Arf GAP SMAP2 is necessary for organized vesicle budding from the trans-Golgi network and subsequent acrosome formation in spermiogenesis. Mol Biol Cell 24(17):2633–2644. https://doi.org/10.1091/mbc.E13-05-0234

Goldberg E, Hawtrey C (1967) The ontogeny of sperm specific lactate dehydrogenase in mice. J Exp Zool 164(3):309–316. https://doi.org/10.1002/jez.1401640302

Gondos B (1977) Testicular development. In: Johnson AD, Gomes WR (eds) The Testis, vol 1. Academic Press, New York, pp 339–411

Görlich D, Henklein P, Laskey RA, Hartmann E (1996) A 41 amino acid motif in importin-alpha confers binding to importin-beta and hence transit into the nucleus. Embo j 15(8):1810–1817

Guidi LG, Holloway ZG, Arnoult C, Ray PF, Monaco AP, Molnár Z, Velayos-Baeza A (2018) AU040320 deficiency leads to disruption of acrosome biogenesis and infertility in homozygous mutant mice. Sci Rep 8(1):10379. https://doi.org/10.1038/s41598-018-28666-6

Hall EA, Keighren M, Ford MJ, Davey T, Jarman AP, Smith LB, Jackson IJ, Mill P (2013) Acute versus chronic loss of mammalian Azi1/Cep131 results in distinct ciliary phenotypes. PLoS Genet 9(12):e1003928. https://doi.org/10.1371/journal.pgen.1003928

Hamilton LE, Acteau G, Xu W, Sutovsky P, Oko R (2017) The developmental origin and compartmentalization of glutathione-s-transferase omega 2 isoforms in the perinuclear theca of eutherian spermatozoa. Biol Reprod 97(4):612–621. https://doi.org/10.1093/biolre/iox122

Hamilton LE, Lion M, Aguila L, Suzuki J, Acteau G, Protopapas N, Xu W, Sutovsky P, Baker M, Oko R (2021) Core Histones Are Constituents of the Perinuclear Theca of Murid Spermatozoa: an Assessment of Their Synthesis and Assembly during Spermiogenesis and Function after Gametic Fusion. Int J Mol Sci 22 (15). https://doi.org/10.3390/ijms22158119

Hamilton LE, Suzuki J, Acteau G, Shi M, Xu W, Meinsohn MC, Sutovsky P, Oko R (2018) WBP2 shares a common location in mouse spermatozoa with WBP2NL/PAWP and like its descendent is a candidate mouse oocyte-activating factor. Biol Reprod 99(6):1171–1183. https://doi.org/10.1093/biolre/ioy156

Hamilton LE, Suzuki J, Aguila L, Meinsohn MC, Smith OE, Protopapas N, Xu W, Sutovsky P, Oko R (2019) Sperm-borne glutathione-S-transferase omega 2 accelerates the nuclear decondensation of spermatozoa during fertilization in mice†. Biol Reprod 101(2):368–376. https://doi.org/10.1093/biolre/ioz082

Han F, Liu C, Zhang L, Chen M, Zhou Y, Qin Y, Wang Y, Chen M, Duo S, Cui X, Bao S, Gao F (2017) Globozoospermia and lack of acrosome formation in GM130-deficient mice. Cell Death Dis 8(1):e2532. https://doi.org/10.1038/cddis.2016.414

Hansson V, Calandra R, Purvis K, Ritzen M, French FS (1976) Hormonal regulation of spermatogenesis. Vitam Horm 34:187–214. https://doi.org/10.1016/s0083-6729(08)60076-x

Hansson V, McLean WS, Smith AA, Tindall DJ, Weddington SC, Nayfeh SN, French FS, Ritzen EM (1974) Androgen receptors in rat testis. Steroids 23(6):823–832. https://doi.org/10.1016/0039-128x(74)90056-7

Harbuz R, Zouari R, Pierre V, Ben Khelifa M, Kharouf M, Coutton C, Merdassi G, Abada F, Escoffier J, Nikas Y, Vialard F, Koscinski I, Triki C, Sermondade N, Schweitzer T, Zhioua A, Zhioua F, Latrous H, Halouani L, Ouafi M, Makni M, Jouk PS, Sèle B, Hennebicq S, Satre V, Viville S, Arnoult C, Lunardi J, Ray PF (2011) A recurrent deletion of DPY19L2 causes infertility in man by blocking sperm head elongation and acrosome formation. Am J Hum Genet 88(3):351–361. https://doi.org/10.1016/j.ajhg.2011.02.007

Hecht NB (1987) Gene expression during spermatogenesis. Ann N Y Acad Sci 513:90–101. https://doi.org/10.1111/j.1749-6632.1987.tb25000.x

Hecht NB, Liem H (1984) Mitochondrial DNA is synthesized during meiosis and spermiogenesis in the mouse. Exp Cell Res 154(1):293–298. https://doi.org/10.1016/0014-4827(84)90688-8

Ho HC, Wey S (2007) Three dimensional rendering of the mitochondrial sheath morphogenesis during mouse spermiogenesis. Microsc Res Tech 70(8):719–723. https://doi.org/10.1002/jemt.20457

Hochereau-de Reviers MT (1970) Etude des divisions spermatogoniales et du renouvellement de la spermatogenie souche chez le taureau. Paris University of Sciences, France

Hochereau MT (1963) Etude comparee dela vague spermatogenetique chez le taureau et chez le rat. Ann Biol Anim Biochim Biophys 3:5–20

Hochereau MT (1967) DNA synthesis during proliferation and regeneration of spermatogonia in the bull. Arch Anat Microsc Morphol Exp 56(3):85–96

Hochereau MT, Courot M, Ortravant R (1964) Duree de la spermatogenese chez le taureau; etude par autoradiograpie testiculaire. . Proc 5th Int Congr Anima Reprod AI 3:541–546

Irons MJ (1983) Synthesis and assembly of connecting-piece proteins as revealed by radioautography. J Ultrastruct Res 82(1):27–34. https://doi.org/10.1016/s0022-5320(83)90094-1

Irons MJ, Clermont Y (1982a) Formation of the outer dense fibers during spermiogenesis in the rat. Anat Rec 202(4):463–471. https://doi.org/10.1002/ar.1092020405

Irons MJ, Clermont Y (1982b) Kinetics of fibrous sheath formation in the rat spermatid. Am J Anat 165(2):121–130. https://doi.org/10.1002/aja.1001650204

Jegou B, Le Gac F, de Kretser DM (1982) Seminiferous tubule fluid and interstitial fluid production. I. Effects of age and hormonal regulation in immature rats. Biol Reprod 27 (3):590–595. https://doi.org/10.1095/biolreprod27.3.590

Jégou B, Le Gac F, Irby DC, de Kretser DM (1983) Studies on seminiferous tubule fluid production in the adult rat: effect of hypophysectomy and treatment with FSH. LH and Testosterone. Int J Androl 6(3):249–260. https://doi.org/10.1111/j.1365-2605.1983.tb00539.x

Juneja SC, van Deursen JM (2005) A mouse model of familial oligoasthenoteratozoospermia. Hum Reprod 20(4):881–893. https://doi.org/10.1093/humrep/deh719

Kang-Decker N, Mantchev GT, Juneja SC, McNiven MA, van Deursen JM (2001) Lack of acrosome formation in Hrb-deficient mice. Science 294(5546):1531–1533. https://doi.org/10.1126/science.1063665

Kierszenbaum AL (2002) Intramanchette transport (IMT): managing the making of the spermatid head, centrosome, and tail. Mol Reprod Dev 63(1):1–4. https://doi.org/10.1002/mrd.10179

Kierszenbaum AL, Tres LL, Rivkin E, Kang-Decker N, van Deursen JM (2004) The acroplaxome is the docking site of Golgi-derived myosin Va/Rab27a/b- containing proacrosomal vesicles in wild-type and Hrb mutant mouse spermatids. Biol Reprod 70(5):1400–1410. https://doi.org/10.1095/biolreprod.103.025346

Knudsen O (1954) Cytomorphological investigations into the spermiocytogenesis of bulls with normal fertility and bulls with acquired disturbances in spermiogenesis. Pathol Microbiol Scand:[Suppl. C1]:12–79.

Komada M, McLean DJ, Griswold MD, Russell LD, Soriano P (2000) E-MAP-115, encoding a microtubule-associated protein, is a retinoic acid-inducible gene required for spermatogenesis. Genes Dev 14(11):1332–1342

Koscinski I, Elinati E, Fossard C, Redin C, Muller J, Velez de la Calle J, Schmitt F, Ben Khelifa M, Ray PF, Kilani Z, Barratt CL, Viville S (2011) DPY19L2 deletion as a major cause of globozoospermia. Am J Hum Genet 88(3):344–350. https://doi.org/10.1016/j.ajhg.2011.01.018

Krausz C, Sassone-Corsi P (2005) Genetic control of spermiogenesis: insights from the CREM gene and implications for human infertility. Reprod Biomed Online 10(1):64–71. https://doi.org/10.1016/s1472-6483(10)60805-x

Kwon YK, Hecht NB (1993) Binding of a phosphoprotein to the 3' untranslated region of the mouse protamine 2 mRNA temporally represses its translation. Mol Cell Biol 13(10):6547–6557. https://doi.org/10.1128/mcb.13.10.6547-6557.1993

Lacroix M, Parvinen M, Fritz IB (1981) Localization of testicular plasminogen activator in discrete portions (stage VII and VIII) of the seminiferous tubule. Biol Reprod 25(1):143–146. https://doi.org/10.1095/biolreprod25.1.143

Lagerlof N (1951) Hereditary forms of sterility in Swedish cattle breeds. Fertil Steril 2(3):230–242

Leblond CP, Clermont Y (1952a) Definition of the stages of the cycle of the seminiferous epithelium in the rat. Ann N Y Acad Sci 55(4):548–573. https://doi.org/10.1111/j.1749-6632.1952.tb26576.x

Leblond CP, Clermont Y (1952b) Spermiogenesis of rat, mouse, hamster and guinea pig as revealed by the periodic acid-fuchsin sulfurous acid technique. Am J Anat 90(2):167–215. https://doi.org/10.1002/aja.1000900202

Leestma JE, Sepsenwol S (1980) Sperm tail axoneme alterations in the Wobbler mouse. J Reprod Fertil 58(1):267–270. https://doi.org/10.1530/jrf.0.0580267

Lerer-Goldshtein T, Bel S, Shpungin S, Pery E, Motro B, Goldstein RS, Bar-Sheshet SI, Breitbart H, Nir U (2010) TMF/ARA160: a key regulator of sperm development. Dev Biol 348(1):12–21. https://doi.org/10.1016/j.ydbio.2010.07.033

Lin YN, Roy A, Yan W, Burns KH, Matzuk MM (2007) Loss of zona pellucida binding proteins in the acrosomal matrix disrupts acrosome biogenesis and sperm morphogenesis. Mol Cell Biol 27(19):6794–6805. https://doi.org/10.1128/mcb.01029-07

Longo FJ, Cook S (1991) Formation of the perinuclear theca in spermatozoa of diverse mammalian species: relationship of the manchette and multiple band polypeptides. Mol Reprod Dev 28(4):380–393. https://doi.org/10.1002/mrd.1080280411

Lyon MF, Glenister PH, Lamoreux ML (1975) Normal spermatozoa from androgen-resistant germ cells of chimaeric mice and the role of androgen in spermatogenesis. Nature 258(5536):620–622. https://doi.org/10.1038/258620a0

Machado de Domenech E, Domenech CE, Aoki A, Blanco A (1972) Association of the testicular lactate dehydrogenase isozyme with a special type of mitochondria. Biol Reprod 6(1):136–147. https://doi.org/10.1093/biolreprod/6.1.136

Means AR, Fakunding JL, Huckins C, Tindall DJ, Vitale R (1976) Follicle-stimulating hormone, the Sertoli cell, and spermatogenesis. Recent Prog Horm Res 32:477–527. https://doi.org/10.1016/b978-0-12-571132-6.50027-0

Meistrich ML, Trostle-Weige PK, Russell LD (1990) Abnormal manchette development in spermatids of azh/azh mutant mice. Am J Anat 188(1):74–86. https://doi.org/10.1002/aja.1001880109

Monesi V, Geremia R, D'Agostino A, Boitani C (1978) Biochemistry of male germ cell differentiation in mammals: RNA synthesis in meiotic and postmeiotic cells. Curr Top Dev Biol 12:11–36. https://doi.org/10.1016/s0070-2153(08)60592-x

Morales C, Clermont Y (1982) Evolution of Sertoli cell processes invading the cytoplasm of rat spermatids. Anat Rec 203(2):233–244. https://doi.org/10.1002/ar.1092030205

Morales CR, Lefrancois S, Chennathukuzhi V, El-Alfy M, Wu X, Yang J, Gerton GL, Hecht NB (2002) A TB-RBP and Ter ATPase complex accompanies specific mRNAs from nuclei through the nuclear pores and into intercellular bridges in mouse male germ cells. Dev Biol 246(2):480–494. https://doi.org/10.1006/dbio.2002.0679

Mountjoy JR, Xu W, McLeod D, Hyndman D, Oko R (2008) RAB2A: a major subacrosomal protein of bovine spermatozoa implicated in acrosomal biogenesis. Biol Reprod 79(2):223–232. https://doi.org/10.1095/biolreprod.107.065060

Mulder E, Peters MJ, van Beurden WM, van der Molen HJ (1974) A receptor for testosterone in mature rat testes. FEBS Lett 47(2):209–211. https://doi.org/10.1016/0014-5793(74)81013-6

Nantel F, Sassone-Corsi P (1996) CREM: a transcriptional master switch during the spermatogenesis differentiation program. Front Biosci 1:d266-269. https://doi.org/10.2741/a131

Odell WD, Swerdloff RS, Jacobs HS, Hescox MA (1973) FSH induction of sensitivity to LH: one cause of sexual maturation in the male rat. Endocrinology 92(1):160–165. https://doi.org/10.1210/endo-92-1-160

Oko R (1977) Normal and defective bovine spermatogenesis. University of Calgary, Canada

Oko R (1984) The effects and possible mechanisms of gossypol action on sperm organelles, spermatogenesis and accessory sex glands of the adult rat. University of Saskatchewan,

Oko R (1998) Occurrence and formation of cytoskeletal proteins in mammalian spermatozoa. Andrologia 30(4–5):193–206. https://doi.org/10.1111/j.1439-0272.1998.tb01161.x

Oko R, Clermont Y (1989) Light microscopic immunocytochemical study of fibrous sheath and outer dense fiber formation in the rat spermatid. Anat Rec 225(1):46–55. https://doi.org/10.1002/ar.1092250108

Oko R, Donald A, Xu W, van der Spoel AC (2011) Fusion failure of dense-cored proacrosomal vesicles in an inducible mouse model of male infertility. Cell Tissue Res 346(1):119–134. https://doi.org/10.1007/s00441-011-1248-9

Oko R, Hermo L, Chan PT, Fazel A, Bergeron JJ (1993) The cytoplasmic droplet of rat epididymal spermatozoa contains saccular elements with Golgi characteristics. J Cell Biol 123(4):809–821. https://doi.org/10.1083/jcb.123.4.809

Oko R, Hrudka F (1984a) Comparison of the effects of gossypol, estradiol-17 beta and testosterone compensation on male rat reproductive organs. Biol Reprod 30(5):1198–1207. https://doi.org/10.1095/biolreprod30.5.1198

Oko R, Hrudka F (1984b) Gossypol-induced early and delayed effects in the seminiferous epithelium of the adult rat. Contracept Deliv Syst 5:22

Oko R, Maravei D (1995) Distribution and possible role of perinuclear theca proteins during bovine spermiogenesis. Microsc Res Tech 32(6):520–532. https://doi.org/10.1002/jemt.1070320605

Oko R, Sutovsky P (2009) Biogenesis of sperm perinuclear theca and its role in sperm functional competence and fertilization. J Reprod Immunol 83(1–2):2–7. https://doi.org/10.1016/j.jri.2009.05.008

Olson GE, Winfrey VP (1990) Mitochondria-cytoskeleton interactions in the sperm midpiece. J Struct Biol 103(1):13–22. https://doi.org/10.1016/1047-8477(90)90081-m

Orgebin MC (1961) Etude du transit epididymaire des spermatozoides de taureau marques a l'aide du 32P. Ann Biol Anim Biochim Biophys 1:117–120

Orth JM (1984) The role of follicle-stimulating hormone in controlling Sertoli cell proliferation in testes of fetal rats. Endocrinology 115(4):1248–1255. https://doi.org/10.1210/endo-115-4-1248

Ortravant R (1958) Le cycle spermatogenetique chez le belier. University of Paris, France

Otani H, Tanaka O, Kasai K, Yoshioka T (1988) Development of mitochondrial helical sheath in the middle piece of the mouse spermatid tail: regular dispositions and synchronized changes. Anat Rec 222(1):26–33. https://doi.org/10.1002/ar.1092220106

Paiardi C, Pasini ME, Gioria M, Berruti G (2011) Failure of acrosome formation and globozoospermia in the wobbler mouse, a Vps54 spontaneous recessive mutant. Spermatogenesis 1(1):52–62. https://doi.org/10.4161/spmg.1.1.14698

Parvinen M (1982) Regulation of the seminiferous epithelium. Endocr Rev 3(4):404–417. https://doi.org/10.1210/edrv-3-4-404

Parvinen M, Ruokonen A (1982) Endogenous Steroids in the Rat Seminiferous Tubules. Comparison of the Stages of the Epithelial Cycle Isolated by Transillumination-Assisted Microdissection. Journal of Andrology 3 (4):211–220. https://doi.org/10.1002/j.1939-4640.1982.tb00672.x

Perey B, Clermont Y, Leblond CP (1961) The wave of the seminiferous epithelium in the rat. American Journal of Anatomy 108(1):47–77. https://doi.org/10.1002/aja.1001080105

Pierre V, Martinez G, Coutton C, Delaroche J, Yassine S, Novella C, Pernet-Gallay K, Hennebicq S, Ray PF, Arnoult C (2012) Absence of Dpy19l2, a new inner nuclear membrane protein, causes globozoospermia in mice by preventing the anchoring of the acrosome to the nucleus. Development 139(16):2955–2965. https://doi.org/10.1242/dev.077982

Protopapas N, Hamilton LE, Warkentin R, Xu W, Sutovsky P, Oko R (2019) The perforatorium and postacrosomal sheath of rat spermatozoa share common developmental origins and protein constituents†. Biol Reprod 100(6):1461–1472. https://doi.org/10.1093/biolre/ioz052

Ritzen EM, Boitani C, Parvinen M, French FC, Feldman M (1982) Stage-dependent secretion of ABP by rat seminiferous tubules. Mol Cell Endocrinol 25(1):25–33. https://doi.org/10.1016/0303-7207(82)90166-6

Robaire B, Ewing LL, Irby DC, Desjardins C (1979) Interactions of testosterone and estradiol-17 beta on the reproductive tract of the male rat. Biol Reprod 21(2):455–463. https://doi.org/10.1095/biolreprod21.2.455

Ross MH (1976) The Sertoli cell junctional specialization during spermiogenesis and at spermiation. Anat Rec 186(1):79–103. https://doi.org/10.1002/ar.1091860107

Russell L (1977a) Desmosome-like junctions between Sertoli and germ cells in the rat testis. Am J Anat 148(3):301–312. https://doi.org/10.1002/aja.1001480302

Russell L (1977b) Movement of spermatocytes from the basal to the adluminal compartment of the rat testis. Am J Anat 148(3):313–328. https://doi.org/10.1002/aja.1001480303

Russell L (1977c) Observations on rat Sertoli ectoplasmic ('junctional') specializations in their association with germ cells of the rat testis. Tissue Cell 9(3):475–498. https://doi.org/10.1016/0040-8166(77)90007-6

Russell L, Clermont Y (1976) Anchoring device between Sertoli cells and late spermatids in rat seminiferous tubules. Anat Rec 185(3):259–278. https://doi.org/10.1002/ar.1091850302

Russell LD (1980) Sertoli-germ cell interrelations: a review. Gamete Res 3(2):179–202. https://doi.org/10.1002/mrd.1120030209

Russell LD, Clermont Y (1977) Degeneration of germ cells in normal, hypophysectomized and hormone treated hypophysectomized rats. Anat Rec 187(3):347–366. https://doi.org/10.1002/ar.1091870307

Russell LD, Griswold MD (1993) The Sertoli cell. Cache River Press, Clearwater, FL

Russell LD, Malone JP, Karpas SL (1981) Morphological pattern elicited by agents affecting spermatogenesis by stimulation. Tissue Cell 13(2):369–380. https://doi.org/10.1016/0040-8166(81)90011-2

Russell LD, Russell JA, MacGregor GR, Meistrich ML (1991) Linkage of manchette microtubules to the nuclear envelope and observations of the role of the manchette in nuclear shaping during spermiogenesis in rodents. Am J Anat 192(2):97–120. https://doi.org/10.1002/aja.1001920202

Sanborn BM, Steinberger A, Tcholakian RK, Steinberger E (1977) Direct measurement of androgen receptors in cultured Sertoli cells. Steroids 29(4):493–502. https://doi.org/10.1016/0039-128x(77)90069-1

Schalles U, Shao X, van der Hoorn FA, Oko R (1998) Developmental expression of the 84-kDa ODF sperm protein: localization to both the cortex and medulla of outer dense fibers and to the connecting piece. Dev Biol 199(2):250–260. https://doi.org/10.1006/dbio.1998.8931

Schneider M, Förster H, Boersma A, Seiler A, Wehnes H, Sinowatz F, Neumüller C, Deutsch MJ, Walch A, Hrabé de Angelis M, Wurst W, Ursini F, Roveri A, Maleszewski M, Maiorino M, Conrad M (2009) Mitochondrial glutathione peroxidase 4 disruption causes male infertility. Faseb j 23(9):3233–3242. https://doi.org/10.1096/fj.09-132795

Setchell BP (1980) The Functional Significance of the Blood-testis Barrier. J Androl 1(1):3–10. https://doi.org/10.1002/j.1939-4640.1980.tb00003.x

Sharpe RM (1983) Local control of testicular function. Q J Exp Physiol 68(3):265–287. https://doi.org/10.1113/expphysiol.1983.sp002723

Sharpe RM, Maddocks S, Kerr JB (1990) Cell-cell interactions in the control of spermatogenesis as studied using Leydig cell destruction and testosterone replacement. Am J Anat 188(1):3–20. https://doi.org/10.1002/aja.1001880103

Skinner MK, Griswold MD (2004) Sertoli cell biology. Elsevier

Solari AJ (1964) THE MORPHOLOGY AND ULTRASTRUCTURE OF THE SEX VESICLE IN THE MOUSE. Exp Cell Res 36:160–168. https://doi.org/10.1016/0014-4827(64)90169-7
Sotomayor RE, Handel MA (1986) Failure of acrosome assembly in a male sterile mouse mutant. Biol Reprod 34(1):171–182. https://doi.org/10.1095/biolreprod34.1.171
Steinberger A, Steinberger E (1965) DIFFERENTIATION OF RAT SEMINIFEROUS EPITHELIUM IN ORGAN CULTURE. J Reprod Fertil 9:243–248. https://doi.org/10.1530/jrf.0.0090243
Steinberger A, Steinberger E (1966) Stimulatory effect of vitamins and glutamine on the differentiation of germ cells in rat testes organ culture grown in chemically defined media. Exp Cell Res 44(2):429–435. https://doi.org/10.1016/0014-4827(66)90449-6
Steinberger A, Walther J, Heindel JJ, Sanborn BM, Tsai YH, Steinberger E (1979) Hormone interactions in the Sertoli cells. In Vitro 15(1):23–31. https://doi.org/10.1007/bf02627076
Steinberger E (1971) Hormonal control of mammalian spermatogenesis. Physiol Rev 51(1):1–22. https://doi.org/10.1152/physrev.1971.51.1.1
Suganuma R, Walden CM, Butters TD, Platt FM, Dwek RA, Yanagimachi R, van der Spoel AC (2005) Alkylated imino sugars, reversible male infertility-inducing agents, do not affect the genetic integrity of male mouse germ cells during short-term treatment despite induction of sperm deformities. Biol Reprod 72(4):805–813. https://doi.org/10.1095/biolreprod.104.036053
Susi FR, Clermont Y (1970) Fine structural modifications of the rat chromatoid body during spermiogenesis. Am J Anat 129(2):177–191. https://doi.org/10.1002/aja.1001290205
Swaminathan J, Baxter EM, Corces VG (2005) The role of histone H2Av variant replacement and histone H4 acetylation in the establishment of Drosophila heterochromatin. Genes Dev 19(1):65–76. https://doi.org/10.1101/gad.1259105
Tang XM, Lalli MF, Clermont Y (1982) A cytochemical study of the Golgi apparatus of the spermatid during spermiogenesis in the rat. Am J Anat 163(4):283–294. https://doi.org/10.1002/aja.1001630402
Thibier M, Rolland O (1976) The effect of dexamethasone (DXM) on circulating testosterone (T) and luteinizing hormone (LH) in young postpubertal bulls. Theriogenology 5(2):53–60. https://doi.org/10.1016/0093-691x(76)90168-0
Thorne-Tjomsland G, Clermont Y, Hermo L (1988) Contribution of the Golgi apparatus components to the formation of the acrosomic system and chromatoid body in rat spermatids. Anat Rec 221(2):591–598. https://doi.org/10.1002/ar.1092210205
Tovich PR, Sutovsky P, Oko RJ (2004) Novel aspect of perinuclear theca assembly revealed by immunolocalization of non-nuclear somatic histones during bovine spermiogenesis. Biol Reprod 71(4):1182–1194. https://doi.org/10.1095/biolreprod.104.030445
Tran MH, Aul RB, Xu W, van der Hoorn FA, Oko R (2012) Involvement of classical bipartite/karyopherin nuclear import pathway components in acrosomal trafficking and assembly during bovine and murid spermiogenesis. Biol Reprod 86(3):84. https://doi.org/10.1095/biolreprod.111.096842
Vihko KK, Suominen JJ, Parvinen M (1984) Cellular regulation of plasminogen activator secretion during spermatogenesis. Biol Reprod 31(2):383–389. https://doi.org/10.1095/biolreprod31.2.383
Vogl AW, Pfeiffer DC, Mulholland D, Kimel G, Guttman J (2000) Unique and multifunctional adhesion junctions in the testis: ectoplasmic specializations. Arch Histol Cytol 63(1):1–15. https://doi.org/10.1679/aohc.63.1
Vogl AW, Pfeiffer DC, Redenbach DM (1991) Ectoplasmic ("junctional") specializations in mammalian Sertoli cells: influence on spermatogenic cells. Ann N Y Acad Sci 637:175–202. https://doi.org/10.1111/j.1749-6632.1991.tb27310.x
Vogl AW, Vaid KS, Guttman JA (2008) The Sertoli cell cytoskeleton. Adv Exp Med Biol 636:186–211. https://doi.org/10.1007/978-0-387-09597-4_11

Vogl AW, Young JS, Du M (2013) New insights into roles of tubulobulbar complexes in sperm release and turnover of blood-testis barrier. Int Rev Cell Mol Biol 303:319–355. https://doi.org/10.1016/b978-0-12-407697-6.00008-8

Waites GM, Gladwell RT (1982) Physiological significance of fluid secretion in the testis and blood-testis barrier. Physiol Rev 62(2):624–671. https://doi.org/10.1152/physrev.1982.62.2.624

Wang H, Wan H, Li X, Liu W, Chen Q, Wang Y, Yang L, Tang H, Zhang X, Duan E, Zhao X, Gao F, Li W (2014) Atg7 is required for acrosome biogenesis during spermatogenesis in mice. Cell Res 24(7):852–869. https://doi.org/10.1038/cr.2014.70

Weddington SC, Hansson V, Ritzen EM, Hagenas L, French FS, Nayfeh SN (1975) Sertoli cell secretory function after hypophysectomy. Nature 254(5496):145–146. https://doi.org/10.1038/254145a0

Wright WW, Frankel AI (1980) An androgen receptor in the nuclei of late spermatids in testes of male rats. Endocrinology 107(1):314–318. https://doi.org/10.1210/endo-107-1-314

Wu AT, Sutovsky P, Manandhar G, Xu W, Katayama M, Day BN, Park KW, Yi YJ, Xi YW, Prather RS, Oko R (2007a) PAWP, a sperm-specific WW domain-binding protein, promotes meiotic resumption and pronuclear development during fertilization. J Biol Chem 282(16):12164–12175. https://doi.org/10.1074/jbc.M609132200

Wu AT, Sutovsky P, Xu W, van der Spoel AC, Platt FM, Oko R (2007b) The postacrosomal assembly of sperm head protein, PAWP, is independent of acrosome formation and dependent on microtubular manchette transport. Dev Biol 312(2):471–483. https://doi.org/10.1016/j.ydbio.2007.08.051

Xiao N, Kam C, Shen C, Jin W, Wang J, Lee KM, Jiang L, Xia J (2009) PICK1 deficiency causes male infertility in mice by disrupting acrosome formation. J Clin Invest 119(4):802–812. https://doi.org/10.1172/jci36230

Xu X, Toselli PA, Russell LD, Seldin DC (1999) Globozoospermia in mice lacking the casein kinase II alpha' catalytic subunit. Nat Genet 23(1):118–121. https://doi.org/10.1038/12729

Yang J, Medvedev S, Yu J, Tang LC, Agno JE, Matzuk MM, Schultz RM, Hecht NB (2005) Absence of the DNA-/RNA-binding protein MSY2 results in male and female infertility. Proc Natl Acad Sci U S A 102(16):5755–5760. https://doi.org/10.1073/pnas.0408718102

Yang J, Morales CR, Medvedev S, Schultz RM, Hecht NB (2007) In the absence of the mouse DNA/RNA-binding protein MSY2, messenger RNA instability leads to spermatogenic arrest. Biol Reprod 76(1):48–54. https://doi.org/10.1095/biolreprod.106.055095

Yao R, Ito C, Natsume Y, Sugitani Y, Yamanaka H, Kuretake S, Yanagida K, Sato A, Toshimori K, Noda T (2002) Lack of acrosome formation in mice lacking a Golgi protein, GOPC. Proc Natl Acad Sci U S A 99(17):11211–11216. https://doi.org/10.1073/pnas.162027899

Yildiz Y, Matern H, Thompson B, Allegood JC, Warren RL, Ramirez DM, Hammer RE, Hamra FK, Matern S, Russell DW (2006) Mutation of beta-glucosidase 2 causes glycolipid storage disease and impaired male fertility. J Clin Invest 116(11):2985–2994. https://doi.org/10.1172/jci29224

Yu Y, Vanhorne J, Oko R (2009) The origin and assembly of a zona pellucida binding protein, IAM38, during spermiogenesis. Microsc Res Tech 72(8):558–565. https://doi.org/10.1002/jemt.20696

Yu Y, Xu W, Yi YJ, Sutovsky P, Oko R (2006) The extracellular protein coat of the inner acrosomal membrane is involved in zona pellucida binding and penetration during fertilization: characterization of its most prominent polypeptide (IAM38). Dev Biol 290(1):32–43. https://doi.org/10.1016/j.ydbio.2005.11.003

Zhou J, Du YR, Qin WH, Hu YG, Huang YN, Bao L, Han D, Mansouri A, Xu GL (2009) RIM-BP3 is a manchette-associated protein essential for spermiogenesis. Development 136(3):373–382. https://doi.org/10.1242/dev.030858

Chapter 3
Bovine Sperm Maturation

Contents

Abstract On completion of spermatogenesis, testicular spermatozoa appear structurally mature but are infertile and must undergo a sequential maturational process in the epididymis to become motile and acquire fertilizing potential. This chapter provides a cell biological overview of the endocytic and secretory activities, along the extratesticular duct system, that provide appropriate conditions for epididymal maturation of bull spermatozoa. The compartmentalization of the bovine epididymis is illustrated and discussed in terms of epithelial cell types and merocrine and apocrine protein secretions by principal cells that influence maturation. Sequential maturational events are followed with examples, first, of testicular proteins associated with spermatozoa that are endocytosed to form a 'clean slate' and then, of epididymal secretory proteins that recondition the sperm milieu and bind to spermatozoa in order to attain its full fertilization potential. Finally, an assessment is made of the potential contributions to epididymal maturation of some well-characterized and identified secretory proteins that interact with the cytoplasmic membrane of spermatozoa.

Keywords Bull · Rete testis · Efferent ducts · Epididymis · Principal cells · Basal cells · Apical cells · Caput · Descending limb · Ascending limb · Corpus · Cauda · Endocytosis · Exocytosis · Merocrine Secretion · Apocrine secretion · Sperm maturation · Fertilization · Proteomics

© The Author(s), under exclusive license to Springer Nature Switzerland AG 2025
A. Barth et al., *Abnormal Morphology of Bovine Spermatozoa*,
Advances in Anatomy, Embryology and Cell Biology 240,
https://doi.org/10.1007/978-3-031-70126-9_3

3.1 Introduction

Even though spermatids released into the lumen of the seminiferous tubules at the end of spermiogenesis appear structurally mature they are infertile. In order to become motile and fertile they must pass through the rete testis, the efferent ductules and at least the proximal portion (species dependent) of the epididymis. During their transit through this part of the excurrent duct system, the acquisition of natural fertilization capability of the spermatozoa is dependent on the coordinated activity of endocytosis and secretion of the epithelial lining cells. Thus, for example, in some parts of the excurrent duct, such as the rete, the efferent ducts and the initial segment, endocytosis appears to play a more prominent role, while in others such as the caput epididymis, secretion appears more dominant. However, overall it is the balance between the secretory and endocytic activities, along the excurrent duct system, that provide the appropriate conditions in the external milieu for sperm maturation. A variety of proteins and other molecules are secreted into the external milieu, some of which condition the sperm milieu, while others associate with the sperm by enzymatic, electrostatic or receptor mediated means. However, the exact contribution to sperm maturation and hence fertilization of many identified epididymal secretory proteins is still in the exploratory stage of research.

Transit of sperm through the rete, efferent ductules and the epididymis is associated with significant maturational changes, which are discussed or alluded to in the forthcoming text. These include:

1. absorption of seminiferous fluid by the rete, efferent ductules and initial segment in preparation for a newly secreted environment in the epididymis
2. gaining the capacity for progressive motility
3. alteration of the nature of the plasma membrane and addition of proteins and glycoconjugates
4. addition of antioxidant enzymes to the milieu and sperm to protect against reactive oxygen species (ROS) and regulate signalling during capacitation
5. gaining the ability to interact more efficiently with the cu mulus, zona pellucida and oolemma
6. cross-linking of structural proteins by disulfide bonds in sperm components such as the perinuclear theca, ODFs and FS
7. final condensation of the nucleus and acrosome – the specific gravity becomes higher
8. migration of the cytoplasmic droplet from a proximal to distal midpiece position of the sperm tail and its eventual separation
9. oxygen deficiency to inhibit metabolism of sperm, as well as secretion of metabolic inhibitors and depletion of substrates for metabolism
10. maintenance of luminal acidification (low pH) to keep spermatozoa in a quiescent, immotile state
11. resorption, phagocytosis, and liquefaction of defective sperm components

The most important maturational changes in sperm occur in the efferent ductules and caput epididymis. Sperm taken from these areas will not fertilize secondary oocytes (Orgebin-Crist 1969; Bedford 1975). Bull and ram sperm taken from the distal corpus and the cauda epididymis already have the potential to fertilize (Amann 1987). The cauda epididymis has properties that allow the sperm to be stored there for several weeks. One property that retains the viability of the sperm is a constant temperature even below that of the testes (Bedford 1978). Transport of the sperm through the efferent ductules is accomplished by ciliated epithelial cells while transport of sperm through the epididymis is accomplished by localized contraction of smooth muscle of the epididymal wall (Bedford 1978). The time taken for bull sperm to pass through the ductus epididymis is 8–11 days. In fact, if the epididymal sperm reserve is depleted by a frequent ejaculation regime covering a 90-day period, passage of bovine spermatozoa takes approximately two days less (Koefoed-Johnsen 1960).

3.2 The Epididymis and Its Cell Types

The bull epididymis consists of one highly coiled duct which can be divided into three major segments (caput, corpus and cauda; Figs. 3.1, 3.2). The caput or head consists of a descending limb and an ascending limb covering the dorsal and caudal surface of the testis (Fig. 3.1). The descending limb includes a short extratesticular rete coming out at the papilla and connecting to the efferent ductules, which in turn connect at several places to the convoluted duct of the epididymis (Amann 1987; Goyal 1985). The ascending limb of the ductus epididymis runs parallel to the descending limb until it turns abruptly downward to continue as the corpus on the medial side of the testis, meeting the bulkier cauda epidididymis at the ventral and medial extremity of the testis (Fig. 3.1). The cauda epididymidis is easily palpated through the ventral and medial-rostral aspect of the scrotum. For a better understanding of the arrangement of the epididymis on the testis and its relationship to the vas deferens and scrotum we recommend viewing an exceptionally well done anatomical dissection of this region in the buck (Campbell 2007), which has almost identical features to the bull. The epididymal epithelium is comprised primarily of principal cells, which have long microvilli that project into the epididymal lumen (Fig. 3.3). Other cell types, including basal, clear, narrow, apical, dendritic, and halo cells, reside sporadically between the principal cells in murids and some other mammalian species (Hermo and Robaire 2002; Robaire and Hinton 2015). However, in the bull only basal and apical cells, macrophages, monocytes and T lymphocytes appear to be associated with the principal cells (Fig. 3.3) (Amann 1987; Goyal 1985). Due to the problem of microscopically differentiating between these leucocytes (i.e., macrophages, monocytes and lymphocytes), they have been put into the category of halo cells (Robaire and Hinton 2015).

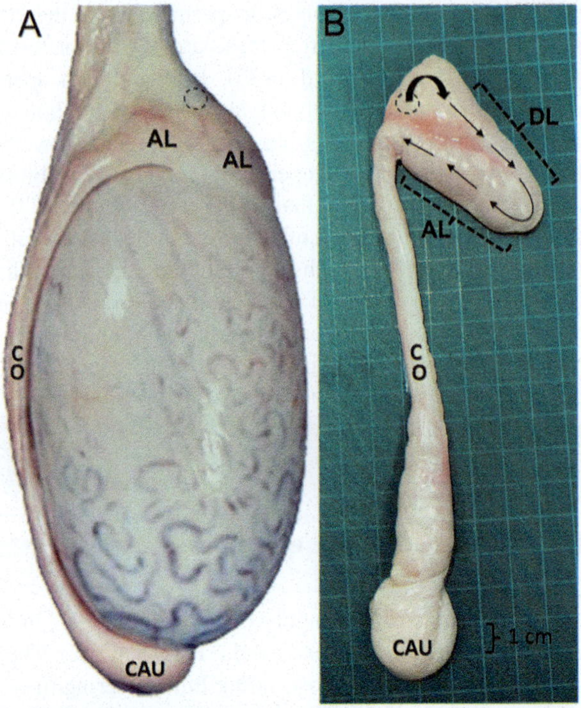

Fig. 3.1 **A.** Right bull testes as viewed from behind (with skin and tunica vaginalis removed). From this view one can see the ascending limb (AL) of the caput epididymidis cross the top of the testis before descending to form the corpus (CO) epididymidis, which runs down the medial side of the testis until it reaches the cauda (CAU) epididymidis, which is suspended ventrally from the testis. **B**. Right dissected epididymidis arranged to show the relationship between the ascending limb (AL) and descending limb (DL) of the caput, which was hidden in the caudal view (A) of the testis. The dashed-lined circle represents the papillary region, underneath the caput, where the testicular rete meets the efferent ducts. The efferent ducts (area represented by the bolded curved arrow) then join the convoluted duct of the descending limb, represented by the downward pointed arrows. Following the arrows, the seminal fluid first descends in the DL and then ascends in the AL before descending in the corpus

The main cell type, the principal cell, has both absorptive and secretory functions. The dominance of one activity over the other is regionally dependent along the epididymis. For example, in the initial segment (IS) of the bull there is more absorption by the principal cell, pumping out excess fluids, proteins and steroids left over from the testis, while in the M1, M2 and M3 regions (Fig. 3.2) there is more secretory activity, presumably providing the right milieu for sperm maturation (Amann 1987). Furthermore, region M3, which constitutes the corpus, is subdivided into a and b because even though the principal cells look similar it is only in M3b and beyond that bull spermatozoa are fertile and result in offspring.

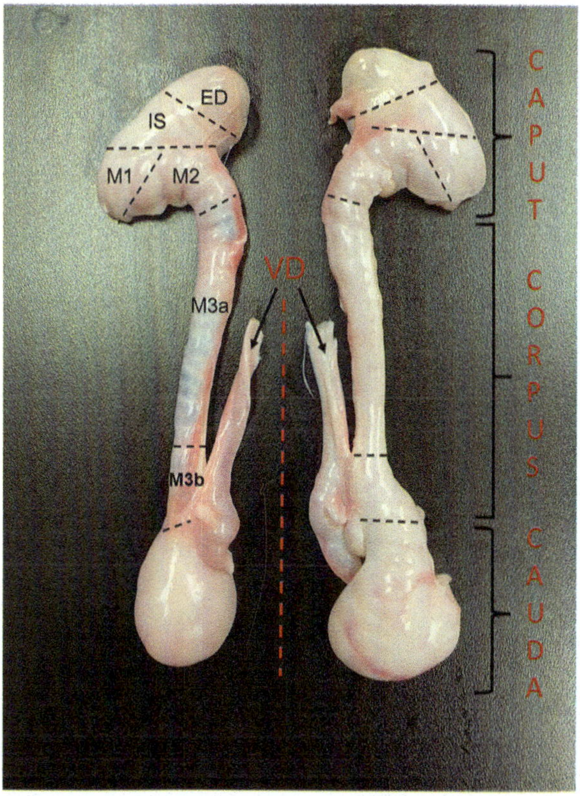

Fig. 3.2 Dissected pair of left and right epididymides placed in position as seen from behind the bull when looking at the scrotum. Note that the vas deferens and corpus of each side face each other medially but are separated from each other in situ by the scrotal septum (represented by the dashed line). The pair of testes would be situated laterally to the each corpus. The designated regions of the epididymis presented here are based on sperm function and electron microscope observations determined by a number of investigators but summarized by Amann (Amann 1987). The initial segment (IS) is involved with adsorption, the middle segments (M1, M2 and M3) in sperm maturation and the cauda in sperm storage. Region M3 is further subdivided into regions M3a and M3b, which although morphologically similar, differ because spermatozoa in M3a have not achieved progressive motility whereas spermatozoa in M3b are motile and fertile

The principal cell height is tallest, microvilli length longest and luminal diameter narrowest in the initial segment of the caput epididymis, a region in the murid which appears equivalent to the mid portion of the descending limb region of the bull caput (Fig. 3.2, IS). Generally, in the murid epididymis, the pseudostratified epithelium becomes progressively shorter and the lumen progressively wider from its proximal to distal end. This is in agreement with what is found in the bull epididymis (Amann 1987) with the exception that the mid-corpus epithelium is slightly taller than the epithelium found in the ascending segment of the caput (Goyal 1985).

Basal cells are small cells located at the base of the epithelium in contact with the basement membrane (Fig. 3.3). They are sporadically interdigitated between the principal cells throughout the epididymis and represent up to 20% of the total number of cells in the epithelium. Although dwarfed by the tall epithelium, basal cells have been shown to have narrow projections that make contact with the lumen (Shum et al. 2009). Furthermore, these cells share extensive interdigitations with the plasma membrane of adjacent principal cells. Therefore, the basal cells are in the position to endocytose or respond to factors coming from the interstitium, lumen or adjacent cells and hence to communicate with and regulate the secretory functions of epididymal epithelial cells. For example, murid basal cells have been shown to express angiotensin II type 2 receptor, which, when activated by angiotensin II, increases proton secretion by adjacent clear cells to maintain the acidity of the luminal fluid (Shum et al. 2009). More recently, Pinel and Cyr provided evidence that epididymal basal cell have properties in common with

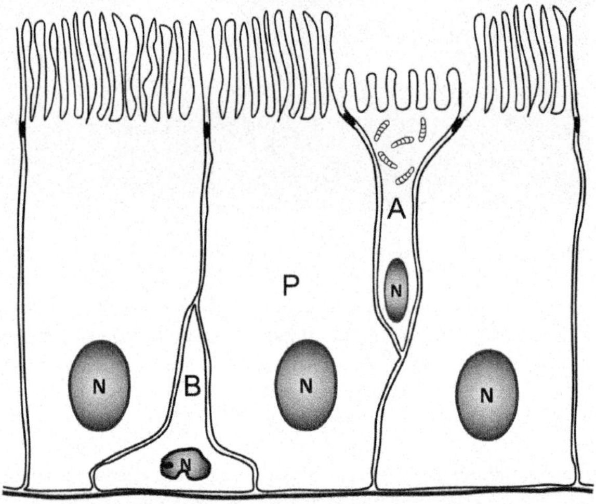

Fig. 3.3 Schematic representation of epithelial cell types in the bull epididymis. The epithelium is comprised primarily of principal cells (P) distinguished by their large size and tall microvilli. Basal cells (B) are less numerous and located at the base of the epithelium in contact with the basement membrane. They have projections that can reach to the lumen and share membrane interdigitations for communication purposes with the neighboring principal cells. The apical cells (A) are located in the apical half of the epithelium and are more sparsely located between the principal cells than basal cells. A distinguishing feature is the wide, apical dome-shaped portion of the apical cell, full of mitochondria. Not shown are the various leucocytes (referred to as halo cells) sparsely dispersed between the principal cells. Clear cells and narrow cells commonly dispersed throughout the murid epididymal epithelium have not been reported to be present in the bull epididymis

stem cells (Pinel and Cyr 2021). The authors isolated an enriched population of rat epididymal basal cells using magnetic bead separation with a CD49f antibody, a biomarker for stem cells. Gene profiling of these isolated basal cells showed numerous transcripts in common with multipotent stem cells. Finally, they demonstrated that these isolated basal cell cultured under 3D conditions can proliferate and differentiate to form organoids, can self-renew and differentiate into other epididymal cell types (i.e., principal cells and clear cells).

In the bull, apical cells are distributed, although sparsely, throughout the epithelium of the caput and corpus, with the exception that they are not found in the initial segment (Goyal 1985), while in murids, they are found primarily in the initial segment. Bull apical cells have a relatively wide dome-shaped cytoplasm close to the lumen, which narrows relatively quickly and incorporates the ovoid nucleus (Fig. 3.3). A distinctive feature is that the wide, apical dome-shaped portion of the cell full of mitochondria. These cells do not appear to contact the basement membrane and are positioned in the apical half of the epididymal epithelium. As for their function, the exclusive localization of carbonic anhydrase activity to this cell (Goyal 1985) suggests that the apical cell promotes hydrogen ion buffering, in this case lowering the pH of the lumen to maintain sperm motility quiescence. Interestingly, in the murid this anhydrase activity was exclusive to the narrow cell (Adamali and Hermo 1996), a cell type not recorded in the bull epithelium. The apical cell also has the ability to endocytose substances and proteins from the lumen and degrade them in lysosomes as revealed by huge buildup of lysosomes in β-hexosaminidase knockout mice (Adamali et al. 1999). In fact, Goyal (Goyal 1985) provided micrographic evidence that bull apical cells are capable of internalizing large sperm components for digestion. Finally, Yf and Yo subunits of glutathione S-transferase-P were immunolocalized in apical cells in murids suggesting that this cell may also be involved in antioxidant defense of spermatozoa (Adamali and Hermo 1996).

Clear cells, not found in the epididymal epithelium of bull or ram but present in murids, are active endocytic cells found distal to the initial segment in increasing numbers from the caput to the cauda (Hermo et al. 1988a; Robaire and Hermo 1988). Furthermore, their endocytic activity is greater than that of adjacent principal cells, especially in the cauda. Murid sperm typically lose their cytoplasmic droplet in the cauda and concurrently the droplet's contents are seen being degraded in endocytic vesicles within clear cells (Hermo et al. 1988a). This suggests that one of the main reasons clear cells are present in the epithelium, and especially numerous in the cauda, is to clean up the debris of the cytoplasmic droplets. It is therefore tempting to speculate that the reason clear cells are not present in the ram and bull is that the extra endocytic activity provided by the clear cells to clean up the cytoplasmic droplets is not required because the majority of bull and ram spermatozoa lose their cytoplasmic droplet during ejaculation (Amann 1987). Phospholipid binding protein (PBP), synthesized in bull ampullae and seminal vesicles, was demonstrated to be the agent for removing cytoplasmic droplets from bull spermatozoa (Matoušek and Kysilka 1980).

3.3 Endocytosis and Secretion of Proteins in the Extratesticular Duct System

The epithelial cells lining the various regions of the male extratesticular duct system have distinct morphological features, reflecting their functions. The rete testis is an anastomotic labyrinth, which in the bull occupies the whole central region of the testis, draining the surrounding seminiferous tubules of mature spermatids, that are diluted in seminiferous tubular fluid (testicular fluid), and transporting them to the efferent ducts (Figs. 3.4, 3.5). The efferent ducts in turn channel the spermatozoa into the convoluted epididymal duct by ciliary action of their epithelium (Fig. 3.4). The anastomotic channel of the rete is lined by a simple cuboidal epithelium, while the efferent duct is lined by a simple columnar epithelium of which the noncilated cell predominates in space and number over the ciliated cell (Figs. 3.5, 3.6). The noncilated cell of the efferent duct and the rete epithelial cell have a well-developed endocytic system when compared to only an adequate 'housekeeping' secretory system, which does not produce dense core secretory granules for export out of the cell (Fig. 3.6). The apical plasma membrane of the rete testis cell displays short microvilli as well as small invaginations or pits (coated or uncoated), which fill by fluid-phase, adsorptive or receptor-mediated endocytosis and pinch off to form small vesicles filled with luminal substances. Hence in the apical cytoplasm these small vesicles fuse with larger empty looking vesicles referred to as endosomes which in turn transform to pale and then dense multi-vesicular bodies (MVBs) and finally form acid phosphatase-positive secondary lysosomes, where degradation of the endocytosed materials occur, thanks to the contribution of hydrolytic enzymes transported from the Golgi apparatus (Figs. 3.6, 3.7A). The nonciliated cell of the efferent duct also has these endocytic features; however, it has a more prominent brush border made up of longer microvilli and tubular, rather than rounded, coated pits which invaginate from the apical membrane, pinch off, and form apical tubules that fuse to endosomes, which transform to MVBs and then secondary lysosomes (Fig. 3.6, 3.7B). These membranous elements comprise the endocytic apparatus of the cell.

Electron-dense tracers injected into the lumen of the rete show a progressive fluid phase incorporation of this material by the endocytic apparatus (i.e., by 2 min in the pits, 5 min in the endosomes, 15–30 min in the MVBs and 1 h in the secondary lysosomes) (Hermo et al. 1994; Morales and Hermo 1983; Morales et al. 1984). Likewise time course studies, following injection of fluid-phase or adsorptive tracers into the lumen of the efferent ducts, show a progressive labeling of the endocytic apparatus such that by 1–2 h, secondary lysosomes are labeled (Hermo and Morales 1984; Hermo et al. 1988b). This demonstrated efficiency of the endocytic system of these epithelial cells, helped explain the observed dramatic decrease in protein concentration of the seminiferous tubular fluid (testicular fluid) as it passes through the rete testis and efferent ducts (Hinton and Keefer 1983; Veeramachaneni et al. 1990). In fact, the latter investigators estimated that more than 85% of water and 65% of bulk protein in rete testis fluid is absorbed between

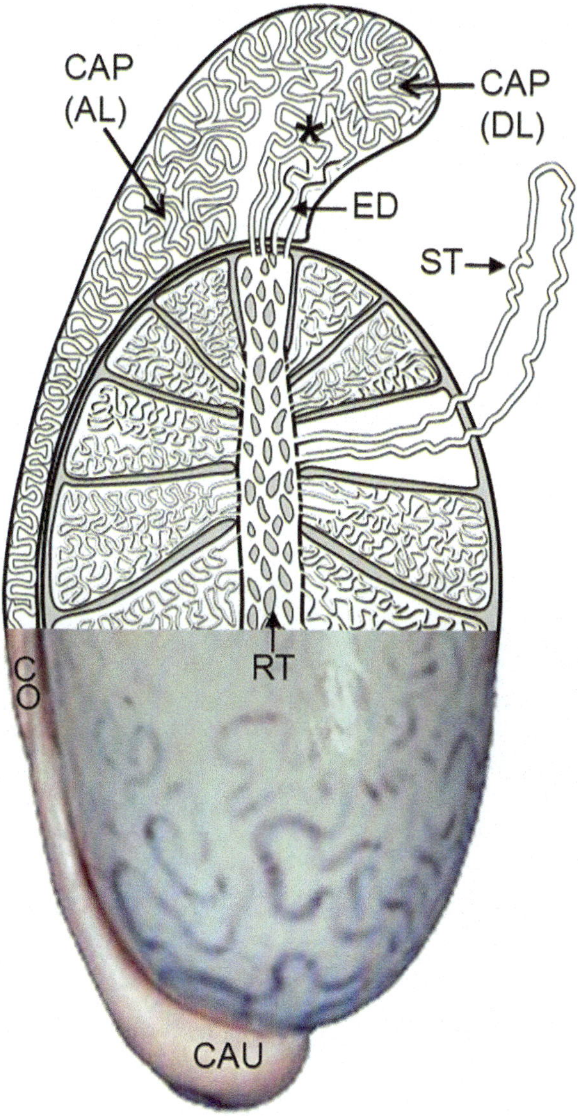

Fig. 3.4 Schematic of the bull testis and extratesticular tract. Spermatozoa in testicular fluid leave the seminiferous tubules (ST) and enter the rete testis (RT), an anastomotic channel that is located centrally in the bull testis. The seminiferous tubules are very long when stretched out but are normally packaged into lobules surrounding the rete testis. At the end of each seminiferous tubule there are two exits leading into the rete. The spermatozoa pass upward through the rete until they reach the top of the testis and enter several tortuous efferent ducts (ED), which drain proximally into the long convoluted duct of the epididymis. The major regions of the epididymis, i.e., the initial segment (asterisk)), caput (CAP), corpus (CO), and cauda (CAU), are indicated. The caput normally sitting flat on the dorsal aspect of the testis has been lifted upwards in this diagram to see its underbelly and thus appreciate the entry of the rete testis into the efferent ducts and the relationship between the descending (DL) and ascending (AL) limbs of the caput

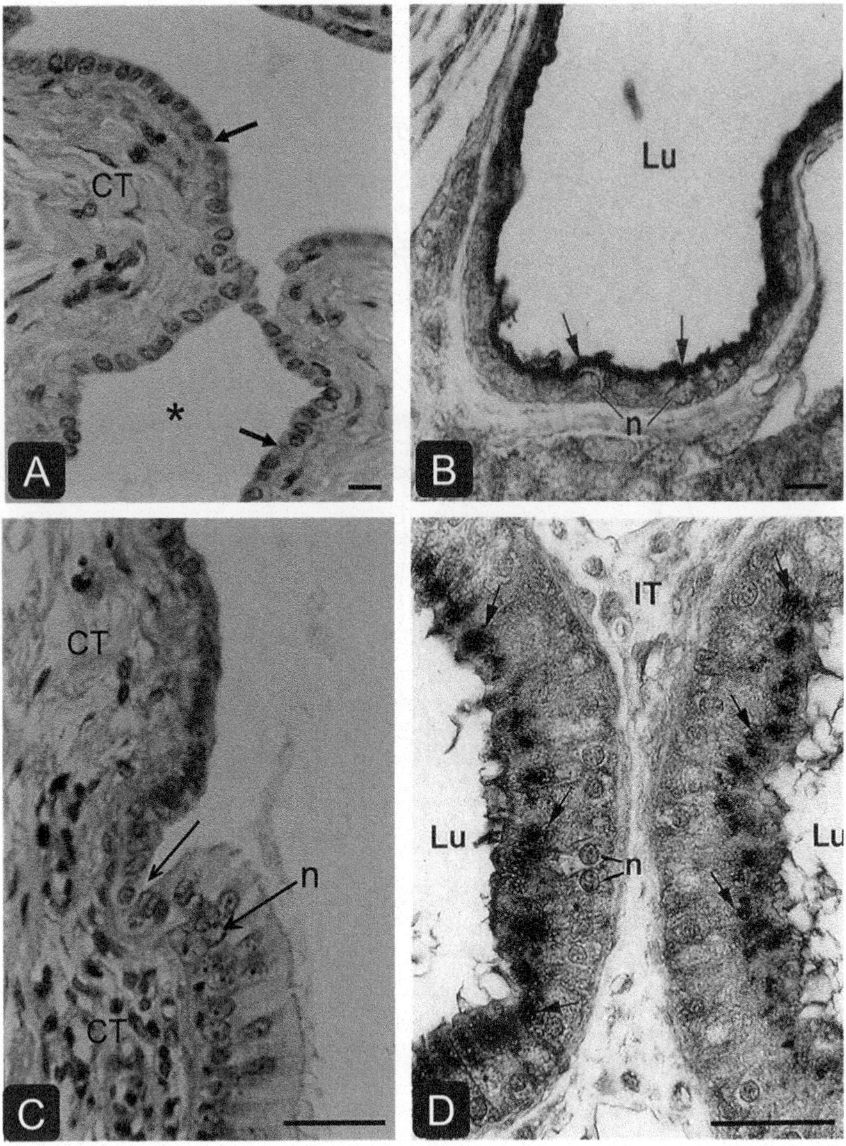
CT
*
A
Lu
n
B
CT
n
CT
C
IT
Lu
n
Lu
D

◀ **Fig. 3.5** Photomicrographs of histological sections through the rete testis and efferent ducts counterstained with methylene blue. **A)** Section through the rete testis illustrating the simple cuboidal epithelium (arrows) lining the anastomotic channel (asterisk) through which the testicular fluid and sperm pass through on their way to the epididymis. CT, connective tissue. Bar = 10 μm. **B)** Section of the rete epithelium immunostained with anti-SGP-2 antibody. The immunoperoxidase reaction product appears as a dense black band over the apical part of the epithelium (arrows). Lu, lumen; n, nuclei. Bar = 10 μm. **C)** Section showing the junction (arrow) between the simple cuboidal epithelium lining the rete testis and the taller simple columnar epithelium lining the efferent duct. CT, connective tissue; n, nuclei of efferent duct. Bar = 20 μm. **D)** Section through two adjacent efferent ducts immunostained with anti-SGP-2 antibody. Unlike the dense continuous immunoperoxidase reaction band over the epithelium of the rete testis (see B), that over the efferent ducts is patchier (arrows), most likely because the ciliated cells found between the noncliated cell do not take part in endocytosis. The strongest immunoreactivity is over the apical and supernuclear regions of the cell. IT, intertubular connective tissue; n, nuclei. Bar = 20 μm. Source: Images B and D reproduced with permission from (Hermo et al. 1991)

the terminal portion of the rete testis and proximal caput of the ram. Importantly, this estimate does not take into account the huge amount of water and protein absorption that occurs in the rete testis. The electron-dense tracer studies above stimulated the question whether proteins found in high concentration in testicular fluid, like SGP-2, and SGP-1, are processed by the endocytic apparatus of the rete and efferent ducts. Sulfated glycoprotein-2 (SGP-2, also known as clusterin or apolipoprotein J) is the major protein secreted by the Sertoli cell into the seminiferous tubular lumen, where it binds to late spermatids (Sylvester et al. 1991). It most likely is involved in the transport of lipids (e. g., cholesterol) to the sperm plasmalemma, thus modifying the surface for further uptake of proteins involved in sperm maturation. Likewise, SGP-1 (or prosaposin, the precursor of saposins) coats the sperm surface and may function in protecting the sperm surface from lysosomal enzymes and/or assisting in the modification of membrane lipids during sperm maturation (Sylvester et al. 1989). Unlike SGP-2, SGP-l can follow two routes: a lysosomal route, occurring from the Golgi apparatus to secondary lysosomes (Morales et al. 1995) and a secretory route, ending in the binding of SGP-l to spermatids in the seminiferous tubules of the testis. The secretory form of SGP-l is in high concentration within the lumen of the rete testis and efferent ducts but decreases significantly by the time the testicular fluid reaches the caput epididymidis (Sylvester et al. 1989).

In the rete testis, immunoperoxide localization of SGP-2 (Hermo et al. 1991) and SGP-1 (Hermo et al. 1992a) revealed a dense reaction product over the apical and supranuclear regions of the rete epithelial cells (Fig. 3.5B). A more detailed electron microscope account by immunogold labeling localized SGP-l and SGP-2 over the various compartments of the endocytic apparatus of these cells (Fig. 3.7A). The endocytosis of SGP-2 and SGP-1 was also coincident with their detachment from spermatozoa during transit through the rete testis and efferent

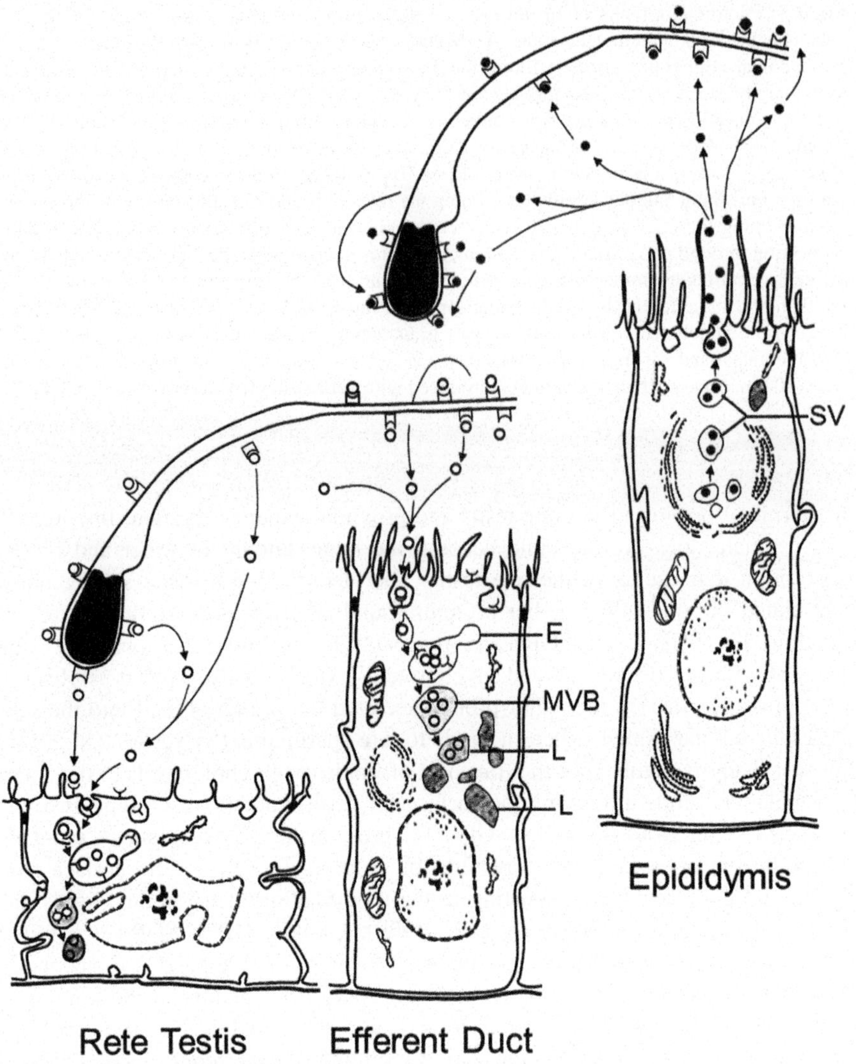
SV
E
MVB
L
L
Epididymis
Rete Testis
Efferent Duct

◀ **Fig. 3.6** Diagram illustrating the endocytosis of testicular SGP-2 and SGP-1by epithelial cells of the rete testis and efferent ducts and secretion of epididymal SGP-2 by principal cells (SGP-1 is not secreted by principal cells) in epididymis. Testicular SGP-2 and SGP-1 (represented as open circles) are secreted by Sertoli cells whereupon they bind to the surface of late spermatids. During transit through the rete testis and efferent ducts, they detach from the sperm and are endocytosed by the cuboidal rete epithelial cells and nonciliated cells of the efferent ducts. Receptors (Y) for testicular SGP- 2 and SGP-1 presumably exist on the apical plasma membrane of these cells. Following their receptor-mediated uptake by invaginations of the cells surface, the receptor-ligand complex appears in endosomes (E), multivesicular bodies (MVB) and secondary lysosomes (L). In the endosome, one may postulate that the receptor and ligand dissociate from one another, with the receptor (Y) being recycled to the apical plasma membrane, while the ligand is destined for degradation in secondary lysosomes. In the epididymis (except for the initial segment), the epithelial principal cells appear to be involved in the secretion of an epididymal SGP-2 (represented as solid black circles) but not of SGP-1. The epididymal SGP- 2 is found in approximately 200—400 nm secretory vesicles of the principal cell next to the Golgi apparatus as well as in the apical region of the cell. Following its secretion, the epididymal SGP-2 binds to the plasma membrane of the sperm especially in the area of its head and principal piece of its tail. Source: Diagram adapted from (Hermo et al. 1991)

ducts, depicted in Fig. 3.6. In the efferent duct, immunoperoxidase localization of SGP-2 (Hermo et al. 1991) and SGP-1 (Hermo et al. 1992a) also revealed a distinct reaction over the apical and supranuclear regions of the nonciliated cells (Fig. 3.5D). This localization corresponded in the electron microscope to an immunogold labeling of SGP-2 and SGP-1 in tubular coated pits, apical tubules, endosomes, MVBs, and secondary lysosomes (Fig. 3.7B). By the time the luminal fluid, including the sperm, arrives in the proximal initial segment region of the epididymis it is nearly devoid of SGP-2 and SGP-1 indicating not only the high efficiency of fluid-phase and adsorptive endocytosis in the rete and efferent ducts, but also of receptor-mediated endocytosis, by which SGP-1 and SGP-2 are most likely internalized. What then is the functional significance of cleansing the spermatozoa of its testicular milieu before reaching the epididymal duct? As will be seen, the 'clean slate' allows for the regional secretory and endocytic activity of the epididymal duct to determine the sperm-milieu and -surface features best suited for the maturation process and hence fertilizing capability of the spermatozoon.

In summary, SGP-2 and SPG-1 are present in fluid and on sperm leaving the testis but are lost from both in the rete testis and efferent ducts, by the efficient endocytic system of the epithelium. Unexpectedly, the sperm is recoated again by an epididymal form of SGP-2 (a lower molecular weight) that reappears in the epididymal fluid from secretions emanating from principal cells that are distal to the initial segment (Fig. 3.6) (Hermo et al. 1991). The strongest secretory activity, as determined by immunocytochemistry, appears in the principal cells of the caput epididymis. The immunoperoxidase staining pattern is characteristically checkerboard-like; some principal cells are intensely stained, some moderately and some unreactive (Fig. 3.8). This is a very typical regional pattern of expression for many

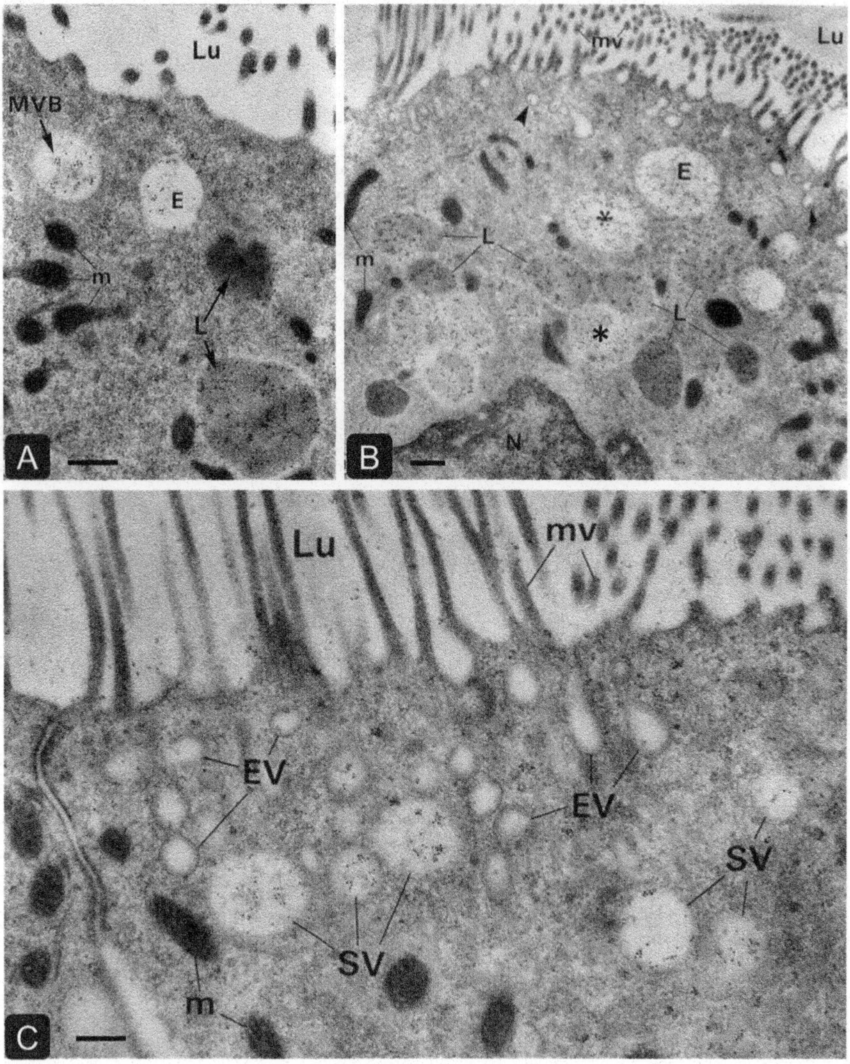

Fig. 3.7 Electron micrographs of the apical regions of rete testis (A), efferent duct (B) and principal cell (C) of the caput epididymis all immunolabelled with anti-SGP-2. **A,B)** Colloidal gold particles, attached to the secondary antibodies, are predominantly found in endocytic structures such as coated pits (arrowheads), endosomes (E), multivesicular bodies (MVB or asterisk*) and secondary lysosomes (L) where the proteins are degraded and resulting amino acids sent for recycling. Lu, Lumen; m, mitochondria; mv, microvilli; N, nucleus. **C)** In the principal cell, on the other hand, colloidal gold particle are predominantly found in the secretory vesicles arising from the Golgi apparatus. The smaller subsurface endocytic vesicles (EV) are virtually empty. Bar = 0.2 μm in all micrographs. Source: Images reproduced with permission from (Hermo et al. 1991)

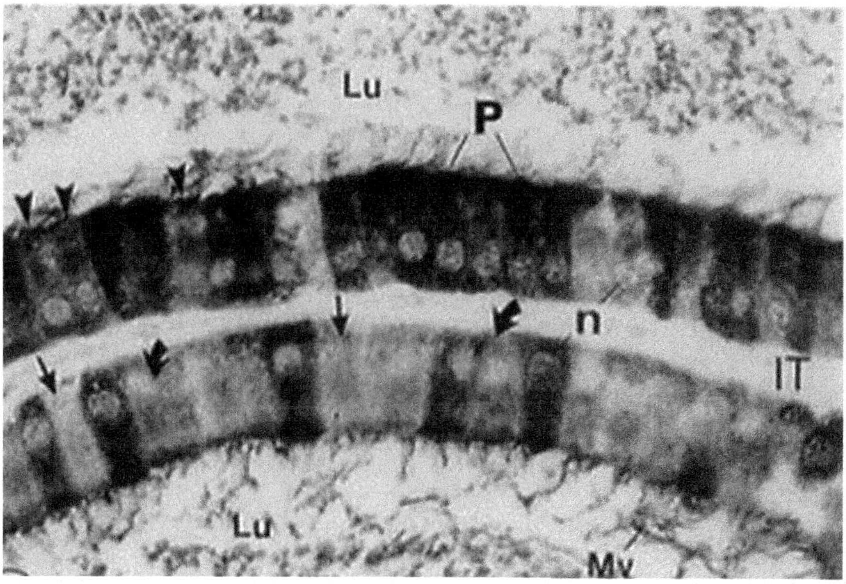

Fig. 3.8 Light micrograph of adjacent epithelia from caput epididymis immunoperoxidase stained with anti-SGP-2. Note the intense staining of some principal cells (P), while others are moderately (arrowheads) or weakly (curved arrowheads) immunoreactive and a few even non-reactive (arrows). The immunostaining extends through the whole cell leaving only the nucleus (n) unstained. Mv, microvilli; Lu, lumen; IT, interstitial connective tissue. Bar = 20 µm. Source: Image reproduced with permission from (Hermo et al. 1991)

proteins secreted by the principal cells (Hermo et al. 1992b). As for immunogold labelling at the ultrastructural level, SGP-2 is localized to the cisternae of rough endoplasmic reticulum in the basal region of the cell, to forming secretory vesicles on the Trans aspect of the Golgi apparatus and to mature apical secretory vesicles, which after fusing with the apical plasma membrane release their contents into the lumen (Figs. 3.6, 3.7C). On the other hand, there is no evidence that SGP-1 (prosaposin) repopulates the epididymal fluid and recoats the sperm from principal cell secretions in the epididymis. Rather, SGP-1 is glycosylated in the Golgi apparatus and processed and transported to MVBs and lysosomes as four saposins involved in the degradation of ingested lipids (Hermo et al. 1992a; Morales et al. 1995).

Principal cells can deliver proteins into the epididymal lumen by either merocine or apocrine secretion. As reviewed above, SGP-2 is an example of a protein delivered to the lumen by merocrine secretion. The protein is translated from mRNA on ribosomes of the rough endoplasmic reticulum (ER), entering the lumen of the ER by means of an N-terminal signal peptide sequence. From the ER the protein is transferred into the saccules of the Golgi apparatus for glycosylation and is finally packaged into secretory vesicles on the Trans aspect of the Golgi apparatus for export out of the cell. Once in the lumen of the duct the glycoprotein normally interacts with the surface of the epithelium, or in this particular case with the sperm surface, bestowing on it various properties that facilitate the maturation process (Fig. 3.6). Apocrine secretion, on the other hand, involves the protrusion of apical blebs (AB) of cytoplasm from between the microvilli of the brush border of the principal cell into the lumen of the duct (Fig. 3.9). These apical blebs eventually detach from the surface of the principal cell and breakdown in the lumen releasing a selective content of cytoplasmic components into the lumen (Aumüller et al. 1999; Hermo and Jacks 2002). The contents are made up of free ribosomes, small vesicles (referred by some researchers as epididymosomes), a few smooth ER cisternae and cytosol with a variety of proteins, incapable of crossing membranes as they lack the N-terminal signal peptide sequence. The contents of the apical blebs are unlike the apical cell cytoplasm of the principal cell that contains endosomes, vesicles of different sizes, lysosomes, and mitochondria indicating that a segregation process is involved in selecting the content of the apical blebs. This segregation process is exquisitely captured in a series of ultrastructural figures featured in a definitive article on this subject (Hermo and Jacks 2002). It suggests that there may be specific functions in sperm maturation for the contents of the apical blebs whose proteins are synthesized on free ribosomes and normally remain in the cytosolic compartment of the cell. Two such proteins localized to the apical blebs by Hermo and Jacks are ubiquitin and the glutathione-S-transferase-Ya subunit, suggesting that the former, is used in ubiquinating sperm- and luminal-breakdown proteins for degradation (Sutovsky et al. 2001) and the latter, protects the lumen and sperm from harmful electrophiles.

Hales et. al. (Hales et al. 1980) were the first to show that the epididymis contains a very high glutathione-S-transferase (GST) activity and that there is a variety of transferases differentially localized along its length. Immuno-localization of members of three GST gene families substantiated this finding (Papp et al. 1995; Veri et al. 1993, 1994), the point being that all these GST members within the principal cells have potential access to the epididymal lumen via apocrine secretion and thus can protect the sperm from reactive oxygen species (ROS). As discussed in the next subsection some antioxidant enzymes that are released by apocrine secretion even bind permanently to the sperm plasmalemma, insuring long lasting antioxidant protection for the sperm on their journey through the epididymis and the female reproductive tract.

In an excellent review, on the characteristics of the epididymal luminal environment responsible for sperm maturation and storage, a list of 20 genes is provided, whose ablation leads to the dysregulation of the intraluminal epididymal environment resulting in subfertile or infertile sperm (Zhou et al. 2018). The expression products of these genes are involved in the regulation of osmolality, acidity/alkalinity, lipid homeostasis/transport, passive water transport, levels of ROS, absorption/secretion ratios, or other physiological roles. For example, male mice lacking estrogen receptor-alpha (ESR1KO) are infertile due to a hypo-osmotic epididymal environment (Joseph et al. 2010b) and failure in luminal acidification

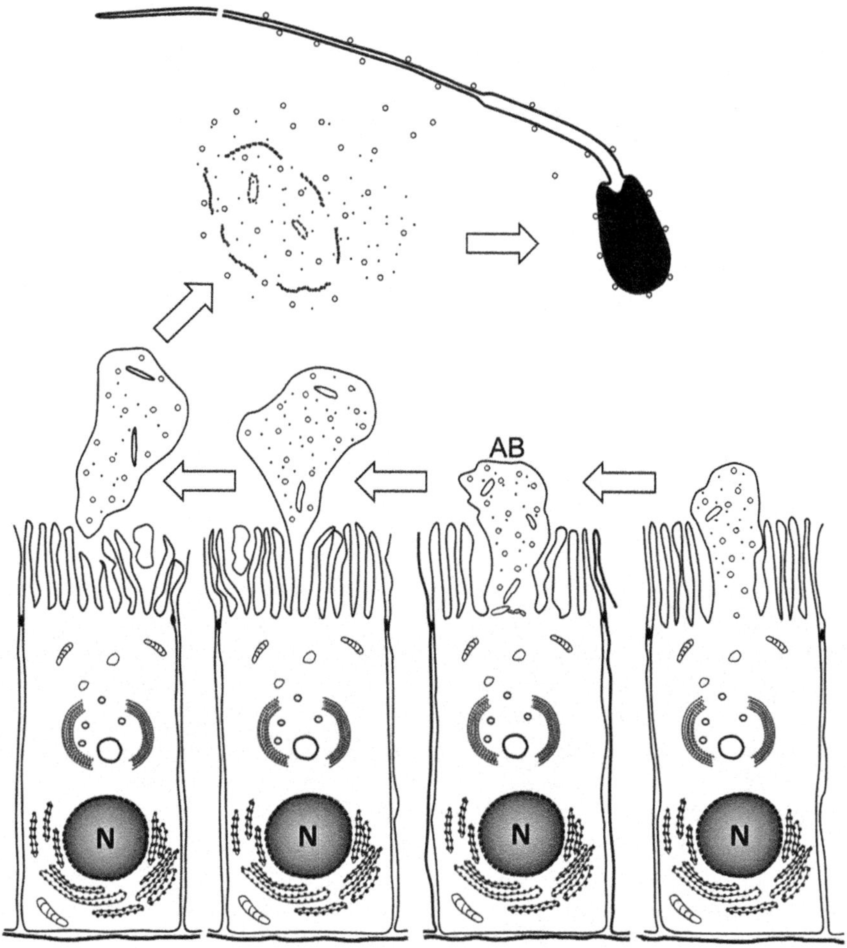

Fig. 3.9 Illustration of the process of apocrine secretion in epididymal principal cells (follow the direction of the arrows). The process begins with the protrusion of an apical cytoplasmic bleb (AB) from the principal cell towards the lumen, followed by a gradual detachment from the cell and finally a breakdown of the AB within the lumen. Of particular importance is that when the membrane surrounding the AB breaks down it releases its segregated cargo of cytosolic proteins and small membrane vesicles (circles) that have the potential of conditioning the milieu and binding to the sperm membrane. Presumably, the small vesicles are the source of epididymosomes, although this has not been directly proven. The liberty is taken in this drawing to show the small vesicles (circles) making contact and interacting with the surface of the bull sperm, for deposition of proteins such as glycosylphosphatidyl inositol anchored proteins, which are essential for sperm maturation. Source: Diagram modified from (Hermo and Jacks 2002)

(Joseph et al. 2010a). This dysregulation of the luminal environment contributed to sperm tail coiling, immotility and a significant increase in spontaneous acrosome reactions. The sperm tail coiling has similarities to the distal midpiece reflex defect seen in Fig. 5.20 of chapter 5 and differs from the Dag defect (Chap. 5 in Fig. 5.26) as there appears to be no axonemal filament breaking. A large part of the phenotype is due to the failure of the efferent ducts to execute their absorptive function (Hess et al. 1997), which together with alterations in ion and fluid transporters within the epididymis lead to a dilution of downstream fluid and significant changes in the luminal fluid composition. Importantly, Hess and coinvestigators show that normal offspring result from transplanting Esr1KO germ cells into wild-type testes, providing proof that the sperm abnormalities do indeed result from a defective intraluminal environment. Furthermore, incubating Esr1KO sperm in a series of solutions of increasing osmolality, resembling the wild-type-like osmotic environment, significantly reduces the abnormal flagellar coiling.

3.4 Epididymal Sperm Protein Transfer

As reviewed by Dacheax and co-investigators, sperm maturation and hence fertilization capability is dependent on sequential modifications that occur on the sperm surface (Dacheux et al. 2009). As already alluded to in the above section, during epididymal transit immature spermatozoa progressively lose a large number of their testicular surface proteins only to gain new ones or modifications (e. g., glycosylations) of existing ones in various plasma membrane domains located throughout the sperm. Basically there are three potential ways by which the sperm plasma membrane can acquire new glycoconjugates during epididymal transit: (1) by epididymal epithelium secretions, (2) by extrinsic glycosylation, and (3) by intrinsic glycosylation (Hermo et al. 1994). The mechanisms for the latter two ways still require verification, while the detection of apocrine secretion in the epididymis helps explain how proteins lacking a N-terminal peptide signal sequence for crossing membranes can be secreted into the lumen and end up associating with the sperm plasmalemma. Furthermore, the demonstration of the transfer of glycosylphosphatidyl inositol (GPI) anchored proteins from epididymosomes (released by apocrine secretion) to lipid rafts on bull sperm plasmalemma (Sullivan et al. 2007) provide an explanation on how the transfer of peripheral membrane proteins in vivo from one cell to another can be accomplished. A large-scale proteomic profiling study of mouse epididymosomes by Nixon and coinvestigators demonstrated that these vesicles are capable of transferring protein cargo to different regions of the maturing sperm, supporting their selective role in coordinating post-testicular sperm maturation (Nixon et al. 2019). Below is an inexhaustive list and evaluation of important bull sperm surface proteins acquired during epididymal passage, many of which are conveyed to the sperm by apocrine secretion.

3.4.1 Forward Motility Protein

Bovine sperm bind a caput-derived protein, coined forward motility protein, that acts in concert with cyclic AMP to produce progressive motility (Acott and Hoskins 1981). Motility is the most obvious observed change in sperm related to acquisition of fertility. Usually, in most species, the sperm tail beat only involves the end-piece in the caput and then gradually propagates up the tail to incorporate the principal piece in the corpus and finally the mid-piece in the cauda. In other words, this motility 'priming process' is not complete until the sperm reaches the distal end of the corpus.

3.4.2 Acrosome Stabilizing Factor (ASF)

ASF reversibly decapacitates sperm and blocks the acrosome reaction in vitro (Eng and Oliphant 1978). ASF is synthesized and secreted by the corpus, where it interacts with the sperm membrane to stabilize the acrosome for storage in the cauda (Thomas et al. 1984).

3.4.3 Immobilin

Immobilin mechanically immobilizes sperm in the cauda epididymis as a result of its viscoelastic properties (Usselman et al. 1985), and in addition serves to protect sperm from shearing at the time of ejaculation (Cardullo and Cone 1986). Light and electron microscope immunocytochemistry showed that a differential secretory pattern exists between epididymal segments with maximal secretory activity of immobilin occurring in the intermediate zone and proximal caput epididymis, while no secretion takes place in the cauda epididymis (Hermo et al. 1992b). Immobilin must be diluted out from the cauda sperm sample for the sperm to become fully motile.

3.4.4 Adam7

A disintegrin and metalloprotease (ADAM) 7 is expressed specifically in the epididymis and transferred to the sperm surface during epididymal transit. It is secreted to the epididymal lumen in an apocrine manner in association with membranous vesicles known as epididymosomes and transferred directly from these vesicles to sperm plasma membrane (Oh et al. 2009).

3.4.5 CD52

CD52 is a glycosylphosphatidyl inositol (GPI) anchored lymphocyte antigen found over the postacrosomal region and tail of the sperm. Correlation between percentages of CD52- immunostained human spermatozoa with percentages of motile cells suggests its involvement in maturation of sperm function (Yeung et al. 1997). It was suggested by Kirchoff and Hale, (Kirchhoff and Hale 1996) that the transfer of peripheral membrane proteins in vivo from one cell to another (in the case of CD52, from the principal cell to the sperm surface) may be a unique feature of lipid anchored membrane proteins.

3.4.6 PH20/SPAM1

Bull sperm possess two isoforms of SPAM1/PH20 (Morin et al. 2010). One is epididymal in origin, expressed at the surface of the postacrosomal region, which may facilitate hyaluronidase catalysis of hyaluronic acid in the cumulus surrounding the egg during sperm penetration. The second is testicular in origin, which associates with the outer acrosomal membrane and may mediate adhesion with the zona pellucida (ZP) during the initial stages of acrosome exocytosis, much as zonadhesion is proposed to act (Bi et al. 2003). Based on relatively newer evidence in the mouse, which suggests that acrosomal exocytosis occurs during penetration of the cumulus (Jin et al. 2011), it is possible that the resultant hybrid vesicles forming the acrosomal shroud also exhibit extracellular hyaluronidase activity on their surface and aid in the penetration of the cumulus along with the sperm surface SPAM1. Interestingly, mouse sperm lacking SPAM1 are still capable of penetrating the cumulus mass despite a delayed dispersal of cumulus cells (Baba et al. 2002). However, a 55-kDa hyaluronidase isoform (HYAL5) was found in both wild-type and PH-20-deficient mouse sperm and shown that it could adequately compensate as a 'cumulus matrix depolymerase' in place of PH20 (Kim et al. 2005). It is generally agreed that SPAM1in most species is anchored to membranes by a Glycosylphosphatidyl inositol (GPI)-link (Cherr et al. 2001; Zhang et al. 2004), however, intriguingly in the bull, SPAM1 appears to lack a GPI anchor. Instead it contains a transmembrane domain that separates the hyaluronidase activity from the zona pellucida binding sites (Lalancette et al. 2001; Morin et al. 2005, 2010). Epididymal SPAM1 was suggested as a useful biomarker for sperm maturation as its uptake in Spam1-null sperm, exposed to W/T sperm-free epididymal luminal fluid, significantly increases cumulus penetration (Chen et al. 2006).

3.4.7 P25b

P25b is a GPI-anchored protein transferred from the epididymosomes to the plasma membrane in the bull sperm acrosomal region (Frenette and Sullivan 2001; Lessard et al. 2000; Parent et al. 1999). Its orthologs have been described in hamster (P26h), human (P34H), mice (P25m) and cynomologus monkey (P31m) (Bégin et al. 1995; Boué and Sullivan 1996; Lamontagne et al. 2001; Sullivan and Bleau 1985) with a similar subcellular localization (Légaré et al. 1999) and purported function of sperm-zona pellucida binding (Bérubé and Sullivan 1994). However, a word of caution regarding P25b's sperm zona-binding role. During natural fertilization in mice, spermatozoa already undergo the acrosome reaction and release their acrosomal contents in the cumulus oophorus making sperm plasmalemma-zona binding irrelevant during natural fertilization (Jin et al. 2011). If this is proven to be the case in bovine fertilization, then infertility associated with lower levels of P25b (Parent et al. 1999) could **not** be due to a defective plasmalemma-zona interaction. Based on their finding that P25b's interaction with the plasmalemma of spermatozoa increases from the corpus to the cauda epididymis and that lower levels of this interaction are correlated with bull subfertility, Frenette and Sullivan (Frenette and Sullivan 2001) propose P25b as a biomarker for sperm epididymal maturation in bulls.

3.4.8 Antioxidant Enzymes

Mammalian spermatozoa function at the knife's edge of oxidative regulation. They must maintain levels of reactive oxygen species (ROS) low enough to prevent against oxidative damage but high enough to act as a signaling requirement for capacitation and hence successful fertilization (de Lamirande and Gagnon 1993; O'Flaherty et al. 1999). The high proportion of polyunsaturated fatty acids within the lipid bilayer structure of the plasmalemma make spermatozoa particularly vulnerable to oxidative damage during their transit through the epididymis and female reproductive tract. As a result, spermatozoa must equip themselves with a highly strategic antioxidant defense system during this journey. Glutathione-s-transferase (GST) activity is predominately characterized on the plasma membrane of spermatozoa, with isoforms of GSTM, GSTP and GSTO subfamilies identified as cytoplasmic membrane residents, reviewed by (Hamilton et al. 2022). Glutathione peroxidases (GPx), which consist of eight major subclasses, are also major scavengers of free radicals that facilitate the protection of cells against oxidative stress and lipid peroxidation. Many of these subclasses are found associated with the epithelium of the epididymis but one in particular, GPX5, is secreted into the luminal compartment where it loosely interacts with sperm cells. The features of two antioxidant enzymes, GSTO2 and GPX5, both associated with the sperm plasmalemma in the epididymis are summarized below.

3.4.9 *GSTO2*

Glutathione-s-transferase omega 2 (GSTO2) is one of two functionally active enzymes within the Omega class of GST enzymes. It is present on the entire surface of epididymal spermatozoa in mouse, bull, and boar but not on the surface of precursor spermatids (Hamilton et al. 2019). GSTO2 is secreted by the principal cells throughout the epididymis and most likely finds its way to the epididymal lumen and sperm through the apocrine pathway since it is a non-glycosylated cytosolic protein. Equipped with a cysteine residue at their active site, GSTOs are not only able to facilitate glutathione reductase reactions but also dehydroascorbate reductase capabilities. Through the replenishment of antioxidant molecules such as ascorbic acid (vitamin C) and glutathione, which are plentiful in the epididymal lumen and seminal secretions, GSTO2 participates in the regulation of the oxidative environment of the sperm plasmalemma during maturation and perhaps more importantly during capacitation (Hamilton et al. 2019).

3.4.10 *GPX5*

GPX5 is a member of the glutathione peroxidase family. It is secreted into the caput epididymis lumen and a small part of this secretion becomes associated with the sperm plasmalemma in the acrosome region (Rejraji et al. 2002). In the lumen of the epididymis, as a free soluble protein or attached to membrane structures/vesicles including the sperm membrane, it most likely participates in protecting the sperm from oxidative stress. Mouse knockout model of GPx5 were fertile, however, an increased incidence of late developmental defects occurred in embryos conceived with spermatozoa from aged Gpx5$^{-/-}$ mice (Chabory et al. 2009).

3.5 Conclusion

In a proteomic study of human epididymal secretory proteins that were detected by immunocytochemistry, 174 of 408 secretory proteins were identified that bound to the sperm surface in 15 different regional ways (Li et al. 2010). The secretory-sperm located proteins were on the plasmalemma covering either the acrosome, equatorial region, postacrosomal sheath, neck, midpiece, annulus, principal piece, end-piece or combinations of these locations. This study indicates that the size of not only the human epididymal secretory-sperm located protein group, but probably those of bull and other species, is approximately fivefold larger than was previously indicated in the literature. However, the challenge remains in providing these newly discovered epididymal secretory-sperm proteins a specific function during maturation and consequently during fertilization or embryo development. In this context, it is easier to show that infertility or subfertility result from inactivating a protein than determining the defective mechanism or process involved. As emphasized by Dacheax et. al. (Dacheux et al. 2009, 2016) significant differences exists between mammalian species either in the sequential maturation changes in the luminal proteome or sperm surface proteins. The reason for this is that each species appears to have evolved its own strategy for sperm maturation, which most likely reflects subtle differences in the fertilization process between species.

There is no doubt that epididymal and accessory sex gland secretions, that surround and are bound to the sperm after ejaculation, prepare spermatozoa to withstand the physiological challenges of the journey through the female reproductive tract, while retaining their fertilization competence and preventing premature capacitiation. For example, binder of sperm (BSP) proteins, expressed in bull seminal vesicles and murid epididymis, are deemed essential for natural fertilization. They are implicated as molecular chaperones in maintaining sperm motility and viability, in the formation of the oviductal sperm reservoir and in promoting capacitation in the oviduct (Plante et al. 2016). For a full appreciation of this remarkable journey we recommend the review by Miller, entitled, 'The epic journey of sperm through the female reproductive tract' (Miller 2018).

Compliance with Ethical Standards
All applicable international, national and institutional guidelines for the care and use of animals were followed. This current work was supported by the Natural Science and Engineering Research Council of Canada (Grant number RGPIN/05305) to Richard Oko at Queen's University, along with the Agriculture and Food Research Initiative Competitive Grant number 2023-67015-39262 from the USDA National Institute of Food and Agriculture, Animal Breeding, Genetics & Genomics Program and seed funding from the College of Agriculture, Food and Natural Resources of the University of Missouri to Peter Sutovsky. Work for this book was also supported by a 2021–2022 Lalor Foundation Postdoctoral Research Fellowship to Lauren Hamilton. The research support to Viv Perry was funded by Australian Research Council (Grant number LP110100649), English Beef and Lamb Executive a division of the Agriculture and Horticulture Development Board (Grant number RC3815), Queensland Sperm Morphology Laboratories, Ridley AgriProducts Pty Ltd and the Australian cattle industry. The Canadian Association of Animal Breeders provided the foundation for research by Albert Barth. They supported the Development of the Diagnostic Spermatology Laboratory at the Western College of Veterinary Medicine, Saskatoon, SK., along with related support for numerous other research directives. Many grants were also received from Saskatchewan Agriculture, including the Cattle Marketing Deductions Fund, the Horned Cattle Trust Fund, the Cattle Breeders Research Council, the Beef Development Fund and the Agriculture Development Fund. Animal research approval for research or clinical studies summarized in these chapters were given by respective University Animal Care Committees. The

authors declare no conflict of interest as there is no perceived interest in prejudicing the impartiality of the research reported.

References

Acott TS, Hoskins DD (1981) Bovine sperm forward motility protein: binding to epididymal spermatozoa. Biol Reprod 24(2):234–240. https://doi.org/10.1095/biolreprod24.2.234

Adamali HI, Hermo L (1996) Apical and narrow cells are distinct cell types differing in their structure, distribution, and functions in the adult rat epididymis. J Androl 17(3):208–222

Adamali HI, Somani IH, Huang JQ, Mahuran D, Gravel RA, Trasler JM, Hermo L (1999) I. Abnormalities in cells of the testis, efferent ducts, and epididymis in juvenile and adult mice with beta-hexosaminidase A and B deficiency. J Androl 20(6):779–802

Amann RP (1987) Function of the epididymis in bulls and rams. J Reprod Fertil Suppl 34:115–131

Aumüller G, Wilhelm B, Seitz J (1999) Apocrine secretion–fact or artifact? Ann Anat 181(5):437–446. https://doi.org/10.1016/s0940-9602(99)80020-x

Baba D, Kashiwabara S, Honda A, Yamagata K, Wu Q, Ikawa M, Okabe M, Baba T (2002) Mouse sperm lacking cell surface hyaluronidase PH-20 can pass through the layer of cumulus cells and fertilize the egg. J Biol Chem 277(33):30310–30314. https://doi.org/10.1074/jbc.M204596200

Bedford JM (1975) Matutation, transpoert and fate of spermatozoa in the epididymis. In: Greep RO, B. AE (eds) Handbook of Physiology, Section 7: endocrinology, Male Reproductive System, vol 5. American Physiology Society, Washington, D.C., pp 303–317

Bedford JM (1978) Influence of abdominal temperature on epididymal function in the rat and rabbit. Am J Anat 152(4):509–521. https://doi.org/10.1002/aja.1001520405

Bégin S, Bérubé B, Boué F, Sullivan R (1995) Comparative immunoreactivity of mouse and hamster sperm proteins recognized by an anti-P26h hamster sperm protein. Mol Reprod Dev 41(2):249–256. https://doi.org/10.1002/mrd.1080410216

Bérubé B, Sullivan R (1994) Inhibition of in vivo fertilization by active immunization of male hamsters against a 26-kDa sperm glycoprotein. Biol Reprod 51(6):1255–1263. https://doi.org/10.1095/biolreprod51.6.1255

Bi M, Hickox JR, Winfrey VP, Olson GE, Hardy DM (2003) Processing, localization and binding activity of zonadhesin suggest a function in sperm adhesion to the zona pellucida during exocytosis of the acrosome. Biochem J 375(Pt 2):477–488. https://doi.org/10.1042/bj20030753

Boué F, Sullivan R (1996) Cases of human infertility are associated with the absence of P34H an epididymal sperm antigen. Biol Reprod 54(5):1018–1024. https://doi.org/10.1095/biolreprod54.5.1018

Campbell K (2007) Manipulation of the reproductive system of feral goats (Capra hircus) to increase the efficacy of Judas goats: field methods utilising tubal sterilisation, abortion, hormone implants and epididymectomy.

Cardullo RA, Cone RA (1986) Mechanical immobilization of rat sperm does not change their oxygen consumption rate. Biol Reprod 34(5):820–830. https://doi.org/10.1095/biolreprod34.5.820

Chabory E, Damon C, Lenoir A, Kauselmann G, Kern H, Zevnik B, Garrel C, Saez F, Cadet R, Henry-Berger J, Schoor M, Gottwald U, Habenicht U, Drevet JR, Vernet P (2009) Epididymis seleno-independent glutathione peroxidase 5 maintains sperm DNA integrity in mice. J Clin Invest 119(7):2074–2085. https://doi.org/10.1172/jci38940

Chen H, Griffiths G, Galileo DS, Martin-DeLeon PA (2006) Epididymal SPAM1 is a marker for sperm maturation in the mouse. Biol Reprod 74(5):923–930. https://doi.org/10.1095/biolreprod.105.048587

Cherr GN, Yudin AI, Overstreet JW (2001) The dual functions of GPI-anchored PH-20: hyaluronidase and intracellular signaling. Matrix Biol 20(8):515–525. https://doi.org/10.1016/s0945-053x(01)00171-8

Dacheux JL, Belleannée C, Jones R, Labas V, Belghazi M, Guyonnet B, Druart X, Gatti JL, Dacheux F (2009) Mammalian epididymal proteome. Mol Cell Endocrinol 306(1–2):45–50. https://doi.org/10.1016/j.mce.2009.03.007

Dacheux JL, Dacheux F, Druart X (2016) Epididymal protein markers and fertility. Anim Reprod Sci 169:76–87. https://doi.org/10.1016/j.anireprosci.2016.02.034

de Lamirande E, Gagnon C (1993) A positive role for the superoxide anion in triggering hyperactivation and capacitation of human spermatozoa. Int J Androl 16(1):21–25. https://doi.org/10.1111/j.1365-2605.1993.tb01148.x

Eng LA, Oliphant G (1978) Rabbit sperm reversible decapacitation by membrane stabilization with a highly purified glycoprotein from seminal plasma. Biol Reprod 19(5):1083–1094. https://doi.org/10.1095/biolreprod19.5.1083

Frenette G, Sullivan R (2001) Prostasome-like particles are involved in the transfer of P25b from the bovine epididymal fluid to the sperm surface. Mol Reprod Dev 59(1):115–121. https://doi.org/10.1002/mrd.1013

Goyal HO (1985) Morphology of the bovine epididymis. Am J Anat 172(2):155–172. https://doi.org/10.1002/aja.1001720205

Hales BF, Hachey C, Robaire B (1980) The presence and longitudinal distribution of the glutathione S-transferases in rat epididymis and vas deferens. Biochem J 189(1):135–142. https://doi.org/10.1042/bj1890135

Hamilton LE, Oko R, Miranda-Vizuete A, Sutovsky P (2022) Sperm Redox System Equilibrium: implications for Fertilization and Male Fertility. Adv Exp Med Biol 1358:345–367. https://doi.org/10.1007/978-3-030-89340-8_15

Hamilton LE, Zigo M, Mao J, Xu W, Sutovsky P, O'Flaherty C, Oko R (2019) GSTO2 Isoforms Participate in the Oxidative Regulation of the Plasmalemma in Eutherian Spermatozoa during Capacitation. Antioxidants (Basel) 8(12). https://doi.org/10.3390/antiox8120601

Hermo L, Dworkin J, Oko R (1988a) Role of epithelial clear cells of the rat epididymis in the disposal of the contents of cytoplasmic droplets detached from spermatozoa. Am J Anat 183(2):107–124. https://doi.org/10.1002/aja.1001830202

Hermo L, Spier N, Nadler NJ (1988b) Role of apical tubules in endocytosis in nonciliated cells of the ductuli efferentes of the rat: a kinetic analysis. Am J Anat 182(2):107–119. https://doi.org/10.1002/aja.1001820202

Hermo L, Jacks D (2002) Nature's ingenuity: bypassing the classical secretory route via apocrine secretion. Mol Reprod Dev 63(3):394–410. https://doi.org/10.1002/mrd.90023

Hermo L, Morales C (1984) Endocytosis in nonciliated epithelial cells of the ductuli efferentes in the rat. Am J Anat 171(1):59–74. https://doi.org/10.1002/aja.1001710106

Hermo L, Morales C, Oko R (1992a) Immunocytochemical localization of sulfated glycoprotein-1 (SGP-1) and identification of its transcripts in epithelial cells of the extratesticular duct system of the rat. Anat Rec 232(3):401–422. https://doi.org/10.1002/ar.1092320310

Hermo L, Oko R, Morales CR (1994) Secretion and endocytosis in the male reproductive tract: a role in sperm maturation. Int Rev Cytol 154:106–189

Hermo L, Oko R, Robaire B (1992b) Epithelial cells of the epididymis show regional variations with respect to the secretion of endocytosis of immobilin as revealed by light and electron microscope immunocytochemistry. Anat Rec 232(2):202–220. https://doi.org/10.1002/ar.1092320206

Hermo L, Robaire B (2002) Epididymal cell types and their functions. In: Robaire B, T. HB (eds) The Epididymis: from Molecules to Clinical Practice. Plenum Press, New York, pp 81–102

Hermo L, Wright J, Oko R, Morales CR (1991) Role of epithelial cells of the male excurrent duct system of the rat in the endocytosis or secretion of sulfated glycoprotein-2 (clusterin). Biol Reprod 44(6):1113–1131. https://doi.org/10.1095/biolreprod44.6.1113

Hess RA, Bunick D, Lee KH, Bahr J, Taylor JA, Korach KS, Lubahn DB (1997) A role for oestrogens in the male reproductive system. Nature 390(6659):509–512. https://doi.org/10.1038/37352

Hinton BT, Keefer DA (1983) Evidence for protein absorption from the lumen of the seminiferous tubule and rete of the rat testis. Cell Tissue Res 230(2):367–375. https://doi.org/10.1007/bf00213810

Jin M, Fujiwara E, Kakiuchi Y, Okabe M, Satouh Y, Baba SA, Chiba K, Hirohashi N (2011) Most fertilizing mouse spermatozoa begin their acrosome reaction before contact with the zona pellucida during in vitro fertilization. Proc Natl Acad Sci U S A 108(12):4892–4896. https://doi.org/10.1073/pnas.1018202108

Joseph A, Hess RA, Schaeffer DJ, Ko C, Hudgin-Spivey S, Chambon P, Shur BD (2010a) Absence of estrogen receptor alpha leads to physiological alterations in the mouse epididymis and consequent defects in sperm function. Biol Reprod 82(5):948–957. https://doi.org/10.1095/biolreprod.109.079889

Joseph A, Shur BD, Ko C, Chambon P, Hess RA (2010b) Epididymal hypo-osmolality induces abnormal sperm morphology and function in the estrogen receptor alpha knockout mouse. Biol Reprod 82(5):958–967. https://doi.org/10.1095/biolreprod.109.080366

Kim E, Baba D, Kimura M, Yamashita M, Kashiwabara S, Baba T (2005) Identification of a hyaluronidase, Hyal5, involved in penetration of mouse sperm through cumulus mass. Proc Natl Acad Sci U S A 102(50):18028–18033. https://doi.org/10.1073/pnas.0506825102

Kirchhoff C, Hale G (1996) Cell-to-cell transfer of glycosylphosphatidylinositol-anchored membrane proteins during sperm maturation. Mol Hum Reprod 2(3):177–184. https://doi.org/10.1093/molehr/2.3.177

Koefoed-Johnsen HH (1960) Influence of ejaculation frequency on the time required for sperm formation and epididymal passage in the bull. Nature 185:49–50. https://doi.org/10.1038/185049a0

Lalancette C, Dorval V, Leblanc V, Leclerc P (2001) Characterization of an 80-kilodalton bull sperm protein identified as PH-20. Biol Reprod 65(2):628–636. https://doi.org/10.1095/biolreprod65.2.628

Lamontagne N, Légaré C, Gaudreault C, Sullivan R (2001) Identification and characterization of P31m, a novel sperm protein in Cynomolgus monkey (Macaca fascicularis). Mol Reprod Dev 59(4):431–441. https://doi.org/10.1002/mrd.1050

Légaré C, Bérubé B, Boué F, Lefièvre L, Morales CR, El-Alfy M, Sullivan R (1999) Hamster sperm antigen P26h is a phosphatidylinositol-anchored protein. Mol Reprod Dev 52(2):225–233. https://doi.org/10.1002/(sici)1098-2795(199902)52:2<225::Aid-mrd14>3.0.Co;2-m

Lessard C, Parent S, Leclerc P, Bailey JL, Sullivan R (2000) Cryopreservation alters the levels of the bull sperm surface protein P25b. J Androl 21(5):700–707

Li J, Liu F, Wang H, Liu X, Liu J, Li N, Wan F, Wang W, Zhang C, Jin S, Liu J, Zhu P, Liu Y (2010) Systematic Mapping and Functional Analysis of a Family of Human Epididymal Secretory Sperm-Located Proteins*. Mol Cell Proteomics 9(11):2517–2528. https://doi.org/10.1074/mcp.M110.001719

Matoušek J, Kysilka Č (1980) The haemolytic factor (phospholipid-binding protein) in the bull reproductive tract - its synthesis and effect on spermatozoal cytoplasm droplets. Anim Reprod Sci 3:195–205

Miller DJ (2018) Review: the epic journey of sperm through the female reproductive tract. Animal 12(s1):s110–s120. https://doi.org/10.1017/s1751731118000526

Morales C, Hermo L (1983) Demonstration of fluid-phase endocytosis in epithelial cells of the male reproductive system by means of horseradish peroxidase-colloidal gold complex. Cell Tissue Res 230(3):503–510. https://doi.org/10.1007/bf00216196

Morales C, Hermo L, Clermont Y (1984) Endocytosis in epithelial cells lining the rete testis of the rat. Anat Rec 209(2):185–195. https://doi.org/10.1002/ar.1092090206

Morales CR, El-Alfy M, Zhao Q, Igdoura S (1995) Molecular role of sulfated glycoprotein-1 (SGP-1/prosaposin) in Sertoli cells. Histol Histopathol 10(4):1023–1034

Morin G, Lalancette C, Sullivan R, Leclerc P (2005) Identification of the bull sperm p80 protein as a PH-20 ortholog and its modification during the epididymal transit. Mol Reprod Dev 71(4):523–534. https://doi.org/10.1002/mrd.20308

Morin G, Sullivan R, Laflamme I, Robert C, Leclerc P (2010) SPAM1 isoforms from two tissue origins are differentially localized within ejaculated bull sperm membranes and have different roles during fertilization. Biol Reprod 82(2):271–281. https://doi.org/10.1095/biolreprod.109.079582

Nixon B, De Iuliis GN, Hart HM, Zhou W, Mathe A, Bernstein IR, Anderson AL, Stanger SJ, Skerrett-Byrne DA, Jamaluddin MFB, Almazi JG, Bromfield EG, Larsen MR, Dun MD (2019) Proteomic Profiling of Mouse Epididymosomes Reveals their Contributions to Post-testicular Sperm Maturation. Mol Cell Proteomics 18:S91–S108. https://doi.org/10.1074/mcp.RA118.000946

O'Flaherty CM, Beorlegui NB, Beconi MT (1999) Reactive oxygen species requirements for bovine sperm capacitation and acrosome reaction. Theriogenology 52(2):289–301. https://doi.org/10.1016/s0093-691x(99)00129-6

Oh JS, Han C, Cho C (2009) ADAM7 is associated with epididymosomes and integrated into sperm plasma membrane. Mol Cells 28(5):441–446. https://doi.org/10.1007/s10059-009-0140-x

Orgebin-Crist MC (1969) Studies on the function of the epididymis. Biol Reprod 1(Suppl 1):155–175. https://doi.org/10.1095/biolreprod1.supplement_1.155

Papp S, Robaire B, Hermo L (1995) Immunocytochemical localization of the Ya, Yc, Yb1, and Yb2 subunits of glutathione S-transferases in the testis and epididymis of adult rats. Microsc Res Tech 30(1):1–23. https://doi.org/10.1002/jemt.1070300102

Parent S, Lefievre L, Brindle Y, Sullivan R (1999) Bull subfertility is associated with low levels of a sperm membrane antigen. Mol Reprod Dev 52:57–65

Pinel L, Cyr DG (2021) Self-renewal and differentiation of rat epididymal basal cells using a novel in vitro organoid model†. Biol Reprod 105(4):987–1001. https://doi.org/10.1093/biolre/ioab113

Plante G, Prud'homme B, Fan J, Lafleur M, Manjunath P (2016) Evolution and function of mammalian binder of sperm proteins. Cell Tissue Res 363(1):105–127. https://doi.org/10.1007/s00441-015-2289-2

Rejraji H, Vernet P, Drevet JR (2002) GPX5 is present in the mouse caput and cauda epididymidis lumen at three different locations. Mol Reprod Dev 63(1):96–103. https://doi.org/10.1002/mrd.10136

Robaire B, Hermo L (1988) Efferent ducts, epididymis and vas deferens: structure, functions and their regulation. In: Knobil E, Neill J (eds) The physiology of reproduction. Raven press, New York, pp 999–1080

Robaire B, Hinton BT (2015) The Epididymis. In: Plant TM, Zeleznik AJ (eds) Knobil and Neill's Physiology of Reproduction, vol 1. Elsevier, Amsterdam, pp 691–771

Shum WW, Da Silva N, Brown D, Breton S (2009) Regulation of luminal acidification in the male reproductive tract via cell-cell crosstalk. J Exp Biol 212(Pt 11):1753–1761. https://doi.org/10.1242/jeb.027284

Sullivan R, Bleau G (1985) Interaction of isolated components from mammalian sperm and egg. Gamete Res 12:101–116

Sullivan R, Frenette G, Girouard J (2007) Epididymosomes are involved in the acquisition of new sperm proteins during epididymal transit. Asian J Androl 9(4):483–491. https://doi.org/10.1111/j.1745-7262.2007.00281.x

Sutovsky P, Moreno R, Ramalho-Santos J, Dominko T, Thompson WE, Schatten G (2001) A putative, ubiquitin-dependent mechanism for the recognition and elimination of defective spermatozoa in the mammalian epididymis. J Cell Sci 114(Pt 9):1665–1675. https://doi.org/10.1242/jcs.114.9.1665

Sylvester SR, Morales C, Oko R, Griswold MD (1989) Sulfated glycoprotein-1 (saposin precursor) in the reproductive tract of the male rat. Biol Reprod 41(5):941–948. https://doi.org/10.1095/biolreprod41.5.941

Sylvester SR, Morales C, Oko R, Griswold MD (1991) Localization of sulfated glycoprotein-2 (clusterin) on spermatozoa and in the reproductive tract of the male rat. Biol Reprod 45(1):195–207. https://doi.org/10.1095/biolreprod45.1.195

Thomas TS, Reynolds AB, Oliphant G (1984) Evaluation of the site of synthesis of rabbit sperm acrosome stabilizing factor using immunocytochemical and metabolic labeling techniques. Biol Reprod 30(3):693–705. https://doi.org/10.1095/biolreprod30.3.693

Usselman MC, Cone RA, Rossignol DP (1985) Rat cauda epididymal fluid is a mucus. J Androl 6(5):315–320. https://doi.org/10.1002/j.1939-4640.1985.tb00851.x

Veeramachaneni DN, Amann RP, Palmer JS, Hinton BT (1990) Proteins in luminal fluid of the ram excurrent ducts: changes in composition and evidence for differential endocytosis. J Androl 11(2):140–154

Veri JP, Hermo L, Robaire B (1993) Immunocytochemical localization of the Yf subunit of glutathione S-transferase P shows regional variation in the staining of epithelial cells of the testis, efferent ducts, and epididymis of the male rat. J Androl 14(1):23–44

Veri JP, Hermo L, Robaire B (1994) Immunocytochemical localization of glutathione S-transferase Yo subunit in the rat testis and epididymis. J Androl 15(5):415–434

Yeung CH, Cooper TG, Nieschlag E (1997) Human epididymal secreted protein CD52 on ejaculated spermatozoa: correlations with semen characteristics and the effect of its antibody. Mol Hum Reprod 3(12):1045–1051. https://doi.org/10.1093/molehr/3.12.1045

Zhang H, Morales CR, Badran H, El-Alfy M, Martin-DeLeon PA (2004) Spam1 (PH-20) expression in the extratesticular duct and accessory organs of the mouse: a possible role in sperm fluid reabsorption. Biol Reprod 71(4):1101–1107. https://doi.org/10.1095/biolreprod.104.030403

Zhou W, De Iuliis GN, Dun MD, Nixon B (2018) Characteristics of the Epididymal Luminal Environment Responsible for Sperm Maturation and Storage. Frontiers in Endocrinology 9. https://doi.org/10.3389/fendo.2018

Chapter 4
Photomicrographic Glossary of the Features of Bovine Sperm Cell Abnormalities

Abstract This chapter is intended to provide an accurate visual description of normal and abnormal bovine sperm morphology as seen under the light microscope. The photographs were taken × 1000 magnification and enlarged × 1.6, unless specified otherwise, using either a) differential interference microscopy (DIC), b) phase contrast microscopy, c) eosin-nigrosin-stained sperm smears, or d) Feulgen stained smears as indicated in the legends.

When viewed with a critical eye, there is an extremely wide variation in form and structure within the various categories of sperm cell abnormalities. It would be an impossible and fruitless task to attempt to show every minor variation in appearance of each type. Therefore, a series of photographs has been put together showing the main features of each type of abnormality so that any other deviation observed can easily be recognized as to the category in which it belongs.

The nomenclature applied to the various abnormalities is that preferred by us, although generally accepted by other workers. Synonymous nomenclature used by other workers is included in the discussion on interpretation of sperm cell abnormalities in Chapter 5.

Keywords Sperm abnormalities: Knobbed acrosome defect · Ruffled plasmalemma and acrosome · Pyriform heads · Tapered heads · Microcephalic and macrocephalic · Nuclear vacuoles · Abnormal condensation of DNA · Rolled head-nuclear crest-giant head syndrome · Pale Centers · Detached heads · Decapitated · Distal midpiece reflex · Dag defect · Segmental aplasia of the mitochondrial sheath · Pseudodroplet · Corkscrew defect · Bowed midpieces · Bent principal pieces · Coiled principal pieces · Abaxial · Accessory · and Multiple tails · Tail stump defect · Short tail defect · Hypotonic shock · Proximal droplets · Distal droplets · Teratoid forms

© The Author(s), under exclusive license to Springer Nature Switzerland AG 2025
A. Barth et al., *Abnormal Morphology of Bovine Spermatozoa*, Advances in Anatomy, Embryology and Cell Biology 240, https://doi.org/10.1007/978-3-031-70126-9_4

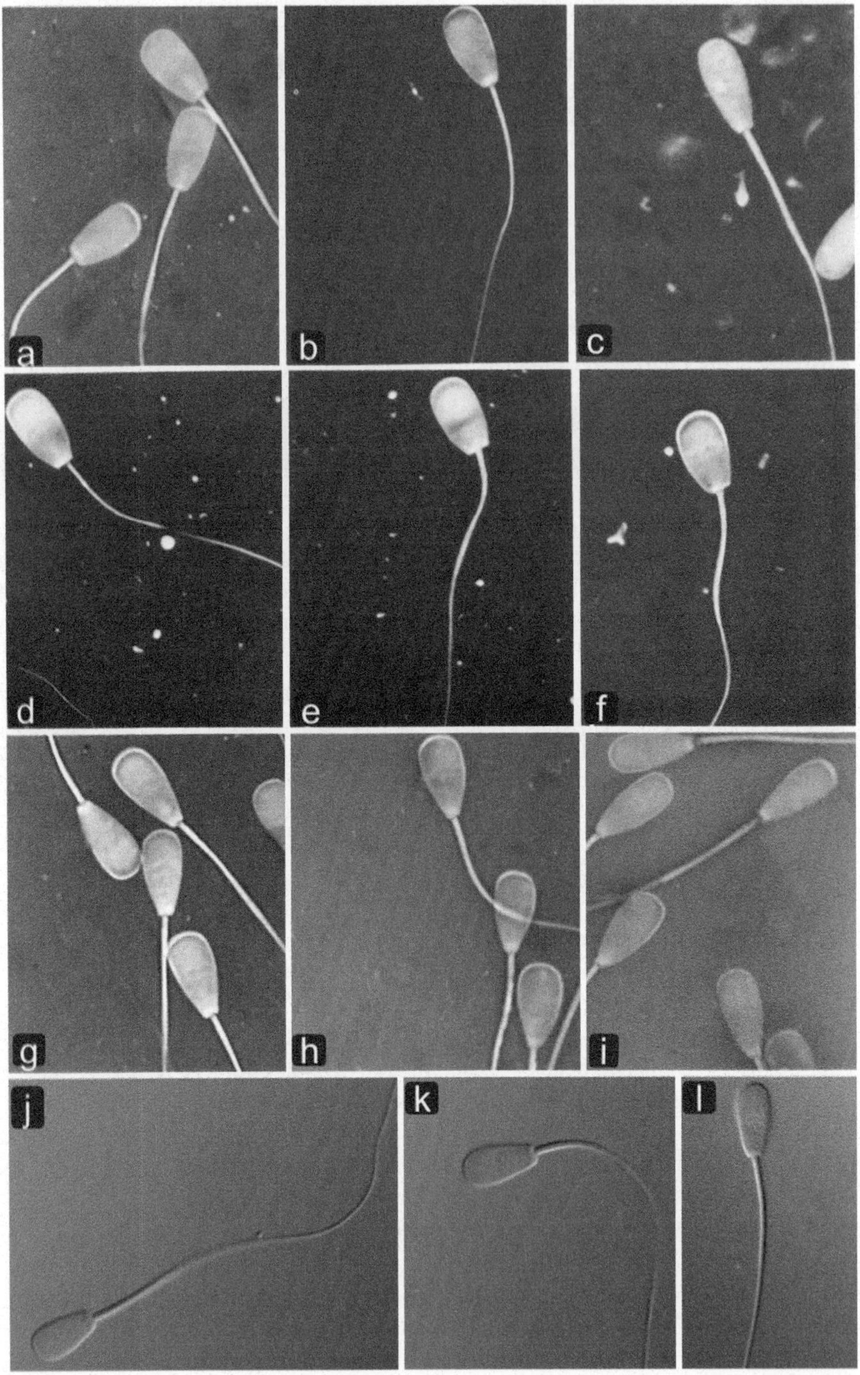
a
b
c
d
e
f
g
h
i
j
k
l

◀ **Fig. 4.1 Normal bovine spermatozoa**. Note the variation in head shape. Views (**a**) and (**b**) show the most common shape of normal bovine spermatozoa; (**c**) shows the longer, narrower profile sometimes observed; (**d**), (**e**), (**f**) show a progression of the shorter, broader variations of normal. (**g**), (**h**), (**i**) show normal sperm intermediate between classical normal shape progressing toward pyriform. These bulls displayed normal fertility indicating the variation in normal where this is the "typical" form for a particular bull. (**j**), (**k**), (**l**) show normal sperm under DIC

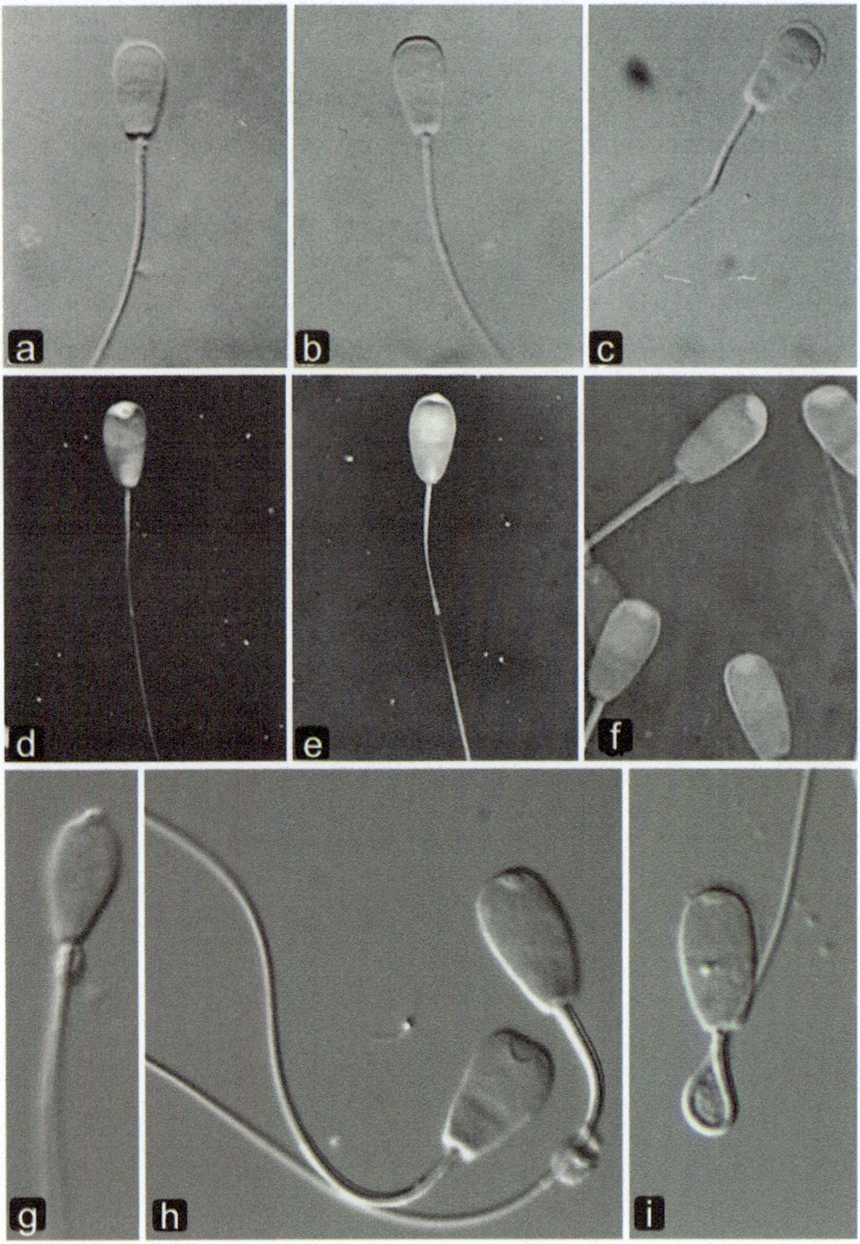
a
b
c
d
e
f
g
h
i

◄ **Fig. 4.2 Knobbed acrosome defect.** Views (**a**)–(**c**) do not have knobbed acrosomes; (**a**) and (**b**) show normal sperm with intact acrosomes as seen by differential interference contrast with the light coming from two different angles, and (**c**) depicts a normal sperm with a swollen acrosome. The swollen appearance is due to break down of the plasmalemma and the acrosomal membrane
Knobbed acrosome defect has two main forms: beaded and indented. In the beaded form, (**d**) (**e**) and (**f**), an enlarged acrosome folds back over the apex of the sperm head and the apex is not indented. This can also be observed in the DIC images in (**g**), (**h**) and (**i**)
Knobbed defect continued. In the indented form which is much more common, affected sperm have an indented appearance as in (**j**), (**k**) and (**l**) or flattened appearance as in (**n**). In contrast (m) shows a beaded form.
View (**o**) is a photomicrograph taken from a Feulgen-stained preparation in which the sperm cell membrane and acrosome have been removed by the caustic action of the staining procedure. Note the dented appearance of the apex in the nucleus caused by a folding back of this portion of the nucleus
The "flattened acrosome" as seen in DIC images (**p**), (**q**), and (**r**). Image (**s**) shows a normal shaped sperm head for comparison

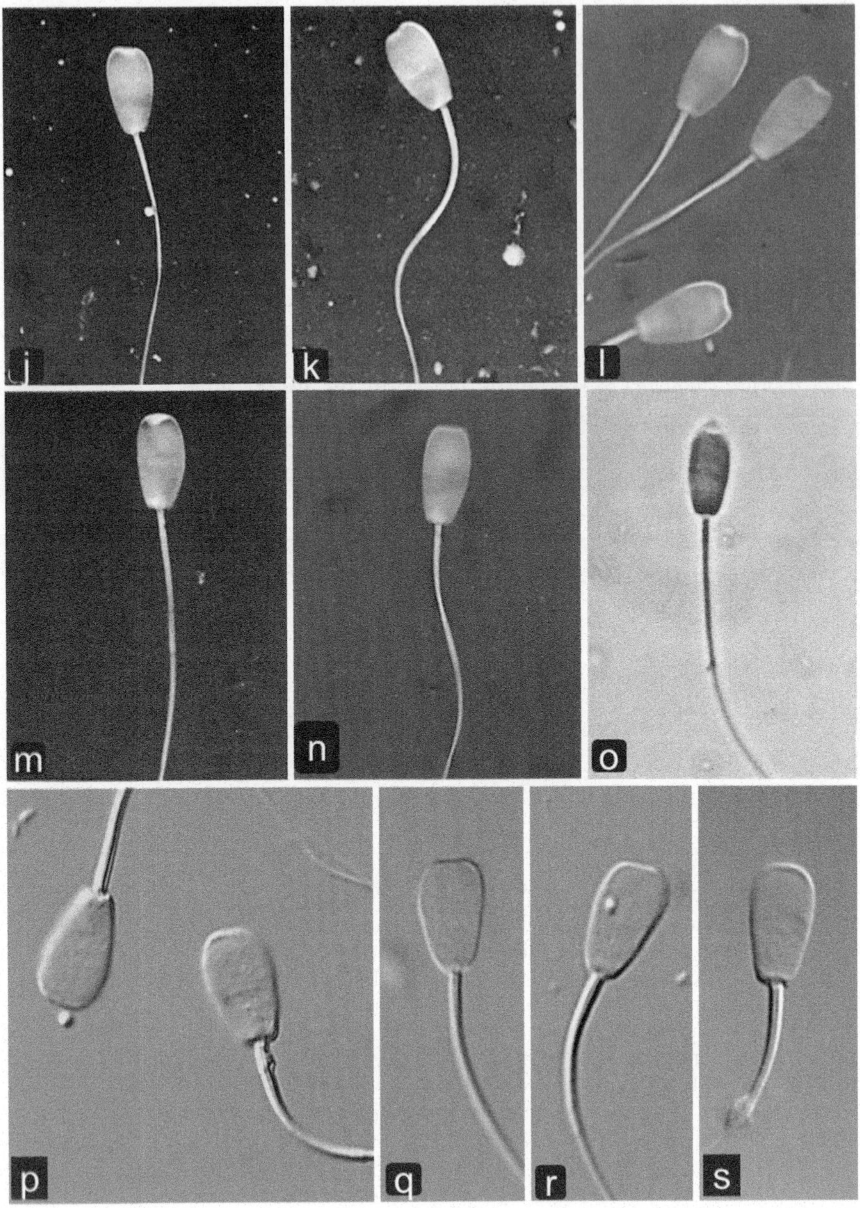

Fig. 4.2 (continued)

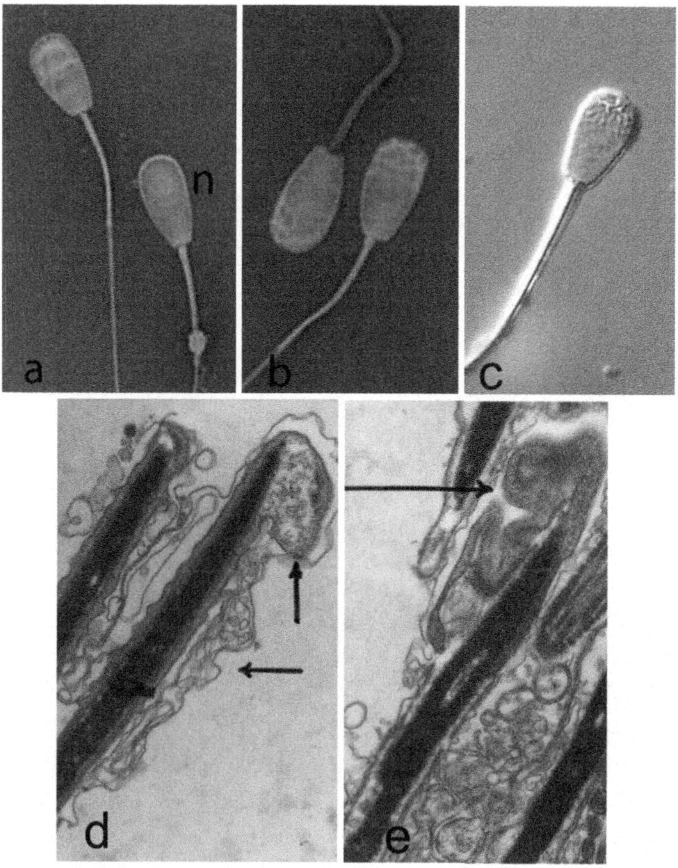

Fig. 4.3 Ruffled plasmalemma and acrosome. A ruffled appearance of the plasmalemma is shown in views (**a**), (**b**) and (**c**). The sperm labelled (n) is normal. View (**d**) shows an EM of the wrinkled plasmalemma (horizontal arrow) and an enlarged acrosome (vertical arrow). The acrosome itself may be wrinkled as shown by a long horizontal arrow in (**e**). It is common to find small percentages of ruffled membranes in bulls with a normal spermiogram. In rare cases, very high percentages of these are observed in otherwise normal spermiograms and such cases have been associated with low pregnancy rates

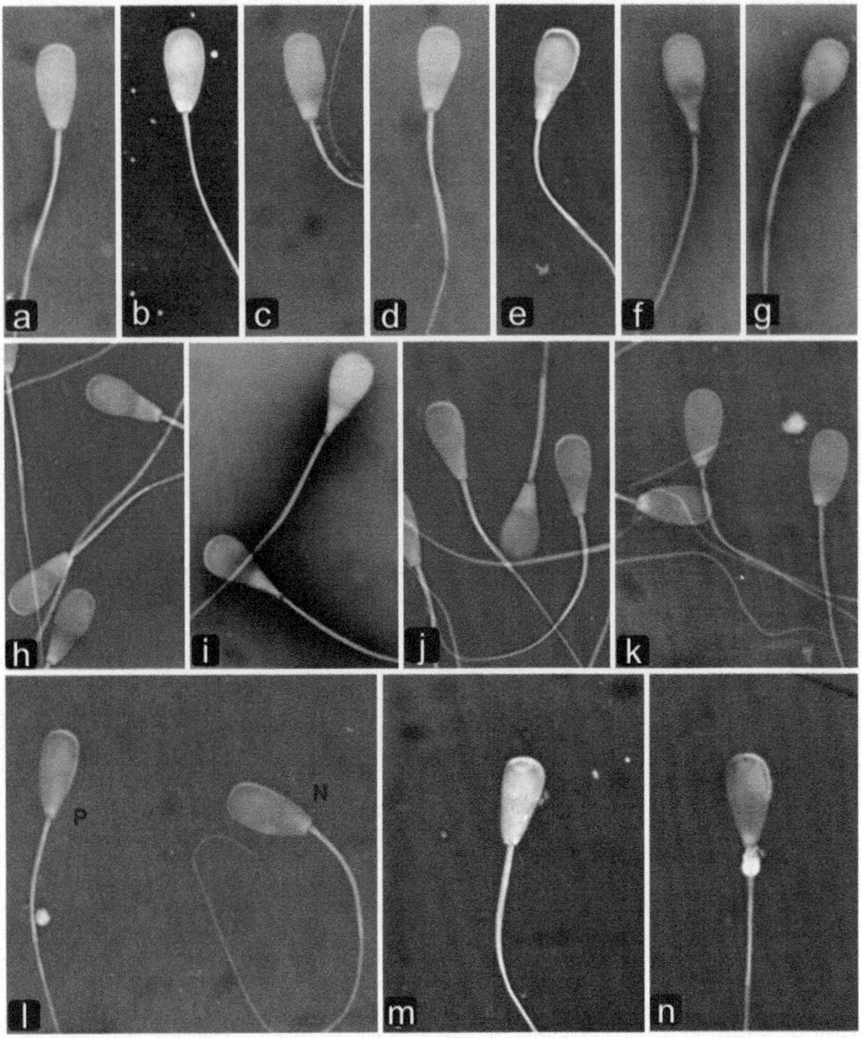

Fig. 4.4 Pyriform heads. These are narrow in the postacrosomal region. Views (**a**)–(**g**) show a progression from normal to severely pyriform. Most evaluators would classify the sperm cells depicted in (**a**) and (**b**) as normal and those in (**e**), (**f**), and (**g**) as abnormal. It has been established that pyriform head shape where the postacrosomal region is concave as in (**e**), (**f**), and (**g**) as well as in images (**h**), (**i**), (**j**) adversely affects fertility, whereas the mild forms as in (**c**) and (**d**) do not

Image (**k**) shows a less common type of pyriform shape in which most of the postacrosomal region appears full and rounded but there is a sharp pinching-in of the base of the nucleus. In a small breeding trial using semen with this morphology fertility was not decreased. View (**l**) demonstrates that pyriform heads (P) often appear smaller than normal sperm heads (N)

The pyriform defect may be accompanied by other defects within the same cell, such as those seen in (**m**) and (**n**). Note the knobbed acrosome and the nuclear vacuoles in the equatorial region of the nucleus in (m) and the proximal droplet and apical nuclear vacuole in (n)

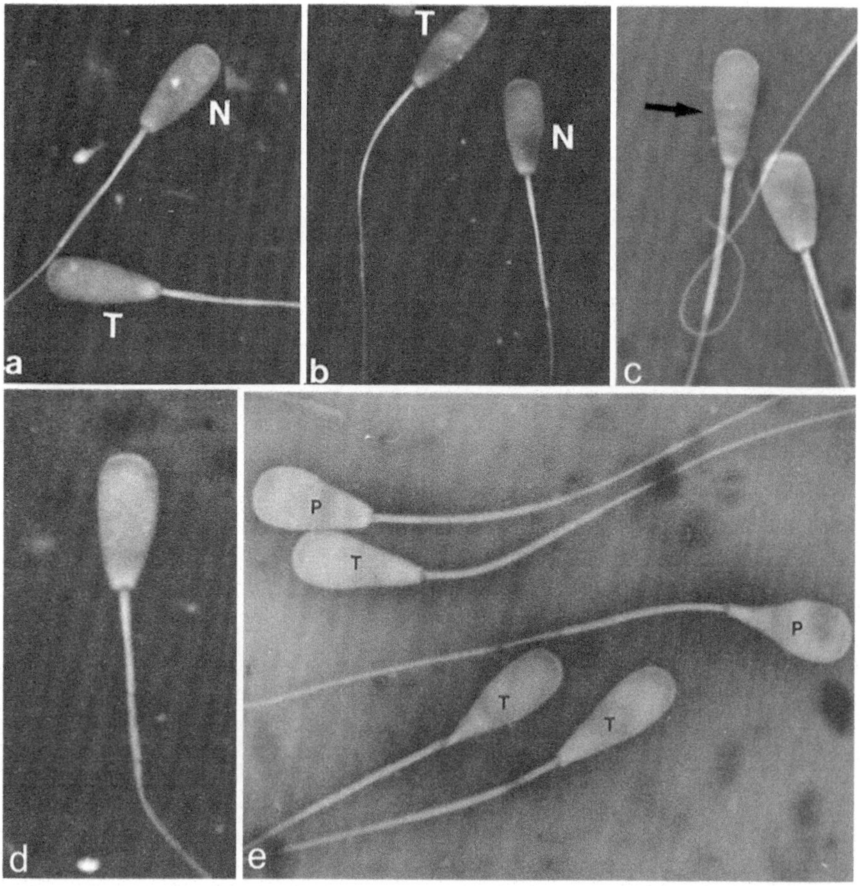

Fig. 4.5 Tapered heads. These are narrow in both the acrosomal and postacrosomal region. Views (**a**), (**b**) and (**c**) compare tapered (T) and normal (N) nuclei. The sperm cell in (**d**) is classified as normal. View (**e**) demonstrates that both tapered (T) and pyriform (P) defects may be found in the same ejaculate, thus the two forms could be considered as a similar problem. However, in one report, nearly all the sperm from a bull of normal fertility had tapered heads

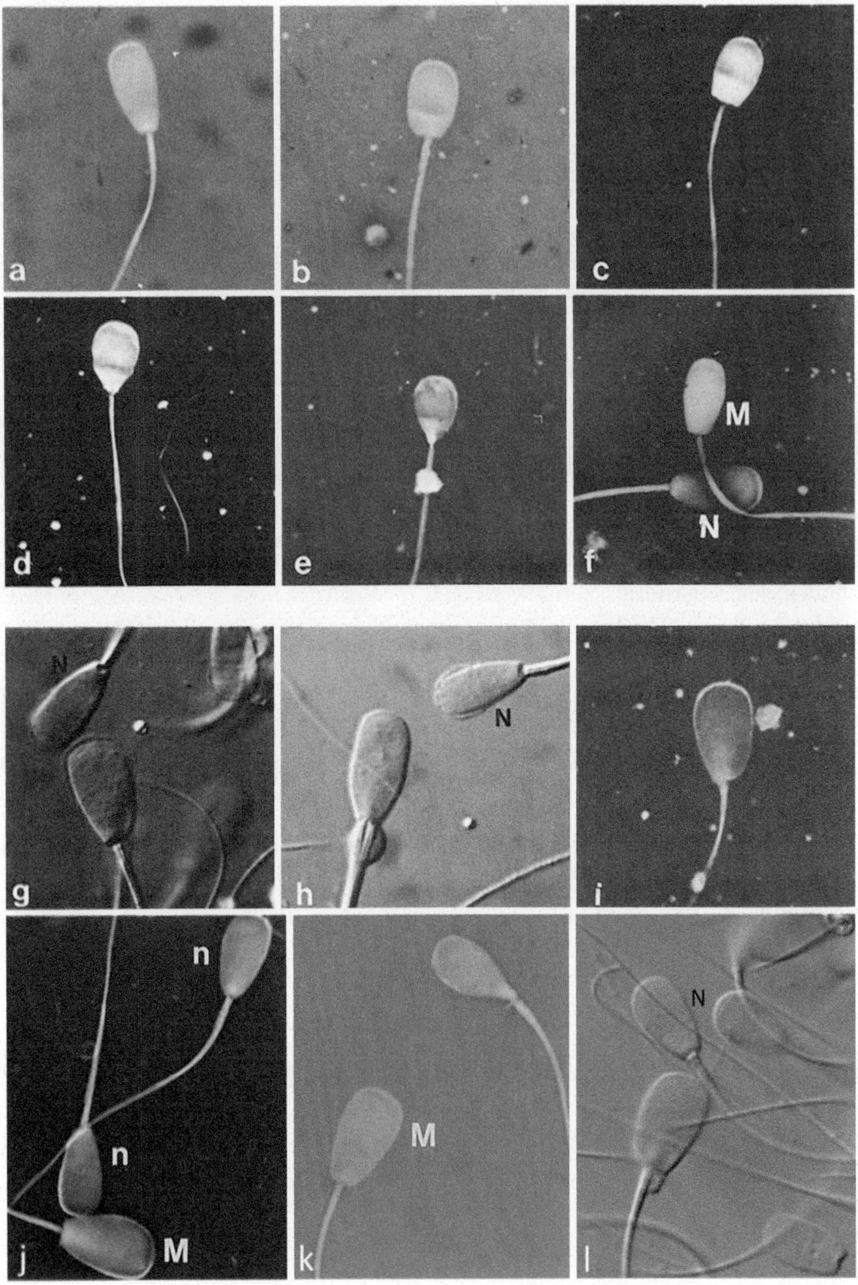
a
b
c
d
e
M
N
f
N
g
N
h
i
n
n
M
j
M
k
N
l

◀ **Fig. 4.6 Microcephalic and macrocephalic sperm.** It appears that these defects arise as a result of an uneven meiotic distribution of chromosomes in spermatocytes. Most of these cells die before reaching the spermatid stage and are phagocytosed by the Sertoli cells. This would explain why microcephalic and macrocephalic sperm are not found in large numbers in semen. The exception to this observation is the Rolled Head-Nuclear Crest-Giant head syndrome
There is a wide variety of shapes and sizes of undersize sperm heads. Although microcephalic sperm are quite common, they rarely are found in more than a very small proportion of sperm. View (**a**) is a normal sperm for comparison. Microcephalic sperm most commonly have short broad-shaped heads as in (**b**) and (**c**) or pyriform heads as in (**d**) and (**e**). Normally shaped microcephalics (M) as in (**f**) are uncommon. Note the normal-sized sperm head (N) in (**f**)
A few macrocephalic sperm are found occasionally in semen smears (**g–l**). Note the normal sperm (N) for size comparison in (**g**), (**h**), (**j**) and (**l**) Macrocephalic sperm frequently display double tails as in (**h**), accessory tails (**l**) or other abnormalities such as droplets or abaxial tail attachments as in (**i**). They are occasionally accompanied by other sperm abnormalities such as pyriform heads (**k**) or nuclear vacuoles (**l**)

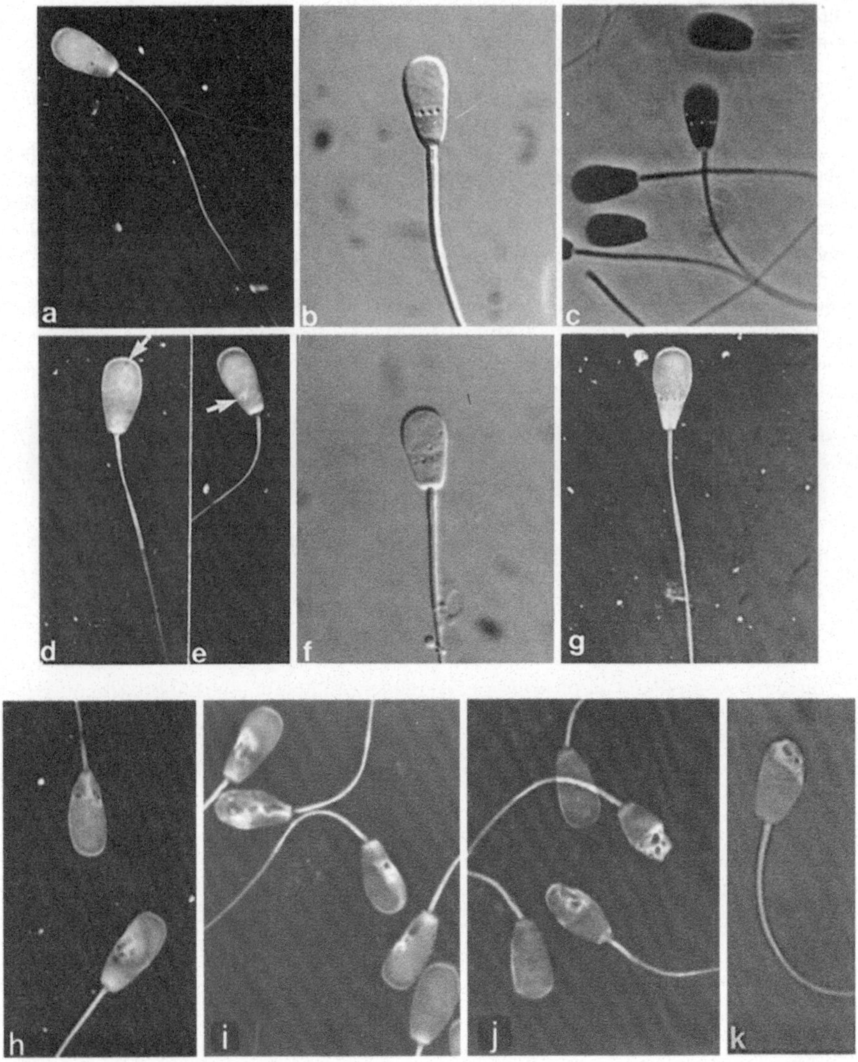

Fig. 4.7 Nuclear vacuoles. These may be considered in three broad categories: **1.** the diadem defect, **2.** the single nuclear vacuole, **3.** the large confluent vacuole
The photomicrographs of (**a**), (**b**), and (**c**) were taken from eosin-nigrosin stained, differential interference contrast microscopy (DIC), and Feulgen-stained preparations respectively. They depict the type of nuclear vacuolation in the equatorial region of the nucleus known as the diadem defect. Figure (b) shows the crater-like appearance of nuclear vacuoles seen with DIC microscopy. This appearance has led to the term 'crater defect', although diadem is still more commonly used
Views (**d**) and (**e**) show single nuclear vacuoles (arrows). The most common location for single vacuoles is in the apical region of the nucleus as seen in (**d**). However, they may occur scattered throughout the nucleus as in image (**f**). Single nuclear vacuoles along with all vacuoles occur more commonly in Bos indicus crossbreds than in Bos taurus breeds. Pyriformity and vacuolation frequently occur together (**g**)
Views (**h**), (**i**), (**j**), and (**k**) show sperm heads with clusters of large confluent vacuoles. Severe vacuolation of the acrosomal region of the nucleus as in (**j**) and (**k**) has sometimes been mistakenly called an acrosome defect

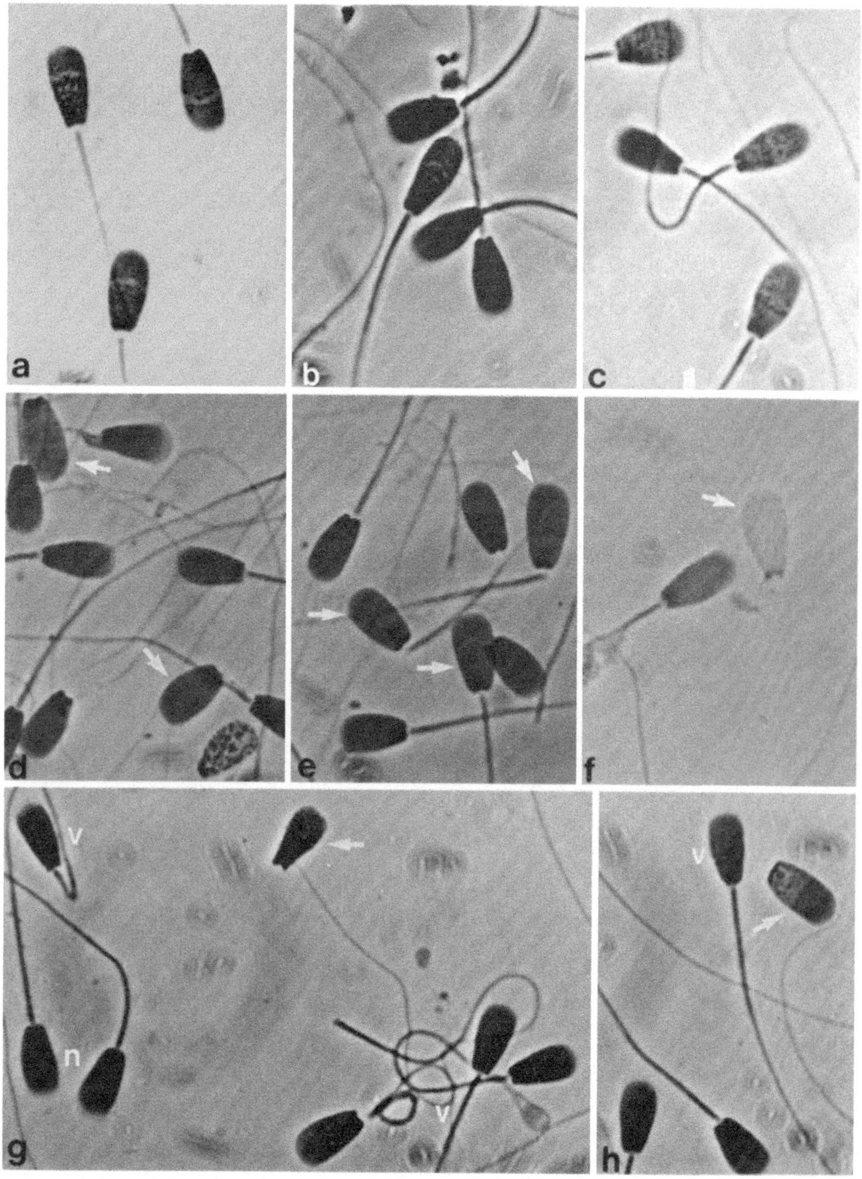

Fig. 4.8 Abnormal condensation of DNA. All photomicrographs in this figure were taken from Feulgen-stained preparations. DNA clumping cannot be seen in wet mounts for Phase or DIC microscopy or in eosin-nigrosin-stained smears

Views (**a**), (**b**), and (**c**) show the most common forms of this type of defect; (**b**) and (**c**) also show uniformly condensed normal nuclei for comparison

Views (**d**) and (**e**) show expanded nuclei (arrows) along with normal nuclei. Note the extremely fine-grained clumping of DNA in the expanded nuclei. Also note the range from very fine to very coarse clumping of DNA in nuclei of (**a**), (**b**), (**c**), and (**d**). View (**f**) shows a ghost cell (arrow). This is a rare form of abnormal DNA condensation in which DNA appears to be virtually absent. Views (**g**) and (**h**) show normal nuclei (n) vacuolated nuclei (v), and nuclei with clumping of the DNA. Figure (**g**) shows coarse clumping of DNA in a pyriform nucleus (arrow); (**h**) shows an expanded nucleus with some coarse clumps of DNA in the postacrosomal region (arrow)

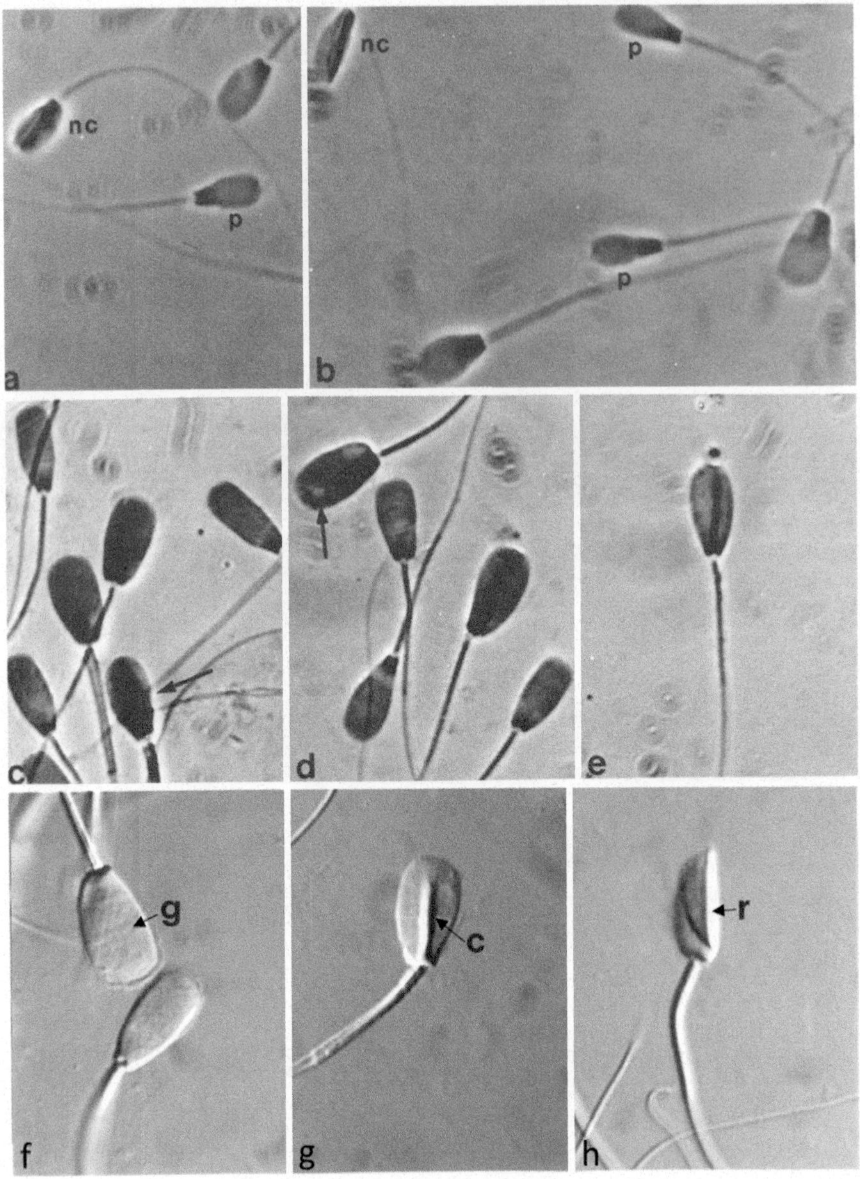

Fig. 4.9 Rolled head-nuclear crest-giant head syndrome. The photomicrographs in (**a**) and (**b**) were taken from wet mounts under phase-contrast microscopy; (**c**), (**d**), and (**e**) are from Feulgen-stained preparations; and (**f**), (**g**), and (**h**) are from wet mounts under differential interference contrast. Figures (**a**) and (**b**) depict sperm cells from a Holstein bull that sired the bull producing the sperm cells shown in (**c**)–(**h**). In (**a**) note the nuclear crest (nc) and a pyriform head with a knobbed acrosome defect (p). In (**b**) note the sperm cell with a nuclear crest (nc), two pyriform heads with knobbed acrosomes (p), and two giant-headed sperm (one of which has a double tail and the other a crest in the postacrosomal region). View (**c**) shows a sperm with a nuclear crest and a double tail (arrow) as well as two giant-headed sperm and some slightly pyriform-headed sperm. In (**d**) note the two giant heads with abaxial tails. One of them also has nuclear vacuoles (arrow). In (**e**) the giant head of this sperm is also crested and appears to have an acrosome defect. In (**f**) note that this sperm has a giant head which may be diploid or polyploid and these are often crested as in (**c**) or rolled as in sperm (r)

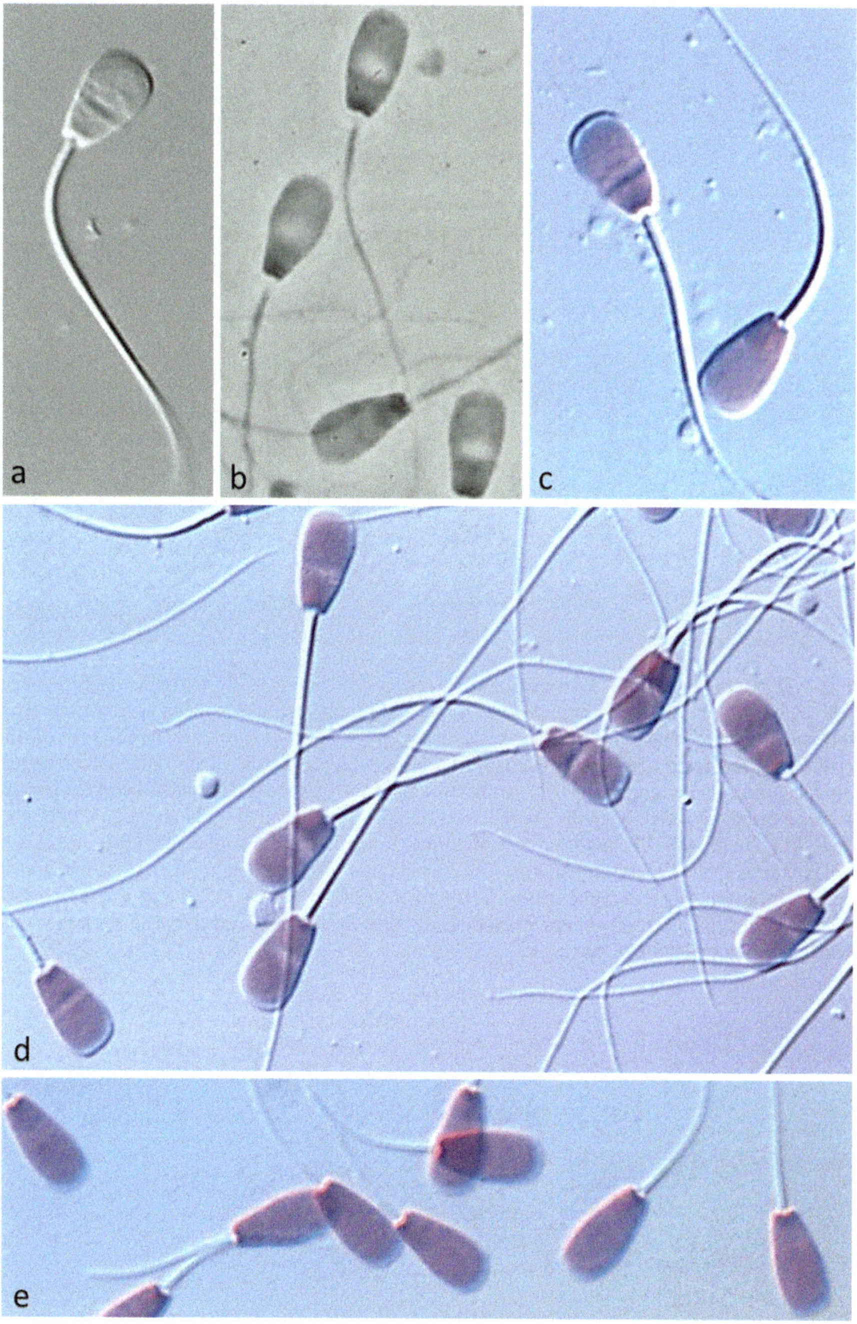

Fig. 4.10 Pale centre. A pale centre appearance is due to narrowing of the nuclear material in the postacrosomal region immediately below the equatorial segment. View (**a**) shows a pale centre under DIC and (**b**) shows a pale centre under phase contrast microscopy. View (**c**) shows a Feulgen stain of a pale centre sperm with the head pointing up compared to a normal sperm with the head pointing down. View (**d**) and (**e**) are also Feulgen stain images with (**d**) exhibiting pale centres compared to normal sperm in (**e**)

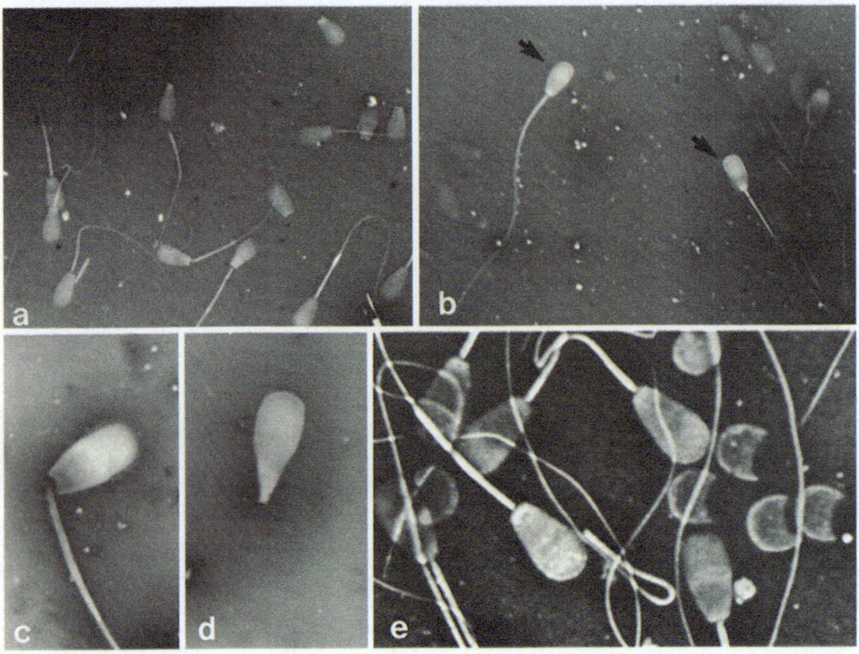

Fig. 4.11 Detached heads. Detached normal sperm heads are usually found in small numbers and are due to senescence. Views (**a**) and (**b**) are from bulls that accumulated spermatozoa from the ampullae and cauda epididymis. In such cases very large proportions of cells take up the eosin stain and large numbers of spermatozoa have detached heads. Eosin-stained cells appear dark in these photomicrographs. Note the two unstained live cells (arrows) in (**b**). View (**c**) shows a detaching head. Note the partial eosin staining of the nucleus and the swollen acrosome. Detached abnormal heads may be due to abnormal formation of the implantation fossa as seen in this pyriform shaped sperm in (**d**). View (**e**) is a photomicrograph taken from an india ink-stained semen smear. The horseshoelike structures are not detached heads but rather acrosomal caps shed by senescent cells. Acrosomal caps are not usually seen in such high numbers and are less visible with other types of stain. (Courtesy of E. Blom)

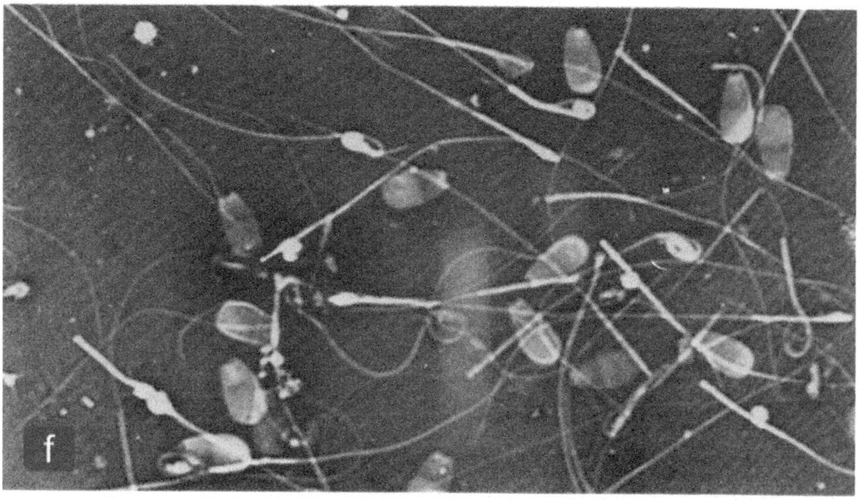

Fig. 4.12 Decapitated sperm defect. This defect affects most of the sperm with separation into loose heads and tails (**f**). Most loose tails are bent around a cytoplasmic droplet and both bent and straight tails are vigorously motile. The cause could be from a shallow implantation fossa, a defective basal plate, or a defective connecting piece. Both the connecting piece and basal plate share structural proteins with the outer dense fibers (ODF). Thus defects in the ODF structure are often associated with decapitations. Gene knockouts of prominent ODF proteins are often associated with sperm decapitations

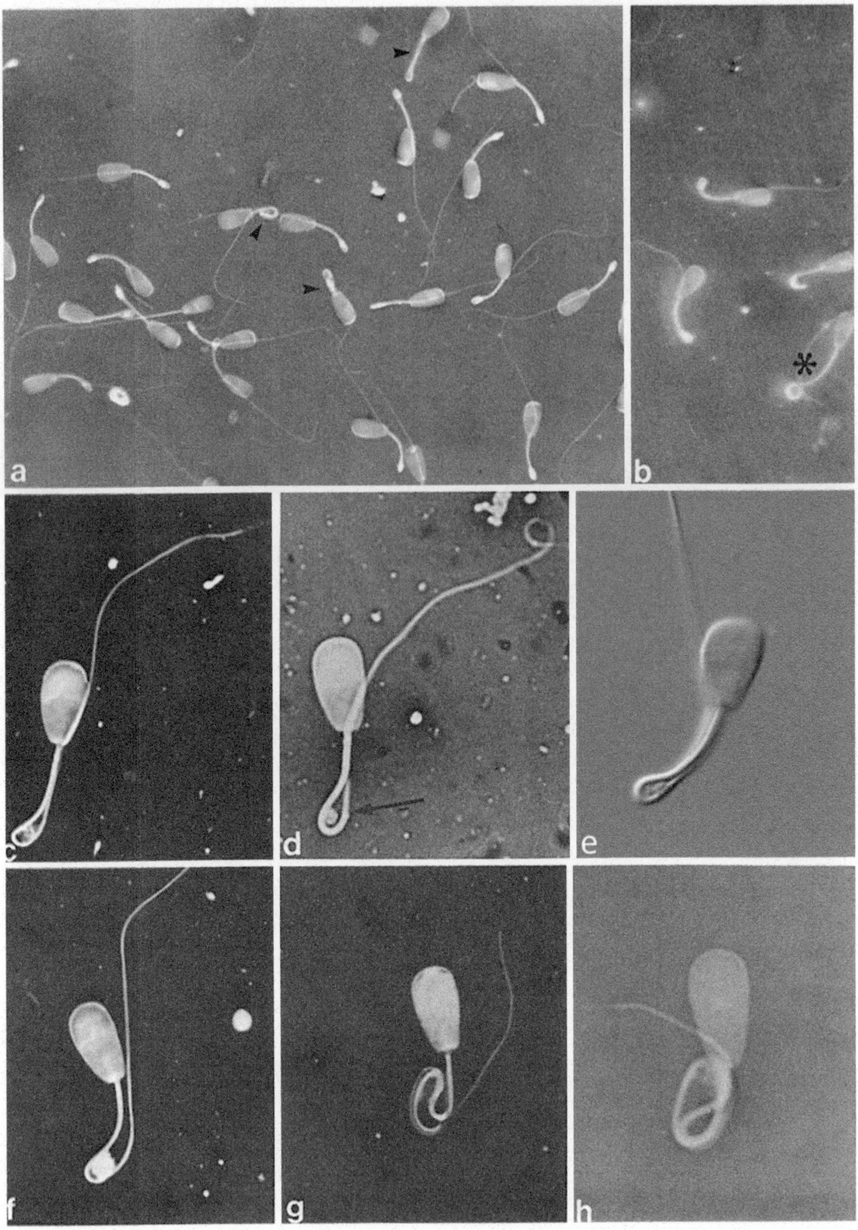
a
b
c
d
e
f
g
h

◀ **Fig. 4.13 Distal midpiece reflex abnormality.** Views (**a**) and (**b**) show a very high incidence of this defect. Notice some of the different forms within the same smear (arrows) and the coiled principal piece (*), which may be found in small numbers concurrent with the distal midpiece reflex abnormality. Views (**c**), (**d**) and (**e**) show the most common form of this defect, the latter being a DIC image; (**d**) very clearly shows the J shape of the midpiece. The arrow indicates the region of the annulus. Views (**f**), (**g**), and (**h**) are less common forms of this defect. Note in (**g**) and (**h**) the double bending of the midpiece, which results in a coiled appearance
Distal midpiece reflex continued—Views (**i**) and (**j**) show an open form of the distal midpiece reflex in which the principal piece does not reflect past the head. Notice that in (**i**) the cytoplasmic droplet has not yet been shed. In both cases the sperm cells appear consistently to retain the kink in the distal midpiece. View (**k**) shows a distal midpiece reflex in association with bent principal pieces. These abnormalities were produced by hypotonicity. Hypotonic solutions most commonly produce bent principal pieces, and distal midpiece reflexes occur less often. Notice the lack of a distal droplet within the reflex bend of the midpiece, which is typical of midpiece reflexes caused by hypotonicity of stains or solutions to which live spermatozoa may be exposed. View (**I**) shows an unusual form associated with this defect. Notice that most of the reflex bend occurs in the region of the principal piece; the arrow indicates the approximate location of the annulus. Views (**l**) and (**m**) show two forms in which fracture at the annular region occurred in different planes with relationship to the sperm head. Thus in (**I**) the principal piece reflects past the head and in (**m**) it reflects over the head. Views (**m**) and (**n**) show two forms of distal midpiece reflexes with fractures at the annulus occurring concurrently with the more typical form of distal midpiece reflex. View (**o**) depicts two types of coiling of the principal piece associated with a single bend (*) or double bend (**) of the midpiece

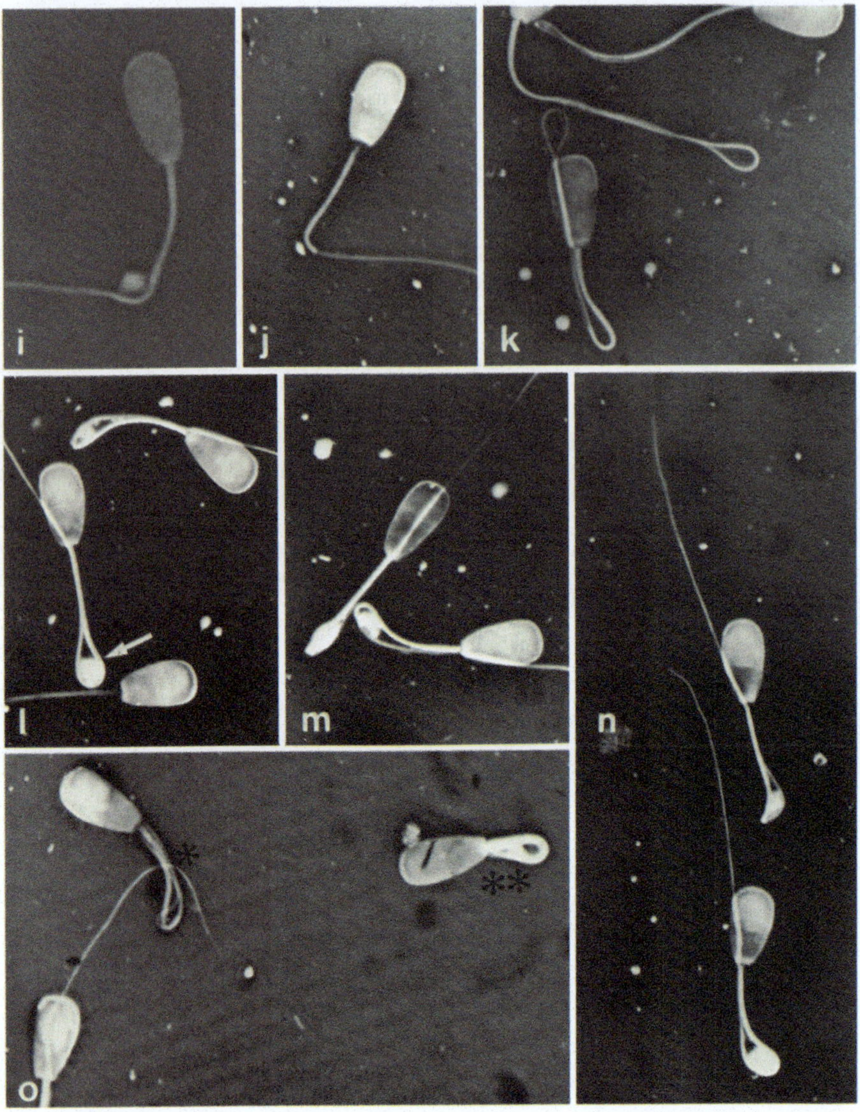

Fig. 4.13 (continued)

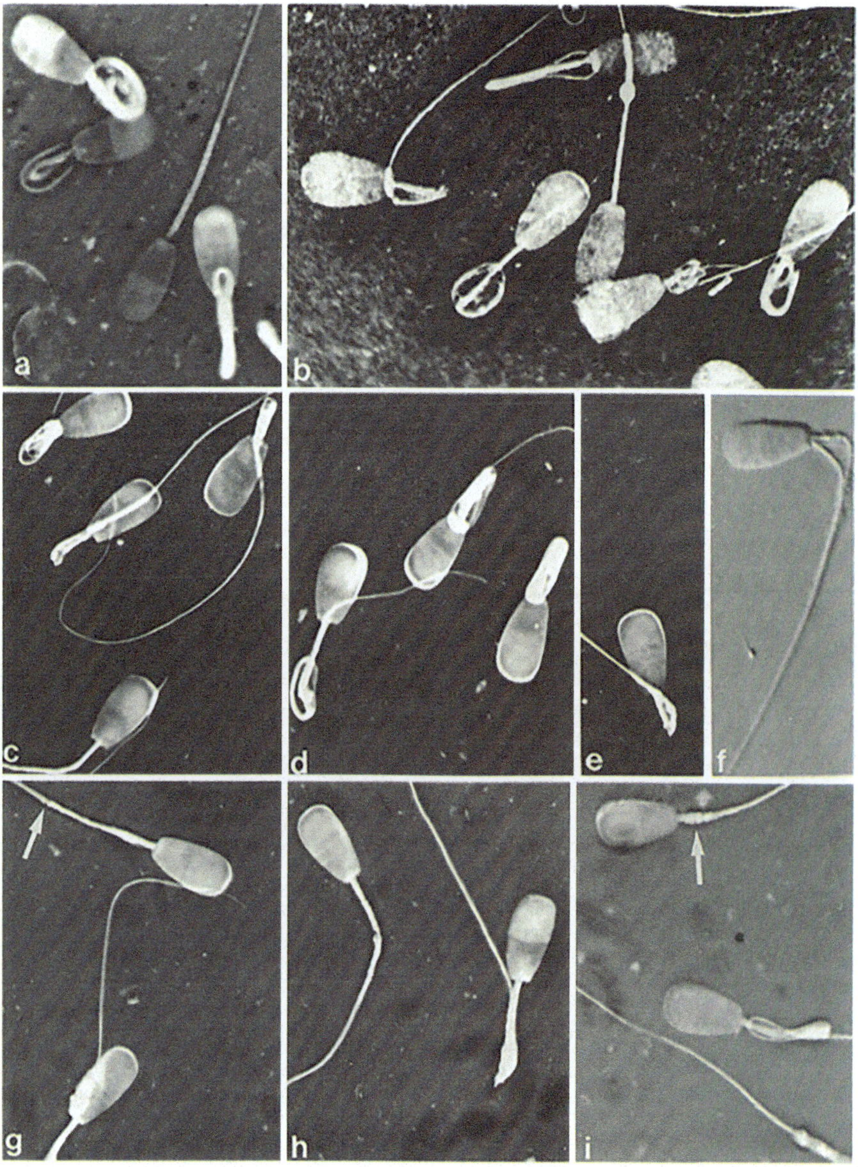

Fig. 4.14 The Dag defect. Views (**a**) and (**b**) are photomicrographs from eosin nigrosin and india ink smears respectively of semen from a Jersey bull producing the Dag defect (Courtesy of E. Blom)

Views (**c**)–(**i**) are photomicrographs of spermatozoa produced by a Hereford bull; (**f**) was produced from a wet mount using differential interference microscopy. Note the disruption of the coarse and fine axonemal fibers in the midpiece region. The lack of outer structural support to these fibers results in the fracturing, shattering, splitting, and coiling of the midpiece seen in these photomicrographs, which is typical of the Dag defect

Views (**g**), (**h**), and (**i**) show the rough, uneven distribution of the mitochondria along the midpiece, which is characteristic of spermatozoa with the Dag defect. Note the gap in the mitochondrial sheath at the annulus in (**g**) (arrow) and the similarity in appearance to the pseudodroplet defect (arrow) in (**i**)

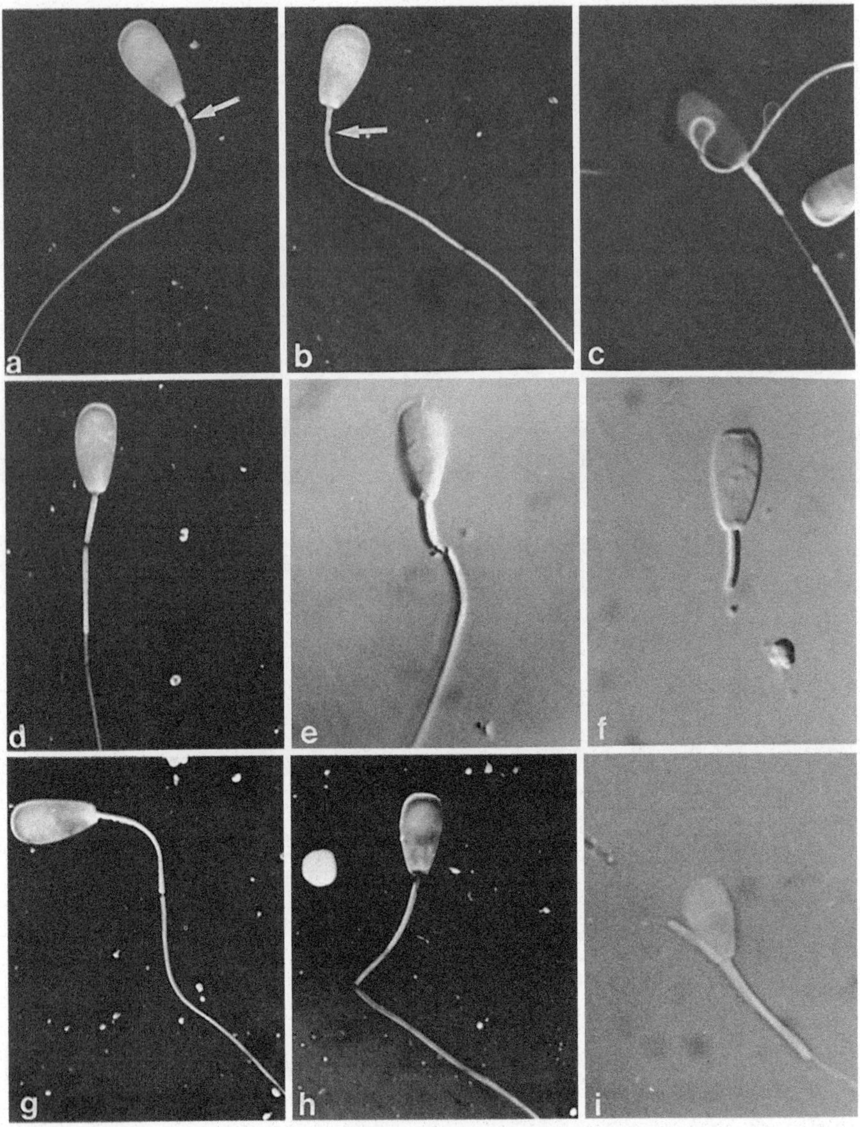

Fig. 4.15 Segmental aplasia of the mitochondrial sheath. Views (**a**) and (**b**) show small gaps in the mitochondrial sheath (arrows). This small defect in the mitochondria does not noticeably affect sperm motility nor affect fertility. We therefore include these small gaps (<3 um) as normal. However, it does represent a weak spot that may predispose the midpiece to fracture. Note also a gap in the fibrous sheath of the principal piece in (**b**). Gaps in the fibrous sheath are rare. A gap such as this may also predispose to fracture of the principal piece. Gaps in the mitochondrial sheath of the magnitude seen in (**c**) are rarely found, since any attempt at motility would result in fracture. This particular cell probably did not possess motility at any time. Views (**d**), (**e**), and (**f**) are a sequence to show how a mitochondrial sheath gap may lead to a fractured midpiece. Figures (**e**) and (**t**) are photomicrographs from wet mounts photographed under differential interference microscopy. Views (**g**), (**h**), and (**i**) represent a special type of segmental aplasia of the mitochondrial sheath at the annulus. This type of defect has been found in over 30% of spermatozoa in the ejaculates of some bulls; it has been produced in rats by gossypol treatment. Spermatozoa with gaps in the mitochondrial sheath at the annulus as seen in (**g**) are not seen as commonly as the type seen in (**i**) because motility of the cell usually results in fracture

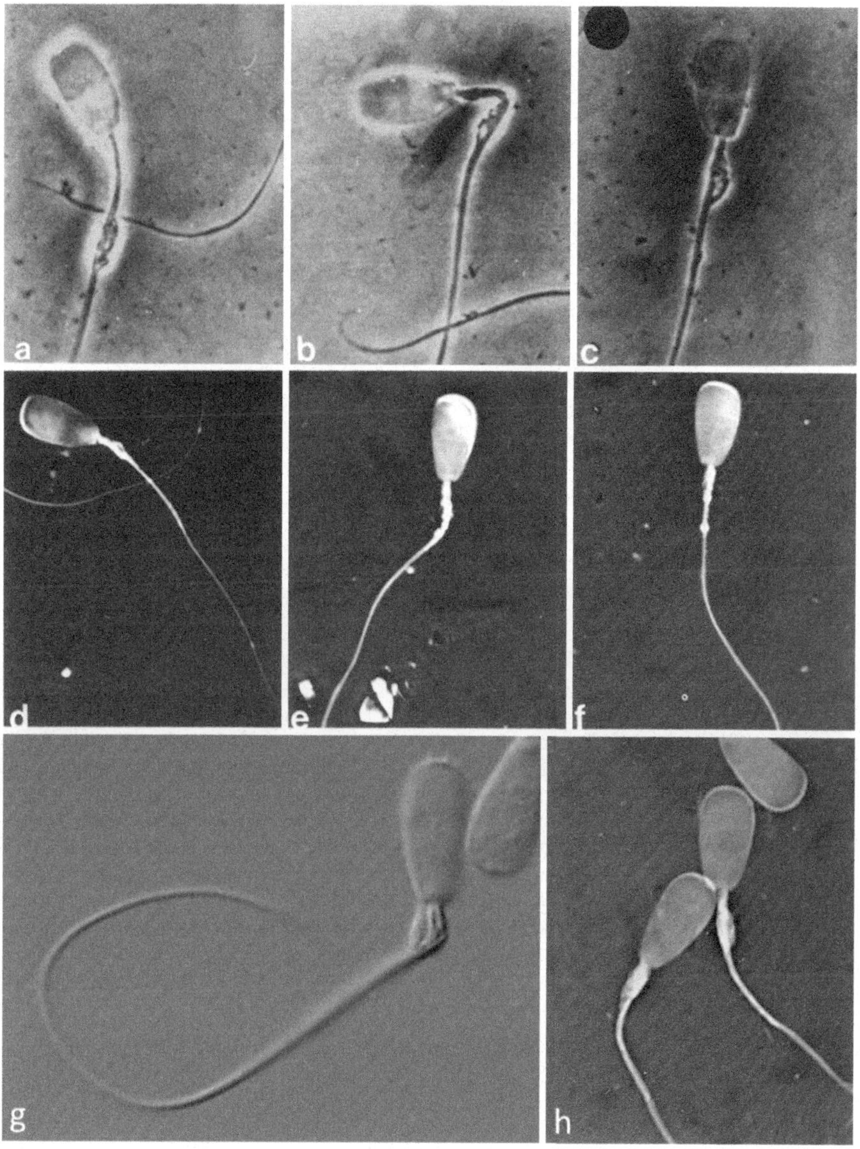

Fig. 4.16 Pseudodroplets. This defect was called a novel segmental aplasia (Rocha et al., 2006). In (**a**)–(**c**) note that the mitochondrial sheath defect may occur at a proximal central or distal midpiece region. Fractures of the midpiece may occur in the vicinity of the mitochondrial sheath defect as seen in (**b**). The light nonstained portions of the swollen region were shown by EM (Blom) to contain granular inclusions surrounded by normal appearing mitochondria. (India ink stain a–c, Courtesy of E. Blom)
Views (**d**), (**e**), and (**f**) show mitochondrial sheath disturbances of spermatozoa from a Canadian Holstein bull. These defects bear some resemblance to the pseudodroplet defect described by Blom and later by Rocha et al., 2006.

It is common to find one or two pseudo droplet-like sperm in bulls with a disturbance of spermatogenesis as in view (**g**). View (**h**) is from a Hereford bull in Manitoba, Canada, 1997

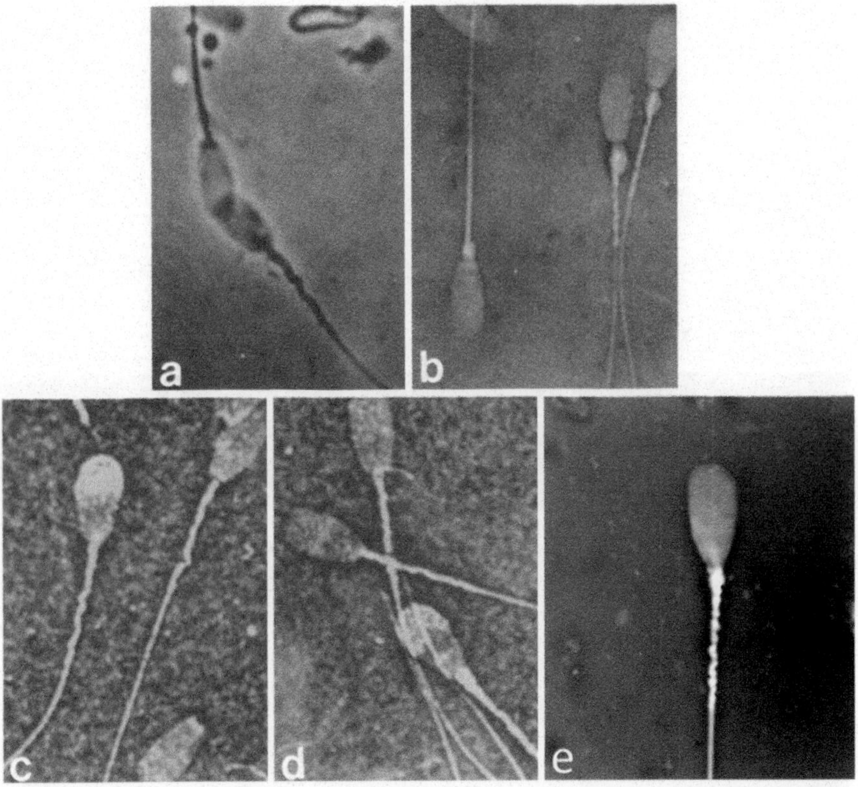

Fig. 4.17 Corkscrew defect. Views (**a**) and (**b**) were produced from eosin-nigrosin stained smears. It was difficult get good photographs because semen from the affected bulls was very dilute. Views (**c**) and (**d**), are india ink smears from semen of Aberdeen Angus bulls (Courtesy of E. Blom). Note the corkscrew like appearance of the midpiece. The defect is characterized by lumpy accumulations of mitochondria along the midpiece and affected sperm are always dead (stained with eosin). View (**e**) is a photomicrograph from one of two Saskatchewan bulls (Charolais and Red Angus) affected with this defect. In these cases, the corkscrew defect was confirmed by electron microscopy, however, the cause was not determined

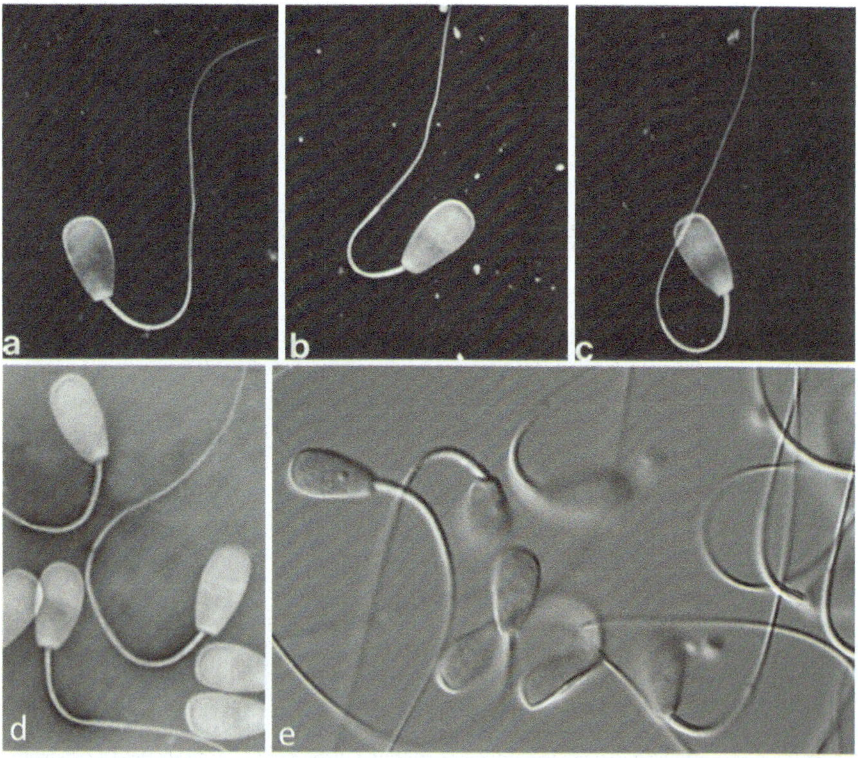

Fig. 4.18 Bowed midpieces. These photomicrographs show the bowlike curving of the midpiece that may occur as the cell gradually loses motility and dies. Rare cases have been observed in which there is a permanent bend in the midpieces of motile sperm. In these cases motion occurs in a tight circle

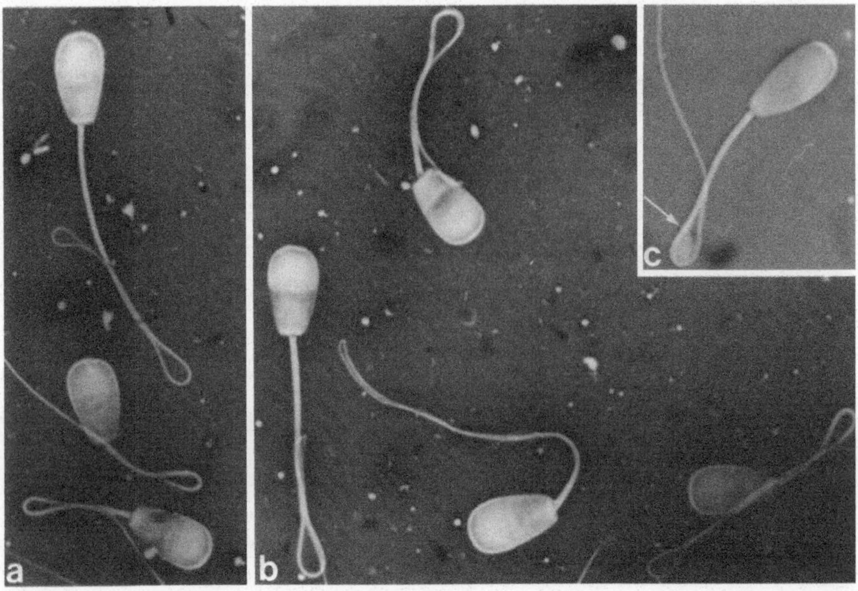

Fig. 4.19 Bent principal pieces. Views (**a**) and (**b**) show a variety of loops and bends primarily in the region of the principal piece, although a reflex bend in the distal midpiece may also occur. These types of bends and loops were produced by exposing normal spermatozoa to a hypotonic solution. Note that none of the spermatozoa in (**a**) and (**b**) have entrapped cytoplasmic droplets. View (**c**) shows a reflex bend in the principal piece with an entrapped droplet. This form is uncommon and primarily occurs in association with the distal midpiece reflex defect originating in the epididymis. The arrow points to the midpiece principal piece junction

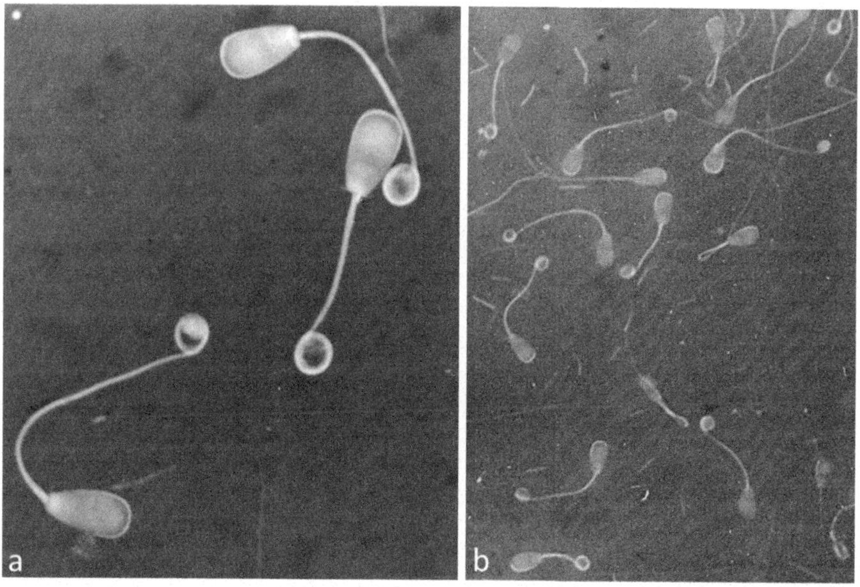

Fig. 4.20 Coiled principal pieces. This defect is not commonly seen in sperm smears. It has never been reported to occur in high proportions of spermatozoa and has not been linked to infertility. View (**a**) is from a semen smear of a Maine Anjou bull that was found to produce spermatozoa with coiled principal pieces at an incidence of 10–15% at each spring prebreeding exam for 2 consecutive years. View (**b**) is from a semen smear of 1 of 4 crossbred bulls that had their scrotums insulated for a 4-day period

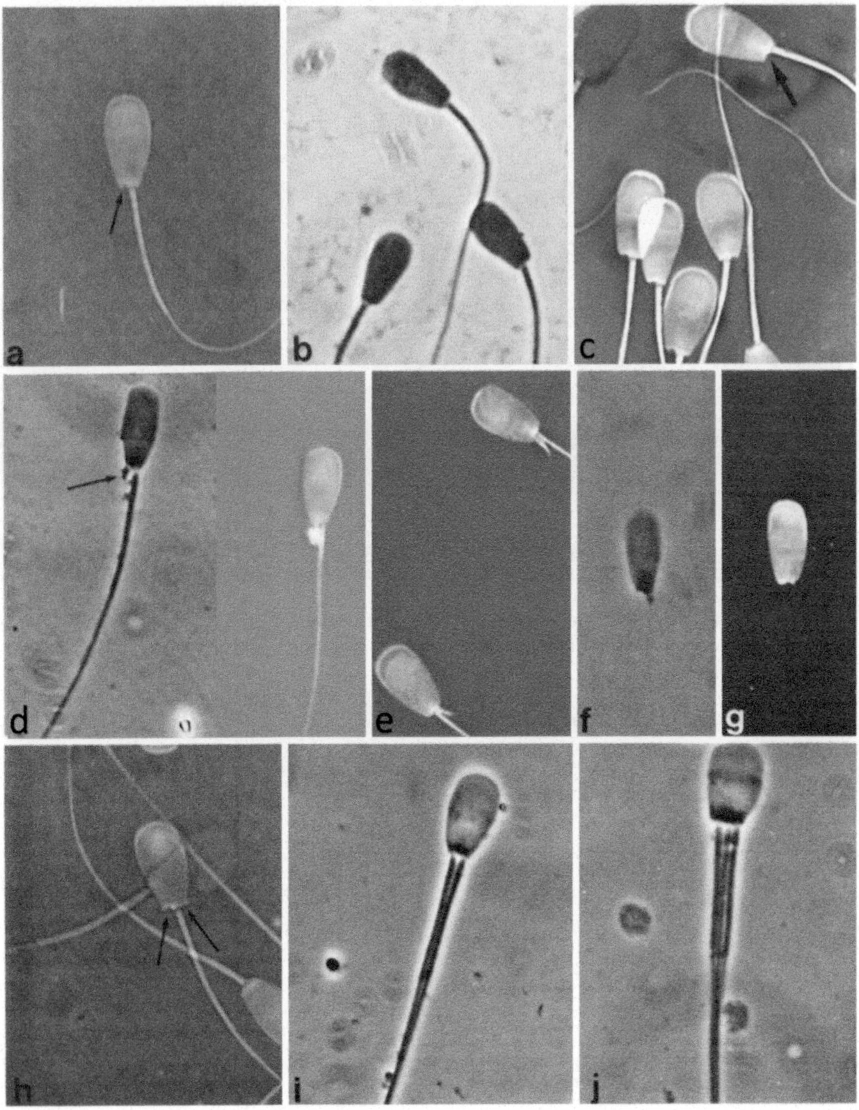
a
b
c
d
e
f
g
h
i
j

◀ **Fig. 4.21 Abaxial, Accessory, and Multiple tails.** These abnormal forms are usually associated with each other; however, abaxial tails (**a–c**) are by far the most common. Abaxial tails alone without multiple tails do not reduce fertility. Views (**b**), (**d**), (**f**), (**i**), and (**j**) are phase-contrast micrographs. View (**a**) depicts the usual appearance of a sperm cell with an abaxial tail attachment. A secondary empty implantation fossa (arrow) is nearly always present. All three sperm cells in (**b**) have abaxial tail attachments. This figure demonstrates how easily the anomaly may be overlooked in routine preparations. Abaxial tails often come off the base of the head at an angle. A slight angle between the axis of the tail and the base of the head is noticeable in most cases with abaxial tail attachment (**a–c**)

View (**d**) depicts the accessory tail defect. Note the short stump (arrow). The stump rarely varies from the size seen here. Often the accessory tail is obscured by the cytoplasmic droplet as seen in the sperm on the right (**d**). In (**e**) the rudimentary tail or filamentous stump can be seen clearly. View (**f**) shows a detached head with an accessory tail. A detached head with two implantation fossas is seen in (**g**). View (**h**) illustrates a relatively normal sperm cell except for extra implantation fossas on either side of the tail (arrows). The implantation fossa at the base of the nucleus is formed before the centriolar complex arrives in preparation for implantation and tail formation. Additional fossas without a tail may occur if a centriolar replication error occurs whereby there are extra centriolar forms moving towards the base. If one or more additional centriolar complexes were to fail to attach or deteriorate en route, empty fossas could result. It is likely that sperm retaining an accessory tail will produce an extra aster within the ovum resulting in uneven chromosome division. Centriole replication results in the multitude of cilia seen in cells of many tissues. In spermatozoa it appears that centriole replication is suppressed; however, perhaps not completely in some cells, resulting in accessory or multiple tails (i) and (j)

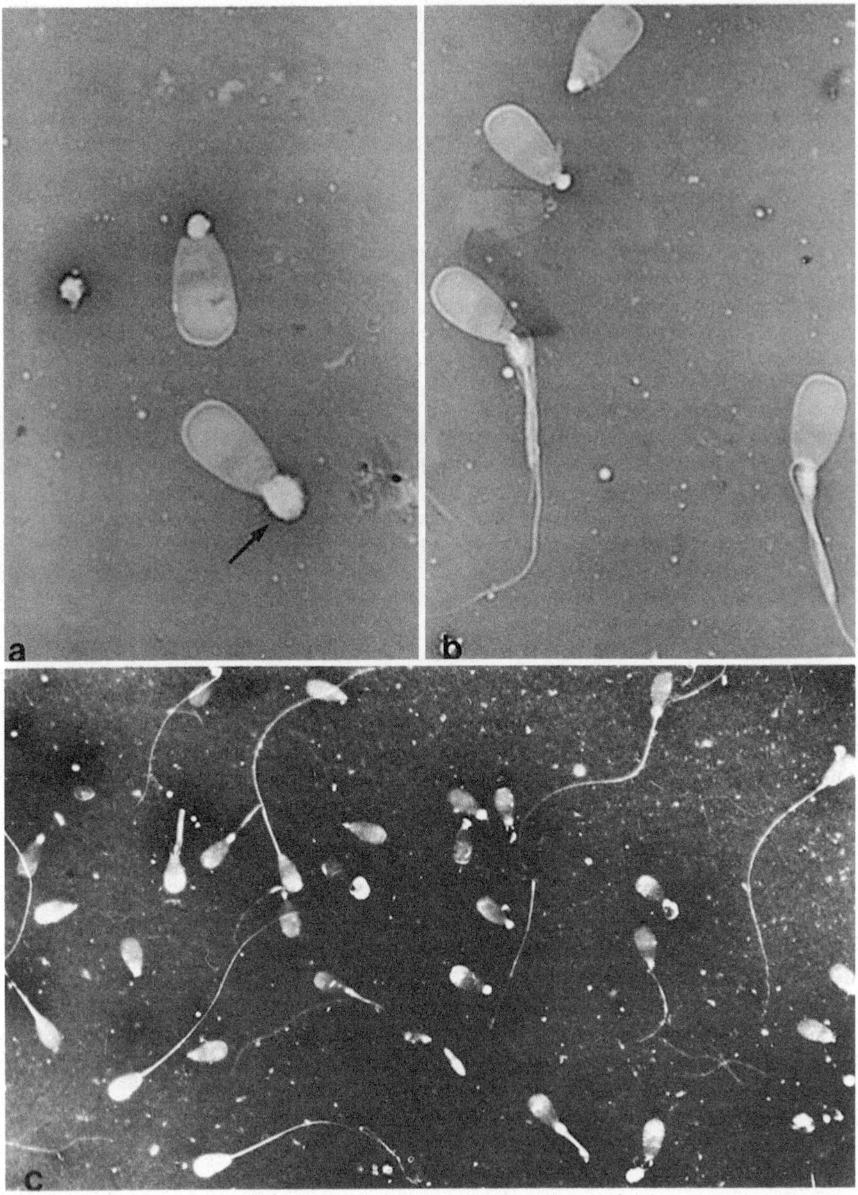

Fig. 4.22 Tail stump defect. This uncommon sperm defect has been reported in many breeds of cattle and in most domestic species. It is most likely inherited as a recessive condition. The typical appearance of this defect is a very short stump attached to the base of the nucleus. Frequently the tail stump is obscured by a retained cytoplasmic droplet (arrow) as seen in (**a**). Other forms of tail defects are often seen in association with the stump tail defect as shown in (**b**) (Courtesy of M. Wenkoff). View (**c**) is a survey micrograph from an india ink smear demonstrating the tail stump defect (Courtesy of E. Blom)

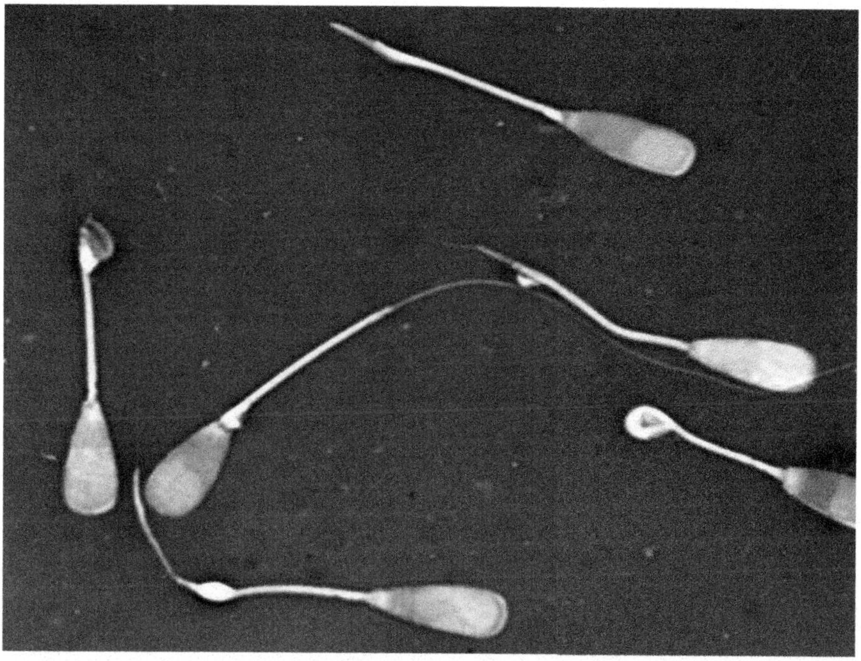

Fig. 4.23 Short tail defect. This defect was first described in boars in 2000. The condition was found in a 2-year-old Limousin bull in Alberta in 2002 as viewed above. In that case, the length of the principle piece was markedly reduced or missing in nearly 100% of sperm resulting in a total lack of sperm motility. Sixty-four percent of sperm had proximal droplets and 19% had distal droplets; otherwise there were very few other sperm defects. The condition was also described in 3 Nelore bulls in 2010 where 34–70% of sperm were affected, proximal droplets occurred in 1–6% of sperm and there were small percentages of other sperm defects. In boars the short tail defect was shown to be due to an autosomal recessive gene and in the case of the Nelore bulls, two were half sibs. Short sperm tails or principal pieces have been induced in mice by gene deletion of Akap4 or Akap3 encoding for two prominent fibrous sheath structural proteins making up the principal piece of the tail

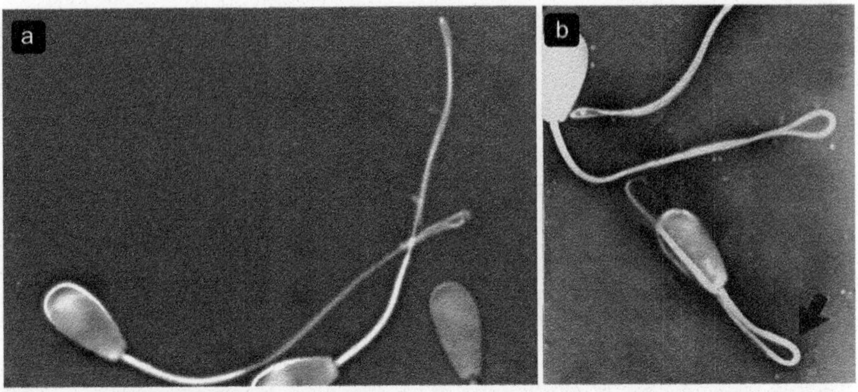

Fig. 4.24 Hypotonic shock. Eosin-nigrosin and eosin-aniline blue stains are very hypotonic. When stain and slides are cold, smears dry slowly resulting in prolonged exposure of sperm to a hypotonic medium. Urine contamination may also induce hypotonic shock. The primary manifestation is a severely contorted distal portion of the principal piece as seen in (a). In severe hypotonic shock, midpieces may also bend distally (arrow in (b)) but without a trapped droplet. It is possible for distal midpiece reflexes to form in the epididymis without a trapped droplet since not all sperm in the cauda epididymis have a distal droplet

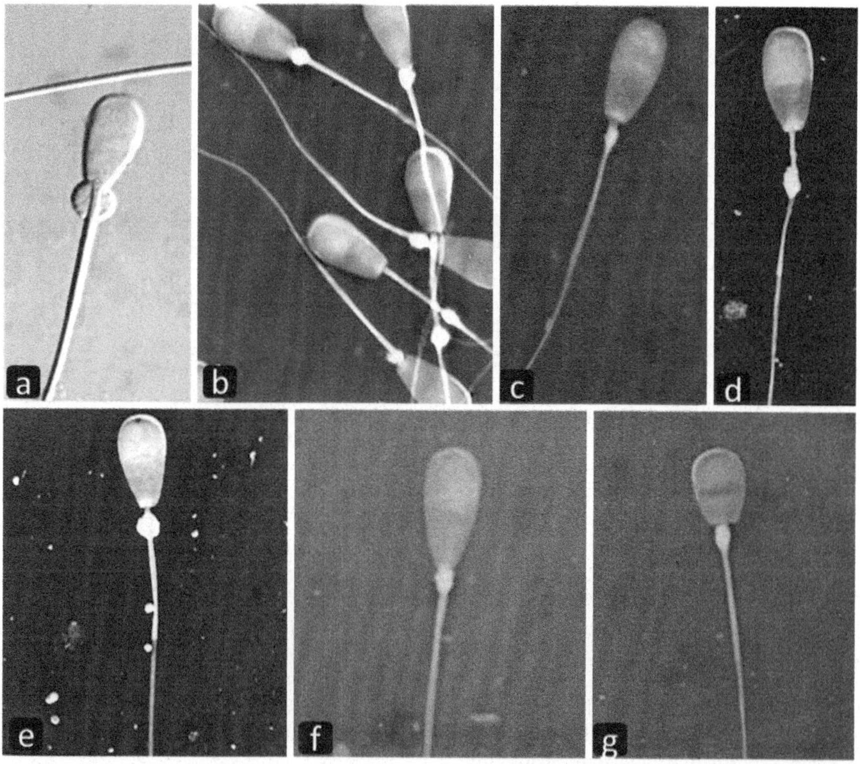

Fig. 4.25 Proximal droplets. View (**a**) shows the large rounded cytoplasmic droplet retained in a proximal midpiece position. This is the view in wet mounts when viewed by DIC or phase microscopy. When the semen is used to prepare smears with morphology stain, droplets usually have a dense opaque appearance (**b–d**). Views (**b**,**c**) show the shrunken refractile appearance of a proximal droplet as it is seen with bright-field microscopy of an eosin-nigrosin-stained smear. View (**d**) shows a droplet in a partially migrated position on the midpiece. The droplet is very rarely found in this position

Views (**e**), (**f**) and (**g**) depict abnormal sperm (nuclear vacuole, pyriform head, and microcephalic sperm respectively), which frequently have retained cytoplasmic droplets in the proximal position

Virtually all sperm entering the caput epididymis have a cytoplasmic droplet at the proximal midpiece position. By the time the sperm have progressed through the caput, approximately 90% of droplets have shifted to the distal midpiece. Proximal droplets in ejaculated sperm are a sign of a disturbance in epididymal, or testicular function. Within 7–10 days after stress, there will be an acute rise in proximal droplets, indicating that normal sperm within the epididymis have been affected. However, the incidence of proximal droplets may also rise 3–4 weeks after an insult to spermatogenesis in association with structurally defective sperm

In vitro studies have shown that sperm with proximal droplets fail to bind to ova. Furthermore, normal appearing sperm associated with such sperm also have a reduced ability to bind to ova. The fundamental problem appears to be an abnormality in sperm surface proteins required for attraction and binding to the zona pelucida

A Red Angus bull that consistently produced about 70% proximal-droplet-sperm, was placed in a breeding trial with 29 females for 21 days. His first service pregnancy rate was 17.2% compared to pregnancy rates of 61.2% and 81.5% for control bulls under similar conditions

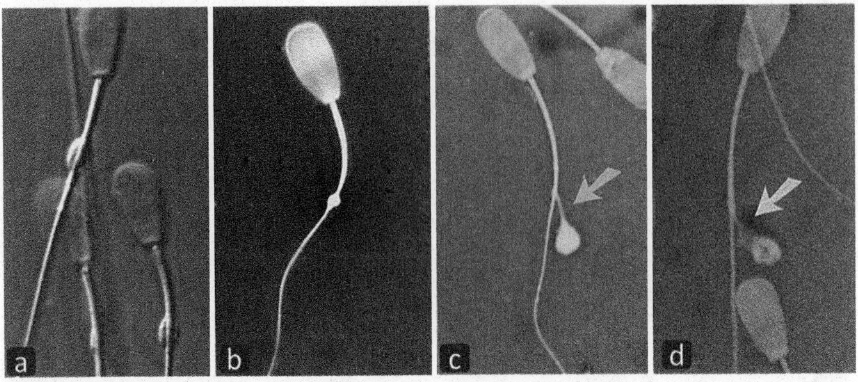

Fig. 4.26 Distal droplets. View (**a**) shows a large rounded cytoplasmic droplet in the distal midpiece position as it normally appears in wet mounts under DIC. View (**b**) shows the shrunken refractile appearance of the droplet in a distal midpiece position as seen in eosin-nigrosin stained semen smears examined with bright-field microscopy
Views (**c**) and (**d**) show the droplet being shed by the formation of long cytoplasmic stalk (arrows). Most of the sperm that are stored in the cauda epididymis have a cytoplasmic droplet at the distal midpiece position. Immediately after mixing with seminal fluids after ejaculation, the droplet is shed. A hemolytic factor present in vesicular fluid is necessary for shedding to occur. Shed droplets tend to break up quickly leaving debris in the seminal plasma. Distal droplets have not been associated with infertility in bulls, therefore, sperm with normal morphology retaining a distal droplet should not be considered abnormal. In one case, a 16-month-old bull produced normal sperm with 51% distal droplets; however, 2 h later, the same bull produced normal sperm with only 1% distal droplets. The bull was used in pasture mating immediately after the breeding soundness evaluation and had normal fertility

Fig. 4.27 Teratoid forms. Views (**a**)–(**f**) show some of the variations of teratoid forms of spermatozoa that may appear in the ejaculate when there is a disturbance of spermatogenesis. These abnormal products of spermatogenesis may be mistaken for debris or inflammatory cells. They are severely defective sperm usually with abnormal DNA condensation, often with a deformed nucleus and, in most cases, a deformed midpiece and principal piece wrapped around the nucleus. Views (**a**), (**b**), and (**c**) show a common type of teratoid form that appears when the disturbance in spermatogenesis is moderately severe. Note that the midpiece usually lies over the sperm head in a bent or partially coiled form. The principal piece may be coiled over the sperm head (arrow heads) or project away from it. Less commonly, the sperm tail is straight, bending acutely at the neck region and lying across the sperm head as in (**c**) (arrow)
Views (**d**), (**e**), and (**f**) show some of the many variations of teratoid forms associated with severe disturbances of spermatogenesis (arrows). Other sperm abnormalities are always present in increased numbers in association with teratoid forms. Note the sperm head abnormalities in (**a**), (**d**), and (**f**) and the proximal droplets on sperm cells in (**a**) and (**f**). When many teratoid forms appear in semen smears, there is usually an increased amount of debris. This is especially noticeable in (**a**) and (**f**)
Teratoid forms are often found at an incidence of 0–2% in normal semen. In disturbances of spermatogenesis, the incidence is usually in the range of 5–10%. Over a period of 20 years azoospermia or oligospermia was observed in 18 bulls of several breeds. Three of the bulls were closely related. All had well-developed firm testicles, but ejaculated white semen-like fluid which contained an abundance of teratoid forms and only a few complete sperm. None showed any sign of recovery over observation periods of 6 to 24 months

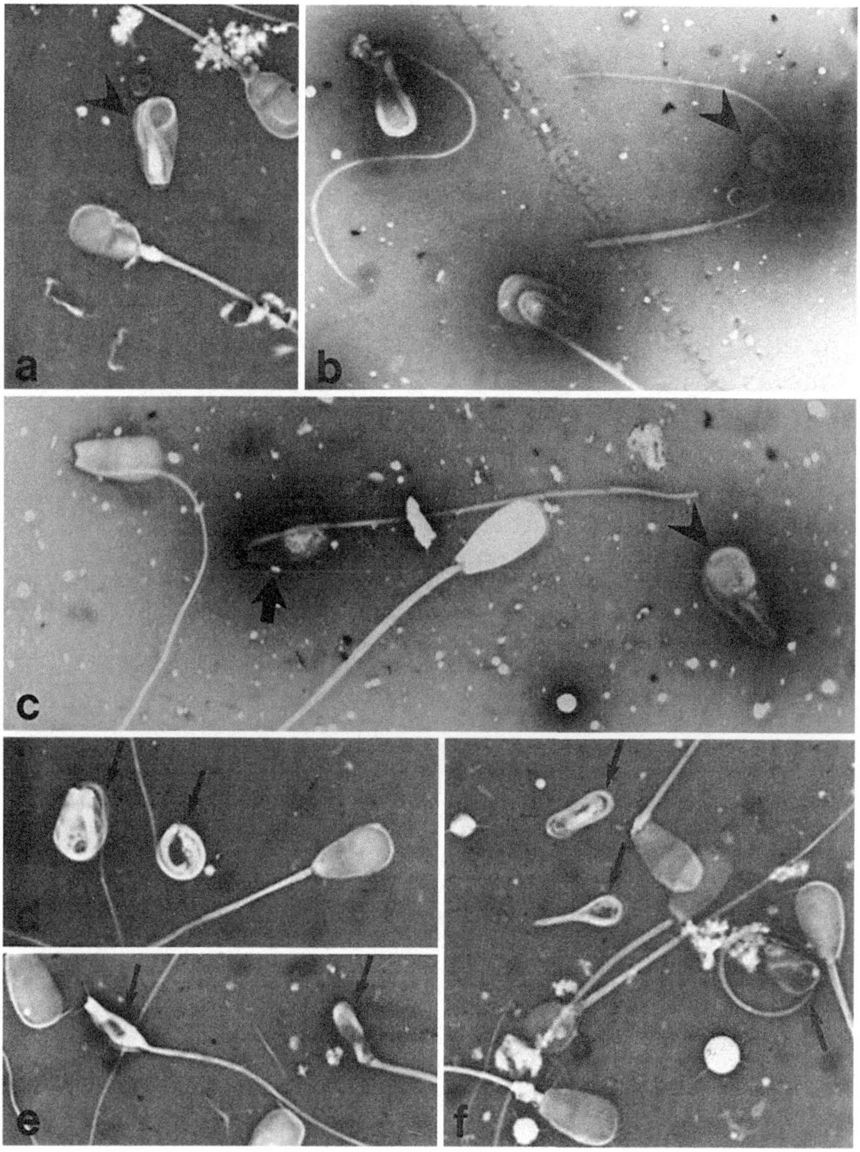
a
b
c
d
e
f

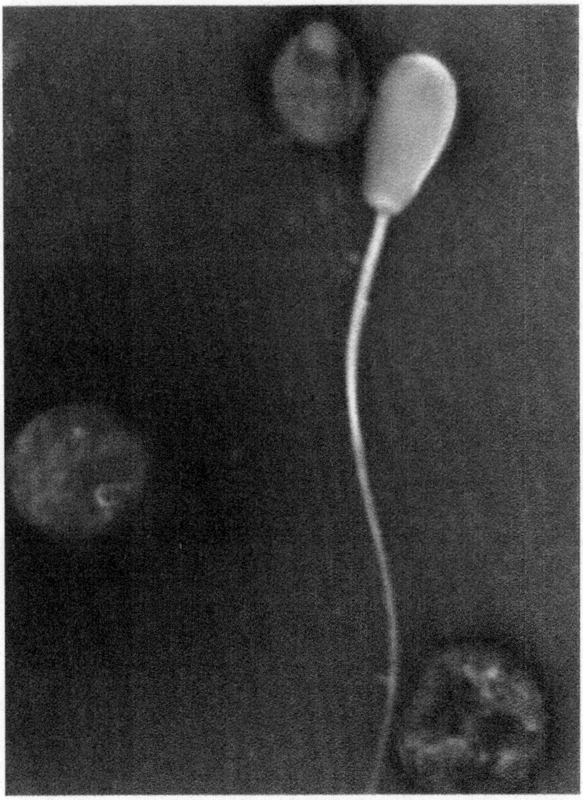

Fig. 4.28 Spheroids. This view shows a normal spermatozoa among three spheroids. Spheroids are not usually seen in large numbers in semen smears even during testicular degeneration, and it may be difficult to determine their origin. Some types contain nuclear material and appear to be degenerating spermatogenic cells, but in most cases they do not contain any nuclear material and they are derived from disintegrating white blood cells in the aftermath of a vesicular gland infection. Occasionally, large numbers of spheroids occur with few if any white blood cells present. Within a month or less, the spheroids disappear from the semen and fertility is unaffected. Sperm morphology may be normal when the spheroids develop in association with inflammatory cells

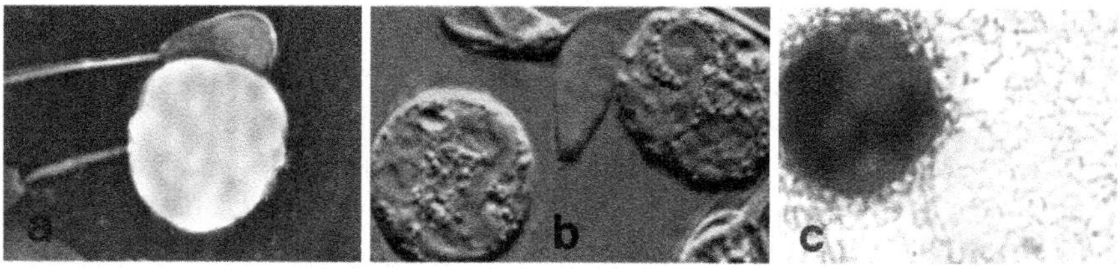

Fig. 4.29 White blood cells (WBCs) in semen may be due to inflammation of the ampullae, accessory sex glands, or epididymides. WBCs may also be contaminants from preputial fluids, particularly when the glans penis and prepuce are mildly abraded or injured. WBCs with intact membranes do not take up eosin stain and therefore they appear as white, somewhat irregular bodies, 1–2 times the length of a sperm head (**a**). Those with broken membranes appear as pink disintegrating bodies in eosin nigrosin stained smears. Positive identification can be made with differential interference contrast (**b**) or phase contrast microscopy. New methylene blue, Wright-Giemsa, or Diff-Quik stains will reveal the typical lobulated appearance of neutrophils (**c**)

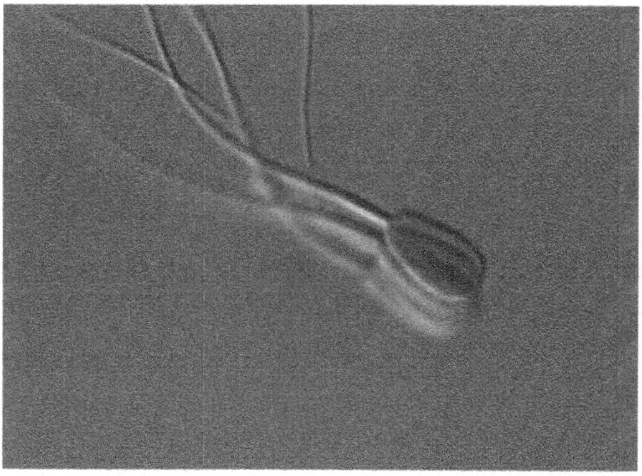

Fig. 4.30 Sperm agglutination. This view show several agglutinated bull sperm. Circulating anti-sperm antibodies are quite common in populations of various domestic species, but sperm agglutination in semen samples is very uncommon. Spermatids and spermatozoa are isolated from the immune system within the seminiferous tubules due to a blood testis barrier. When this barrier is disrupted, exposure of the immune system to spermatic antigens would stimulate the formation of antisperm antibodies (ASA). ASA formation can be induced by trauma, infectious and noninfectious inflammation, or by obstruction followed by rupture in the testicular efferent duct system. Experiments have shown that the presence of ASA is capable of reducing fertility; however, there are no reports of naturally occurring cases definitively linking anti-sperm antibodies, or sperm agglutination with infertility. Therefore, when sperm agglutination is encountered during semen assessment, there can be no certainty of an effect on fertility. Furthermore, it is not known how long a condition of sperm agglutination might be expected to persist

Compliance with Ethical Standards

All applicable international, national and institutional guidelines for the care and use of animals were followed. This current work was supported by the Natural Science and Engineering Research Council of Canada (Grant number RGPIN/05305) to Richard Oko at Queen's University, along with the Agriculture and Food Research Initiative Competitive Grant number 2023–67015-39262 from the USDA National Institute of Food and Agriculture, Animal Breeding, Genetics & Genomics Program and seed funding from the College of Agriculture, Food and Natural Resources of the University of Missouri to Peter Sutovsky. Work for this book was also supported by a 2021–2022 Lalor Foundation Postdoctoral Research Fellowship to Lauren Hamilton. The research support to Viv Perry was funded by Australian Research Council (Grant number LP110100649), English Beef and Lamb Executive a division of the Agriculture and Horticulture Development Board (Grant number RC3815), Queensland Sperm Morphology Laboratories, Ridley AgriProducts Pty Ltd and the Australian cattle industry. The Canadian Association of Animal Breeders provided the foundation for research by Albert Barth. They supported the Development of the Diagnostic Spermatology Laboratory at the Western College of Veterinary Medicine, Saskatoon, SK., along with related support for numerous other research directives. Many grants were also received from Saskatchewan Agriculture, including the Cattle Marketing Deductions Fund, the Horned Cattle Trust Fund, the Cattle Breeders Research Council, the Beef Development Fund and the Agriculture Development Fund. Animal research approval for research or clinical studies summarized in these chapters were given by respective University Animal Care Committees. The authors declare no conflict of interest as there is no perceived interest in prejudicing the impartiality of the research reported.

Chapter 5
Bull Sperm Abnormalities in Practice

Contents

Abstract In this chapter we discuss spermatozoal abnormalities which are associated with reduced bull fertility in practice as assessed by published case studies or research experiments. Each abnormality follows the standardized form found in Chap. 9. An initial summary of the abnormality is given including the suggested tolerated threshold for the abnormality in an ejaculate. This is followed by detail of research into the abnormality and its associated affect upon fertility in the bull. Images of the abnormalities are presented and the reader is further directed to photomicrographs of Chap. 4 where appropriate.

Keywords Proximal droplets · Pyriform heads · Knobbed acrosomes · Ruffled plasmalemma and acrosome · Nuclear vacuoles · Abnormal DNA condensation · Pale centres · Microcephalic heads macrocephalic heads · Decapitated sperm · Rolled nuclear crest · Distal reflex midpiece · Segmental aplasia · Corkscrew defect · Pseudodroplet · Dag defect · Abaxial tails · Short tail defect · Detached heads · Coiled principal piece · Azoospermia · Oligospermia · Multinuclear-multiflagellar · Sperm accumulation

© The Author(s), under exclusive license to Springer Nature Switzerland AG 2025

A. Barth et al., *Abnormal Morphology of Bovine Spermatozoa*,
Advances in Anatomy, Embryology and Cell Biology 240,
https://doi.org/10.1007/978-3-031-70126-9_5

5.1 Abnormalities Associated with the Head of the Sperm

5.1.1 *Proximal Cytoplasmic Droplets (PD)*

Summary: PDs are normally observed in the pubertal bull with incidence decreasing with age (Amann et al. 2000; Copping et al. 2018). In the mature bull they indicate abnormal spermiogenesis (and/or epididymal function). They were observed 7–10 days following a temperature or stress event (Barth and Bowman 1994; Copping et al. 2018) and 15 days following ruminal acidosis (Callaghan et al. 2016).

Counts of 10–15% proximal droplets (Barth et al. 2008) have been associated with decreased fertility. This trait is considered uncompensable as the sperm fail to bind to the ova and, furthermore, other sperm associated with high numbers of proximal droplet sperm have impaired ability to bind with the ova (Thundathil et al. 2001a). In bulls with > 30% proximal droplets, Amann et al. (2000) reported that the associated apparently normal spermatozoa displayed immaturity and reduced ability to fertilize ova. PDs are also associated with decreased membrane integrity and increased chromatin damage post-thaw (Carreira et al. 2012). We consider this defect should have a threshold of 20% as studies show proximal droplets are associated with poor pregnancy rates (Söderquist et al. 1991).

Detail: As spermatids within the seminiferous epithelium change from spherical to elongated cells, the cytoplasm is drawn toward the tail region. As development progresses further, the Sertoli cell shapes the superfluous cytoplasm into a lobule called a residual body. The residual body remains attached to the spermatid by a slender stalk, and at spermiation, the stalk is severed, leaving a small droplet in the neck region of the sperm (Bloom and Nicander 1961). This is a selective process, as the organelle content of the droplet is not made up of a random selection of superfluous organelles as in the residual body but rather a select group of saccular elements originating from the Trans aspect of the Golgi apparatus (Oko et al. 1993). All sperm entering the caput epididymis have a cytoplasmic droplet at the sperm neck position containing these flattened saccular membranous elements. By the time sperm have progressed through the caput approximately, 90% of droplets have shifted to the distal midpiece (Barth 1989). Droplets are normally removed at ejaculation in the bull (Cooper 2011).

Clinically when droplets on sperm are viewed in wet mounts (physiological solution) they appear round and translucent; however, when the same semen is used to prepare smears with morphology stain, droplets have a dense opaque appearance. Often the ragged opaque appearance of a proximal droplet viewed using morphology stain is suggestive of a mitochondrial helix defect. Whether it is droplet or mitochondrial material can be determined by viewing the semen in a wet mount with phase contrast. On phase contrast, the droplet matrix appears as an accumulation of granular material and the midpiece can be seen clearly through the droplet. On electron microscopy, the droplet is limited by the sperm plasmalemma and consists of flattened saccular elements as the dominant membranous component (Oko et al. 1993; Bloom and Nicander 1961; Garbers et al. 1970). The flattened saccular elements are loose or in stacked arrangements and between them are occasional vesicles and a few spherical bodies with a double membrane (Figs. 5.1, 5.2).

The proposed functions of the droplet need further elucidation. There are components of organelle degradation pathways within the droplet material, including 15-LOX and various components of the ubiquitin–proteasome pathway (Fischer et al. 2005). These contribute to organelle membrane breakdown and degradation in other cellular systems (van Leyen 1998) and may work within the droplet to

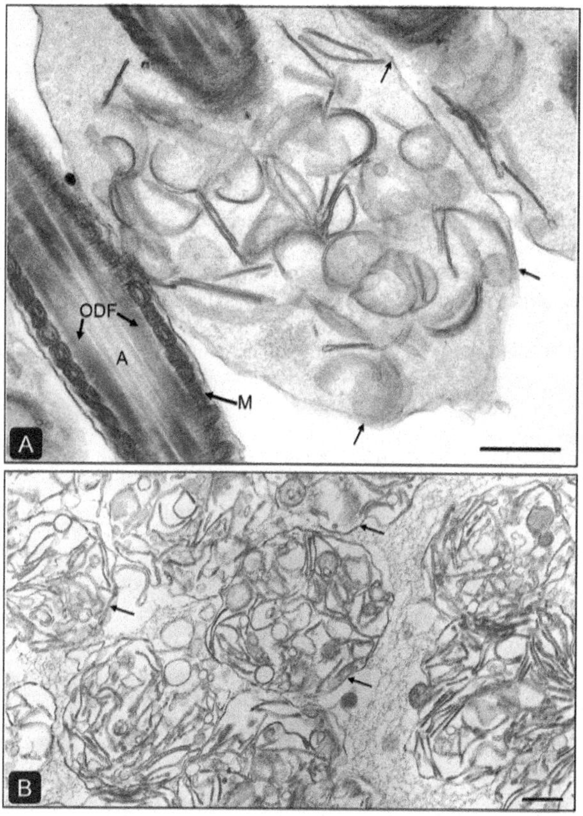

Fig. 5.1 (**A**) Oblique section through the tails of rat spermatozoa at the level of the cytoplasmic droplet in the lumen of the epididymis showing loosely stacked, straight, or C-shaped flattened saccular elements, selectively derived from the Trans face of the Golgi apparatus. The flattened saccular elements show areas of close apposition but large gaps between elements are also observed. The edges of the saccular elements are occasionally observed in close proximity to the plasma membrane (arrows). M, mitochondrial sheath; ODF, outer dense fiber; A, axoneme. (**B**) Morphological analysis of isolated droplets showing numerous flattened straight or C-shaped saccular elements enclosed within a remnant plasma membrane (arrows). Bars = 0.5 μm. For detail see The Journal of Cell Biology, Volume 123, Number 4, November, 1993: 809–821

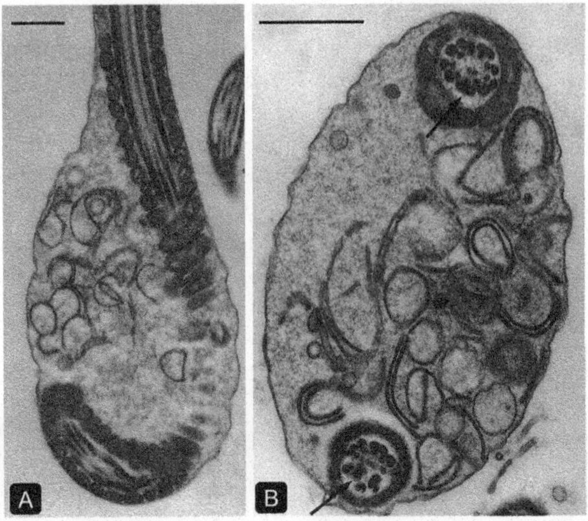

Fig. 5.2 **(A)** Oblique section through sperm tail of bull affected with the distal reflex defect. The tail bends at the distal end of the midpiece around the distal cytoplasmic droplet, which is laden with saccular elements. **(B)** Cross section through the cytoplasmic droplet of bull sperm tail affected with the distal reflex defect. Note that the saccular elements make up the majority of organelle-content in the cytoplasmic droplet and have a similar structure and disposition as found in the cytoplasmic droplet of rat spermatozoa (see Fig. 5.1A). The arrows point to the disorganization of axial fibers in one half of the axial fiber bundle of the principal piece (lower arrow) and the unsymmetrical positioning of axial fibers in the midpiece (upper arrow). Bars = 0.5 μm

promote its movement from the proximal to distal midpiece position (Fischer et al. 2005). The evaluation of the content of acid phosphatase within the isolated droplet fraction was found to be very low (Oko et al. 1993), confirming previous negative cytochemical reactivity tests for acid phosphatase in the cytoplasmic droplet (Hermo et al. 1988). It is therefore unlikely that the droplet represents a modified lysosome, a view proposed by Dott and Dingle (1968). The identification of Golgi/TGN glycosylating activities in the saccules of the droplet may be related to plasma membrane modifications which occur during epididymal sperm maturation (Oko et al. 1993). The passage of the droplet along the midpiece may aid lipid diffusion and hence sperm maturation (Cooper 2011).

Proximal droplets in ejaculated sperm are a sign of a disturbance in epididymal or testicular function. In normal bulls, within 7–10 days after stress, there will be an acute rise in proximal droplets on otherwise normal appearing sperm, indicating that normal sperm within the epididymis have been affected (Barth and Bowman 1994). However, the incidence of proximal droplets may also rise 2–4 weeks after a relatively severe insult to spermatogenesis on sperm that are abnormally formed within the testis (Callaghan et al. 2016). Abnormal sperm, e.g., those with pyriform heads, seem more predisposed to retain the droplet in a proximal position on the midpiece. Pubertal bulls often have elevated numbers of sperm with proximal droplets, as well as a large variety of other defects (Felton-Taylor et al. 2020). Usually by the time a bull is 14 to 15 months old, spermatogenesis becomes normal and the elevated numbers of proximal droplets, as well as other types of sperm defects, decline, this may take longer in Bos indicus breeds (20 months). In northern regions of Canada, it is not uncommon to find a small percentage of mature bulls in spring breeding soundness exams to have elevated percentages of proximal droplets. In the UK, elevations in proximal droplets are more likely to occur in summer (Sekoni and Gustafsson 1987). In Australia, however, no effect of season upon proximal droplets was observed although increased PD were found in the hot Northern regions (Felton-Taylor et al. 2020). In this study it was also found that some breeds were more susceptible to PD than others. Taken together, over all breeds, 8.7% of bulls were observed to have greater than 20% PD in their ejaculate thereby failing Bull Breeding Soundness Evaluation (BBSE) on this category.

Proximal droplets must be considered as a serious abnormality even when sperm look structurally normal. Although the cytoplasmic droplet's functions still need to be elucidated, the droplet's abnormal retention or positioning along the tail may simply be a manifestation of abnormal epididymal function resulting in a deficiency of sperm membrane remodeling. There are some very important changes that take place in the sperm membrane during sperm passage through the epididymis that are necessary for normal motility, capacitation and fertilizing ability (Jones 1989).

IVF studies have shown that sperm with proximal droplets fail to bind to ova. Furthermore, even normal sperm without droplets that are associated with high numbers of proximal-droplet-sperm have an impaired ability to bind to ova (Thundathil et al. 2001a).

A Red Angus bull that consistently produced about 70% proximal-droplet-sperm, (otherwise sperm were structurally normal) was placed in a breeding trial with 29 females for 21 days. His first service pregnancy rate was 17.2% compared to pregnancy rates of 61.2% and 81.5% for control bulls under similar conditions (Barth 1997). Others have also shown that high percentages of proximal droplets are related to poor non-return rates on AI and pregnancy rates in natural service (Söderquist et al. 1991; Nöthling and Arndt 1995).

In our experience, most bulls that have little evidence of any serious disturbance of spermatogenesis, but have a high percentage of proximal droplets, even as high as 80%, usually recover within a few weeks. In rare cases, possibly due to due to subclinical epididymitis, elevated numbers of sperm with proximal droplets may persist for many months.

5.1.2 *Distal Cytoplasmic Droplets*

Summary: Although present along the midpiece rather than adjacent to head they are considered here to clearly differentiate them from proximal droplets. Unlike the boar, there are no reports of distal droplets being associated with infertility in the bull (Barth 2013; Sutovsky 2015). Case studies using bulls with high numbers of distal droplets in natural service achieve normal pregnancy rates (Barth 2013). Sperm with distal droplets will lose the droplet if left in a water bath for 15–30 min or if gently agitated. The number of sperm within the ejaculate with distal droplets, also vary widely between sequential ejaculates (McAuliffe et al. 2010). For this reason distal droplets in isolation are not generally considered to be a defect by the authors or by other researchers (Johnson 1997; Barth 2013) and are placed in the normal category.

Detail: A hemolytic factor, phospholipid binder of sperm protein (BSP) with a molecular weight of 80,000, was found in bovine vesicular glands and purified by (Kysilka 1973). The protein was demonstrated, by indirect immunofluorescence, to bind to the sperm membrane over the acrosome and over the distal cytoplasmic droplet (Matousek and Kysilka 1984). It was shown that, besides its hemolytic activity, BSP was able to remove cytoplasmic droplets from bull epididymal sperm as well as from ram, boar and rabbit sperm (Matousek and Kysilka 1984; Matoušek and Kysilka 1980). It appears that, with the mixing of sperm with seminal fluid and the onset of motility, droplets are shed from the distal midpiece due to increased fluidity of the sperm membrane over the droplet region. At the distal midpiece region the membrane stretches, forms a stalk and ruptures releasing a non-membrane bound remnant of spermatid cytoplasm. Much of the debris seen in semen smears is from disintegrating shed droplets.

BSP was also found in the ampullae and, not surprisingly, many ampullary sperm were found to be free of distal cytoplasmic droplets (Matousek and Kysilka 1984). Since this time many other proteins have also been found within the bovine accessory gland secretions which are essential to the fertility of sperm. For this reason, massage of the seminal vesicles during BBSE should not be ignored.

The retention of a cytoplasmic droplet on sperm, whether in the proximal or distal position, is an indication that the sperm membrane may not have acquired these essential binding proteins. In the case of proximal droplets, missing membrane proteins are likely those that should have been acquired during epididymal passage and which are needed for sperm to bind to the zona pellucida of ova (Jones 1989). Sperm membrane proteins that are acquired from seminal plasma are those needed for droplet shedding and for binding of sperm to the oviductal epithelium in order to establish a sperm reservoir within the isthmus of the oviduct (Suarez 2015). Establishment of an oviductal sperm reservoir is essential for fertility. Frozen boar semen containing a high percentage of distal-droplet-sperm had impaired fertility (Waberski et al. 1994) confirming the theory that membrane-binding proteins might be deficient in such sperm. As far as we are aware there are no reports of distal droplets being associated with bull infertility.

In one case, a 15-month-old bull produced normal sperm with 51% distal droplets; however, 2 h later the same bull produced normal sperm with only 1% distal droplets. Perhaps the first ejaculate, collected by electroejaculation, lacked sufficient vesicular gland fluid for droplet shedding. Interestingly, sperm from the first ejaculate, after incubation for 2 h, were in the process of shedding their droplets.

The bull was placed alone with 38 cows immediately after the breeding soundness evaluation for 70 days resulting in a pregnancy rate of 87%. Occasionally, the author, Albert Barth, observed incidences of distal droplets as high as 90% in mature bulls with otherwise satisfactory sperm morphology. Upon re-examination of semen collected from those bulls on the same day, or 2 to 4 weeks later, the incidence of distal droplets was less than 5%. It appears that electroejaculation may sometimes fail to expel sufficient vesicular gland fluid to allow droplet shedding.

In light of present information, sperm with normal morphology, but retaining a distal droplet should usually be placed in the normal category when constructing spermiograms for bulls to be used in natural service. When the percentage of sperm with distal droplets is very high, it would be wise to collect more semen samples on the same day, or within a few weeks, to determine whether the bull has a potential fertility problem.

5.1.3 Pyriform Heads and Tapered Heads (PY)

Summary: Narrow in the postacrosomal region. This condition is particularly seen in young over fat bulls (Barth 2013). It is important to note, that there is variation in the degree of this abnormality: In a series of trials (Barth et al. 1992) reported that fertility was related to the severity of pyriformity of the head. A moderate degree of pyriformity, in the absence of other signs of disturbed spermatogenesis, is not detrimental to fertility. However, extreme tapering in the postacrosomal region results in significant reductions in fertility. Pyriformity is considered only partially compensable (Thundathil et al. 1998a). This study also showed that even though some pyriform sperm were able to fertilize oocytes they had a reduced ability to cleave. The threshold of not more than 20% is therefore applied to this abnormality.

Pyriform heads are induced in bulls following stress such as dexamethasone treatment and scrotal insulation (Barth and Bowman 1994). The difference in severity of response between bulls, 3 to 4 weeks after initiation of these treatments, is suggestive of a genetic predisposition to develop pyriform heads in response to adverse environmental influences. Pyriform heads are differentially excluded from advancement in the female reproductive tract and specifically precluded from the cervix, uterus, utero-tubule junction (Mitchell et al. 1985).

Detail: In cattle, the most common abnormality; in the shape of sperm heads, is the pyriform (pear-shaped) defect characterized by a fully rounded acrosomal region and a narrow post-acrosomal region (Barth and Oko 1989). Tapered heads are less common and are narrow in both the acrosomal and post acrosomal region. They may coexist with pyriform heads. The overall incidence of affected bulls used in natural service or artificial insemination has been reported to be 9% and 16%, respectively; the percentage of pyriform sperm in samples from these bulls varied from 10% to>50% (Barth and Oko 1989). A recent Australian study of

11,387 range bulls, however found bulls failing BBSE on the pyriform head defect to be below 1% (Felton-Taylor et al. 2020) a much lower incidence than in the Canadian study. It is important to note, however, that, unlike earlier papers, this paper included only extreme forms of pyriformity based on the series of experiments completed by Barth et al. (1992): The broad range of severity within and between bulls in the Pyriform deformity category make it difficult to decide the point at which a sperm is normal or abnormal. The use of the pictogram available from this unique study assists practitioners in this differentiation (Barth et al. 1992). In this study seven experiments were conducted to determine the importance of various degrees of sperm-head narrowness to fertility (Barth et al. 1992). Sperm head shape varied among the study bulls from tapered to pyriform. Frozen semen used in these experiments was selected to be normal in all respects except for very high percentages (65–95%) of this single specific type of sperm head aberration. Semen with this sperm aberration and control semen were selected to be as similar as possible in dose and post thaw viability so that differences in fertility would be attributable to the morphological variant under study. Fertilization rates were determined by collecting embryos from the reproductive tracts of superovulated heifers that had been slaughtered 7 days after insemination. Pregnancy rates and rates of embryonic loss were studied in estrus-synchronized heifers by repeated transrectal ultrasound examinations from day 22 to day 55 after insemination. The results indicated that a moderate degree of sperm head narrowness, in the absence of other seminal signs of a disturbance of spermatogenesis, is not detrimental to fertility. However, extreme narrowness of the postacrosomal region of the sperm head of most spermatozoa, as was found in 2 bulls, resulted in significantly reduced fertility. These experiments helped to define the pyriform head abnormality at least in bulls that had few other kinds of sperm defects. Such bulls are actually rare and in most cases bulls with pyriform sperm have signs of a disturbance of spermatogenesis such as an increase in tail defects, detached heads, and proximal droplets.

The cause of the pyriform head defect appears to be due mainly to adverse environmental conditions. Early workers found that increased levels of pyriform heads were frequently present in clinical investigations of bull infertility and in cases of infertility due to disturbed spermatogenesis (Lagerlof 1938; Hancock 1959; Williams and Savage 1925). Treatment of bulls with dexamethasone to simulate stress, or with scrotal insulation, resulted in an increase in pyriform sperm 3 to 4 weeks after initiation of these treatments (Barth and Bowman 1994). Since there was a difference in severity of response between bulls with similar treatments it appeared that some bulls may be genetically predisposed to develop pyriform heads in response to adverse environmental influences (Barth 1986; Barth and Oko 1989).

The effect of pyriform heads on fertility has been studied with both in vivo and in vitro experiments. Thundathil et al. (1999) reported on two breeding trials. In the first trial heifers were superstimulated with FSH and artificially inseminated to a bull with 85% pyriform sperm or a control bull with 91% normal sperm. Fertilization rates of ova/embryos were lower for the semen with pyriform heads than for the control bull, 68.5% vs. 84.4% ($P<0.05$). In a second trial, estrous-synchronized heifers were artificially inseminated with semen from the same bulls. Pregnancy rates at day 60 were 37% and 61% for the bull with pyriform sperm and the control bull, respectively; ($P=0.22$) and the rate of embryo/fetal loss between Days 22 and 60 of pregnancy were 23% and 8% for the pyriform and control semen, respectively ($P=0.55$).

Failure of fertilization by pyriform sperm may be due to failure of sperm transport in the female reproductive tract or an inability of sperm to interact normally with the ovum (Hawk 1987, 1988). Transport of sperm with abnormal heads was impaired at the cervix (Mortimer et al. 1991; Karabinus and Saacke 1987; Kot and Handel 1987), uterus (Mitchell et al. 1985), and utero-tubal junction (Krzanowska 1974; Mortimer et al. 1991). Exclusion of spermatozoa with tapered or pyriform heads by the female reproductive tract was related to the severity of the deformity; the more severe the distortion from the normal shape, the greater the chances for exclusion. More severely deformed spermatozoa were also apparently excluded at the level of the zona pellucida (Mitchell et al. 1985).

One method of assessing sperm fertilizing competence has been to compare the morphology of accessory sperm (sperm trapped in the zona pellucida after ovum fertilization) with those of the inseminate. This comparison would evaluate sperm transport in the female tract as well as ovum binding and penetration. Such studies have shown that there is discrimination against pyriform and tapered heads. Accessory sperm classified as either tapered or pyriform were quite subtle in their deformity, suggesting that those with severe deformity did not gain access to the ovum (Saacke et al. 1998).

A series of IVF and culture studies was done to determine the cause of reduced fertility by specifically investigating the stage(s) during fertilization and early embryonic development in which pyriform sperm failed (Thundathil et al. 1999). Pyriform sperm had a reduced ability to bind to and penetrate the zona pellucida, and zygotes resulting from the fertilization of oocytes by pyriform sperm appeared to have a reduced ability to initiate cleavage.

In vitro and in vivo experiments such as this show that sperm head pyriformity is not compatible with fertility, however, the tolerable level of pyriform heads in ejaculates of breeding bulls or in frozen semen used for AI, has not been determined. The general recommendation that not more than 20% of abnormal sperm heads should be allowed for a satisfactory classification should apply to pyriform heads as well.

5.1.4 Knobbed Acrosomes

Summary: This abnormality may be heritable or arise following a stress event (Callaghan et al. 2016) and is often observed in the peripubertal bull (Copping et al. 2018; Callaghan et al. 2020) prior to the adult spermiogram. It was observed to rise 30d after a single acidotic event (Callaghan et al. 2016).

Two forms are regularly observed; beaded and indented, however a third form; flattened, a subcategory of the indented form (Thundathil et al. 2000, 2001b; Meyer and Barth 2001), is essential to differentiate due to the fertility prognosis of each:

i) The beaded form is considered inherited by an autosomal recessive gene (Barth 1986; Hancock 1953). More recently, a homozygous point mutation was shown to give to this form of the defect (Nogueira et al. 2022). The beaded form is often associated with sterility and usually occurs as a high percentage of the ejaculate.
ii) The indented form is described as an enlargement of the apical ridge that then folds back on the apex of the sperm head and is much more common than the beaded form.

As stated, indented or flattened acrosomes vary in their effect upon fertility. In natural service non-competitive matings such bulls may achieve near normal fertility, however this may reflect that normal sperm coexisting with these sperm are in sufficient numbers to achieve conception, as sperm with the flattened or indented form were unable to penetrate the zona pellucida (Thundathil et al. 2000). This abnormality is therefore considered a compensable defect and is given a 30% threshold in the BBSE. This is supported by work by Andersson et al. (1990) who reported that when present in less than 25% of sperm there was no decrease in fertility. In bulls with a high percentage of this abnormality (>80%) the indented acrosome defect may not be compensable as such sperm did not bind to the zona pellucida, and furthermore, other sperm present in the ejaculate that appeared normal and bound to the zona, resulted in zygotes with a reduced ability to cleave (Thundathil et al. 2001b).

Abnormal developmental morphology: Fig. 5.3 shows the characteristic appearance of the beaded for of the knobbed acrosome defect. It compares light and electron micrographs of bull spermatozoa affected with this defect. There appears to be two different modes of origin of the beaded form of the knobbed acrosome defect, both having similar consequences of swelling of the apical ridge of the acrosome and indenting or flattening the apex of the acrosome and the nucleus. One of the modes of origin is the entrapment of the membranous vesicles in the developing acrosome during the cap phase of spermiogenesis (Fig. 5.4 A), eventually causing a swelling of the acrosomic granule in step 9 spermatids (Fig. 5.4 B, C) and henceforth an abnormal development of the apical crest and hence apical segment of the acrosome (Fig. 5.4 D). These vesicles arise from invaginations of the acrosome membrane, which incorporate cellular components associated with the Golgi apparatus. Attesting to this, multivesicular bodies, normally associated with the Glogi apparatus, are sometimes found entrapped in the acrosomal membrane invaginations (Fig. 5.4 E) Another mode of origin occurs later in the elongation phase of spermiogenesis (steps 11–12). It appears as if the tongue-like acrosomal projection or apical crest normally found at these steps, folds back on itself entrapping membrane components of the Sertoli cytoplasm, such as the endoplasmic reticulum, that normally surround the projections (Figs. 5.5, 5.6 and

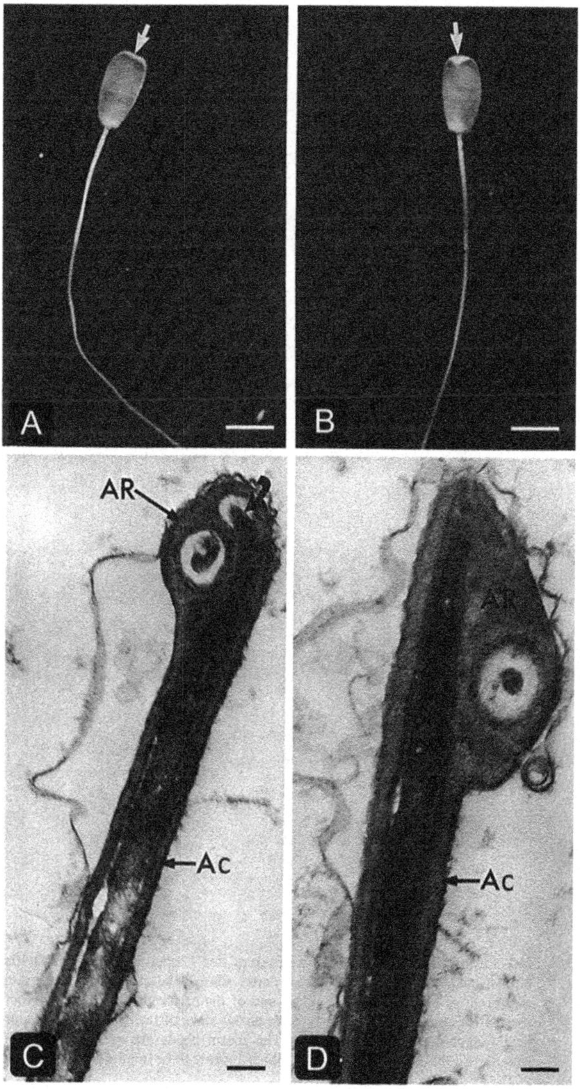

Fig. 5.3 **(A, B)** Light micrographs of eosin-nigrosin stained sperm of a bull affected with the knobbed sperm defect, showing the two faces of the sperm head. Bars = 5 μm. The arrow in **(A)** points to the flattening of the back side of the apical ridge of the acrosome as a result of the collapse and swelling of the apical ridge (AR) seen in **(B, arrow and D)** and the resultant bending of the apical portion of the nucleus **(C, curved arrow)**. **(C, D)** Electron micrographs of ejaculated bull sperm show the characteristic knob appearance of the apical ridge (AR) of the acrosome (Ac). Bars = 0.5 μm. The knobs usually contain membrane-bound vacuoles and inclusions. The bending of the apical end of the nucleus (curved arrow) in **(C)**, contributes to the flattening of the anterior end of the sperm head seen in **(A)**

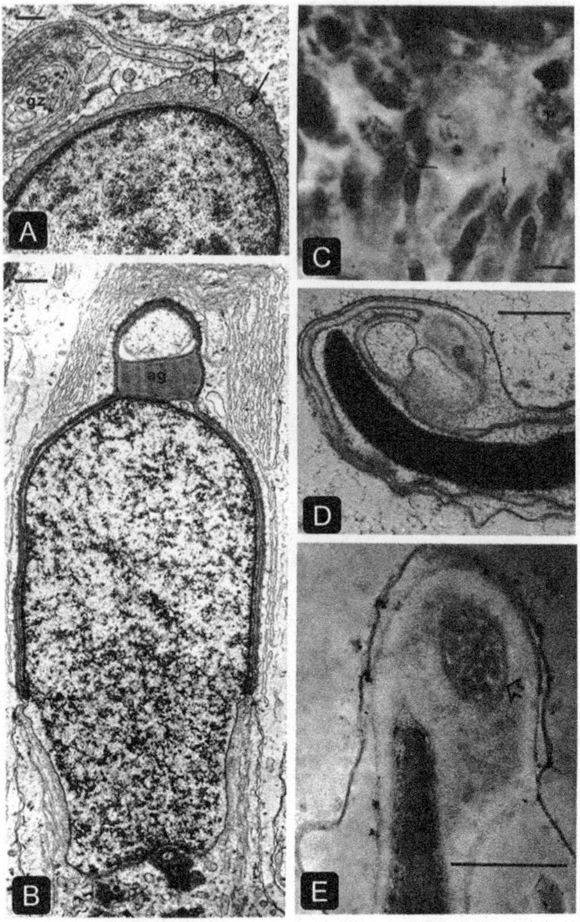

Fig. 5.4 **(A)** Electron micrograph of a step 7 spermatid from a bull affected with the knobbed sperm defect. This form of the knobbed defect (Fig. 5.4) has a different origin than described for other bulls in the legend of Figs. 5.5 and 5.6. Here the acrosome has incorporated membrane-bound vesicles (arrows) presumably derived from the Golgi apparatus (gz). Bar = 0.5 μm. **(B)** Electron micrograph of a step 9 spermatid with a membrane-bound vacuole now incorporated in the acrosomic granule (ag). Bar = 0.5 μm. Compare with (C). This abnormally located vacuole impairs the development of the apical crest of the acrosome and also leads to the knobbed defect as seen in (D). **(C)** Histological section through part of a tubule in Stage IX (S9) of development. Arrows indicate swelling and vacuolation of the acrosomic granule of step 9 spermatids. Compare with (B). Bar = 5 μm. **(D)** Bovine spermatid at spermiation showing characteristic feature of the knobbed sperm defect with mode of origin described in (A-C). Bar = 0.5 μm. **(E)** Knobbed sperm acrosome of the same origin as the spermatid in (D), which has incorporated a multivesicular body (arrow) normally found in association with the Golgi apparatus in the Golgi and cap phase of spermiogenesis. The presence of the multivesicular body indicates that the origin of the defect occurred early in acrosomic development. Bar = 0.5 μm

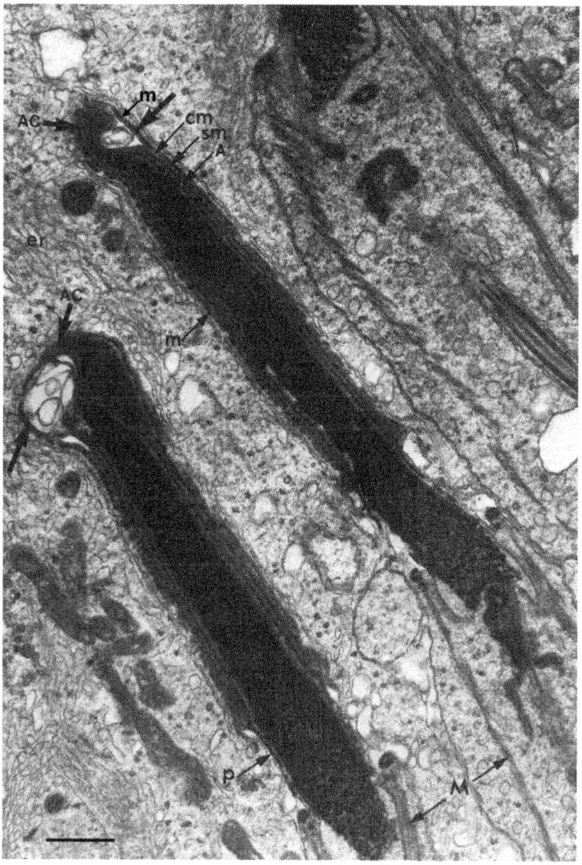

Fig. 5.5 Electron micrograph of a group of step 11 spermatids from the testis of a bull affected with the knobbed sperm defect. This form of the knobbed defect has a different origin than described for other bulls in Fig. 5.4. The tongue-like apical crest (AC) appears to fold over the membranous inclusions (arrows) incorporating them into the acrosome. The membranous inclusions in the upper and lower spermatid appears to be derived from the Sertoli membrane (see large arrows). Note the Sertoli cytoplasmic membrane (sm), mantle (m), endoplasmic reticulum (er), spermatid cytoplasmic membrane (cm), acrosome (A), postacrosomal sheath (p), and manchette (M). Bar = 0.5 μm

5.7). However, in this case it is difficult to determine if the abnormal developmental process resulted from a defective Sertoli cytoplasm or a defective acrosome. In the pig the KA defect has been linked to genes associated with ubiquitination; a prerequisite for both chromatin remodelling and acrosome formation (Sironen et al. 2010). Further, another study has associated the defect with a point mutation in EML5 (Nogueira et al. 2022). The KA defect is actively selected against in the female reproductive tract such that normal sperm within the ejaculate are more likely to reach the ova (Mitchell et al. 1985).

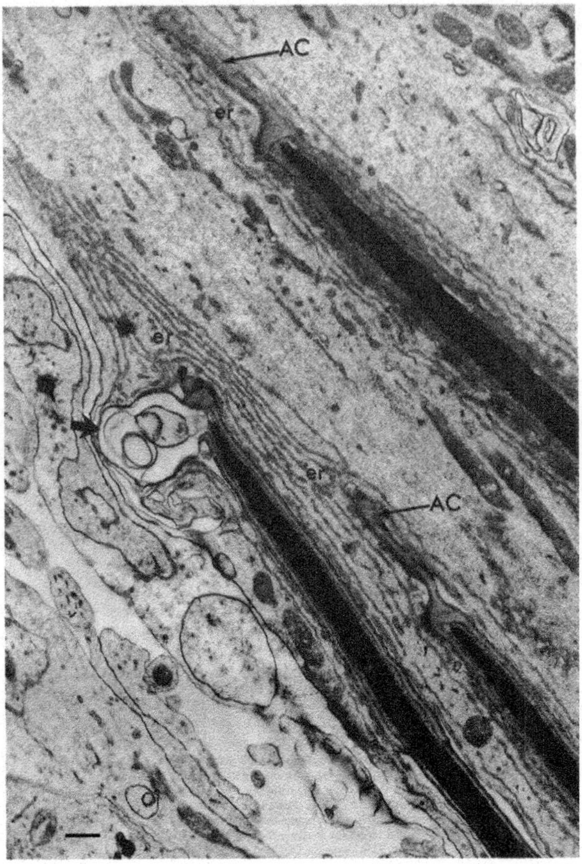

Fig. 5.6 Electron micrograph of a group of step 12 spermatids from the testis of a bull affected with the knobbed sperm defect. One of the spermatids is incorporating membranous inclusions at the apical end of the acrosome (arrow). The other two spermatids appear to develop normally at this stage of the cycle and have a prominent tongue-like apical crest (AC). The sperm heads are surrounded by Sertoli smooth endoplasmic reticulum (er). Bar = 0.5 μm

***Detail*:** The two separate types of knobbed acrosome conditions in bulls, the indented form and a beaded form, have been recognized for at least 20 years (Barth 1997; Hancock 1953; Thundathil et al. 2001b). The indented form also includes a flattened form. Early publications of photomicrographs and electron micrographs of sperm from infertile Friesian bulls showed that, in the beaded form, the defect was characterized by an enlargement of the apical ridge that became folded back on the apex of the sperm head. Often the swollen apical ridge contained vesicles with inclusions (Blom and Birch-Andersen 1962; Hancock 1953). On light microscopy, the majority of sperm had a rounded apex; however, indentations were also present. In the indented form, the majority of sperm have a flattened or indented apex with thickening of the acrosomal ridge. Often

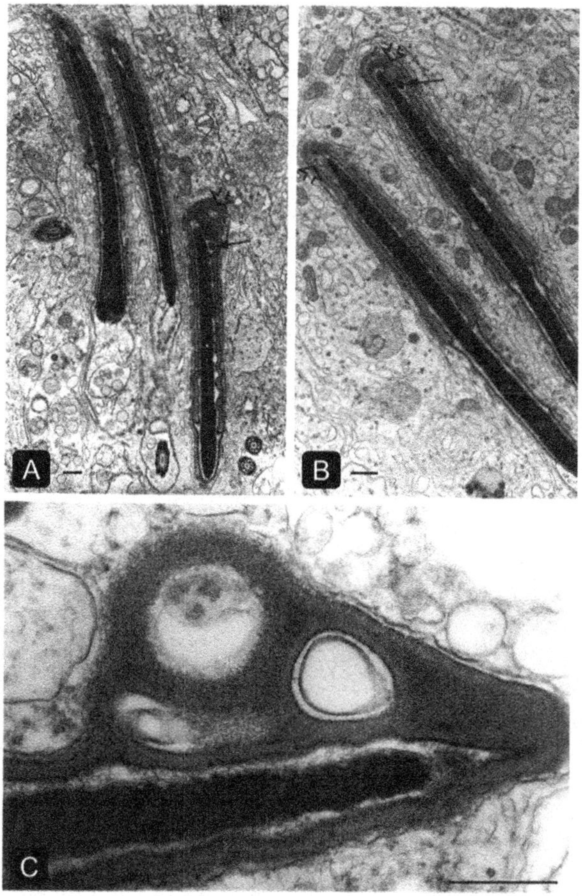

Fig. 5.7 **(A, B)** Spermatids in step 13 of development from the testis of a bull affected with the knobbed sperm defect. Note the knobbed acrosomes (open arrows) and the pieces (arrows) that have broken off the apical end of two of the nuclei and are squeezed between the apical crest and nucleus. Bar = 0.5 μm. **(C)** Step 14 bovine spermatid showing the characteristic knobbed sperm defect. Bar = 0.5 μm. Compare with Fig. 4 D

small number of sperm may have a bead like enlargement with or without apical indentation.

There is no particular breed disposition for either type of acrosomal defect, as both of these types of acrosomal defects occur in many common breeds of western Canada (Barth and Oko 1989) and no influence of breed was observed in an extensive study of range bulls in Australia (Felton-Taylor et al. 2020). This latter study did however show an increased number of bulls failing BBSE on the KA defect in the summer months.

The beaded form is much less common than the indented. Sterility was reported in Friesian bulls affected with the beaded form (Hancock 1953) and sterility is also the experience of the author for other breeds. With the beaded form, when only sperm that do not take up vital stain are counted (hence membrane intact sperm), the percentage of beaded sperm will be extremely high, often approaching 100%. However, when the indented form is present, the percentage of affected sperm may range from < 10 to 100%. With the indented form the effect upon fertility of many bulls varies and is altered by whether the bull is single sire or multiple sire mated. This is due to the inability of some forms of KA to compete with normal sperm and conversely the ability of normal sperm within the ejaculate to fertilize.

The beaded form in Friesian bulls was reported to be inherited by an autosomal recessive gene (Hancock 1953). A pedigree analysis of eight Charolais bulls severely affected with indented acrosomes also indicated a recessive inheritance of this defect (Barth 1986).

Many cases of infertility due to the knobbed sperm defect have been reported in bulls (Barth 1986; Hancock 1953; Meyer and Barth 2001). The following two cases studies by Albert Barth represent a common history of infertility seen with knobbed sperm. A Charolais bull was examined for breeding soundness in the spring at 16 months of age. Although his scrotal circumference was only 30 cm, he was classified as satisfactory. The bull was placed with 36 purebred cows and was observed to have good serving capacity. After a breeding season of 105 days, the cows were pregnancy tested in late November and 53% of cows were found to be open. In December, the bull was re-evaluated. His scrotal circumference was 32 cm, a semen sample had good concentration and good gross motility. There were few sperm abnormalities, but 76% of his sperm had flattened or indented acrosomes.

In another case, a Horned Hereford bull was rented from a neighbour who had used the bull in two previous years in multiple sire breeding groups. Since calf crops were normal, it was assumed that the bull was fertile. The bull was placed alone with 50 cows and was observed to be a good breeder; however, after a 90-day breeding season, only 8 cows (16%) were pregnant. The bull was evaluated in November after pregnancy testing. He had a scrotal circumference of 34 cm and the semen sample had good concentration and good motility. The sperm morphologic features of his spermiogram were normal, except that 77% of sperm had beaded acrosomes.

These cases seem to indicate that the effect of knobbed sperm on fertility is quite straight-forward, however, other anecdotal reports indicated that some bulls that were quite severely affected with flattened or indented acrosomes were unexpectedly fertile. Therefore, several in vivo and in vitro experiments were done at the Western College of Veterinary Medicine to further study the indented form of knobbed acrosomes.

Four bulls with the indented form of knobbed acrosomes and 2 control bulls were used in single sire natural mating (Meyer and Barth 2001). The four affected bulls each consistently produced good quality semen, except that 65–85% of sperm had flattened and indented acrosomes. Pregnancy rates after 21 days of breeding ranged from 61–85% for the bulls with indented form of knobbed acrosomes and 52 to 65% for the control bulls. Frozen semen from 2 of these bulls producing 61 and 85% flattened and indented sperm and from a normal control bull was used to artificially inseminate 98 heifers (32–33 per bull). The number of sperm per insemination dose was 11 million and the percent motile sperm per dose was similar for each bull used. Pregnancy rates for the bulls with flattened and indented sperm were 48.5 and 50% and 46.3% for the control bull. Therefore, with single-sire natural service and artificial insemination, it appears that fertility was unaffected by a high percentage of the indented form of the knobbed acrosome defect. One explanation for the unexpected fertility levels was that there were sufficient numbers of sperm with normal acrosomes present in both the natural service and artificial insemination experiments to result in normal fertility.

Two Charolais bulls producing high percentages of flat or indented sperm were placed with normal control bulls in competitive mating situations at pasture (Meyer and Barth 2001). A competitive mating was considered to be one in which more than one bull bred a cow at the same estrus period. Breeding was observed twice daily and chin ball markers were used to confirm normal breeding activity. Calves produced were identified by phenotype and blood-group typing.

In the first competitive mating trial, a Charolias bull that produced about 90% flat or indented sperm, and two normal bulls (Simmental and Brown Swiss) were placed with 49 cows for 52 days. The Charolais bull produced only 2 calves, winning 8.7% of 23 competitions, compared to 50–55% of competitions won by the control bulls.

In the second competitive mating trial, another Charolais bull with flat or indented sperm was placed with 4 control bulls (2 Hereford and 2 Charolais) and 118 cows for 63 days. The affected Charolais bull produced only 3 calves winning 13.6% of 22 known competitions compared to 41–50% of competitions won by the control bulls.

These trials indicate that in noncompetitive mating situations bulls with the indented form of knobbed sperm can achieve normal or nearly normal fertility. In-vitro fertilization experiments indicated that sperm with indented and flattened acrosomes generally did not attach to ova and the few that did were unable to penetrate the zona pellucida (Thundathil et al. 2001b). Therefore it would appear that normal sperm coexisting with the knobbed sperm most likely were available in sufficient numbers after natural service to accomplish good levels of fertility. However, in competitive mating situations, both knobbed sperm and coexisting normal sperm may compete poorly with sperm from unaffected bulls. This concept is supported by a study designed to compare the differential retention of morphologically abnormal sperm within the cow genital tract after artificial insemination (Mitchell et al. 1985). The proportions of sperm with knobbed acrosomes recovered from all tract regions, cervix, uterus, uterotubal junction and isthmus were reduced compared to the inseminate. Substantially fewer knobbed sperm were retained in the genital tract when compared with those of normal sperm even though the abnormal sperm did not appear to differ from the normal sperm in the quality or type of motility. The mechanisms responsible for differential retention are not known; however, the proportion of normal sperm achieved in the oviductal sperm population after a competitive mating would be much lower for the bulls with knobbed sperm than from normal bulls. Thus, the chances of fertilizing ova favored the control bulls during competitive mating.

5.1.5 *Ruffled Plasmalemma and Acrosome*

Summary: A ruffled appearance of outer acrosomal sperm membranes is shown in Fig. 4.3 a-c (Chap. 4.4). Figure 4.3 d shows by EM that the outer acrosomal membrane is wrinkled and the acrosomal apex may be enlarged. The acrosome itself may be wrinkled (Chap. 4 in Fig. 4.3 e). It is common to find a small percentage of sperm with ruffled membranes in bulls with normal or abnormal spermiograms. In rare cases a very high percentage of sperm with ruffled membranes may be found with otherwise normal spermiograms and such cases have been associated with low pregnancy rates (Barth 2013). The author (VP) has also observed that semen from a bull, with greater than seventy percent ruffled-membrane sperm, had a reduced ability to produce viable embryos.

5.1.6 *Swollen Acrosomes (SA)*

Swelling and sloughing of the acrosome is a normal progression during sperm aging and it is therefore not a morphological abnormality. It is, however, a criterion of viability as these sperm are not functional: the loss of the acrosome necessitates an inability to fertilize ova.

In the Australian classification system for bull breeding soundness evaluation swollen acrosomes are enumerated in a category separate from sperm abnormalities. In the assessment of frozen thawed sperm the percent intact acrosomes (PIA) is enumerated as a measure of viability. In the frozen thawed sperm sample, acrosome retention may be altered by iatrogenic, rather than animal effects. This PIA count is correlated with fertility in the frozen thawed sample (Berndtson et al. 1981).

Swollen acrosomes can be associated with sperm accumulation during abnormal sperm transport in the male reproductive tract (McAuliffe et al. 2010). Ageing of these sperm causes the acrosome to undergo a similar reaction to capacitation resulting in swelling or loss of the acrosome, disabling the sperm to attach to the oocyte. In disturbances of spermatogenesis the percentage of swollen acrosomes may be increased along with an increase in other sperm abnormalities such as knobbed acrosomes. This is because the abnormality may cause premature initiation of the acrosome reaction (Thundathil et al. 2001b). The swollen acrosome in these cases may hide the knobbed defect in initial observations. Swollen acrosomes seldom occur in very high numbers except in accumulator bulls which store senescent sperm in the ampullae and cauda epididymis (Barth 2007).

5.1.7 Nuclear Vacuolation

Summary: Nuclear vacuolation occurs during spermiogenesis and may be caused by an environmental stress although there is some evidence for the heritability of this trait (Callaghan et al. 2016; Felton-Taylor et al. 2020). The abnormality is more commonly observed in Bos indicus content bulls than in Bos taurus breeds (Felton-Taylor et al. 2020). Some bulls are predisposed to this condition (perhaps due to a hormone imbalance in the testis) following a stress event (Callaghan et al. 2016; Barth 2013). Vacuoles appeared 30d after ruminal acidosis (Callaghan et al. 2016) or 20d after dexamethasone treatment (Barth and Bowman 1994). Three forms of vacuolation occur: large confluent vacuoles, the diadem defect and, small apical vacuoles.

i) Large confluent vacuoles (LCV) can be so large as to be a "bite" size piece missing from the side of the head. This abnormality may occur after disruption to spermiogenesis, for example following ruminal acidosis (Callaghan et al. 2016). It has also been reported as an inherited abnormality in a Santa Gertrudis herd (Olley 2001). Smaller craters were also seen in other sperm in the ejaculate. Bulls with a high percentage of this abnormality were infertile. Canadian studies concur with this effect finding levels of>20% reduce pregnancy rates (Barth 2013).
ii) The diadem defect, an arrangement of vacuoles along the equatorial region of the sperm is a serious cause of infertility in the bull (Blom 1973). Fluctuations in the prevalence of this defect occur between ejaculates (Larsen and Chenoweth 1990) with stress being a predisposing factor.
iii) The occurrence of small single vacuoles in the apex of the of the sperm nucleus is common, however, the importance of this type of vacuolation to fertility remains unresolved. Small apical vacuoles are commonly associated with the diadem defect or with multiple small vacuoles throughout the nucleus. Unlike LCV or diadems, small single vacuoles appear to be more transient than the other forms. Ejaculates with high numbers of apical vacuoles (80%) have reduced conception rates and in an IVF study no sperm with these vacuoles were observed inside the zona pellucida (Barth 2013).

A high incidence of nuclear vacuoles >60% (Pilip et al. 1996; Thundathil et al. 1998b) is known to cause severe reduction in fertility. There is debate as to whether this abnormality is compensable by increasing dose size of inseminations as some sperm with this defect were shown to bind to the zona and initiate fertilization, but it could not be determined if this fertilization produced viable zygotes (Thundathil et al. 1998b). Further, most of the sperm with this abnormality did not reach the zona. However, Pilip et al. (1996) reported that sperm with multiple nuclear vacuoles had a reduced ability to fertilize ova. In view of this uncertainty nuclear vacuoles are given a threshold level of not more than 20% (Thundathil et al. 2001b) in BBSE.

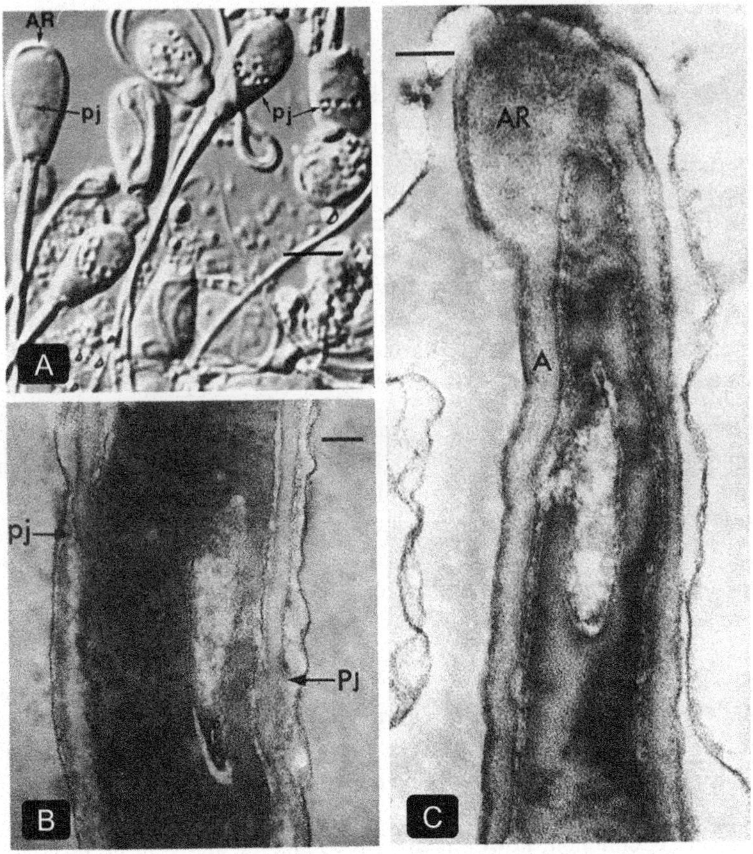

Fig. 5.8 **(A)** Cratered ejaculated sperm as seen under differential interference microscopy. Nuclear vacuoles are usually located just posterior to the apical ridge (AR) or at the postacrosomal sheath junction (pj). Bar = 5 μm. **(B)** Electron micrograph of a portion of an ejaculated sperm head showing membrane-bound nuclear invagination at the postacrosomal sheath junction (Pj). Bar = 0.1 μm. **(C)**) Electron micrograph of the apical portion of an ejaculated sperm head showing membrane-bound nuclear invagination just posterior to the apical ridge (AR) or the acrosome (A). Bar = 0.1 μm

***Abnormal developmental morphology*:** Fig. 5.8 A shows the characteristic arrangement under differential interference contrast (DIC) microscopy of the diadem defect in the equatorial region of the bull sperm head. Note the craterlike appearance of each nuclear vacuole. Occasionally, a single to several apical nuclear vacuoles are found in sperm of bulls affected with the diadem defect By electron microscopy the crater like appearance of the nuclear vacuoles seen by DIC is due to narrow mouth invagination of the nuclear membrane into the nucleoplasm, leading to the term nuclear pouches (Bane and Nicander 1965). These nuclear pouches distribute equatorially and sometimes apically in the sperm nucleus (Fig. 5.8 B, C). The nuclear vacuoles are membrane bound and develop from invagination of the nuclear envelope in the spermatid nucleoplasm during the cap phase of spermiogenesis, just before the condensation and elongation of the spermatid nucleus (Fig. 5.9 A, B). The vacuoles, once formed, are ubiquitously arranged (Fig. 5.10 A-C) until condensation of the chromatin is almost completed (Figs. 5.10 D, 5.11), at which time the vacuoles become positioned in one of the two predominant locations found in the ejaculated sperm.

Fig. 5.12 shows that nuclear vacuolation and the knobbed defect can develop together. In this case it happened in a bull in which stress was simulated by

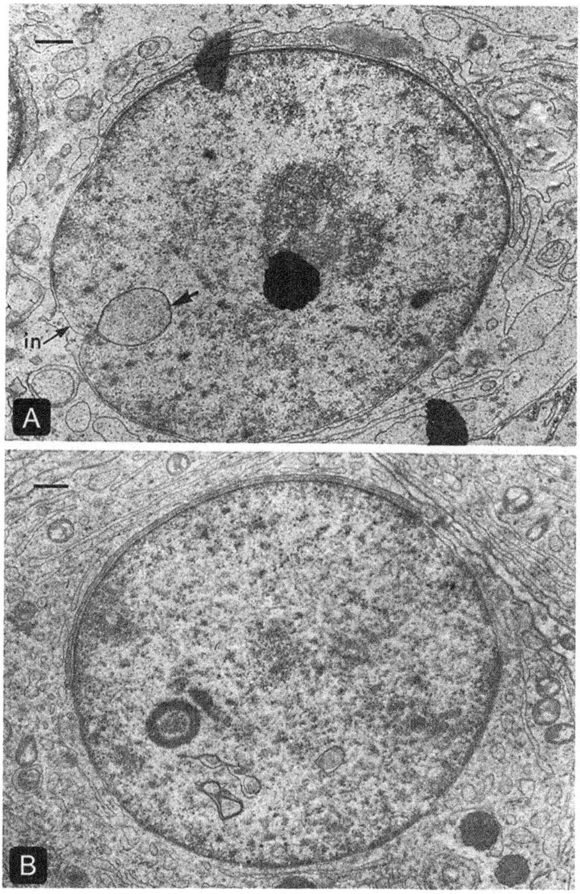

Fig. 5.9 **(A)** Electron micrograph of a late step 6 or early step 7 spermatid. Note that the nuclear defect (arrow) in this case originates from the infolding of the inner nuclear membrane (in) that forms this membrane-bound vacuole. Bar = 0.5 um. **(B)** Electron micrograph of a step 7 spermatid. Note that often several nuclear membrane vacuoles, having various arrangements and forms, are found in the nuclei. Bar = 0.5 μm

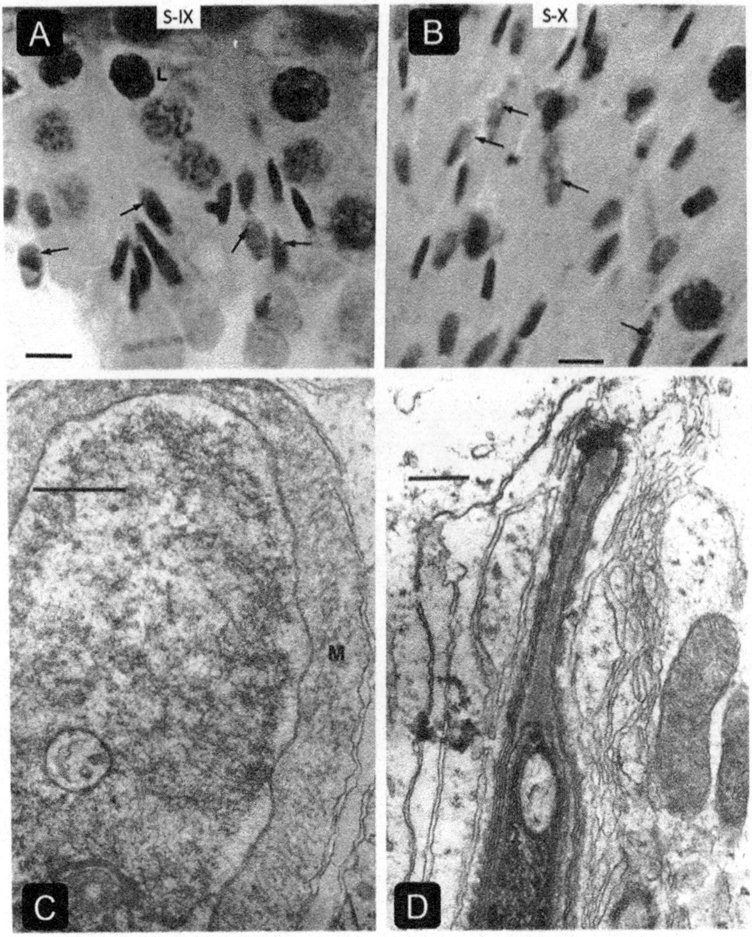

Fig. 5.10 **(A, B)** Histological sections through the seminiferous epithelium of stages (S) IX and X of the bovine cycle, respectively. At these stages the nuclei of step 9 and 10 spermatids are condensed enough that visualization of the nuclear vacuoles (arrows) is possible by light microscopy. PAS-hematoxylin stain. Bars = 5 μm. **(C)** Electron micrograph of a step 9 spermatid containing a membrane-bound nuclear vacuole in the distal part of the nucleus. This part of the nucleus is surrounded by the microtubular manchette (M). At this stage of development the condensation of nuclear material becomes more apparent, allowing this defect to be visualized on hematoxylin-stained histological sections. Bar = 0.5 μm. **(D)** Electron micrograph of step 10 -11 spermatids. By this stage of development the vacuoles become positioned either at the apical end of the nucleus as seen here, or at the postacrosomal sheath junction as seen in Figs. 5.8 B, 5.11. Bar = 0.5 μm

dexamethasone treatments. The other bulls on this treatment regimens only had an increase in nuclear vacuolation without the acrosome effect.

Detail: Sperm nuclear vacuolation is observed commonly in the cattle as well as in other mammals (Bonet et al. 1993; Truitt-Gibert and Johnson 1980; Fawcett

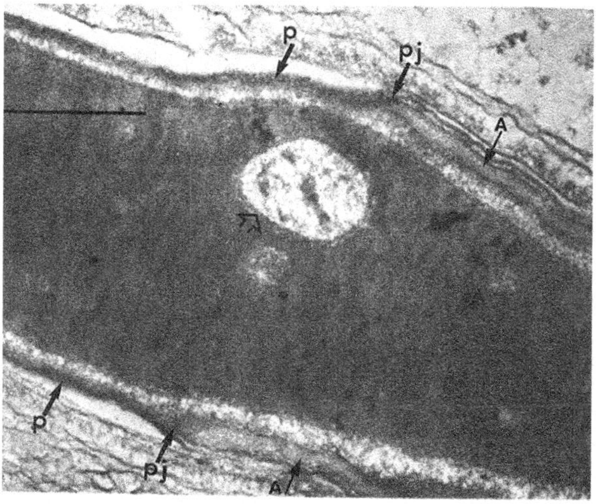

Fig. 5.11 Electron micrograph of a portion of a step 12 spermatid head containing a membrane-bound nuclear vacuole (arrow) at the postacrosomal sheath junction (pj). Note the acrosome (A) and the postacrosomal sheath (p). Bar = 0.5 μm

et al. 1971). Submicroscopic membrane-unlined spaces are normally found within the sperm nucleus and do not cause infertility. However, larger vacuoles visible with light microscopy at 1000× magnification, were associated with infertility as early as 1965 (Bane and Nicander 1965). Large confluent vacuoles (LCV) were shown by electron microscopy to be narrow mouthed pouches formed by invagination of the nuclear envelope (Bane and Nicander 1965; Coulter et al. 1978). Later work showed that the pouches/vacuoles developed from invaginations of the inner nuclear membrane into the spermatid nucleoplasm (Barth and Oko 1989; Larsen and Chenoweth 1990; Oko 1977) (Fig. 5.9). This period coincides with the beginning of condensation of the spermatid nucleoplasm and lengthening and flattening of the nucleus. The membrane-bound vacuoles usually contain inclusions that appear to be derived from the spermatid cytoplasm and/or remnants of nuclear membrane. Theories for the formation of vacuoles in sperm heads can be developed based on either; abnormal chromatin condensation or abnormal nuclear membrane assembly, in response to hormonal- or toxin-induced alterations of enzyme systems (Larsen and Chenoweth 1990). When observed by differential interference microscopy, vacuoles resemble craters and the term craters has been used in several reports (Coulter 1976; Coulter et al. 1978; Foote 1999; Larsen and Chenoweth 1990).

With eosin-nigrosin morphology stain, vacuoles appear as dark spots in the nucleus that must be differentiated from artifacts such as stain crystals or other debris. Vacuoles can easily be overlooked during semen evaluations, even with good preparations and good microscope equipment, if very careful examinations are not done. Vacuoles may appear singly or in multiples anywhere in the sperm nucleus; however, usually specific locations are observed.

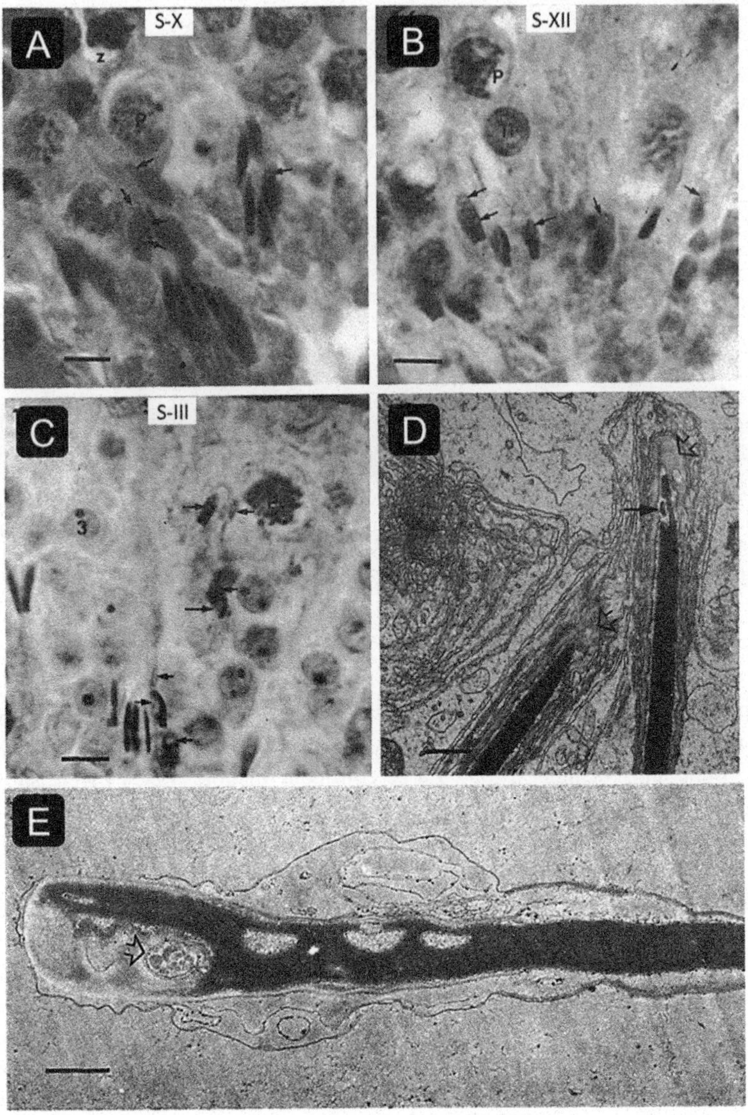

Fig. 5.12 **(A-C)** Histological sections through parts of bovine seminiferous tubules in stages (S) X, XII, and III, respectively, showing development of both the knobbed and crater defect together (arrows) in elongating step 10, 12, and 13 spermatids, respectively. Zygotene spermatocyte (z), pachytene spermatocyte (P), secondary spermatocyte (II), step 3 spermatid (3). The knobbed defect has a similar mode of origin as described in the legend of Fig. 5.4, while the crater defect (nuclear vacuoles) has the same origin as described in Figs. 5.8, 5.9, 5.10 and 5.11. Bars = 5 μm. **(D)** Electron micrograph of a portion of two step 14 spermatids, displaying knobbed acrosomes (open arrows) and a nuclear vacuole (arrow). Bars = 0.5 μm. **(E)** Ejaculated bovine sperm displaying both the knobbed defect and nuclear vacuoles or craters. The knobbed acrosome contains an inclusion (arrow) reminiscent of a multivesicular body. See Fig. 5.4 E. Bar = 0.5 μm

The incidence of vacuoles in the sperm of affected bulls varies from less than 1% to nearly 100% (Bane and Nicander 1965). In one study in Czechoslovakia, of 411 breeding bulls, 43% had no vacuoles, 54% had from < 5% to 20%, and 2.7% of bulls had > 20%. In a survey of western Canadian bulls, in 316 ejaculates from 46 AI stud bulls, representing a wide variety of ages and breeds, all bulls produced some craters, ranging from < 1% to > 60% of sperm affected and of these, 28% had vacuoles in > 20% of spermatozoa (Coulter et al. 1978). In a more recent study (Foote 1999) only 3 of 100 bulls had a substantial incidence of craters (15 to 23%), whereas, on average, the other 97 bulls had 1 to 3% vacuolated sperm. A 2020 study of 12,000 range bulls (Felton-Taylor et al. 2020) found the percentage of bulls with greater than 20% vacuolated sperm (the threshold for BBSE) to be 11.9%. Bulls with Bos indicus content were significantly more affected than Bos taurus bulls, as previously suggested (Barth 2013), with more bulls failing the BBSE on vacuolation in the summer. Similarly, bulls with Bos indicus content seem predisposed to vacuolation of the sperm after a stress event, rather than any other type of sperm abnormality. Indeed, in this study, where > 60% of the bulls had Bos indicus content, vacuolation was the abnormality most commonly observed (Felton-Taylor et al. 2020).

The proportion of sperm affected by vacuolation may vary with time, or may remain relatively stable. In one of the first reports describing sperm nuclear vacuolation (Bane and Nicander 1965), the percentage of affected sperm in 4 bulls changed a great deal in a little as 2 weeks. Decreases from 100 to 5% and 0% were observed in 29 days and 33 days, respectively, in two of the bulls. However, in another study involving 46 beef bulls in AI centers, the proportion of vacuolated sperm did not vary significantly over a period about 30 days (Coulter et al. 1978). In the above reports, the cause of vacuolation was considered to be a disturbance of spermatogenesis, since many other types of sperm defects were present. However, other reports indicate that sperm nuclear vacuolation may occur without concurrent signs of a disturbed spermatogenesis (Barth and Oko 1989; Larsen and Chenoweth 1990).

Large changes in the prevalence of diadem defects in sperm of two related Angus bulls occurred over a period of about 4 months (Larsen and Chenoweth 1990). The bulls were not outwardly suffering from any disease or stressful condition at the time of examination. In another report, in 36 ejaculates from a Simmental bull, that were cryopreserved over the period of a year, the percentage of vacuolated sperm fluctuated frequently over the entire year from as low as 10% up to 90% (Barth and Oko 1989). In this case there were no physical signs or items in his past history that might account for the variation in percentages of sperm affected.

A variety of causes of nuclear vacuolation have been postulated including stress, inadequate thermoregulation of the testis, viral infection, toxins and inheritance. It has been demonstrated that dexamethasone treatments, induced to mimic stress in bulls, caused an acute increase in vacuoles, followed by a decline a few weeks later (Barth and Oko 1989; Barth and Bowman 1994; Coulter 1976). Maximum levels of vacuoles occurred 2–4 weeks after dexamethasone treatment at 20 mg per day for 7 days. Scrotal insulation also resulted in sperm vacuolation in some bulls (Barth and Bowman 1994). Courtens et al. (1980) treated bulls and Hrudka and Eljack (1979) treated rams with ethylene dibromide, a fumigant for insect control in grain. This resulted in vacuole formation in sperm nuclei often associated with inadequate condensation of the spermatid nucleoplasm and a disturbance in acrosome formation. In one study, a bull with many sperm cells with vacuoles was slaughtered, and the epididymal spermatozoa were examined. The high incidence of sperm cells with craters was limited to one side, with the testis on that side having 2 Sertoli cell tumors (Foote 1999). This also suggests that impairment of testis function is a cause of nuclear vacuolation. However, other reports indicate that sperm nuclear vacuolation may occur in healthy unstressed bulls and without concurrent seminal signs of a disturbed spermatogenesis (Barth and Oko 1989; Larsen and Chenoweth 1990). Furthermore, in an unpublished study, 2 bulls with an incidence of approximately 35% nuclear vacuolation—mainly the diadem defect, neither insulation nor dexamethasone treatments caused an increase in the percentage of vacuolated sperm in spite of an obvious severe disturbance of spermatogenesis caused by these treatments. It appears therefore, that a heritable predisposition for vacuolation must also be considered.

The use of sires that produced large percentages of sperm with the diadem defect have been associated with infertility (Larsen and Chenoweth 1990; Miller et al. 1982; Bane and Nicander 1965). Vacuolated sperm were shown to be transported normally to the oviduct (Saacke 1988), and there is evidence from in-vitro fertilization that some sperm with the diadem defect were able to penetrate ova, but were incompatible with embryonic development (Pilip et al. 1996). Thus, this type of defect may be considered non compensable, i.e., fertility can't be improved by increasing sperm dosage (sperm numbers within the oviduct). However, IVF studies indicated that the ovum vestments selected against diadem sperm and therefore vacuolation of sperm may be at least a partially compensable trait (Pilip et al. 1996; Thundathil et al. 1998b).

In the in vitro fertilization experiments done by Pilip et al. 1996, the effect of sperm nucleus vacuolation on sperm-ova interactions, after Percoll washes was examined: 73.5 ± 3.0% of sperm from one bull (A) contained vacuoles and these sperm were compared to normal spermatozoa from a second bull (B). The percentage of normal sperm and vacuolated sperm bound to the zona pellucida did not differ, however, the proportion of vacuolated sperm penetrating the zona pellucida was significantly lower. In one third of the oocytes fertilized by vacuolated spermatozoa, sperm head decondensation was observed to be incomplete and normal male pronucleus formation did not occur; however, all oocytes fertilized by normal sperm had normally decondensed sperm heads. Mean fertilization percentage was significantly lower in bull A (19.7%) than in bull B (67.6%).

In the case reported by Miller et al. 1982, a bull with a history of very low fertility in both natural service and with artificial insemination had been evaluated for breeding soundness by two veterinarians on separate occasions. The bull had large, grossly normal, testicles and produced concentrated semen with very good motility. In addition, morphological evaluation of fresh sperm using routine methods did not reveal any abnormalities that could be related to the low conception rate. However, upon referral of the case, further evaluation revealed that at least 80% of sperm had abnormal nuclei and the predominant finding was nuclear vacuolation of the diadem type. Two controlled breeding trials utilizing a total of 15 super ovulated females were conducted to evaluate the bull's fertilization rate. Combined data demonstrated an 18% fertilization rate of recovered ova. At the same time, the fertilization rate of seven bulls classified as satisfactory potential breeders was 72%. Field results supported the superovulation trial. A seven-week natural breeding exposure to 42 normal cycling females resulted in only two pregnancies (5%), diagnosed at 150 days. Similarly, only 5 pregnancies (8%) were achieved when 180 females at different locations were inseminated with frozen semen. In addition, the use of frozen semen to inseminate 8 superovulated females in 2 embryo-transfer units resulted in a fertilization rate of 41% (30/73). However, of the 30 fertilized ova, 11 were degenerate and were not transferred. A pregnancy rate of 57% (11/19) resulted from the transfer of 19 fertilized ova.

Two reports indicate that when the percentages of vacuolated sperm fluctuate dramatically, infertility occurs during periods when large numbers of vacuolated sperm are present in semen. During periods when of low levels of vacuolated spermatozoa, fertility improves (Larsen and Chenoweth 1990; Barth and Oko 1989).

It appears that the diadem defect may be heritable. In the case report of (Miller et al. 1982), a Simmental bull with poor fertility produced 4 sons that were similarly affected. Bulls also differ in their response to dexamethasone treatments or scrotal insulation (Barth and Bowman 1994) in terms of the percentages of vacuolated sperm produced. Some bulls did not produce vacuolated sperm and some produced large percentages of vacuolated sperm. Similarly, 2 bulls that produced $34 \pm 4.0\%$ vacuolated sperm over a period of 3 months were subjected to dexamethasone treatments for 7 days. Despite the evident severity in disturbance of spermatogenesis as seen in the spermiograms, nether bull had an increase in the percentage of vacuolated sperm in response to the dexamethasone treatments. This was contrary to the finding of (Coulter 1976) in which dexamethasone treatments in 3 Chianina bulls in which surges and declines of vacuolated sperm corresponded to repeated dexamethasone treatments.

The combined results of available reports indicate that likely there is a genetic predisposition to nuclear vacuolation; however, levels of vacuolated sperm can fluctuate dramatically over short intervals without any apparent reason and the response to stressors is very unpredictable.

Single apical nuclear vacuoles (SAV) are commonly associated with the diadem defect or with multiple vacuoles scattered throughout the nucleus. However, we have observed that in a few bulls at certain times 81–100% of sperm had single apical vacuoles in the absence of, or associated with, very low numbers of spermatozoa with multiple vacuoles. A breeding trial was conducted using frozen semen from a bull that had produced varying levels of single apical nuclear vacuoles. In one of the freeze dates that was used, only 4% of spermatozoa had single apical vacuoles, whereas, the other freeze date used had 81% single apical vacuoles. There were few other sperm defects present in either of the freeze dates and the number of motile sperm per dose for both freeze dates was similar. Fifty-two cows were artificially inseminated, 26 with 4% vacuole semen and 26 with 81% vacuole semen. Pregnancy rates were 50% for the 4% single vacuoles vs 32% for the 81% vacuoles. An IVF study done in our lab to determine the ability of sperm with single apical vacuoles to bind, penetrate and fertilize oocytes and to determine the developmental competence of zygotes resulting from fertilization by single vacuole sperm. Semen from bull A with 38% single apical vacuoles was compared to normal control semen from bull B. Single apical vacuoles did not reduce binding to ova; however, the percentage penetration with single vacuoles was lower and no sperm with vacuoles were found inside the zona pellucida. In vitro embryo production was lower for bull A than for bull B.

The proportion of bulls that produce sperm with large confluent vacuoles (LCV) exclusively without diadem and SAV, have not been reported and there appear to be no reports in the literature of infertility due to this defect alone. The author also has observed that in 1760 beef bulls aged 1–10 years, the incidence of LCV was 1.0% for those with 10–19% LCV and 0.58% of bulls with 20–65%. However, all of these bulls had been previously evaluated for semen quality one or more times, therefore since these bulls would have been subjected to previous culling, the true incidence of bulls with LCV could be much higher. In 452 cases of infertility submitted to the author by veterinarians from western Canada, LCV were present at levels of 20–65% of sperm in 6 cases. In these cases, the diadem defect coexisted with LCV ranging from 0–18% of sperm. The presence of very high percentages of LCV in the semen of unstressed, physically normal bulls without concurrent signs of a disturbance of spermatogenesis suggests that the defect is genetic in origin. A genetic origin is further supported when a large proportion of sperm are affected with LCV over a prolonged period of time without other signs of a disturbed spermatogenesis. In one such case, a bull whose semen was used to artificially inseminate 80 cows produced a pregnancy rate of 20%. The bull was housed in 2 different AI centers over a period of 16 months and consistently produced semen with little variation in numbers.

Overall, the studies on the effect of sperm nuclear vacuolation on fertility indicate that affected sperm have a greatly reduced ability to fertilize ova and if fertilization does occur, embryonic development is unlikely. The level of tolerance of this partially compensable trait is not clear; however, most bulls that produce > 20% vacuolated sperm would be expected to have reduced fertility. Those bulls that have shown good fertility in spite of elevated percentages of vacuolated sperm might have had an undetected decline in the percentage of vacuoles during the breeding period. Once a bull is found to produce semen affected by nuclear vacuolation, he should be either culled or closely monitored for the rest of his life. Some severely affected bulls may make dramatic recoveries within a short time, but are likely to revert to periods of producing high percentages of vacuolated sperm.

5.1.8 Abnormal DNA Condensation

Summary: This abnormality cannot be observed under the light microscope without staining. It appears as coarse or fine clumping of the nuclear material. This abnormality can be detected by SCSA- a flow cytometric assay that uses the metachromatic properties of acridine orange to measure the susceptibility of chromatin to denaturization (Barth 2013) or by Feulgen staining under × 1000 phase contrast or DIC microscopy. Feulgen and SCSA methods correlate and both methods enable assessment of affected sperm (Dobrinski et al. 1994). Advances in DIC microscopy also now permit the examination of abnormal DNA condensation to a limited extent. When this or pale centres are considered the major reason for bull sperm falling below threshold, Feulgen staining is recommended to completed as a check.

Detail: During the final stages of spermatogenesis there are marked changes in the structure of sperm chromatin. In early stages of spermatogenesis, sperm DNA is associated with large basic histone nucleoproteins similar to other somatic cells. As spermiogenesis proceeds, histones are replaced by transition proteins which are subsequently replaced by small protamines (Ward and Coffey 1991). This allows DNA chains to lie parallel to each other forming a compact nucleus resistant to denaturation. Incomplete condensation and partial retention of histones results in sperm malfunction.

When healthy bulls have good serving capacity and semen quality, but cowherd infertility occurs, abnormal DNA condensation may be the explanation. A range of sophisticated techniques are available to assess chromatin status in sperm (Agarwal et al. 2004), including the sperm chromatin structure assay (SCSA). The SCSA is a flow cytometric assay that uses the metachromatic properties of acridine orange to measure the susceptibility of sperm chromatin to acid-induced denaturation.

Feulgen staining allows simple visual microscopic detection of abnormal DNA condensation. In the Feulgen reaction, aldehyde groups in the deoxyribose of the DNA are unmasked by acid hydrolysis and are then exposed to Schiff's reagent (basic fuchsin in sodium metabisulfite). Normal sperm nuclei appear homogeneously purplish red with bright field microscopy and intensely violet with phase contrast microscopy at 1000X. Abnormal DNA condensation is usually observed as coarse or fine clumping of nuclear material with a generalized reduced stainability of the nucleus. Often affected nuclei have an expanded appearance. In some cases, nuclei have a slightly mottled pale appearance and nuclear vacuolation may be associated with the abnormal chromatin condensation. Control samples should always be used concurrently to confirm abnormal DNA staining with the Feulgen and other techniques. The existence of clumped granular chromatin has also been shown by electron microscopy (Fig. 5.13 A,B). The Feulgen and SCSA methods have been compared with both methods showing good correlations for the proportion of affected sperm and for fertility (Dobrinski et al. 1994).

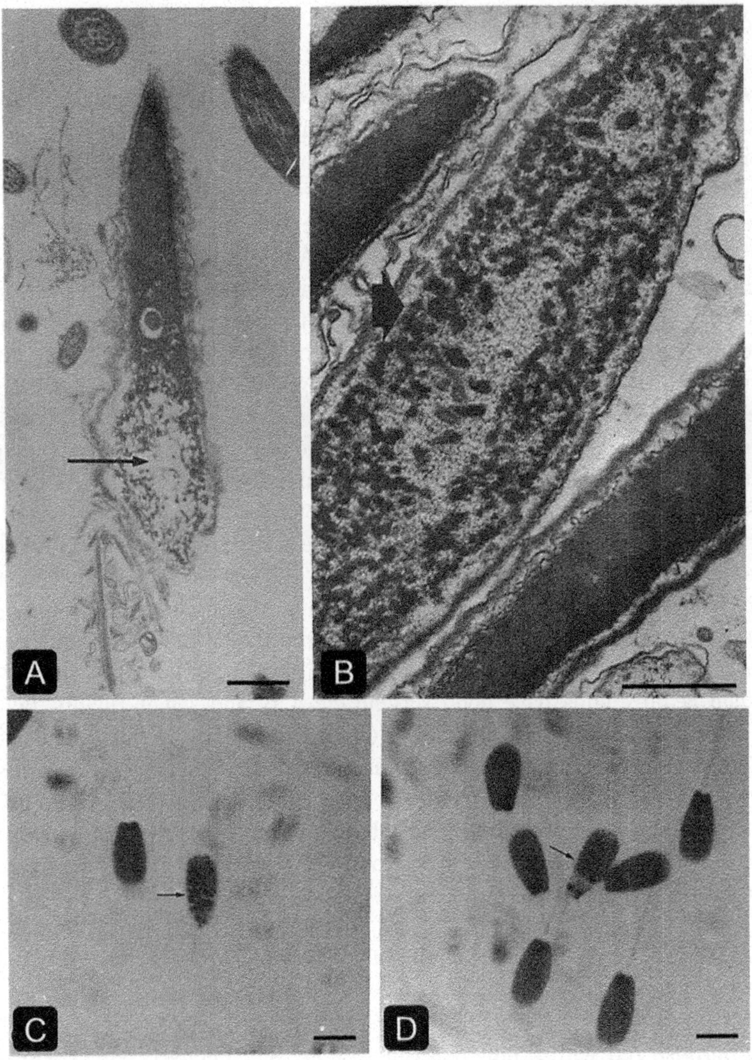

Fig. 5.13 Abnormal DNA condensation. **(A)** An electron micrograph of ejaculated sperm with loosely compacted chromatin, which is more typical of spermatids in the elongation phase of spermiogenesis. In sperm with abnormal DNA condensation the degree of chromatin packing can be quite uniform throughout the apical half of the nucleus, however, it is common for the postacrosomal region to be extremely rarefied as seen here and in (C) and (D) (arrows). Bar = 0.5 µm. **(B)** An electron micrograph of sperm, showing a portion of a nucleus (arrow) that has extremely loosely packed chromatin. The degree of chromatin condensation may vary considerably between affected sperm. Note the homogeneous condensation of DNA in adjacent normal nuclei. Bar = 0.5 µm. Views **(C)** and **(D)** are light micrographs that show variation in the density of chromatin packing in different regions of the nucleus (arrows). Note the homogeneous condensation of DNA in adjacent normal nuclei. Bars = 5 µm

It is quite rare to find bulls that produce a high percentage of sperm with abnormal chromatin condensation. We have seen no breed predisposition. In most bulls, the number of affected sperm is under 5%, but a few bulls may produce > 80% normal sperm as seen on eosin nigrosin stains, but have as high as 100% of nuclei with poor chromatin condensation. The author received approximately 3 referral cases per year of abnormal DNA condensation from veterinarians in western Canada. Typically, in these cases, the spermiogram of affected bulls was normal upon examination with eosin nigrosin stains. The bulls were confidently recommended for use in breeding only to incurr low pregnancy rates. After the finding of low pregnancy rates, re-examination of affected bulls still showed normal spermiograms with routine staining methods; however, with Feulgen staining abnormal DNA condensation was discovered. In one such a case, a Simmental bull was observed to breed 17 cows over a period of 6 weeks, but all cows returned to estrus. Feulgen staining showed that 100% of sperm nuclei had course, grainy chromatin. Upon joining the cows with a second bull, normal first service conception rates occurred. A year later, the same bull produced sperm with mostly fine grainy chromatin. In a breeding trial that year, he remained sterile. Later a project was done using 6 bulls that had abnormal chromatin in 2% to 75% of their sperm (Dobrinski et al. 1994). The results indicated increasing infertility with increasing levels of abnormal DNA condensation. It was shown in one bull that scrotal insulation exacerbated the problem, however, clearly some bulls produce high percentages of the abnormality without any known physical cause for abnormal testis function.

5.1.9 Pale Centres

Analogous to DNA condensation this abnormality may be observed under Feulgen staining at × 1000 using phase contrast microscopy and might be considered the gold standard for the observation of this abnormality. However, it may also be initially observed under DIC at × 1000, with Feulgen used to confirm. The abnormality is displayed as a narrowing of the sperm nucleus in the PAS region with vesiculation of the nuclear envelope. This abnormality has been reported to cause decreased fertility in a case study in Canada by the author (AB) and in Australia by the author (VP). This condition is currently under study by the authors and colleagues. The abnormality may appear in conjunction with vacuole abnormalities or as the only abnormality present. (Fig. 5.14).

5.1.10 Teratoid Sperm

These are sperm that are so grossly abnormal in structure as to be barely recognisable as a sperm cell. The sperm nucleus varies from normal to grossly misshapen, may be vacuolated and the tail is often coiled up completely and lies superimposed on the head (Fig. 5.15). These sperm, in high numbers, are indicative of severe

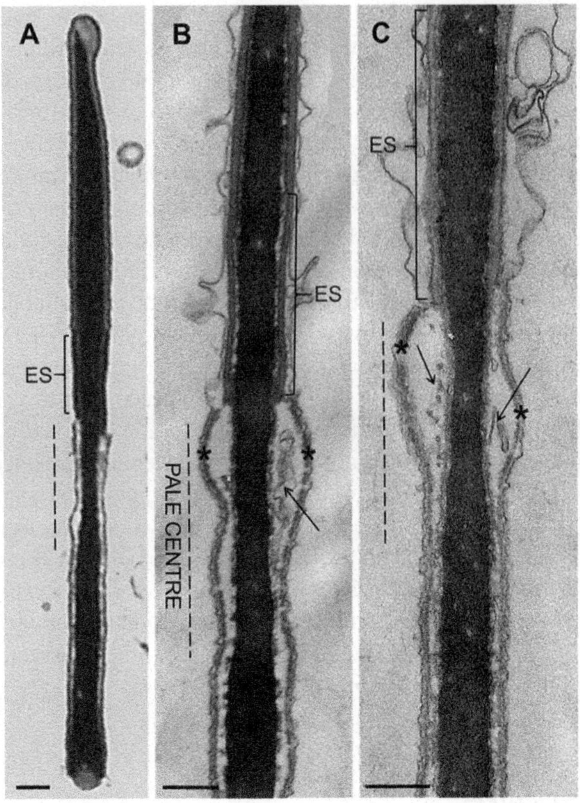

Fig. 5.14 A-C. EM micrographs of sagittal sections through bull spermatozoa displaying the pale centre defect. Note the confined area of thinning (dashed lines) of the nucleus in the postacrosomal region just below the equatorial segment (ES). The nuclear envelope tends to vesiculate and become disorganized (arrows) in this area of thinning and the postacrosomal sheath tends to bulge out (asterisks). Compare with normal bull spermatozoa in Chap. 1 in Figs. 1.3, 1.4. Bars = 0.2 μm

disturbance to spermatogenesis and spermiogenesis. They often occur at very low levels in the spermiogram (1%) but when seen at higher levels the prognosis is poor. There should be no more than 15% of this type of sperm in an ejaculate and they should be associated with at least 70% normal sperm. In addition, there are rare cases in which nearly all sperm are teratoid forms. One of the authors (AB) has seen 19 such cases over a 20 year period in six breeds of bulls including Charolais, Angus, Hereford, Simmental, Holstein and Braford. Three of the Charolais bulls were very closely related. The condition was also seen in a ram and a dog. None of these cases improved over observation period spanning 3 months to 1.5 years. All produced white coloured semen and had large, firm testes. In one bull, treatments spanning 1 month each, first with GnRH and then testosterone, had no effect on the condition. The condition appears to be analogous to Y-chromosomal azoospermia factors in men.

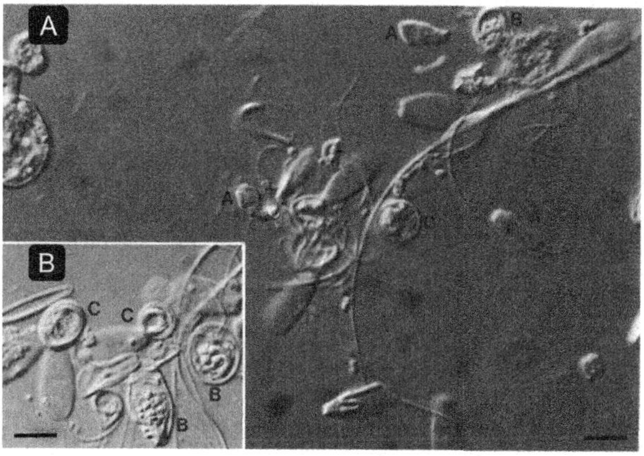

Fig. 5.15 Teratoid spermatozoa in a severely affected Charolais bull. These figures show some of the variety of severely malformed spermatozoa consistently produced by this bull. Both sperm heads and tails were affected. **(A)** Although some of the detached heads appeared grossly normal, very few spermatozoa were completely normal. **(B)** Note the Dag-like forms in the inset micrograph. A, severely malformed heads; B, severely malformed heads with tails coiled around them; C, Dag-like forms. Bars = 5 μm

5.1.11 Microcephalic and Macrocephalic Sperm

Summary: Sperm both smaller and larger than the norm for a particular bull are relatively frequently observed in the ejaculate in very small numbers. Infertility due to this abnormality has not been reported.

Detail: The term microcephalic is used to denote any sperm head that, when compared to the normal population of sperm in a preparation, is clearly smaller. Similarly, the term macrocephalic is used to denote any sperm head that is clearly larger than the heads of the normal population of sperm. There is little information in the literature on micro- and macrocephalic sperm. This might be because these forms, though commonly found in sperm preparations, have never been found in large numbers in bull semen. Nor are there any reports of bull infertility related specifically to either micro- or macrocephalic sperm. The only exception appears to be the Rolled head-Nuclear crest-Giant head syndrome (Barth and Oko 1989).

(Williams and Savage 1925), measured the length of sperm heads in bulls of good, questionable, and poor fertility with a microprojection apparatus that projected the sperm image onto a screen at a magnification of 3000 diameters. They observed that the sperm of fertile bulls were remarkably uniform in length. Bulls with a large variation in sperm head length had lower fertility.

(Salisbury and Baker 1966), studied incidence and size of diploid (macroephalic) sperm in single Feulgen stained smears from 80 bulls. One thousand sperm were counted on each smear. Most of the bulls were used in an inbreeding experiment and included inbred Hereford lines, line outcrosses, some breed crosses and a few dairy bulls. Diploid sperm were found in 46 of 80 samples; the overall incidence was 0.104%. The percentage for inbreds was 0.123 vs 0.106% for the line crosses. In one of the lines of bulls, the incidence of diploid sperm was significantly higher than in the other lines. The highest incidence of diploid sperm was 24/1000 in one individual. Salisbury and Baker interpreted their data as indicating that diploid sperm production is genetically influenced. Although there are no data reported for microcephalic sperm, it seems likely that a genetic influence occurs in this regard as well. The size of the sperm head might be due to a deficiency or excess of nuclear chromatin. There is some evidence that uneven chromosome division occurs during meiosis in spermatogenesis (Gustavsson 1969; Knudsen 1954). In most cases, the cells resulting from uneven chromosome division would not progress forward in spermatogenesis and were lost. Thus, although uneven chromosome division may occur frequently, apparently few such cells survive to enter the sperm population of ejaculates. This would explain why microcephalic and macrocephalic sperm are not found in large numbers in semen.

In scrotal insulation studies, some bulls responded with a mild increase in microcephalic sperm in their ejaculates 21–28 days after the start of insulation, but many did not (Barth and Bowman 1994; Brito et al. 2003).

It is unlikely that micro- or macrocephalic sperm are able to participate in ovum fertilization and embryonic development. However, since these forms have not been found in high percentages in semen, bull infertility due to these forms has not been reported.

5.1.12 The Decapitated Sperm Defect

Summary: The decapitated head defect is rarely observed. It can be differentiated from detached heads due to senesce by the presence of very high percentages of detached heads and large numbers of vigorously moving tails in the fresh specimen and the presence of the proximal droplet still attached to the tail. This trait when it occurs affects 80–100% of sperm in the ejaculate.

Detail: This is a rare sterilizing defect of hereditary nature that was first reported in Guernsey and Hereford bulls (Alun Jones 1962; Blom 1977; Hancock 1949; Williams 1965). Characteristically, separation into loose heads and tails involves most of the spermatozoa. Many of the loose tails are bent around a cytoplasmic droplet and often both bent and straight tails are vigorously motile. The cause of head separation was shown to be a shallow implantation fossa and a defective basal plate. Head and tail are kept together by the cytoplasmic membrane, but head separation occurs when motility develops. The three main characteristics of the decapitated defect were summarized as: (a) 80–100% of sperm are affected; (b), a high percentages of headless tails show active progressive motility; and (c) proximal cytoplasmic droplets with or with out the tail bending around the droplet is are often seen in swimming tails and may be mistaken for small heads (Blom 1977). In the classical cases described for sterile Guernsey bulls, there is little doubt that the condition is inherited, probably as a recessive gene through male and female carriers (Alun Jones 1962).

5.1.13 Rolled Head-Nuclear Crest-Giant Head Syndrome (RCGH)

Summary: This abnormality is included under the vacuole/teratoid category in the Australian standardized assessment of sperm. It may be observed in combination with vacuolated sperm but is uncommon. It is thought to be an inherited condition. The prognosis for recovery is very poor (Barth and Oko 1989). The number tolerated in the ejaculate is at 20% because of the ability to penetrate the zona pellucida but the inability to produce a viable embryo. Reports upon its effect on conception rates when present at 20–30% of the ejaculate vary between 27–74% (Barth 2013).

Detail: This rare defect has been reported in Brown Swiss (Bertschinger 1975), South Devon (Blom 1980) and Friesian (Cran et al. 1982) bulls. The authors have observed the defect in Santa Gertrudis and Droughtmaster bulls. It was also observed in a Jersey bull and in three Canadian Holstein bulls (Albert Barth pers. com). The incidence of RCGH affected sperm has ranged from 8–80% among these cases. The giant heads, which may be diploid or polyploid (Bertschinger 1975), are often crested, or rolled (Figs. 5.16, 5.17). A high percentage of these forms have double tails. Other abnormal forms such as pyriform heads, knobbed acrosomes, double heads and a variety of bizarre head shapes may be seen concurrently. A genetic cause is supported by the work of (Bertschinger 1975), in which a high incidence of RCGH sperm was found in a family of Brown Swiss bulls. Pedigree studies of a Canadian Holstein bull, his son and his grandson also suggested a hereditary origin for this defect.

Four Brown Swiss bulls with 20–32% RCGH sperm were used in natural service and the 90-day non-return rate varied from 27–74% (Bertschinger 1975). A South Devon bull whose semen sample appeared normal except for a high percentage of RCGH sperm was nevertheless quite fertile as 79% of 24 cows artificially inseminated were reported to be pregnant as judged by their 60–90 day non return rate. Presumably, normal sperm associated it defective sperm were able to fertilize ova.

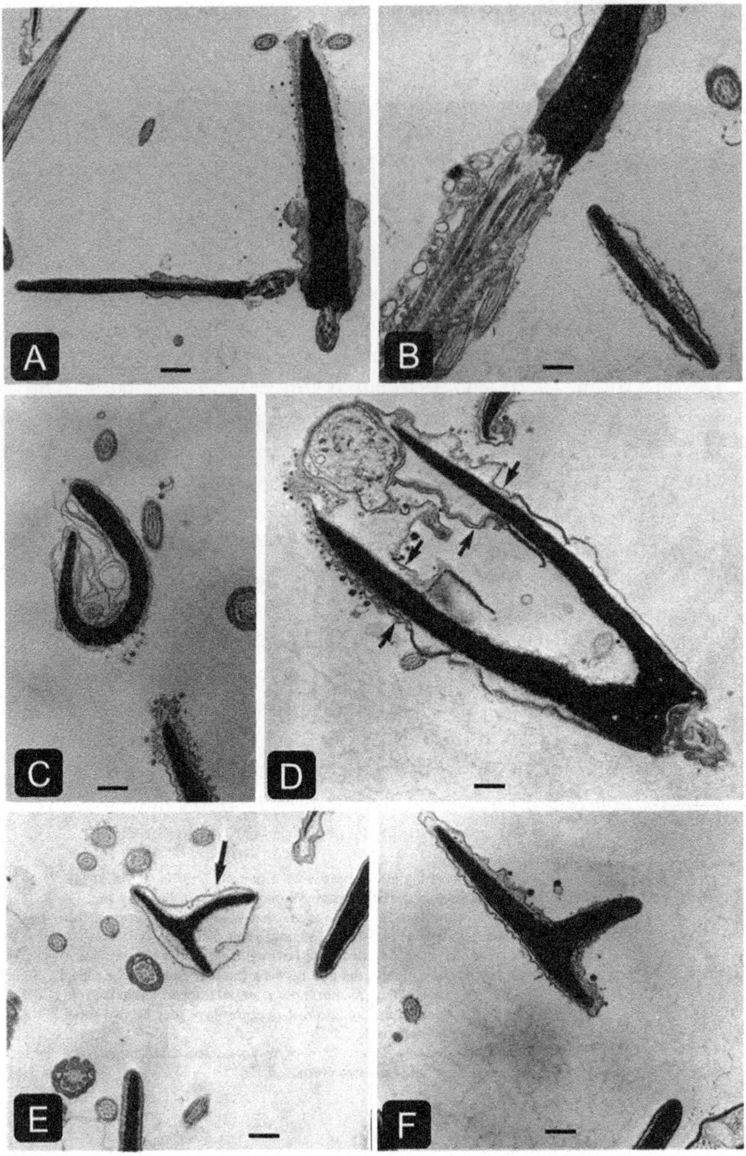

Fig. 5.16 Rolled head-nuclear crest-giant head syndrome. All electron micrographs in this figure and the next are all taken at the same magnification. Bars = 0.5 μm. Note the tremendous variation in size of nuclear sections. Note the difference in size of the normal head and a giant head in (**A**). View (**B**) shows a longitudinal section of a giant head with two implantation fossas for the double tail. A cross section of a rolled head is shown in (**C**). The cell membrane appears to at least partially follow the inner contour of the curvature. Membraneous material is trapped inside the curvature of the nucleus. View (**D**) shows a longitudinal section of a rolled head, resulting in a forked appearance. Note the acrosomal membranes (arrows) and a vesicle filled with granular material trapped in the curvature of the rolled head. View (**E**) shows a cross section of a normal-size nucleus with a crest (arrow) while (**F**) shows a cross section of a crested giant nucleus

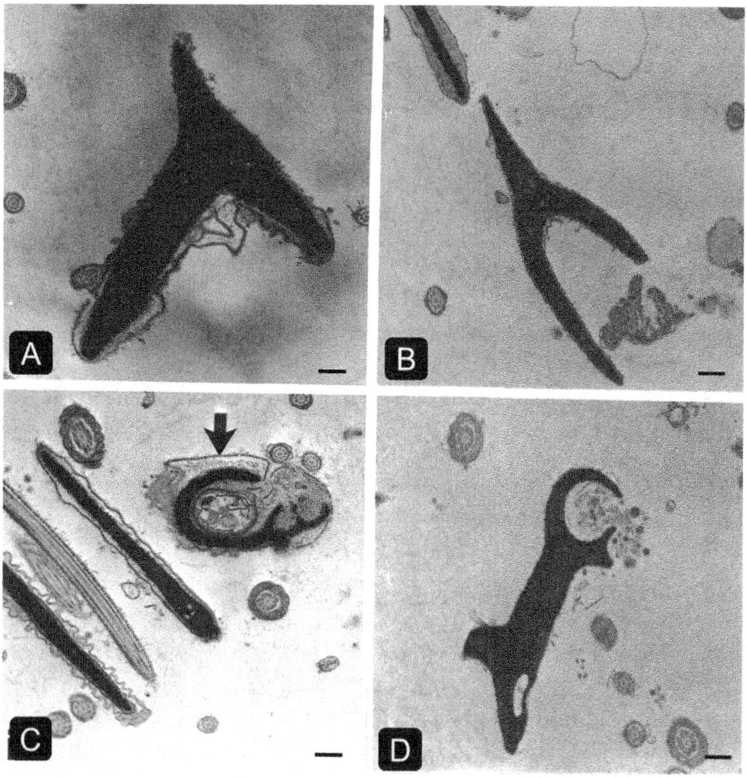

Fig. 5.17 (**A**) Cross section of a giant head with a nuclear crest. Compare this with the sizes of the crested nuclei of the same magnification in (E) and (F) in the previous Fig. 5.16. (**B**) Shows a cross section of a nucleus that has all three forms of this syndrome combined, i.e., a rolled, crested, and giant form. (**C**) A cross section of a rolled head (arrow) in which the cell membrane appears to almost completely bridge the gap between the two edges of the rolled head. Note also the vesicle containing membranous material trapped within the fold caused by rolling of the nucleus. In addition, it appears there may be two small crests on at least part of the nucleus. (**D**) A cross section of a bizarre form of a giant nucleus that appears to be partly rolled and vacuolated and has two crests. Bars = 0.5 μm

5.2 Midpiece Abnormalities

5.2.1 Distal Midpiece Reflexes (DMRs)

Summary: This may be the most common sperm tail defect seen in bull ejaculates (Perry et al. 2002; Menon et al. 2011), not to be confused with a simple bent tail, as the midpiece is also involved in the bend. This defect can occasionally arise as an artefact due to prolonged contact with a hypotonic solution (e.g. Nigrosin-Eosin stain), cold-shock, or solutions >pH 7. It is one of the first abnormalities to appear after a stress event as it develops in the cauda epididymis. An increase in DMRs was observed 16 days after an acidotic event (Callaghan et al. 2016) or only 4–11 days after dexamethasone treatment to stimulate stress (Barth and Bowman 1994).

Dissections and aspirations of sperm have shown that DMRs develop mainly in the distal one-third of the cauda epididymis. The prognosis varies with circumstance and the types of other abnormalities present. When it occurs with abnormalities such as a fracture at the tail bend, aplasia of the midpiece or Dag-like defects there may be an underlying cause such as disturbed spermiogenesis. Some bulls have a predisposition for this defect with fluctuations in the percentage of affected sperm throughout the year. Up to 30% of this abnormality is tolerated in the ejaculate as these cells display reverse motility and would therefore be unable to penetrate the zona pellucida so other normal cells would be able to participate in ovum fertilization (Barth and Oko 1989).

Abnormal developmental Morphology: Fig. 5.18 compares light and electron microscope micrographs of the characteristic J-shaped bend of the distal midpiece reflex defect. A cytoplasmic droplet is nearly always trapped in the bend (Figs. 5.18 A-D, 5.19 A-D and 5.20 A, C). Electron microscope sections through different regions of the reflex show the midpiece and principal piece enclosed in a common membrane (Figs. 5.18 E, 5.19 B and 5.20 A-C). This may be regarded as evidence that when parts of the sperm tail contact each other during extreme bending the opposing sperm cell membranes fuse. Alternatively, the sperm cell membranes may simply stretch around the outside curvature of the bend and contract on the inside curvature, thus encapsulating both midpiece and principal piece (Fig. 5.21). A major difference between the distal reflex and the Dag defect is that in the former the mitochondrial sheath is smooth and complete as opposed to the rough and disrupted mitochondrial distribution seen in the latter (Figs. 5.19 A-C, 5.20 A-C versus Figs. 5.24, 5.27 B, C, 5.28). This difference can also be seen by DIC microscopy (Fig. 5.18 A versus Fig. 5.24).

Detail: The frequency of this abnormality is illustrated in one study (Barth and Oko 1989), of 1214 bulls in western Canada evaluated in May and June, 8.2% had elevated numbers of DMRs in their semen ranging from 15 to>51%. It is not uncommon for some bulls to produce 60 to 80% DMRs due to the effects

Fig. 5.18 **(A)** Light micrograph of ejaculated sperm from a bull affected with the distal midpiece reflex defect showing the characteristic folding of the sperm tail. Folding of the tail takes place in the distal midpiece and is associated with the position of the distal cytoplasmic droplet. Bar=5 μm. **(B)** Same as in (A), but the tail is only partially bent around the cytoplasmic droplet. This happens less commonly but indicates the close association of the bending with the position of the cytoplasmic droplet. It is doubtful that the droplet plays a significant causative role in the development of this defect. Bar=5 μm. **(C)** Same as in (A), but the cytoplasmic droplet appears to be densely packed with cellular components. The lower arrow depicts the level at which the sperm tail was cut in (D) and Fig. 5.20 A, while the upper arrow depicts the level at which the sperm tail was cut in (E) and Fig. 5.20B. Bar=0.5 μm. **(D)** Electron micrograph of the distal part of the folded tail cut longitudinally at the level depicted by the lower arrow in (C). Note that the bend involves the distal end of the midpiece and that the cytoplasm is packed with membranous organelles mostly made up of the saccular elements of the cytoplasmic droplet. Bar=0.5 μm. **(E)** Electron micrograph of the proximal part of the folded tail cut longitudinally at the level depicted by the upper arrow in (C). The mid- and principal-pieces appear to be fused. Bar=0.5 μm ▶

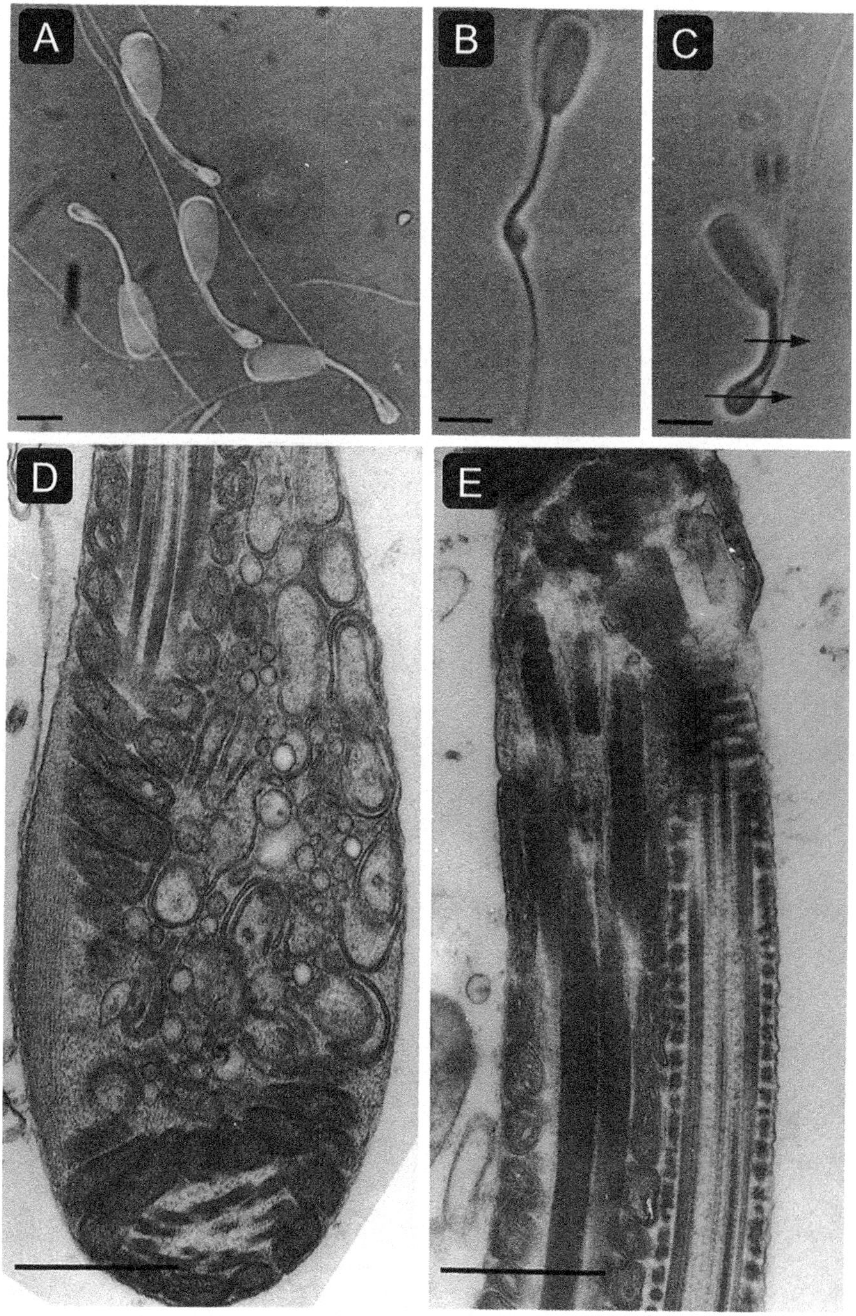
A
B
C
D
E

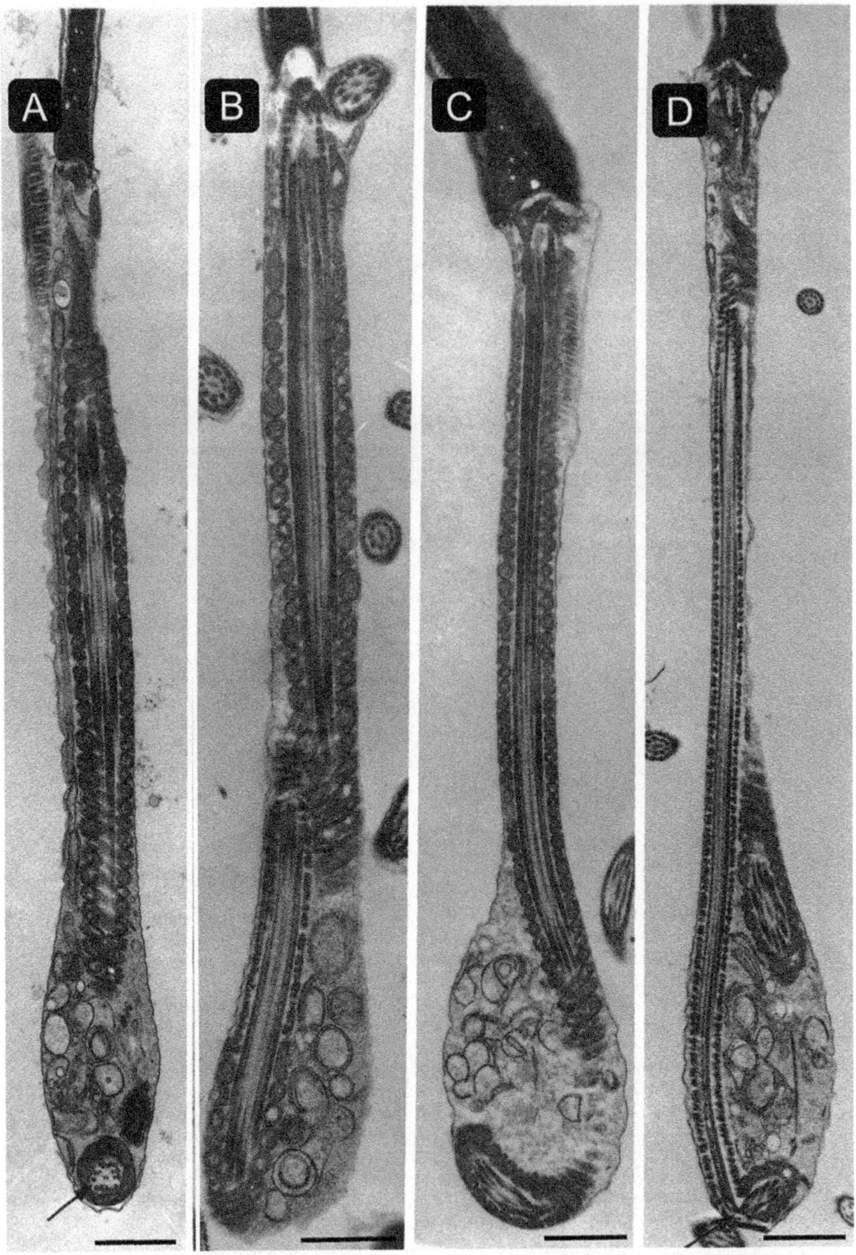

Fig. 5.19 Longitudinal sections through sperm tails of bulls affected with the distal reflex defect. Note that the bending occurs at the distal end of the midpiece around the distal cytoplasmic droplet, the principal pieces fuse with the midpieces, and relatively few axial fibers are disrupted. In **(A)** there is a slight disorganization of axial fibers at the level of the fold (arrow). In **(B and C)** the fracture or bend occurred in the distal midpiece, while in **(D)** there is a breakage of axial fibers at the midpiece-principal piece junction (arrow). Bending usually takes place proximal to this junction. Bars = 0.5 μm

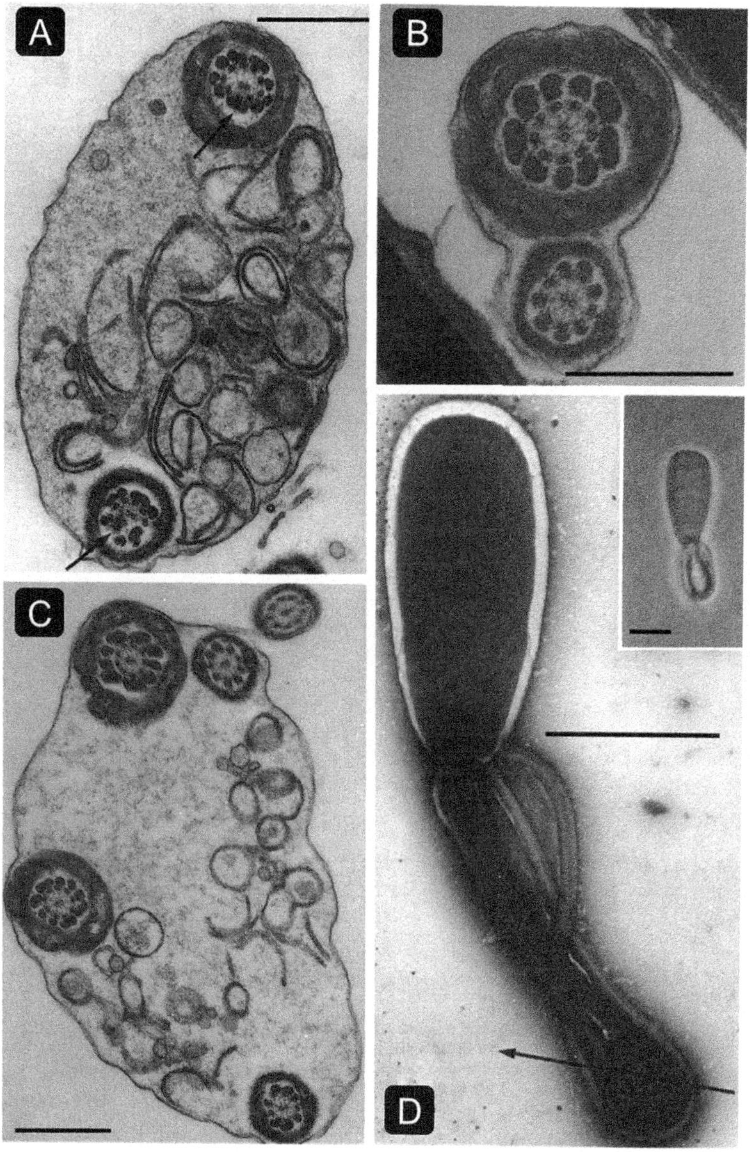

Fig. 5.20 **(A)** Cross section through the fold at the level of the distal cytoplasmic droplet as depicted by the lower arrow in Fig. 5.18 C. Note the disorganization of axial fibers in one of the halves of the axial fiber bundle of the principal piece (lower arrow) and the unsymmetrical positioning of axial fibers in the midpiece (upper arrow). The distal cytoplasmic droplet is packed with saccular elements. Bar = 0.5 µm. **(B)** Cross section through the fold just below the neck piece as depicted by the upper arrow in Fig. 5.18 C, showing the midpiece and principal piece within a common membrane. Bar = 0.5 µm. **(C)** Cross section through the lower part of a coiled tail as depicted by the arrow in (D). Coiling occasionally occurs with the distal reflex defect. However, as compared to the Dag defect (see Fig. 5.26 C), it is not associated with mass disorganization, dislocation, and disruption of axial fibers. Bar = 0.5 µm. **(D)** Electron micrograph of a negatively stained bull sperm displaying a coiled tail, which is occasionally found with the distal reflex defect. Bar = 5 µm. There is no evidence of axial fiber disruption. Inset—light micrograph showing similar features as in (D). Bar = 5 µm

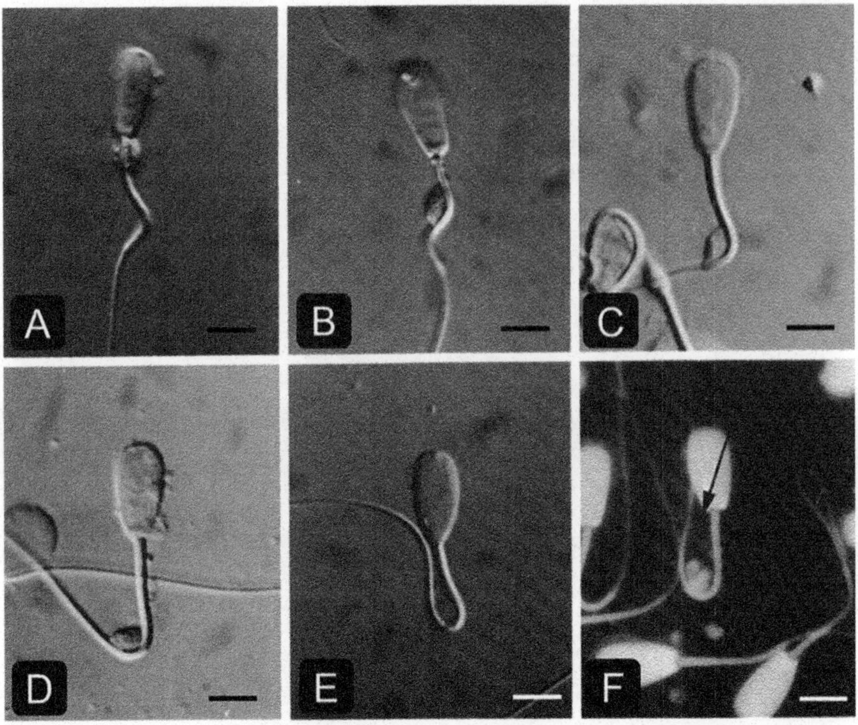

Fig. 5.21 Photomicrographs demonstrating the development of the distal midpiece reflex. Note in **(A)** and **(B)** that bending in the distal midpiece may occur prior to descent of the cytoplasmic droplet from a proximal position. The sequence of **(C)**-**(F)** shows how the sperm cell membrane may stretch around the greater curvature of the bend and contract in the inner curvature (arrow, F), resulting in the incorporation of the descending and ascending parts of the tail and the cytoplasmic droplet within a saclike common membrane. Bars = 5 μm

of stress and, or a seasonal low in testosterone concentrations. Angus bulls seem more prone to DMRs and in western Canada they were a great deal more likely to harbor the problem into late spring. A few Angus bulls maintained the problem throughout the year. Thus, there appears to be a hereditary predisposition in Angus bulls. The Jersey breed also appears to have a hereditary predisposition for sperm DMRs. Pedigree analysis done by the author showed that DMRs occurred in familial lines in this breed. Some Jersey bulls have virtually 100% of sperm affected year-round.

On light microscopy, the typical appearance is that of a bend in the distal midpiece in the shape of the letter J. Usually a cytoplasmic droplet is trapped in the bend of the midpiece and the principle piece projects past the sperm head; in the case of live sperm, reverse motility will be present. Variations in the bending result in different manifestations of the same defect, including an L-shaped bend and a second bend in the opposite direction higher-up in the midpiece. In the latter case, the midpiece may appear to be coiled. On electron microscopy, (Barth and Oko 1989) found that occasionally a section cut through the distal midpiece bend of affected sperm revealed fracturing in the axial fiber bundle of this region but the majority of sections did not show this (Barth and Oko 1989)(Figs. 19 D, 20). (Gustafsson et al. 1974) did not report fractures of fibers in the bend of DMRs. The majority of sperm in the cauda epididymis, where the DMR is produced (Swanson and Boyd 1962), normally have a cytoplasmic droplet in the distal position of the midpiece. Thus when the DMR forms in the cauda epididymis, it entraps the droplet. The defect can be induced experimentally with hypotonic solutions (Lindahl and Drevius 1964; Drevius 1963); however, when the defect is produced in vitro the distal droplet will already have been shed, thus a cytoplasmic droplet will not be present in the J-shaped bend. The finding of a large number of DMRs without entrapped droplets on semen smears should alert the examiner to the possibility the defect was artificially induced during semen handling or smear preparation.

It appears that the cause of the DMR defect is an abnormal environment in the cauda epididymis. Very high concentrations of testosterone are necessary for normal epididymal function (Goyal 1983). A stress-induced rise in cortisol decreases the secretion of LH and consequently testosterone and may thus interfere with epididymal function (Welsh et al. 1979). It seems plausible that epididymal malfunction associated with the production of DMRs may be induced by a common mechanism of low testosterone secretion due to a variety of adverse conditions. For example, estradiol treatment (Cupps et al. 1960; Cupps and Briggs 1965), low thyroid activity (Swanson and Boatman 1953), ephemeral fever (Burgess and Chenoweth 1975), scrotal insulation, or the stress associated with a snowstorm (Barth and Oko 1989), all suppress testosterone.

Sperm enter the epididymis normally formed and make most of their way through the epididymis before developing a reflex bend. The numbers of DMRs suddenly increase in the cauda epididymis, particularly in the distal 1/3 of the cauda epididymis (Fig. 5.21) (Barth and Oko 1989; Purswell et al. 1983). This is also supported by the finding that when normal bulls are subjected to stress, the percentages of DMRs in ejaculates increase rapidly within a week after the onset of the stress. Thus sperm, which take approximately 10 days to pass through the epididymis and some additional days to pass through the vas deferens and ampulla, most likely develop DMR in the distal epididymis in order for the defect to appear so quickly in ejaculates after the onset of stress. The length of exposure of sperm to an abnormal epididymal environment also appears to be important, since successive ejaculates in exhaustion tests had progressively reduced percentages of DMRs (Swanson and Boyd 1962).

Gustafsson et al. 1974 considered that epididymal dysfunction resulted in alteration to the normal intracellular and extracellular ionic balances of the spermatozoa. This caused fluid uptake or loss by the spermatozoa resulting in distortion of the cell membrane, leading to bending and coiling of the sperm tail. Hypotonic solutions (Drevius 1963; Lindahl and Drevius 1964) also cause a marked increase in this defect in vitro. In sperm flushed from the epididymides of bulls that were predisposed to producing sperm with DMRs, egg yolk prevented sperm tail coiling, whereas, isotonic sodium citrate seemed to promote coiling (Swanson and Boyd 1962). Lecithins in egg yolk have may reduce permeability of the sperm membrane to ions in semen diluent, or to leaching of ions from the sperm.

DMRs are a seasonal problem for many bulls in northern regions (Barth and Waldner 2002). This may be related to photoperiodic changes in testosterone and/or the stress of winter conditions. Testosterone concentrations are low in fall and winter, increasing as spring approaches and highest in June (Barth 2010). Most affected bulls recover from DMRs by April or early May. DMRs are easily produced in all bulls by stress. Stress results in reduced testosterone concentrations (Barth and Bowman 1994). If the stress is mild and short lived, the DMRs will appear over a period of about 10 days and then soon regress. If the stress was more severe and prolonged, other types of defects will follow the appearance of DMRs and recovery will take 6 weeks to 2 months, or more. Increased testis temperature due to some abnormality in thermoregulation of the testis also leads to reduced testosterone secretion (Prabhakar et al. 1990).

There is anecdotal evidence that nutritional deficiency of selenium may be involved in elevated levels of DMRs. A Canadian AI center experienced elevated levels of DMRs in many of their bulls. Investigations revealed that selenium supplementation had ceased in the preceding 6 months. After selenium supplementation was reintroduced, the problem dissipated. However, since selenium deficiency causes segmental aplasia of the mitochondrial sheath, resulting in a characteristic sharp bend formation at the midpiece-principal piece junction (Olson et al. 2007, 2005) it is possible that the sperm defect was misdiagnosed.

There are few reports of studies on the tolerable levels of this defect in semen of bulls used for natural service, or in semen frozen for artificial insemination. Herman and Swanson (1941), reported that two bulls with 26% and 55% bent tails had poor fertility on AI, but another bull with 31% bent tails had a 1.55 service per conception rate. Studies which have examined the effect of sperm defects on sperm transport in the female tract have shown that spermatozoa with abnormal tails are filtered by the cervix; thus DMRs would likely be compensated by increasing sperm dosage (Saacke et al. 1994). Experimentally, we have shown that even when DMRs were at levels as high as 39%, whether in frozen semen, or in bulls used in natural service, pregnancy rates were normal. In the Australian standardized format of assessment the DMR threshold is < 30%.

5.3 Mitochondrial Sheath Abnormalities; Segmental Aplasia, Corkscrew Defect, Pseudodroplet Defect and Dag Defect

5.3.1 Segmental Aplasia

Summary: Gossypol in the diet (Chenoweth et al. 2000) and a viral disease (Bovine Ephemeral fever), (Chenoweth and Burgess 1972) have both been shown to effect the mitochondrial sheath. If gaps in the midpiece are larger than 3 microns these may result in fractures of the midpiece and sperm showing such severe segmental aplasia are considered under midpiece abnormalities, however, sperm observed with slight gaps are considered under the normal category (McAuliffe et al. 2010). In a case study a bull with 90% segmental aplasia was reported to have normal fertility over 3 breeding seasons (Barth 2013) indicating that small gaps do not affect fertilizing ability of sperm. The condition can be permanent or transient; if the defect is seen to occur in two tests done ten weeks apart it suggests a permanent condition.

A report of a new abnormality of the mitochondrial sheath which did not affect bull fertility (Rocha et al. 2006) is similar to the previously described pseudodroplet both under light microscope and TEM images (Blom 1968; Barth and Oko 1989; Barth 2013). The difference between the observations is that the effect upon the mitochondrial sheath is apical (Rocha et al. 2006) compared to points along the midpiece (Barth 2013) and that, at least in the Blom study, this defect did affect fertility via an effect on sperm motility. The defect is observed as a thickening of the midpiece often associated with a bend or fracture. TEM reveals an accumulation of dense granules within these thickened areas. The light microscope images of the apical defect (Rocha et al. 2006) are more similar to a broken neck appearance.

Detail: The mitochondrial sheath is laid down in the final steps of spermiogenesis, just before sperm are released into the epididymis (see Chap. 2.2). The mitochondria are laid down first in the sperm neck region and then progress distally toward the annulus at the midpiece-principal piece junction. Occasionally, the last mitochondria are not placed, leaving a gap at the annulus and a weak section at the end of the midpiece. Fractures at that site result in a sharp reflex of the principal piece (Fig. 1.32 Chap. 1.1). This type of midpiece fracture is often seen along with DMRs. It is common to find a few spermatozoa with one or two small gaps in various regions of the midpiece in normal fertile bulls. Rarely, a bull may produce a high percentage of affected sperm.

In some cases, swollen supernumerary mitochondria result in a ragged enlarged area on the midpiece. It is often difficult to distinguish between mitochondrial disruptions and proximal cytoplasmic droplets with light microscopy; however, with the use of phase contrast microscopy droplet material is distinguishable from mitochondrial disruption. In the majority of cases, ragged swollen areas on the midpiece are likely to be cytoplasmic droplet material.

Although mitochondrial sheath segmental aplasia in low percentages of sperm is common in bulls, little is known about its cause. In our experience, it is very uncommon to find bulls that produce very high percentages of sperm with this abnormality. Barth reported (Barth 2013), a Limousin bull that consistently produced 80–90% sperm with small mitochondrial gaps, but otherwise sperm morphology and motility were good. Usually, one or several gaps in the mitochondrial helix were located randomly along the midpiece. Most of the gaps ranged in size from 1–1.5 μm although much larger gaps (2-4μm) occurred in a few sperm and these were more likely to result in a midpiece fracture. There was no indication of the cause of the abnormality in this bull, since his herd mates (n = 6) that were overwintered under the same conditions did not have this sperm defect. There were no abnormal physical findings or any indication of a health problem that might predispose him to this condition. A year later, the bull still produced a similar spermiogram and was used for further breeding. The bull was used in 2 controlled breeding trials, one using frozen semen and a second using natural mating. Pregnancy rates were similar to those of the control bulls.

Gossypol toxicity has been shown to cause sperm mitochondrial aplasia in bulls (Chenoweth et al. 2000). Young Brahman bulls received either cottonseed meal delivering 8.2 g free gossypol/bull/d (n = 4) or isonitrogenous soybean meal (control, n = 4) for 11 wk. Those receiving gossypol developed elevated levels of midpiece defects consisting primarily of gaps in the mitochondrial helix that appeared to be similar to those seen in the case of the Limousin bull described in the previous paragraph.

In rats, an induced dietary deficiency of selenium was associated with defects of the mitochondrial sheath (Olson et al. 2004). Selenium in rat sperm is associated with a cysteine-rich structural protein of the mitochondrial capsules (Calvin et al. 1981). Late spermatids of selenium-deficient rats displayed normal structural organization of the axoneme, outer dense fibers, fibrous sheath and annulus, but they exhibited a premature termination of the mitochondrial sheath due to swelling and degeneration of a proportion of mitochondria. The size of the gap reflects the proportion of selenium deficient and hence swollen mitochondria, which did not take part in the construction of the sheath. Cauda epididymal sperm exhibited extensive flagellar disorganization resulting from the apparent sliding and extrusion of specific outer dense fiber-doublet microtubule complexes from the missing segment of the mitochondrial sheath. A characteristic feature of selenium deprived murid sperm is a hairpin-like bend formation just above the midpiece-principal piece junction where segmental aplasia is most severe (Olson et al. 2007, 2005). These findings in murids suggest that selenium deficiency may play a role in the development of the mitochondrial sheath abnormalities of bull sperm as well.

5.3.2 *Corkscrew Defect*

***Summary*:** The corkscrew defect was first described by Blom in two sterile Red Danish bulls (Blom 1959). He later reported it in 60 Danish bulls of the Red Danish, Holstein–Friesian, Jersey, and other breeds (Blom 1978). The defect is characterized by an irregular distribution of mitochondria along the length of the mitochondrial sheath (Fig. 5.22 A). Electron microscopy shows sporadic mitochondrial swelling, degeneration, and aplasia along the entire length of the sheath (Fig. 5.22 B). Frequently, affected sperm have a proximal droplet Fig. 5.22 B). In spite of the weakened condition of the midpiece, folding and coiling of the tail does not occur in the corkscrew defect. The classic appearance of the defect is found only in dead cells, as in living cells the midpiece would fracture during motility. Since affected sperm are always dead (red on eosin-nigrosin smears) and

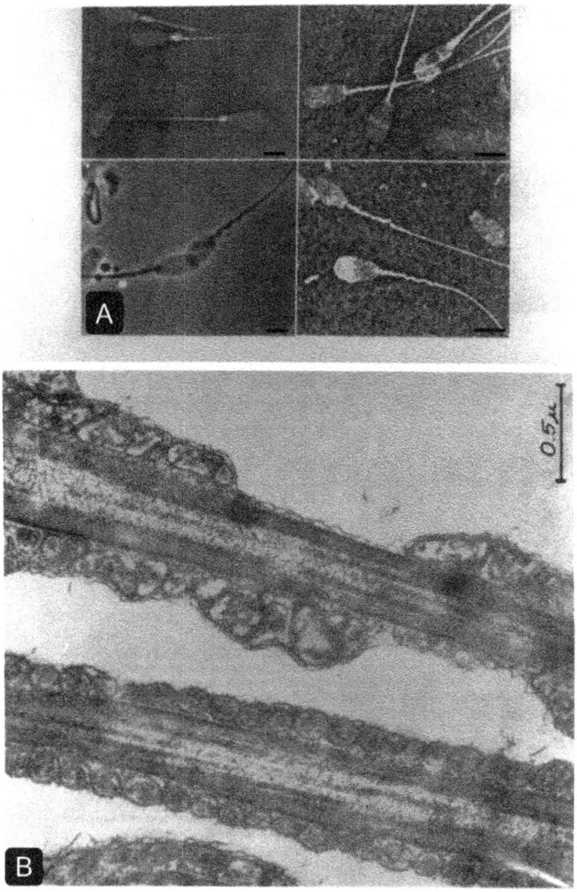

Fig. 5.22 (**A**) Eosin-nigrosin and india ink smears from bulls showing the cork-screw sperm defect. Bars = 5 μm. (Courtesy of E. Blom) (**B**) Electron micrograph of longitudinal sections of the midpiece of a normal sperm below and a typical corkscrew sperm above. Note in the latter the large irregular lumping and swelling of mitochondria replacing the regular mitochondrial helix. 0.5 μm. (Courtesy of E. Blom)

completely immotile, midpiece breakage does not occur. Oko and Hrudka (1982) found that dislocations, foldings, and breakages of the axial fiber complex, which normally accompany gossypol-induced segmental aplasia of the mitochondrial sheath, did not occur once the sperm became immotile after extended treatment with this drug. Thus folding and coiling that occurs in the epididymis of affected sperm is dependent on the cell's capacity for motility or, in other words, on some degree of respiratory activity.

The reported maximum occurrence of this defect was between 1960 and 1965. During this period it was also reported to occur in other countries in Europe (Munro et al. 1961; Smidt 1965). The occurrence of the defect was linked to a progressive degeneration of the testis and coincided quite closely with high levels of nuclear fission products, strontium-90, cesium-137, and iodine-131. This led Blom to speculate that the defect, which is now rare, may have been caused by nuclear fallout from atmospheric testing of nuclear bombs in the early 1960s (Blom 1978). The fact that, among Danish breeds of cattle, a much higher proportion of the cases was observed in the Red Danish breed may indicate an inherited cause (Blom 1978). A similar type of defect, accompanied also by other major sperm defects and testicular regression, was experimentally induced in rams by feeding ethylene dibromide, a grain fumigant (Eljack and Hrudka 1979). This experiment and others (Chenoweth et al. 2000; Oko and Hrudka 1982) are reminders that, of the cellular organelles, mitochondria are probably the most sensitive indicators of toxicity.

We have seen the corkscrew sperm defect in healthy Red Angus and Charolais bulls maintained in a normal environment with other fertile bulls. When the Red Angus bull was first examined in April of 1996, 99.5% of sperm were dead and 96% had the corkscrew defect. The same bull was examined in May of 1997 and 76% of sperm were dead with 74% corkscrew sperm. In the case of the Charolais bull, only 18% of sperm had the corkscrew defect and there were many other types of sperm defects indicating a fairly severe disturbance of spermatogenesis of unknown cause. The sperm problem did not improve over a period of 6 months.

5.3.3 *Pseudodroplet Defect*

The pseudodroplet defect was reported in Denmark by (Blom 1968) in 5 Friesian bulls. The incidence of the defect in ejaculates of these bulls ranged from 7 to 26%. The frequency of the defect appeared to increase with age, and at the same time sperm motility and fertility declined. The defect appeared on light microscopy as a local thickening of the midpiece somewhere along its length and was often associated with a bend or fracture at the same site (Fig. 5.23 A). On eosin-nigrosin or India ink semen smears the thickening appeared to have vacuoles and granules within it (Fig. 5.23.B). On electron microscopy the thickened areas were shown to be an accumulation of dense granules on the surface of the mitochondrial helix surrounded by a single or several layers of mitochondria of normal appearance. The origin of the dense granules was unknown. The defect was demonstrated in 2 half-brothers and, since the condition was neither transient in nature nor associated with testicular or epididymal impairment, Blom suggested that it might be inherited in some way.

(Fig. 5.23 B). The dense granules resemble granular bodies involved in the assembly of the mitochondrial sheath in the last steps of spermiogenesis (Fig. 162.16 in Chap. 2). It appears that the granular bodies lay down the cement substance that attaches mitochondria to the outer dense fibers. Normally, once the assembly of the mitochondrial sheath is complete the superfluous granular bodies, often surrounded by superfluous mitochondria, are extruded with the spermatid

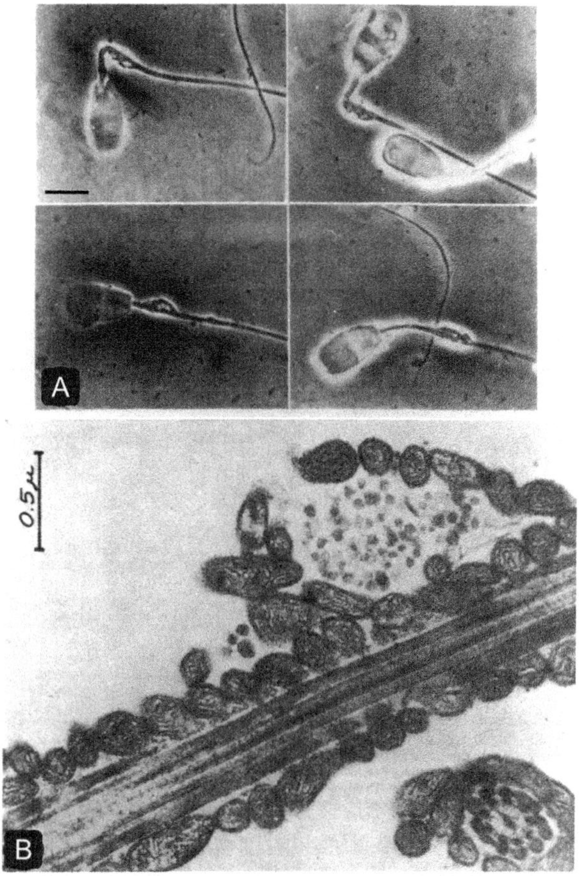

Fig. 5.23 **(A)** Eosin-nigrosin and india ink smears from bulls with impaired fertility showing typical pseudodroplet defect. Note the locations of the defects and the light vacuoles or inclusion bodies within the thickenings. Bar = 5 μm in all light micrographs. (Courtesy of E. Blom). **(B)** Electron micrograph of a longitudinal section from a midpiece showing a characteristic pseudodroplet. Note the accumulation of electron-dense granules on the mitochondrial helix surrounded by a single or double layer of mitochondria of normal appearance. Bar = 0.5 μm. (Courtesy of E. Blom)

cytoplasm as the residual body which is digested by the Sertoli cytoplasm. It appears that in the case of the pseudodroplet defect these granular bodies remain on the sheath of mature sperm This defect was also reported in a Droughtmaster bull that had 41 ± 5% pseudodroplet-sperm (Bongso et al. 1981). Barth (2013) found the pseudodroplet defect in two polled Hereford bulls, both aged 2 years. The spermiograms of these bulls (A and B) were as follows:

Bull A: 3% head defects, 31% pseudodroplet defect (19% unfractured and 12% fractured at the site of the pseudodroplet), 1% detached heads, 0.5% knobbed or accessory tails, and 68% normal sperm. The proportion of sperm staining alive was 75%.

Bull B: 2% head defects, 37% pseudodroplet defect (12% fractured), 10% irregular mitochondrial helixes, 30% distal midpiece reflexes, 1% detached heads and 20% normal sperm. In both cases there were 12 to 25 pen mates that were maintained under similar conditions but did not produce any pseudodroplet sperm.

In both of these cases, the defect persisted at approximately the same level over periods of 1–4 months and then the bulls were slaughtered.

A similar defect to the pseudodroplet defect, although occurring in a more apical position (Rocha et al. 2006), appeared to be caused by a loss of the cementing substance that links mitochondria to the plasma membrane, in association with mitochondrial aplasia at the neck region of the midpiece. This latter defect did not affect fertility or sperm motility.

5.3.4 The Dag Defect

Summary: The characteristic features of this defect are multiple fractures of the axonemal fibers in the midpiece with mitochondrial sheath disruption. 'Dag-like' defects may be present in a few sperm due to a disturbance of spermatogenesis in any breed; however, the term Dag Defect should be reserved for those cases where the defect is present in >60% of sperm and is inherited. The Dag Defect was shown to be inherited via a recessive gene in the Jersey breed and appears to have a similar heritable basis in the Swedish Red and White, Galloway and Hereford breeds. When a bull has both recessive genes, 60–100% of sperm will be affected (Barth and Oko 1989; Koefoed-Johnsen et al. 1980; Hellman et al., 1980; Perry et al. 2002).

Abnormal Developmental Morphology: A developmental perspective of the Dag defect through the excurrent duct and epididymis is presented in Figs. 5.24 and 5.25. In the rete testis folding and coiling of the sperm tail is almost negligible. However, the mitochondria sheath appears rough and swollen and occasionally small gaps are seen (Fig. 5.24 A). With the onset of sperm motility and translocation of the cytoplasmic droplet in the caput epididymis the mid-pieces begin to bend, stretching and incorporating the droplet in the bend (Figs. 5.25 B). In the corpus epididymis there is a high incidence of excessive bending of the midpiece with hairpin turns being a prominent feature, while coiling associated with breakage and dislocation of axial fibers is less frequent (Figs. 5.25 A). In the cauda epididymis and ejaculate, coiling of the tail associated with splitting, fracturing, retraction and dislocation of axial fibers becomes the characteristic feature of the Dag defect (Figs. 5.25 B-D, 5.26). In the process of tail folding, the sperm membranes of different regions of the tail come together and appear to fuse with each other, mimicking what occurs in the Distal Midpiece Reflex defect; as a result, separate parts of the tail (i.e., mid- and principal-pieces) become encapsulated by a common membrane (Figs. 5.27). However, the Dag defect distinguishes itself from Distal Midpiece Reflex defect by the tail folding on itself excessively

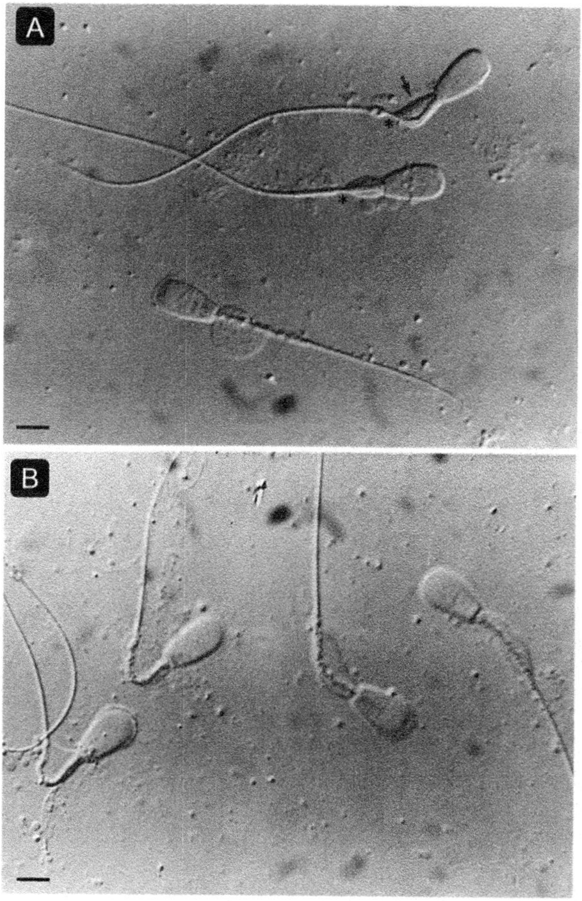

Fig. 5.24 **(A)** Sperm extracted from the rete testis of a bull affected with a Dag defect. Note the roughness of the mitochondrial sheath in the lower sperm, segmental aplasia in the upper two (asterisks) and slight bending associated with the translocation of the cytoplasmic droplet in the top sperm (arrow). Differential interference microscopy. Bar = 5 μm. **(B)** Sperm extracted from caput epididymis of a bull affected with the Dag defect. Bending of the tail occurs within the cytoplasmic membrane and incorporates the cytoplasmic droplet. The mitochondrial sheaths are rough and disorganized. Tail bends are now more prominent than in the rete testis as capacity for motility gradually increases during epididymal transit. Differential interference microscopy. Bar = 5 μm

(i.e., coiling), such that in cross sections through the tail multiple parts of the tail can be seen (e.g., 2 mid-piece parts and 4 principal piece parts) encapsulated by a common membrane (Figs. 5.26 C). Additionally, within the disorganized, membrane encapsulated compartments many of the axial fibers (including both microtubular doublets and corresponding outer dense fibers) retract themselves from the middle and principal pieces of the tail and are found dislocated in the common cytoplasm (Fig. 5.26 C). Ultrastructural analysis of spermatids at the end of

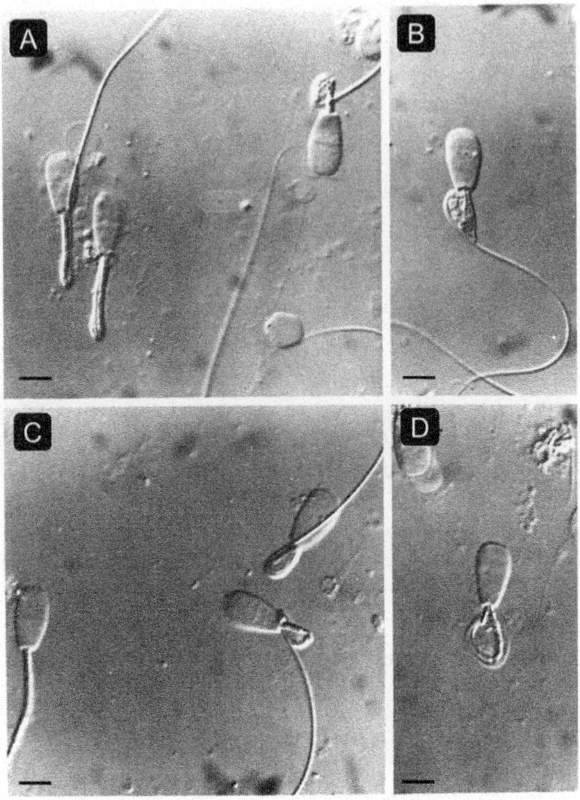

Fig. 5.25 **(A)** Sperm extracted from the corpus epididymis of a bull affected with the Dag defect. In two of the sperm the principal pieces are tightly juxtaposed against the midpieces in a hairpin configuration, whereas in another sperm the tail has coiled and broken in the midpiece. The phenotype very much resembles the gossypol and selenium-deficiency induced midpiece sperm defects in murids (Oko and Hrudka 1982; Olson et al. 2005). Differential interference microscopy. Bar = 5 µm. **(B-D)** Sperm extracted from the cauda epididymides showing characteristic coiling and breakage of the tails. One tail (D, upper right corner) has completely disintegrated. In the majority of the sperm the mitochondrial sheath is rough and disorganized. The cytoplasm of the sperm in (B) is packed with superfluous mitochondria, granular material and saccular elements of the cytoplasmic droplet. Differential interference microscopy. Bars = 5 µm

spermiogenesis suggests that the predisposing factor to the folding and coiling and dislocation of axial fibers is a retardation in the alignment, condensation and arrangement of mitochondria that make up the mitochondrial sheath (Fig. 5.28).

***Detail*:** The name of this uncommon defect was taken from a Jersey bull named Dag who shared the sperm problem with a full brother (Blom 1966). The Dag defected has been reported to occur in Jersey (Blom 1966), Swedish Red and White (Hellmen et al. 1980) and Hereford bulls (Wenkoff 1978). Recently it was seen in 4 sibling, Galloway bulls obtained from a single flush of bovine embryos (Barth, unpublished).

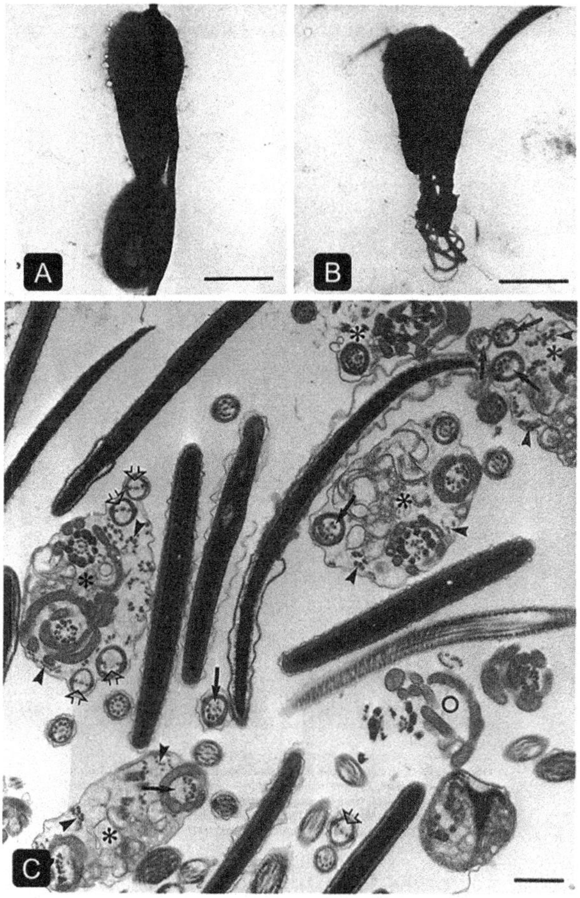

Fig. 5.26 **(A, B)** Electron micrographs of negatively stained ejaculated sperm of a bull affected with the Dag defect. In (A) the coiled tail is enclosed by the plasmalemma, while in (B) the axial fibers of the coiled tail have burst the plasmalemma and are freed. Bars = 5 μm. **(C)** Survey electron micrograph of ejaculated sperm of a bull affected with the Dag defect. Note the cross sections (asterisks) through coiled tails where the different parts of the flagellum have reflected over each other and are encapsulated by a common membrane. Many of the axial fibers (including both microtubular doublets and corresponding outer dense fibers) are missing from the middle and principal pieces of the tail and are found dislocated in the common cytoplasm (arrowheads). Retraction of axial fibers usually involves the larger half of the axial fiber bundle (arrows), namely, fibers 4, 5, 6, and 7; although frequently the smaller half (fibers 9, 1, and 2) is also retracted (hollow arrows). Occasionally, the plasmalemma has disintegrated and its contents are freed (circle). Bar = 1 μm

The Dag defect is inherited via a simple recessive gene (Koefoed-Johnsen et al. 1980) and as such, when the condition is present, nearly all sperm will be affected, especially if sperm with straight tails, but irregular, swollen mitochondria

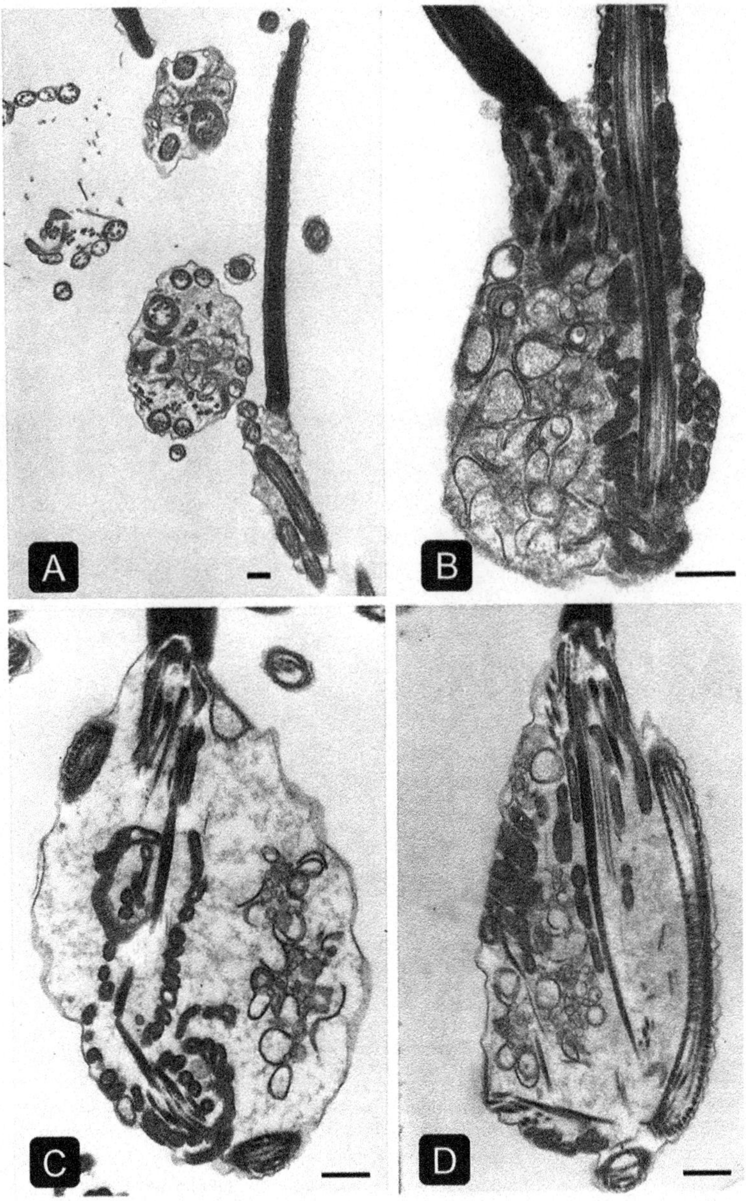

Fig. 5.27 (A-D) Electron micrographs of ejaculated sperm of a bull affected with the Dag defect. These oblique to longitudinal sections of the tails show the characteristic folding in the midpiece. Note the disorganization and irregularity of mitochondria surrounding the axial fiber bundle. The residual cytoplasm is full of saccular elements of the cytoplasmic droplet as well as extracted axial fibers. In (D), note how the principal piece has encroached on the midpiece and both tail pieces or parts are encapsulated by the same membrane and cytoplasm. Bars = 0.5 μm

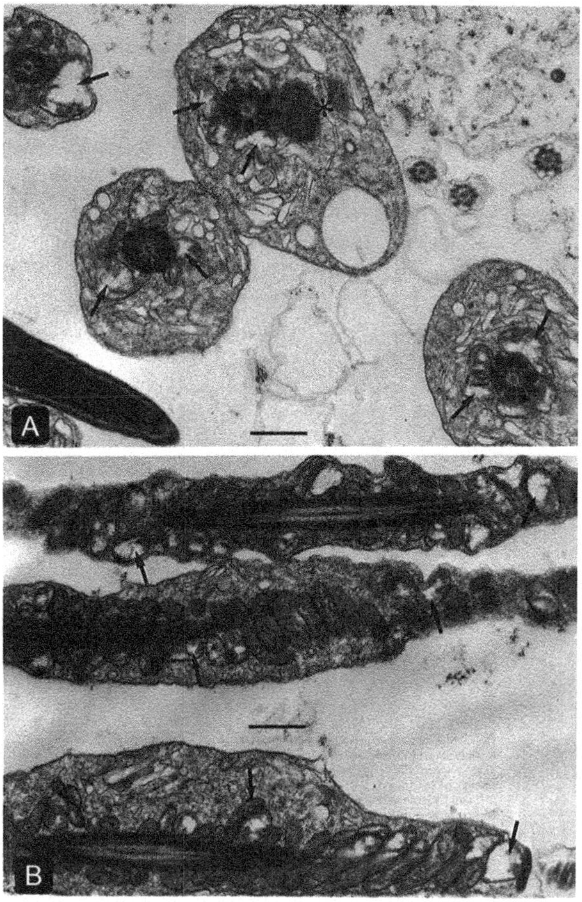

Fig. 5.28 (**A**) Electron micrograph of cross sections through spermatid tails, in the very last step of spermiogenesis (just before spermiation), of a bull affected with the Dag defect. The tails cut through the cytoplasmic droplet region contain a large amount of swollen saccular elements. The mitochondria (arrows) surrounding the axial fiber bundle of the midpiece are also swollen and misaligned. A granular deposit (asterisk) is situated on the outer dense fibers of one of the midpieces and is reminiscent of the substance seen in rat spermatids after gossypol treatment (see Chap. 1 in Fig. 1.31). (**B**) Longitudinal sections through midpieces of late spermatids showing similar characteristics as in (A). Note the swollen mitochondria (arrows). Bars = 0.5 μm.

are counted as defective. A very high percentage of affected sperm distinguishes the Dag defect from similar midpiece disruptions that might be seen in only small numbers of sperm produced by bulls with a disturbance of spermatogenesis. In the latter case, the defects should likely not be called Dag defects, but rather Dag-like defects.

The cardinal features of the Dag defect condition are a swollen disrupted mitochondrial helix, multiple fractures of the midpiece often with coiling of the principal piece, and nearly all sperm are affected (Figs. 5.25, 5.26, 5.27 and 5.28). Although we have chosen to categorize the Jersey and other breeds as similar with respect to pathogenesis, some caution is warranted. In contrast to other breeds, the severity of midpiece disruption appears to be less severe in Jerseys and may have a separate mode of inheritance (Barth and Oko 1989). Development of multiple fractures may be due to structural weakness in the mitochondrial sheath and cementing material of the midpiece (Barth and Oko 1989). Sperm leaving the testis and entering the caput epididymis have an irregular mitochondrial sheath, but the midpiece is intact; when motility begins, breakage of the axonemal fibers occurs.

Affected bulls are sterile and, since the defect is inherited, recovery does not occur.

5.4 Other Sperm Tail Abnormalities

5.4.1 Abaxial and Accessory Tails

Summary: The prognosis for this abnormality is determined by the presence or absence of an accessory tail. Ejaculates containing 60–100% spermatozoa with abaxial tails alone (Barth 1989) cause no decrease in fertility. However, when accessory tails (Aughey and Renton 1968) are present there may be a significant drop in fertility. The cause of this difference lies in the formation of the tail within the spermatid. Tail formation begins with the migration of the proximal and distal centriole to the base of the nucleus. The distal centriole gives rise to the tail with the proximal centriole forming the neck of the midpiece. Normally in spermatids replication of the centrioles is suppressed so that one flagella is formed. Lack of this suppression may allow the formation of additional centrioles and thus additional tails or partial tails. The proximal centriole within the capitulum of the sperm midpiece is contributed to the oocyte at fertilization and is critical for the separation of chromosomes during the first cleavage of the oocyte. This being the case, abaxial tails alone contribute only one proximal centriole and would not affect fertility. However, abaxial tails together with an accessory tail might contribute two proximal centrioles and thus disrupt normal chromosome separation. When there were >17% accessory tails bulls were considered to be of questionable fertility (Barth and Oko 1989).

Detail: Abaxial attachments of the sperm tail to the base of the nucleus may occur in low to very high proportions of sperm, alone or in conjunction with accessory tails. Sperm with abaxial tails usually have an empty secondary implantation fossa (Fig. 5.29 A-C). Accessory tails are 2–3 μm long stumps coming from

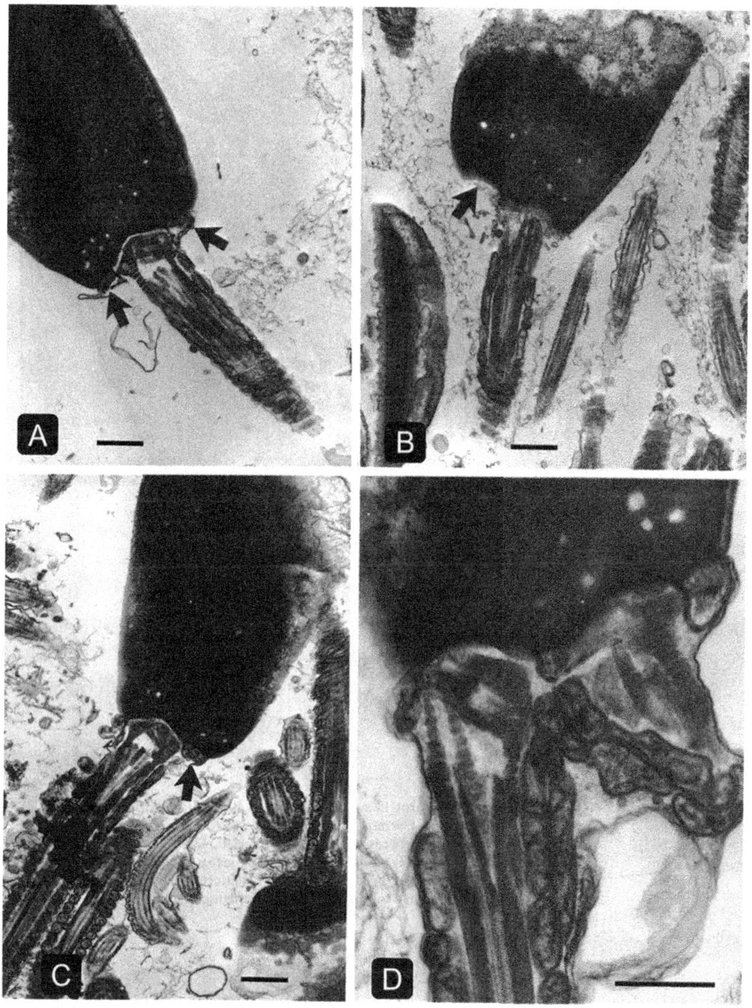

Fig. 5.29 Abaxial and accessory tails. View **(A)** is an electron micrograph of a normal spermatozoan cut through the neck region. Note the basal knobs (arrows), which are a normal feature. View **(B)** is an electron micrograph of a sperm cell with an abaxial tail attachment. Most spermatozoa with abaxial tails have an empty secondary implantation fossa (arrow). The empty fossa is usually distinguishable with ordinary light microscopy at×1000 on good semen smears. View **(C)** is an electron micrograph of a sperm cell with an abaxial tail but no secondary implantation fossa. A basal knob (arrow) on the right side is similar to those seen in (A). A small, seemingly distorted basal knob is present on the left as well. It is common for abaxial tails to come off at an angle to the base of the nucleus, as seen here as well as in (B). View **(D)** is an electron micrograph of a sperm cell with an accessory tail. Mitochondria partly surround the vestigial accessory tail. Bars = 0.5 μm

a secondary implantation fossa lateral to the main tail (Fig. 5.29 D). Accessory tails rarely occur in more than 35% of bull sperm and may occur alone or in association with abaxial tails. When accessory tails are present, often there are a few sperm with complete double tails (Barth and Oko 1989).

In normal spermatogenesis, tail formation begins with migration of the proximal and distal centrioles to the base of the round spermatid nucleus. Importantly, the nucleus forms an implantation fossa *in advance* of the arrival of the pair of migrating centrioles. The proximal centriole arranges itself perpendicular to the distal centriole and becomes part of the neck of the midpiece (capitulum) and the distal centriole gives rise to the sperm tail. In ciliated cells of other tissues, the centrioles replicate many times and give rise to multiple cilia (flagella); however, in spermatids centriolar replication is suppressed so that only one flagellum develops. Failure of suppression of centriolar division could give rise to the development of accessory tails or double tails.

The proximal centriole is a very important structure involved in embryonic development. It forms a sperm aster, which together with a second aster coming from the oocyte are critical to chromosome separation during the first cleavage after fertilization (Long et al. 1993; Navara et al. 1994; Rawe et al. 2002; Sathananthan et al. 1996). During bovine gametogenesis, the spermatozoon retains its proximal centriole while losing centrosomal (pericentriolar) proteins, whereas the oocyte loses its centrioles while retaining centrosomal proteins (Sathananthan et al. 1997; Sun and Schatten 2007). Biparental centrosome contributions to the zygote are typical for most species with some exceptions such as the mouse in which centrosomes are maternally inherited and centrioles are assembled de novo during the blastocyst stage (Manandhar et al. 2005).

Accessory tails and abaxial tails do not appear to originate from extrinsic factors that have an adverse effect on spermatogenesis. If they did, one would expect to see these defects much more commonly. In addition, abnormal spermatogenesis is manifested by marked changes in the spermiogram as an individual succumbs to, and subsequently recovers from, adverse conditions. The spermiograms of bulls affected by abaxial tails and, or accessory tails have remained quite constant over long periods of observation (Barth and Oko 1989). We have found that bulls producing abaxial and, or accessory tails generally have grossly normal testes and produce semen of normal density and progressive motility. Furthermore, sperm with these defects may have good post thaw viability following cryopreservation.

It is unknown whether sperm with accessory tails can fertilize ova; however, if such sperm did enter the ovum an accessory tail would potentially supply an extra aster resulting in abnormal chromosome separation at first cleavage. An Ayrshire bull with 17.8% abaxial tails, 2.9% accessory tails and 12.7% swollen midpieces was reported to be sterile (Aughey and Renton 1968). The fertility of 3 bulls with 17–22% accessory tails was 15–25% below the average non-return rate of the AI centers where the bulls were used (Barth and Oko 1989). In these cases post-thaw viability and other semen parameters were otherwise normal; therefore, the reduced fertility might have been due to the accessory tail defect.

Bulls may produce sperm with up to 100% abaxial tails without a reduction in fertility (Barth 1989). Therefore, abaxial sperm tails may be considered normal within the spermiogram. However, in rare cases, abaxial tails may come off the sperm head at an acute angle and affected sperm swim in a tight circle. In such cases, it is presumed that affected sperm are incapable of fertilization.

5.4.2 *Tail Stump Defect*

Summary: This condition is inherited via a recessive gene. It is a compensable defect, as the affected sperm are immotile and cannot journey to the fertilization site. Bulls with 30–40% of this defect have been found to be fertile (Barth 2013). A cytoplasmic droplet may cover and hide the accessory tail. From an ultrastructural developmental analysis, Blom and Birch-Anderson (1980) concluded that the tail stump is the result of a dysfunctional distal centriole that fails to form the axoneme during spermiogenesis.

Detail: The Tail Stump Defect is uncommon, but most veterinarians will encounter cases from time to time. It has been reported in many countries and many breeds of cattle (Blom 1976; Blom and Birch-Anderson 1980; Coubrough and Barker 1964). A similar defect has been reported in a variety of species, including the mouse, rabbit, dog, stallion and man (Barth and Oko 1989).

The semen produced by severely affected bulls usually has a low sperm concentration and the percent of motile cells is very low as most sperm lack a functional tail. On light microscopy, semen smears may give an initial impression of a very high incidence of detached heads. Closer examination reveal that all that remains of the tail is a vestige of the connecting piece (the stump) at the base of the sperm head. Importantly, the stump is often veiled by a cytoplasmic droplet and may appear as single or double stump 2–3 μm long. Some of the affected sperm have coiled axonemal remnants of a diminutive tail. Although the proportion of tail stump sperm may vary from 15 to 100%, most affected bulls produce a very high percentage of sperm with the defect and are sterile. The tail stump defect is often associated with other defects such as Dag-like midpieces, nuclear vacuoles and pyriform heads. Very few loose tails will be found in sperm smears, as no tail is formed for most cells.

The tail stump defect is most likely inherited via a recessive gene. Affected bulls are usually completely sterile and would not be expected to recover.

5.4.3 *Detached Heads*

Summary: Detached normal heads are most likely due to senescence , whereas, detached abnormal heads may be due either to senescence or to abnormal formation of the basal plate and/or implantation fossa. The basal plate connects the base of the sperm nucleus to the capitulum of the sperm midpiece. Severe stress or abnormal thermoregulation of the testis usually results in the early appearance of sperm head separation, possibly due to abnormal development of the implantation region. Large numbers of detached heads associated with a high percentage of dead sperm appear to be due to abnormal sperm accumulation in the ampullae and cauda epididymides. Normally, peristalsis throughout the epididymides and ampullae constantly moves sperm into the urethra ensuring a reserve of new sperm for ejaculation. Failure of this transport mechanism results in sperm accumulation and senescence.

Detail: Bulls producing semen with a high percentage of detached heads have been associated with low fertility and in some cases sterility (Cooper and Peet 1983). Detached normal heads are most likely due to senescence, whereas, detached abnormal heads may be due either to senescence, or to abnormal formation of the basal plate, and/or implantation fossa. As explained in Chap. 6, detached heads can also arise from weaknesses in the connecting piece (CP) due to targeted mutations in mice of major outer dense fiber genes, which encode the structural proteins of the CP. However, whether such cases exist in the bovine has neither been explored nor documented. When detached heads stain alive, suggesting a recent head separation from a live sperm, one might presume that the head to CP attachment was weak. The basal plate is a thickening of a part of the caudally directed nuclear envelope that connects the base of the sperm nucleus to the capitulum of the sperm CP. Abnormal morphogenesis of the distal nucleus and basal plate was associated with elevated levels of a variety of head detachments (Blom and Birch-Andersen 1970; Thilander et al. 1985). Thus, abnormal basal plate formation and an elevation in the proportion of detached heads may often be associated with disturbances of spermatogenesis caused by a variety of factors including stress, abnormal thermoregulation of the testis and toxicity.

Large numbers of detached heads may be associated with a high percentage of dead sperm due to an abnormal accumulation of senescent sperm in the cauda epididymis and ampullae (Barth 2007). However, sperm accumulation of dead sperm may occur without large numbers of detached heads. Normally, a continuous peristaltic process moves spermatozoa out of the cauda epididymis into the urethra ensuring a reserve of new sperm for ejaculation (Lino et al. 1967). Failure of this transport mechanism results in sperm accumulation and senescence. This defect is considered compensable as the sperm do not have forward motility and is therefore given a $<30\%$ threshold.

5.4.4 Coiled Principal Pieces

Summary: These are seldom seen in high percentages in semen samples and the same factors that cause a rise in distal midpiece reflexes may cause a rise in coiled principal pieces in bulls with a predisposition (Barth and Bowman 1994). However, it is likely that the defect arises in the seminiferous tubules. Levels of 30% are acceptable with 70% normal sperm as this is a compensable abnormality.

Detail: The coiled principal piece defect is characterized by a tight coiling of the principal piece at various points along its length. Bulls that produce sperm with coiled principal pieces are not very common and, in most cases, the percent of affected sperm is under 10%. In such cases, the spermiogram may still have more than 70% normal sperm and satisfactory overall sperm motility. However, in certain lines of bulls the proportion of bulls affected, and the percentage of affected sperm that they produce, is elevated. In such cases, the percentage of sperm with coiled principal pieces may reach as high as 50% or more. Thus, it appears there may be a hereditary predisposition for coiled principal pieces. Various breeds are affected with recovery from the defect occurring within a month or two after late winter to spring prebreeding-season testing, in northern regions, for breeding soundness (Barth 2013). Therefore, it appears that season may be a factor involved in the pathogenesis of the defect. Often many epididymal defects such as distal midpiece reflexes are seen in association with coiled principal pieces. However, based on a scrotal insulation study, it is likely that the defect develops within the seminiferous tubules. A high incidence (22%) of the defect was produced in 1 out of 4 sixteen-month-old crossbred bulls that had their scrotums insulated for 4 days. The incidence of coiled principal pieces suddenly increased from 0 to 22% nineteen days after the insulation was removed and then declined back to 0% over the following 8 days (Barth and Oko 1989). Overall, it appears that low testosterone due to season or stress and poor testis thermoregulation may result in an increase in coiled principal pieces in bulls predisposed to this sperm defect.

5.4.5 Short Tail Defect

In this defect the midpiece is present, but the length of the principle piece is markedly reduced or missing in all sperm resulting in a total lack of sperm motility (Andersson et al. 2000; Siqueira et al. 2010). Transmission electron microscopy was done in boars showing that the midpiece microtubular components are also seriously affected. The author saw the condition in a 2-year old Limousin bull in 2005. In that case, 64% of sperm, with and without short tails, had proximal droplets and 19% had distal droplets; otherwise there were very few other sperm defects. Thus, except for the diminutive tails, spermatogenesis appeared to be normal and in addition, the history of the case did not suggest a stress, nutritional, or toxicity basis for this sperm defect. The condition has also been described in 3 Nelore bulls, two of which were half sibs (Siqueira et al. 2010). In these bulls, proximal droplets occurred in 1–6% of sperm and there were small percentages of other sperm defects, but overall, there were no indications of an environmentally based disturbance of spermatogenesis. In boars, the short tail defect was shown to be due to an autosomal recessive gene and it appears likely that the problem is also inherited in bulls (Sironen et al. 2006). As discussed in Chap. 6, targeted mutations in mice of two fibrous sheath genes presents a similar phenotype with the consequence that the principal piece of the sperm tail is shortened substantially.

5.5 Azoospermia/Oligospermia

There are two conditions in which semen may contain mainly teratoid sperm, one is with severe testicular degeneration and the second appears to be of genetic origin. Sperm morphology may be quite similar in both of these cases, making differentiation difficult if it is based on the semen sample alone. With testicular degeneration usually the testes are soft and the history and physical exam may indicate a significant loss in scrotal circumference as well as reasons for degeneration to occur. When bulls are in good health and body condition, physically sound and have large firm testes, but have azoospermia/oligospermia, a genetic origin might be suspected. Recovery often occurs from testicular degeneration after 3 to 4 months when the inciting cause is removed, however, when the cause is genetic recovery never occurs. Over a 20-year period, Albert Barth on of the authors has recorded Azoospermia/Oligospermia of apparent genetic origin in 19 bulls of 6 breeds including Charolais, Angus, Hereford, Simmental, Holstein, and Brahford. All bulls were approximately 2 years old. Azoospermia/Oligospermia of apparent genetic origin was also seen in a ram and a dog. Two of the Charolais bulls were maintained for more than one year from the time of diagnosis without change in their spermiogram. In all cases, the animals were in good health, there were no clinically detectable physical abnormalities and the testes were of normal size and consistency. Usually, the cauda epididymides felt small and flaccid. In all cases, semen samples were opaque-white, but had few, or no normal sperm.

Three azoospermic Charolais bulls were closely related; two were half-brothers and the sire of those bulls was the grandsire of the third bull. All three bulls also had close relatives on the female side of the pedigree. This suggests a hereditary cause for azoospermia/oligospermia in some bulls.

One of the Charolais bulls was studied over a period of 1.5 years. A karyotype was done and found to be normal. Serial blood samples taken every 30 min over an 8-h period showed normal concentrations of LH and testosterone. Two types of treatment were tried based on a hypothesis that the cause of azoospermia might be a deficiency of gonadotrophin stimulation of the Leydig and Sertoli calls, or a lack of receptors to gonadotrophins in target cells. In the first summer, experimental GnRH implants were placed subcutaneously. In the second summer, 2 g of depot testosterone enanthate was injected subcutaneously every two weeks for three treatments. Neither of the treatments resulted in improvement in the spermiogram. Testis histology was done on two bulls after slaughter and in both cases spermatogenesis progressed poorly up to the spermatid stage. Most tubules showed sloughing of spermatids, and severely malformed sperm into the lumens of seminiferous tubules. In a few tubules, there appeared to be a small number of mature spermatids.

In another case, a 14-month-old Charolais bull with a scrotal circumference of 36 cm received a satisfactory BSE classification in a March prebreeding test. He was placed with 14 cows from May 15 to July 31. Although the bull had been seen to mate normally, none of the cows were pregnant at autumn pregnancy testing. He was retested in October and classified as 'decision deferred' due to inability to obtain a semen sample by electroejaculation. In December the bull was brought to the Western College of Veterinary Medicine for a second opinion. The bull was moderately thin, but otherwise physically sound. The testicles felt firm and the scrotal circumference measurement was 37 cm. A 2 mL semen sample was collected and it had an opaque, white appearance. Wet mounts and stained semen smears revealed that there were virtually no normal sperm in the semen. The semen consisted almost entirely of severely malformed sperm with very small, shriveled nuclei, often with the tail coiled around it (teratoids). At first glance, a semen smear from such bulls may appear to contain mostly white blood cells and debris. It is believed in this case there might have been an error in the practitioner's examination.

Deletions of Y-chromosomal regions are known as a common cause of infertility in men (Silber et al. 1998; Krausz et al. 1999). The DAZ (deleted in azoospermia) gene cluster is located at the AZFc (azoospermia factor) region of the Y chromosome and is deleted in many azoospermic and severely oligospermic men. Three distinct AZF loci have been identified on the Y chromosome (AZF a, b and c) that are required for normal fertility (Vogt et al. 1996). Other studies suggest that entire AZFb and/or AZFb-c deletions may influence the ability of Y chromosome to pair and consequently cause azoospermia (Yogev et al. 2004). There is also an autosomal copy of the DAZ gene (DAZ like, or DAZL). The human DAZL gene was found on chromosome 3 (Silber et al. 1998) and in mice on chromosome 17 (Reijo et al. 1996; Cooke et al. 1996). The autosomal copy, DAZ-like (DAZL) is homologous to the DAZL gene found throughout the vertebrate lineage (Saxena et al. 1996).

5.5.1 Multinuclear- Multiflagellar Sperm Defect

An oligoasthenoteratozoospermic and sterilizing sperm defect, which based on its morphological characteristic was named the multinuclear-multiflagellar sperm (MNMFS) defect (Kopp et al. 2007). This defect is a sterilizing sperm defect encompassing all sperm in the ejaculate. Similar looking multiflagellar cell defects have been reported in bull ejaculates known as medusa cells (Blom 1944) but these are ciliated epithelial cells originating from the ductuli deferens without a distinct head. Medusa like multiflagellar sperm cells with abnormal nuclear material are also occasionally found amongst normal sperm (Sutovsky and Oko 2011). This MNMFS defect indicates a major disturbance (genetic or environmental) in spermiogenesis. This defect has not been reported again to our knowledge.

5.6 Sperm Agglutination and Antisperm Antibodies

The incidence of antisperm antibodies (ASA) in bulls is quite high; however, it is not clear whether there is a significant association between ASA and fertility in animals. In one study (Zralý et al. 2002), blood sera from 612 bulls of different age and breed, and randomly selected in artificial insemination stations and bull rearing stations, were examined by ELISA for the presence of ASA. At serum dilutions of 1:40 and higher, the incidence of positive bulls was 35.5%. The first occurrence of antibodies was recorded at the age of 5 to 6 months (13.8%) with a significant increase (58.1%) at 9 to 10 months of age. In a study of bulls tested for breeding soundness (Purswell et al. 1983), 25 satisfactory and 25 unsatisfactory bulls were examined for the presence of sperm agglutinating antibodies in their serum and seminal fluid. No correlation was found between breeding soundness classification or age and the presence of sperm agglutinating antibodies in serum or seminal fluid.

Agglutination may be head-to-head or tail-to-tail, indicating that different regions of the cell membrane can be affected and bind to each other. Agglutination in the head region might be in the acrosomal, or post-acrosomal region. Perhaps the location of sperm membrane attraction might give a clue as to the cause of the agglutination; for example, agglutination in the acrosomal region might indicate a lack of deposition of decapacitation protein during epididymal passage or premature removal of such proteins within the seminal medium. Post-acrosomal binding, or binding in the tail region, might implicate ASA have developed due to exposure of sperm to the immune system.

Autoimmunity to sperm can occur because sperm and their antigens are first produced long after the fetal period when immunologic self-tolerance occurs. Spermatids and spermatozoa are isolated from the immune system within the seminiferous tubules due to a blood testis barrier created by tight junctions between neighbouring Sertoli cells. When this barrier is disrupted, autoimmunization can occur[92]. Exposure of the immune system to spermatic antigens would stimulate a high titer of ASA that could persist for many years (Purvis and Christiansen 1993; Hinting et al. 1996).

ASA formation can be induced by trauma, infectious and noninfectious inflammation, or by obstruction followed by rupture in the testicular efferent duct system (Zralý et al. 2002). The formation of sperm granulomas has been associated with infertility in men with epididymal-orchitis of infectious or traumatic origin, and in vasectomized men (Hinting et al. 1996). The formation of ASA has also been induced by breeding accidents in a stallion (Papa et al. 1990; Zhang et al. 1990), testicular biopsies, sperm granulomas, or epididymal aspirates in dogs (Kalaydjiev et al. 2007), cryptorchidism in boars (Pinart et al. 2000), and vasectomy in rams (Jessop and Ladds 1995).

Infectious agents associated with granulomas may act as adjuvants in developing antisperm autoimmunity and infertility (Purvis and Christiansen 1993). Male Beagles infected with Brucella canis for 3 months developed serum antibodies that agglutinated normal canine spermatozoa. Titres were highest in dogs that had been infected for 4 to 6 months. Lower sperm agglutinin titres were detected in serum collected 10 months after inoculation. Antibodies were also observed in the seminal plasma of chronically infected dogs. Seminal plasma from infected, but not from healthy dogs, caused head-to-head agglutination of normal sperm (George and Carmichael 1984). *Brucella ovis* infection causes chronic epididymitis in rams and may result in the formation of spermatic granulomas and reduction in fertility. Sera from affected rams, when mixed with semen, agglutinated or immobilized $\geq 50\%$ of sperm (Paolicchi et al. 2000).

It is common for bulls to have one or several blind ending efferent ductules that may become impacted with sperm and then rupture, causing granuloma formation in the caput epididymis. In one report, 8 of 25 bulls were affected (Hemeida et al. 1978). In some bulls there is palpable enlargement and hardness in the caput, but in others, the amount of fibrosis is too small to be palpated. In a 2-year old dog that had a small nodule in the left cauda epididymis, ejaculates had sperm agglutination (Kawakami et al. 2003). Mixing of the dog's seminal plasma or serum with the semen of 2 normal dogs also resulted in agglutination of their sperm. Histological examination showed that the nodule was a sperm granuloma and many lymphocytes were present in the stroma surrounding it.

The presence of natural ASA in fertile humans (Kurpisz and Alexander 1995), virgin girls (Eggert-Kruse et al. 1993) and boys before puberty (Lenzi et al. 1991), may put the authenticity of serum ASA in question. It has been suggested that there may be molecular similarities between bacteria and sperm and antibacterial antibodies might result in anti-sperm immunological reactions (Kurpisz and Alexander 1995). It has been shown that natural ASA cross-react with bacteria such as E. coli and *Salmonella sp* in humans (Kalaydjiev et al. 2007) as well as other pathogens in rams (Paolicchi et al. 2000), dogs (George and Carmichael 1984) and bulls (Hegazi and Ezzo 1995). Tests for antibodies against sperm are usually based on agglutination reactions; however, there are reports that sperm can also be agglutinated by some mycoplasmas, viruses and non-antibody serum components (Rose et al. 1976).

The effect of ASA on fertility in mammalian species is not well documented. It appears there are very few reports of studies that evaluated the effect anti-sperm antibodies on bovine fertility. Perez and Carrasco (Perez and Carrasco 1964), reported two naturally occurring cases of ASA in bulls associated with head-to head agglutination of ejaculated sperm, reduced sperm motility and reduced fertility. However, in that report the affected bulls had clinical vesicular adenitis, therefore, the presence of inflammatory products in their semen is a confounding factor in regard to the effect of ASA on fertility.

Anti-sperm antibodies were generated in three 13-mo-old Holstein bulls by auto-immunization to determine whether ASA interfere with in vitro fertilization. Sperm were incubated with serum from the immunized and control bulls and then placed with bovine oocytes in vitro. Incubation of sperm with post-immunization serum resulted in reduced in vitro fertilization rates, demonstrating that naturally occurring ASA could potentially affect in vivo fertility in bulls (Kim et al. 1999). These studies did not indicate whether there was an association between ASA and sperm agglutination.

In another experiment, bovine sperm heads were separated by ultrasonic treatment and centrifugation (Gou et al. 1998). Anti-bull sperm IgG was produced by immunizing rabbits with acrosome-reacted bull sperm. The main membrane proteins on acrosome-reacted bull sperm head were shown to be of the sp18 family. Anti-sp18 IgG caused agglutination of bull and rabbit spermatozoa. In another experiment, lactoferrin was purified from pig cauda epididymal fluid and anti-lactoferrin antisera were produced. Lactoferrin, which is synthesized in the regions from the distal caput to the cauda epididymis, appeared to protect sperm from agglutination. Anti-lactoferrin antiserum induced tail-to-tail agglutination of mature spermatozoa (Jin et al. 1997). It seems possible that anti-sp18, or anti-lactoferrin antibodies could develop in naturally occurring cases in bulls; however, the question that remains is whether these antibodies would be secreted into the seminal secretions. It appears that this might occur in view of an experiment conducted with rabbits. Male rabbits immunized with homologous spermatozoa were mated. Mating with females after reimmunization of the males significantly reduced conception rate and litter size. Mating of the females by the males after the antibody titre had decreased resulted in improved conception rate and litter size.

The epididymis is responsible for adding decapacitating proteins to the sperm membrane; these are later removed in the female tract at sperm capacitation. Vesicular fluid also supplies some of the protein molecules that attach to the surface membrane. The environment in the female reproductive tract can be simulated by incubating sperm in a capacitation medium in vitro (Yanagimachi 1994), and it has been observed that many sperm become agglutinated at the acrosome during incubation in capacitation media (Harayama et al. 1999; Bedford and Yanagimachi 1991). Thus head-to-head agglutination in semen samples might indicate a lack of deposition of decapacitation proteins in the epididymis or premature removal of such proteins within the seminal medium.

In summary, circulating anti-sperm antibodies are likely quite common in populations of various domestic species. These antibodies are not necessarily due to exposure of the immune system to sperm, but might often be due to interaction of the immune system and certain bacterial antigens that may mimic those of sperm membranes. Experiments have shown that the presence of seminal anti-sperm antibodies are capable of reducing fertility; however, there are no reports of naturally occurring cases definitively linking anti-sperm antibodies, or sperm agglutination with infertility. Therefore, when sperm agglutination is encountered during semen assessment, there can be no certainty of an effect on fertility. Furthermore, it is not known how long a condition of sperm agglutination might be expected to persist. Therefore, clinicians would need to be cautious in predicting the potential fertility of affected individuals and in the formulation of a prognosis for the resolution of the condition.

5.7 Sperm Accumulation

This condition was reported in 1.1% of western Canadian range bulls (Barth and Oko 1989). In the normal male, spermatozoa are continuously moved out of the cauda epididymis, up into the ampullae and into the urethra by peristalsis (Lino et al. 1967). This ensures that there is a constant fresh supply of new spermatozoa even in sexually rested bulls. In sperm accumulators, sperm accumulate in the ampulla and the cauda epididymis. The characteristics of sperm accumulation are:

(1) The ability to collect 25 to 40 mL of concentrated semen at one time by electroejaculation or artificial vagina.

(2) Usually sperm morphology is normal as the problem is not related to a disturbance of spermatogenesis, but rather, abnormal sperm transport.

(3) Most sperm are dead and often the percentage of detached heads is elevated, presumably due to senescence.

(4) Repeated ejaculation over about one week results in a return to normal percentages of live sperm. This could be accomplished by electroejaculation until the semen density is dilute, every second day for a period of six days (3 collection days).

In cases of a severe disturbance of spermatogenesis, there may also be a high percentage of detached heads and dead spermatozoa; however, there will be a great elevation in the percentage of sperm abnormalities and the sperm output is low. These are not cases of sperm accumulation.

Yearling bulls as well as older bulls have been affected and sperm accumulation in bulls appears to be a permanent condition (Barth 2007). There are several mechanisms involved in contractility of the epididymis and the vas deferens including factors altering electrical conductivity through the smooth muscle fibers of these structures and hormonal influences such oxytocin and vasopressin that are present in the mammalian male reproductive tract and play a role in regulating contractility of the excurrent duct and probably sperm transport (Omu et al. 2001). Innervation of the excurrent duct system plays a crucial role in sperm transport as well. It is possible that a disturbance of the sympathetic nerves emanating from the inferior mesenteric ganglion would result in sperm accumulation in some animals, since these nerves innervate the epididymis and vas deferens. In rats, removal of the inferior mesenteric ganglion resulted in excessive sperm accumulation of spermatozoa within the cauda epididymis (Ricker and Chang 1996). Sperm accumulation in bulls has been associated with spinal disorders (Barth and Oko 1989) and surgical vesiculectomy performed to resolve seminal vesiculitis (Nöthling and Volkmann 1997). When bulls are "cleaned out" by frequent ejaculation, re-accumulation of senescent sperm occurs over approximately 1 month of sexual rest (Barth 2007).

In a breeding trial, 2 bulls with sperm accumulation were placed in pens with 30 heifers each for 21 days (Barth 2007). Breeding was observed for 2 h twice daily and bulls had chin ball markers. Pregnancy was determined by ultrasonography 27 days after the end of the breeding period. From these observations it was possible to determine the order in which heifers were bred and whether pregnancy occurred. The bulls bred 8 and 11 heifers over a period 7 and 11 days before the first pregnancy occurred. By 14 days into the breeding period nearly every heifer that was bred became pregnant. Semen collected on day 10 and at the end of the experiment had about 50% and 85% live sperm, respectively. Pregnancy rates were 50 and 56% from affected bulls, whereas control bulls had pregnancy rates of 72 and 77%.

Compliance with Ethical Standards
All applicable international, national and institutional guidelines for the care and use of animals were followed. This current work was supported by the Natural Science and Engineering Research Council of Canada (Grant number RGPIN/05305) to Richard Oko at Queen's University, along with the Agriculture and Food Research Initiative Competitive Grant number 2023-67015-39262 from the USDA National Institute of Food and Agriculture, Animal Breeding, Genetics & Genomics Program and seed funding from the College of Agriculture, Food and Natural Resources of the University of Missouri to Peter Sutovsky. Work for this book was also supported by a 2021–2022 Lalor Foundation Postdoctoral Research Fellowship to Lauren Hamilton. The research support to Viv Perry was funded by Australian Research Council (Grant number LP110100649), English Beef and Lamb Executive a division of the Agriculture and Horticulture Development Board (Grant number RC3815), Queensland Sperm Morphology Laboratories, Ridley AgriProducts Pty Ltd and the Australian cattle industry. The Canadian Association of Animal Breeders provided the foundation for research by Albert Barth. They supported the Development of the Diagnostic Spermatology Laboratory at the Western College of Veterinary Medicine, Saskatoon, SK., along with related support for numerous other research directives. Many grants were also received from Saskatchewan Agriculture, including the Cattle Marketing Deductions Fund, the Horned Cattle Trust Fund, the Cattle Breeders Research Council, the Beef Development Fund and the Agriculture Development Fund. Animal research approval for research or clinical studies summarized in these chapters were given by respective University Animal Care Committees. The authors declare no conflict of interest as there is no perceived interest in prejudicing the impartiality of the research reported.

References

Agarwal A, Nallella K, Allamaneni S, Said T (2004) Role of antioxidants in treatment of male infertility: an overview of the literature. Reprod Biomed Online 8(6):616–627

Alun Jones W (1962) Abnormal Morphology of the Spermatozoa in Guernsey Bulls. British Veterinary Journal 118(6). https://doi.org/10.1016/S0007-1935(17)43005-3

Amann R, Seidel G, Mortimer R (2000) Fertilizing potential in vitro of semen from young beef bulls containing a high or low percentage of sperm with a proximal droplet. Theriogenology 54(9):1499–1515

Andersson M, Peltoniemi O, Makinen A, Sukura A, Rodriguez—Martinez H (2000) The Hereditary 'Short Tail' Sperm Defect—A New Reproductive Problem in Yorkshire Boars. Reprod Domest Anim 35(2):59–63. https://doi.org/10.1046/j.1439-0531.2000.00196.x

Andersson M, Vierula M, Alanko M (1990) Three types of acrosomal aberrations of bull spermatozoa and their relation to fertility. Acta Vet Scand 31(2):175–179

Aughey E, Renton JP (1968) Abnormal spermatozoa in an Ayrshire bull. Veterinary Record 82:129–131

Bane A, Nicander L (1965) Pouch formations by invaginations of the nuclear envelope of bovine and porcine sperm as a sign of disturbed spermiogenesis. Nord Vet Med 17:628–632

Barth A (1997) Analysis of the bovine spermiogram. Proceedings of the Ann Mtg Soc for Theriogenology, Montreal:123–132

Barth A (2013) Bull Breeding Soundness. Third Edition edn. WCABP, Western Canadian Association of Bovine Practitioners, Saskatoon, Canada

Barth AD (1986) The Knobbed Acrosome Defect in Beef Bulls. Can Vet J 27(10):379–384

Barth AD (1989) Abaxial tail attachment of bovine spermatozoa and its effect on fertility. The Canadian veterinary journal = La revue veterinaire canadienne 30(8):656–662

Barth AD (2007) Sperm accumulation in the ampullae and cauda epididymides of bulls. Anim Reprod Sci 102(3–4):238–246. https://doi.org/10.1016/j.anireprosci.2006.11.005

Barth AD The pathogenesis of abnormal sperm production. In: Proc Bull Mgt Fertility Conf, Manhattan, Kansas, 2010 2010.

Barth AD, Bowman PA (1994) The sequential appearance of sperm abnormalities after scrotal insulation or dexamethasone treatment in bulls. Can Vet J 35(2):93–102

Barth AD, Bowman PA, Bo GA, Mapletoft RJ (1992) Effect of narrow sperm head shape on fertility in cattle. The Canadian veterinary journal = La revue veterinaire canadienne 33(1):31–39

Barth AD, Brito LFC, Kastelic JP (2008) The effect of nutrition on sexual development of bulls. Theriogenology 70(3):485–494

Barth AD, Oko RJ (1989) Abnormal morphology of bovine spermatozoa, 1st edn. Iowa State University Press, Iowa, US

Barth AD, Waldner CL (2002) Factors affecting breeding soundness classification of beef bulls examined at the Western College of Veterinary Medicine. The Canadian veterinary journal = La revue veterinaire canadienne 43(4):274–284

Bedford JM, Yanagimachi R (1991) Epididymal storage at abdominal temperature reduces the time required for capacitation of hamster spermatozoa. J Reprod Fertil 91(2):403–410

Berndtson WE, Olar TT, Pickett BW (1981) Correlation between post-thaw motility and acrosomal integrity of bovine sperm. J Dairy Sci 64(2):346–349. https://doi.org/10.3168/jds.S0022-0302(81)82572-6

Bertschinger HJ (1975) The heriditary occurrence of diploid spermatozoa in the semen of Brown Swiss bulls. Zurich

Blom E (1944) On the Occurrence in Bull Sperm of Certain »Medusa Formations« Derived from the Epithelium of the Efferent Ducts of the Testis. Acta Pathologica Microbiologica Scandinavica 21(5):713–720. https://doi.org/10.1111/j.1699-0463.1944.tb04971.x

Blom E (1959) A rare sperm abnormality: corkscrew-sperms associated with sterility in bulls. Nature 183(4670):1280–1281. https://doi.org/10.1038/1831280a0

Blom E (1966) A New Sterilizing and Hereditary Defect (the 'Dag Defect') located in the Bull Sperm Tail. Nature 209:739. https://doi.org/10.1038/209739a0

Blom E (1968) A new sperm defect—"pseudo-droplets"—in the middle piece of the bull sperm. [State Veterinary Serum Laboratory], [Copenhagen]

Blom E (1973) Ultrastructure of some characteristic sperm defects and a proposal for a new classification of bull spermiogram. Nordisk Veterinaer Medicin 25(7–8):383-+

Blom E (1976) A sterilizing tail stump sperm defect in a Holstein-Friesian bull. Nord Vet Med 28(6):295–298

Blom E (1977) A decapitated sperm defect in two sterile Hereford bulls. Nord Vet Med 29:119–123

Blom E (1978) The corkscrew sperm defect in Danish bulls: a possible indicator of nuclear fallout? Nord Vet Med 30:1–8

Blom E (1980) Rolled head and nuclear crest sperm defects in a rare case of presumed diploidy in the bull. Nord Vet Med 32:409–416

Blom E, Birch-Andersen A (1970) Ultrastructure of the "decapitated sperm defect" in Guernsey bulls. J Reprod Fertil 23(1):67–72. https://doi.org/10.1530/jrf.0.0230067

Blom E, Birch-Anderson A (1980) Ultrastructure of the tail stump sperm defect in the bull. Acta Path Microbiol Scand 88:397–405

Blom E, Birch-Andersen A (1962) Ultrastructure of the sterilizing knobbed sperm defect in the bull. Nature 194:989–990

Bloom G, Nicander L (1961) On the ultrastructure and development of the protoplasmic droplet of spermatozoa. Z Zellforsch Mikrosk Anat 55(6):833–844. https://doi.org/10.1007/BF00381652

Bonet S, Briz M, Fradera, A (1993) Ultrastructural abnormalities of boar spermatozoa. Theriogenology 40:383–396

Bongso TA, Jainudeen MR, Dass S (1981) Examination of Droughtmaster bulls for breeding soundness. Theriogenology 15(4):415–425. https://doi.org/10.1016/0093-691x(81)90008-x

Brito LF, Silva AE, Barbosa RT, Unanian MM, Kastelic JP (2003) Effects of scrotal insulation on sperm production, semen quality, and testicular echotexture in Bos indicus and Bos indicus x Bos taurus bulls. Anim Reprod Sci 79(1–2):1–15. https://doi.org/10.1016/s0378-4320(03)00082-4

Burgess GW, Chenoweth PJ (1975) Mid-piece abnormalities in bovine semen following experimental and natural cases of bovine ephemeral fever. Br Vet J 131(5):536–544. https://doi.org/10.1016/s0007-1935(17)35185-0

Callaghan MJ, McAuliffe P, Rodgers RJ, Hernandez-Medrano J, Perry VEA (2016) Subacute ruminal acidosis reduces sperm quality in beef bulls1. J Anim Sci 94(8):3215–3228. https://doi.org/10.2527/jas.2015-0235

Callaghan MJ, Rodgers RJ, Perry VEA (2020) Supplementation of rangeland primiparous Bos indicus x Bos taurus beef heifers during lactation. 2. Effects upon the reproductive development of bull calf progeny. Theriogenology 152:83–93. https://doi.org/10.1016/j.theriogenology.2020.04.032

Calvin HI, Cooper GW, Wallace E (1981) Evidence that selenium in rat sperm is associated with a cysteine-rich structural protein of the mitochondrial capsules. Gamete Res 4(2):139–149. https://doi.org/10.1002/mrd.1120040208

Carreira JT, Mingoti GZ, Rodrigues LH, Silva C, Perri SH, Koivisto MB (2012) Impact of proximal cytoplasmic droplets on quality traits and in-vitro embryo production efficiency of cryopreserved bull spermatozoa. Acta Vet Scand 54(1):1. https://doi.org/10.1186/1751-0147-54-1

Chenoweth PJ, Burgess GW (1972) Mid-piece abnormalities in bovine semen following ephemeral fever. Aust Vet J 48(1):37–38

Chenoweth PJ, Chase CC Jr, Risco CA, Larsen RE (2000) Characterization of gossypol-induced sperm abnormalities in bulls. Theriogenology 53:11193–12003

Cooke HJ, Lee M, Kerr S, Ruggiu M (1996) A murine homologue of the human DAZ gene is autosomal and expressed only in male and female gonads. Hum Mol Genet 5(4):513–516

Cooper A, Peet R (1983) Infertility in a Hereford bull associated with increased numbers of detached sperm heads in his ejaculate. Aust Vet J 60(7):225–226

Cooper T (2011) The epididymis, cytoplasmic droplets and male fertility. Asian J Androl 13(1):130–138

Copping KJ, Ruiz-Diaz MD, Rutland CS, Mongan NP, Callaghan MJ, McMillen IC, Rodgers RJ, Perry VEA (2018) Peri-conception and first trimester diet modifies reproductive development in bulls. Reprod Fertil Dev 30(5):703–720. https://doi.org/10.1071/RD17102

Coubrough R, Barker C Spermatozoa: an unusual middle piece abnormality associated with sterility in bulls. In: Proc, 1964. pp 219–229

Coulter G Effect of dexamethasone on the incidence of the crater defect of bovine sperm. In: International Congress on Animal Reproduction and Artificial Insemination, 1976. pp 694–697

Coulter G, Oko R, Costerton J (1978) Incidence and ultrastructure of "crater" defect of bovine spermatozoa. Theriogenology 9(2):165–173

Courtens J-L, Amir D, Durand J (1980) Abnormal spermiogenesis in bulls treated with ethylene dibromide: an ultrastructural and ultracytochemical study. J Ultrastruct Res 71(2):103–115

Cran D, Dott H, Wilmington J (1982) The structure and formation of rolled and crested bull spermatozoa. Gamete Res 5(3):263–269

Cupps P, Laben R, Rahlmann D (1960) Effects of estradiol benzoate injections on the characteristics of bovine semen. J Dairy Sci 43(8):1135–1139

Cupps PT, Briggs J (1965) Changes in the epididymis associated with morphological changes in the spermatozoa. J Dairy Sci 48(9):1241–1244

Dobrinski I, Hughes HPA, Barth AD (1994) Flow cytometric and microscopic evaluation and effect on fertility of abnormal chromatin condensation in bovine sperm nuclei. J Reprod Fertil 101(3):531–538. https://doi.org/10.1530/jrf.0.1010531

Dott HM, Dingle JT (1968) Distribution of lysosomal enzymes in the spermatozoa and cytoplasmic droplets of bull and ram. Exp Cell Res 52(2):523–540. https://doi.org/10.1016/0014-4827(68)90493-X

Drevius L-O (1963) Spiralization in tails of mammalian spermatozoa in hypotonic media. Nature 197(4872):1123–1124

Eggert-Kruse W, Böckem-Hellwig S, Doll A, Rohr G, Tilgen W, Runnebaum B (1993) Antisperm antibodies in cervical mucus in an unselected subfertile population. Hum Reprod 8(7):1025–1031. https://doi.org/10.1093/oxfordjournals.humrep.a138187

Eljack A, Hrudka F (1979) Pattern and dynamics of teratospermia induced in rams by parenteral treatment with ethylene dibromide. J Ultrastruct Res 67(2):124–134

Fawcett DW, Anderson WA, Phillips DM (1971) Morphogenetic factors influencing the shape of the sperm head. Dev Biol 26(2):220–251

Felton-Taylor J, Prosser KA, Hernandez-Medrano JH, Gentili S, Copping KJ, Macrossan PE, Perry VEA (2020) Effect of breed, age, season and region on sperm morphology in 11,387 bulls submitted to breeding soundness evaluation in Australia. Theriogenology 142:1–7. https://doi.org/10.1016/j.theriogenology.2019.09.001

Fischer K, Van Leyen K, Lovercamp K, Manandhar G, Sutovsky M, Feng D, Safranski T, Sutovsky P (2005) 15-Lipoxygenase is a component of the mammalian sperm cytoplasmic droplet. Reproduction 130(2):213–222

Foote R (1999) Bull sperm surface "craters" and other aspects of semen quality. Theriogenology 51(4):767–775

Garbers DL, Wakabayashi T, Reed PW (1970) Enzyme profile of the cytoplasmic droplet from bovine epididymal spermatozoa. Biol Reprod 3(3):327–337. https://doi.org/10.1093/biolreprod/3.3.327

George L, Carmichael L (1984) Antisperm responses in male dogs with chronic Brucella canis infections. Am J Vet Res 45(2):274–281

Gou KM, Shang LJ, An XR, Lin AX, Ren LM, Chen YF (1998) Extraction of sp18 family membrane proteins on posterior head of bull sperm and the effect of anti-sp18 IgG on sperm motility and murine in vitro fertilization. Anim Biotechnol 9(2):149–153. https://doi.org/10.1080/10495399809525901

Goyal H (1983) Histoquantitative effects of orchiectomy with and without testosterone enanthate treatment on the bovine epididymis. Am J Vet Res 44(6):1085–1090

Gustafsson B, Einarsson S, Nicander L, Holtman M, Soosalu O (1974) Morphological, physical and chemical examination of epididymal contents and semen in a bull with epididymal dysfunction. Andrologia 6(4):321–331

Gustavsson I (1969) Cytogenetics, distribution and phenotypic effects of a translocation in Swedish cattle. Hereditas 63(1–2):68–169

Hancock J (1949) A seminal defect associated with sterility of Guernsey bulls. Vet Rec 61:742–743

Hancock J (1959) The morphologic characteristics of spermatozoa and fertility. Int J Fertil 4:347–359

Hancock JL (1953) The spermatozoa of sterile bulls. J Exp Biol 30:50–58

Harayama H, Magargee SF, Kunze E, Shidara O, Iwamoto E, Arikawa S, Miyake M, Kato S, Hammerstedt RH (1999) Changes in epididymal protein anti-agglutinin on ejaculated boar spermatozoa during capacitation in vitro. Reprod Fertil Dev 11(4–5):193–199. https://doi.org/10.1071/rd99056

Hawk H (1987) Transport and fate of spermatozoa after insemination of cattle. J Dairy Sci 70(7):1487–1503

Hawk H (1988) Gamete transport in the superovulated cow. Theriogenology 29(1):125–142

Hegazi A, Ezzo O (1995) Sperm Auto Antibodies in Buffaloes and Cattle Naturally Infected with IBRV and Brucella. Buffalo Journal 11:325–330

Hellmen E, Ploen L, Settergren I, Nicander L (1980) Middle piece defects of testicular origin in bull sperm. Nord Vet Med 32(10):423–426

Hemeida N, Sack W, McEntee K (1978) Ductuli efferentes in the epididymis of boar, goat, ram, bull, and stallion. Am J Vet Res 39(12):1892–1900

Herman HA, Swanson EW (1941) Variations in dairy bull semen with respect to its use in artificial insemination. University of Missouri, Agricultural Experiment Station

Hermo L, Dworkin J, Oko R (1988) Role of epithelial clear cells of the rat epididymis in the disposal of the contents of cytoplasmic droplets detached from spermatozoa. Am J Anat 183(2):107–124. https://doi.org/10.1002/aja.1001830202

Hinting A, Soebadi DM, Santoso RI (1996) Evaluation of the immunological cause of male infertility. Andrologia 28(2):123–126. https://doi.org/10.1111/j.1439-0272.1996.tb02767.x

Hrudka F, Eljack AH (1979) The effect of ethylene dibromide on differentiation of the acrosome, nucleus, and transient nuclear appendages in ram spermatids. J Ultrastruct Res 67(2):135–151. https://doi.org/10.1016/S0022-5320(79)80003-9

Jessop TS, Ladds PW (1995) The immunopathology of unilateral vasectomy in the ram. Vet Immunol Immunopathol 47(1):123–133. https://doi.org/10.1016/0165-2427(94)05385-6

Jin Y-Z, Bannai S, Dacheux F, Dacheux J-L, Okamura N (1997) Direct evidence for the secretion of lactoferrin and its binding to sperm in the porcine epididymis. Molecular Reproduction and Development 47(4):490–496. https://doi.org/10.1002/(SICI)1098-2795(199708)47:4<490::AID-MRD16>3.0.CO;2-#

Johnson WH (1997) The Significance to Bull Fertility of Morphologically Abnormal Sperm. Veterinary Clinics: food Animal Practice 13(2):255–270. https://doi.org/10.1016/S0749-0720(15)30339-X

Jones R (1989) Membrane remodelling during sperm maturation in the epididymis. Oxf Rev Reprod Biol 11:285–337

Kalaydjiev S, Dimitrova D, Mitov I, Dikov I, Nakov L (2007) Serum sperm antibodies after diarrhoeal diseases. Andrologia 39(3):101–108. https://doi.org/10.1111/j.1439-0272.2007.00772.x

Karabinus D, Saacke R (1987) Selection of spermatozoa across the bovine cervix. J Dairy Sci 70(Suppl 1):248

Kawakami E, Koga H, Hori T, Tsutsui T (2003) Sperm granuloma and sperm agglutination in a dog with asthenozoospermia. J Vet Med Sci 65(3):409–412. https://doi.org/10.1292/jvms.65.409

Kim CA, Parrish JJ, Momont HW, Lunn DP (1999) Effects of experimentally generated bull antisperm antibodies on in vitro fertilization. Biol Reprod 60(6):1285–1291. https://doi.org/10.1095/biolreprod60.6.1285

Knudsen O (1954) Cytomorphological investigations into the spermiocytogenesis of bulls with normal fertility and bulls with acquired disturbances in spermiogenesis. Acta Pathol Microbiol Scand Suppl (1926) 101:1–79

Koefoed-Johnsen H, Andersen J, Andresen E, Blom E, Philipsen H (1980) The Dag defect of the tail of the bull sperm. Studies on the inheritance and pathogenesis. Theriogenology 14(6):471–475

Kopp C, Sukura A, Tuunainen E, Gustavsson I, Parvinen M, Andersson M (2007) Multinuclear-Multiflagellar Sperm Defect in a Bull – a New Sterilizing Sperm Defect. Reprod Domest Anim 42(2):208–213. https://doi.org/10.1111/j.1439-0531.2006.00754.x

Kot MC, Handel MA (1987) Binding of morphologically abnormal sperm to mouse egg zonae pellucidae in vitro. Gamete Res 18(1):57–66. https://doi.org/10.1002/mrd.1120180107

Krausz C, Bussani-Mastellone C, Granchi S, McElreavey K, Scarselli G, Forti G (1999) Screening for microdeletions of Y chromosome genes in patients undergoing intracytoplasmic sperm injection. Hum Reprod 14(7):1717–1721. https://doi.org/10.1093/humrep/14.7.1717

Krzanowska H (1974) The passage of abnormal spermatozoa through the uterotubal junction of the mouse. J Reprod Fertil 38(1):81–90. https://doi.org/10.1530/jrf.0.0380081

Kurpisz M, Alexander NJ (1995) Carbohydrate moieties on sperm surface: physiological relevance. Fertil Steril 63(1):158–165

Kysilka C (1973) Hemolytic factor from bull seminal vesicle fluid: isolation and some properties. J Reprod Fertil 32:185–192

Lagerlof N (1938) Infertility in Male Domestic Animals. Rep 13th Int. Vet. Congr., 1938. 1:214–289

Larsen RE, Chenoweth PJ (1990) Diadem/crater defects in spermatozoa from two related angus bulls. Mol Reprod Dev 25(1):87–96. https://doi.org/10.1002/mrd.1080250115

Lenzi A, Gandini L, Lombardo F, Cappa M, Nardini P, Ferro F, Borrelli P, Dondero F (1991) Antisperm antibodies in young boys. Andrologia 23(3):233–235. https://doi.org/10.1111/j.1439-0272.1991.tb02545.x

Lindahl PE, Drevius LO (1964) Observations on bull spermatozoa in a hypotonic medium related to sperm mobility mechanisms. Exp Cell Res 36:632–646. https://doi.org/10.1016/0014-4827(64)90319-2

Lino BF, Braden AW, Turnbull KE (1967) Fate of unejaculated spermatozoa. Nature 213(5076):594–595. https://doi.org/10.1038/213594a0

Long CR, Pinto-Correia C, Duby RT, Ponce de Leon FA, Boland MP, Roche JF, Robl JM (1993) Chromatin and microtubule morphology during the first cell cycle in bovine zygotes. Mol Reprod Dev 36(1):23–32. https://doi.org/10.1002/mrd.1080360105

Manandhar G, Schatten H, Sutovsky P (2005) Centrosome reduction during gametogenesis and its significance. Biol Reprod 72(1):2–13. https://doi.org/10.1095/biolreprod.104.031245

Matousek J, Kysilka C (1984) The phospholipid-binding protein of the reproductive tract of the bull — Effect on the removal of spermatozoal cytoplasm droplets in other species and influence of antibodics on its reactivity. Anim Reprod Sci 7(5):433–440. https://doi.org/10.1016/0378-4320(84)90045-9

Matoušek J, Kysilka č (1980) The haemolytic factor (phospholipid-binding protein) of the bull reproductive tract—Its synthesis and effect on spermatozoal cytoplasm droplets. Anim Reprod Sci 3(3):195–205. https://doi.org/10.1016/0378-4320(80)90016-0

McAuliffe P, Johnston P, Johnston H, Perry VEA (2010) Ejaculators, Morphology and Microscopes. Australian Cattle Veterinarian, Barton ACT

Menon AG, Barkema HW, Wilde R, Kastelic JP, Thundathil JC (2011) Associations between sperm abnormalities, breed, age, and scrotal circumference in beef bulls. Can J Vet Res 75(4):241–247

Meyer RA, Barth AD (2001) Effect of acrosomal defects on fertility of bulls used in artificial insemination and natural breeding. Can Vet J 42(8):627–634

Miller DM, Hrudka F, Cates WF, Mapletoft RJ (1982) Infertility in a bull with a nuclear sperm defect: a case report. Theriogenology 17(6):611–621. https://doi.org/10.1016/0093-691X(82)90059-0

Mitchell JR, Senger PL, Rosenberger JL (1985) Distribution and Retention of Spermatozoa with Acrosomal and Nuclear Abnormalities in the Cow Genital Tract2. J Anim Sci 61(4):956–967. https://doi.org/10.2527/jas1985.614956x

Mortimer RG, Seidel GE, Amann RP, Grotelueschen DM, Olson SE (1991) Use of in vitro fertilization to evaluate spermatozoa with proximal droplets in young beef bulls. Theriogenology 35(1):247. https://doi.org/10.1016/0093-691X(91)90223-Z

Munro IB, Stewart DL, Mel-Rose DR (1961) Middle piece defect in bovine spermatozoa. Vet Bee 73:577–578

Navara CS, First NL, Schatten G (1994) Microtubule organization in the cow during fertilization, polyspermy, parthenogenesis, and nuclear transfer: the role of the sperm aster. Dev Biol 162(1):29–40. https://doi.org/10.1006/dbio.1994.1064

Nogueira E, Tirpák F, Hamilton LE, Zigo M, Kerns K, Sutovsky M, Kim J, Volkmann D, Jovine L, Taylor JF, Schnabel RD, Sutovsky P (2022) A Non-Synonymous Point Mutation in a WD-40 Domain Repeat of EML5 Leads to Decreased Bovine Sperm Quality and Fertility. Front Cell Dev Biol 10:872740. https://doi.org/10.3389/fcell.2022.872740

Nöthling J, Volkmann D (1997) Dilatation of the Ampullae and an Increased Incidence of Loose Sperm Heads after Bilateral Vesiculectomy in a Bull. Reprod Domest Anim 32(6):321–324. https://doi.org/10.1111/j.1439-0531.1997.tb01303.x

Nöthling JO, Arndt EP (1995) Fertility of two bulls with poor sperm morphology. J S Afr Vet Assoc 66(2):74–76

Oko R (1977) Normal and defective bovine spermatogenesis. Master's thesis, University of Calgary, Canada,

Oko R, Hermo L, Chan PT, Fazel A, Bergeron JJ (1993) The cytoplasmic droplet of rat epididymal spermatozoa contains saccular elements with Golgi characteristics. J Cell Biol 123(4):809–821. https://doi.org/10.1083/jcb.123.4.809

Oko R, Hrudka F (1982) Segmental aplasia of the mitochondrial sheath and sequelae induced by gossypol in rat spermatozoa. Biol Reprod 26(1):183–195. https://doi.org/10.1095/biolreprod26.1.183

Olley T (2001) Nuclear Vacuole Defects in Bovine Spermatozoa. University of Queensland, Brisbane, Honours

Olson GE, Winfrey VP, Hill KE, Burk RF (2004) Sequential development of flagellar defects in spermatids and epididymal spermatozoa of selenium-deficient rats. Reproduction 127(3):335–342. https://doi.org/10.1530/rep.1.00103

Olson GE, Winfrey VP, Nagdas SK, Hill KE, Burk RF (2005) Selenoprotein P is required for mouse sperm development. Biol Reprod 73(1):201–211. https://doi.org/10.1095/biolreprod.105.040360

Olson GE, Winfrey VP, NagDas SK, Hill KE, Burk RF (2007) Apolipoprotein E Receptor-2 (ApoER2) Mediates Selenium Uptake from Selenoprotein P by the Mouse Testis *. J Biol Chem 282(16):12290–12297. https://doi.org/10.1074/jbc.M611403200

Omu AE, Al-Bader AA, Dashti H, Oriowo MA (2001) Effect of extracellular Mg concentration on electrically induced contractions of rat vas deferens in vitro. Arch Androl 46(3):159–167. https://doi.org/10.1080/01485010151096388

Paolicchi FA, Casaro PA, Gimeno EJ, Kortebani LG, Mazzolli AB (2000) Antisperm response in rams experimentally infected with Brucella ovis. Small Rumin Res 36(1):7–15. https://doi.org/10.1016/S0921-4488(99)00108-X

Papa FO, Alvarenga MA, Lopes MD, Campos Filho EP (1990) Infertility of autoimmune origin in a stallion. Equine Vet J 22(2):145–146. https://doi.org/10.1111/j.2042-3306.1990.tb04230.x

Perez T, Carrasco L (1964) Autoimmunization Against Sperm as a Cause of Bull Infertility. Vth Int Congr Anim Reprod Artif Insem 5:527

Perry VE, Phillips N, Fordyce G, Gardiner B, Entwistle K, Chenoweth P, Doogan VJ Semen collection and evaluation. In: Bull Fertility: selection & Management in Australia, 2002. Australian Assoc. of Cattle Veterinarians, pp 1–19

Pilip R, Del Campo MR, Barth AD, Mapletoft RJ (1996) In vitro fertilizing characteristics of bovine spermatozoa with multiple nuclear vacuoles: a case study. Theriogenology 46(1):1–12. https://doi.org/10.1016/0093-691X(96)00136-7

Pinart E, Sancho S, Briz MD, Bonet S, Garcia N, Badia E (2000) Ultrastructural study of the boar seminiferous epithelium: changes in cryptorchidism. J Morphol 244(3):190–202. https://doi.org/10.1002/(sici)1097-4687(200006)244:3<190::Aid-jmor4>3.0.Co;2-b

Prabhakar J, Chimbombi M, Malmgren L, Fredriksson G, Madej A (1990) Effects of testosterone and LH concentrations of induced testicular degeneration in bulls. Acta Vet Scand 31(4):505–507. https://doi.org/10.1186/bf03547536

Purswell BJ, Dawe DL, Caudle AB, Williams DJ, Brown J (1983) Spermagglutinins in serum and seminal fluid of bulls and their relationship to fertility classification. Theriogenology 20(4):375–381. https://doi.org/10.1016/0093-691x(83)90198-x

Purvis K, Christiansen E (1993) Infection in the male reproductive tract. Impact, diagnosis and treatment in relation to male infertility. Int J Androl 16(1):1–13. https://doi.org/10.1111/j.1365-2605.1993.tb01146.x

Rawe VY, Terada Y, Nakamura S, Chillik CF, Olmedo SB, Chemes HE (2002) A pathology of the sperm centriole responsible for defective sperm aster formation, syngamy and cleavage. Hum Reprod 17(9):2344–2349. https://doi.org/10.1093/humrep/17.9.2344

Reijo R, Seligman J, Dinulos MB, Jaffe T, Brown LG, Disteche CM, Page DC (1996) Mouse autosomal homolog of DAZ, a candidate male sterility gene in humans, is expressed in male germ cells before and after puberty. Genomics 35(2):346–352. https://doi.org/10.1006/geno.1996.0366

Ricker DD, Chang TS (1996) Neuronal input from the inferior mesenteric ganglion (IMG) affects sperm transport within the rat cauda epididymis. Int J Androl 19(6):371–376. https://doi.org/10.1111/j.1365-2605.1996.tb00530.x

Rocha A, Oliveira E, Vilhena MJ, Diaz J, Sousa M (2006) A novel apical midpiece defect in the spermatozoa of a bull without an apparent decrease in motility and fertility. A Case Study. Theriogenology 66(4):913–922. https://doi.org/10.1016/j.theriogenology.2006.02.032

Rose NR, Hjort T, Rumke P (1976) Techniques for detection of iso and auto antibodies to human spermatozoa. Clin Exp Immunol 23(2):175–199

Saacke R Sperm transport and accessory sperm evaluation. In: Proc 12th Tech Conf AI and Reprod, 1988, 1988. pp 7–14

Saacke RG, Dejarnette JM, Bame JH, Karabinus DS, Whitman SS (1998) Can spermatozoa with abnormal heads gain access to the ovum in artificially inseminated super- and single-ovulating cattle? Theriogenology 50(1):117–128. https://doi.org/10.1016/S0093-691X(98)00119-8

Saacke RG, Nadir S, Nebel RL (1994) Relationship of semen quality to sperm transport, fertilization, and embryo quality in ruminants. Theriogenology 41(1):45–50. https://doi.org/10.1016/S0093-691X(05)80047-0

Salisbury GW, Baker FN (1966) Frequency of occurrence of diploid bovine spermatozoa. J Reprod Fertil 11(3):477–480. https://doi.org/10.1530/jrf.0.0110477

Sathananthan AH, Ratnam SS, Ng SC, Tarín JJ, Gianaroli L, Trounson A (1996) The sperm centriole: its inheritance, replication and perpetuation in early human embryos. Hum Reprod 11(2):345–356. https://doi.org/10.1093/humrep/11.2.345

Sathananthan AH, Tatham B, Dharmawardena V, Grills B, Lewis I, Trounson A (1997) Inheritance of sperm centrioles and centrosomes in bovine embryos. Arch Androl 38(1):37–48. https://doi.org/10.3109/01485019708988530

Saxena R, Brown LG, Hawkins T, Alagappan RK, Skaletsky H, Reeve MP, Reijo R, Rozen S, Dinulos MB, Disteche CM, Page DC (1996) The DAZ gene cluster on the human Y chromosome arose from an autosomal gene that was transposed, repeatedly amplified and pruned. Nat Genet 14(3):292–299. https://doi.org/10.1038/ng1196-292

Sekoni VO, Gustafsson BK (1987) Seasonal variations in the incidence of sperm morphological abnormalities in dairy bulls regularly used for artificial insemination. Br Vet J 143(4):312–317. https://doi.org/10.1016/0007-1935(87)90064-9

Silber SJ, Alagappan R, Brown LG, Page DC (1998) Y chromosome deletions in azoospermic and severely oligozoospermic men undergoing intracytoplasmic sperm injection after testicular sperm extraction. Hum Reprod 13(12):3332–3337. https://doi.org/10.1093/humrep/13.12.3332

Siqueira JB, Pinho RO, Guimarães SE, Miranda Neto T, Guimarães JD (2010) Immotile short-tail sperm defect in Nelore (Bos taurus indicus) breed bulls. Reproduction in domestic animals = Zuchthygiene 45(6):1122–1125. https://doi.org/10.1111/j.1439-0531.2009.01494.x

Sironen A, Thomsen B, Andersson M, Ahola V, Vilkki J (2006) An intronic insertion in KPL2 results in aberrant splicing and causes the immotile short-tail sperm defect in the pig. Proc Natl Acad Sci U S A 103(13):5006–5011. https://doi.org/10.1073/pnas.0506318103

Sironen A, Uimari P, Nagy S, Paku S, Andersson M, Vilkki J (2010) Knobbed acrosome defect is associated with a region containing the genes STK17b and HECW2 on porcine chromosome 15. BMC Genomics 11:699–699. https://doi.org/10.1186/1471-2164-11-699

Smidt WJ (1965) Changes in the middle piece of the bull sperm. Tijdschr Diergeneeskd 90:1613–1615

Söderquist L, Janson L, Larsson K, Einarsson S (1991) Sperm Morphology and Fertility in A. I. Bulls. Journal of Veterinary Medicine Series A 38(1–10):534–543. https://doi.org/10.1111/j.1439-0442.1991.tb01045.x

Suarez SS (2015) Gamete and zygote transport. Knobil and Neill's Physiology of Reproduction 1:113–145

Sun QY, Schatten H (2007) Centrosome inheritance after fertilization and nuclear transfer in mammals. Adv Exp Med Biol 591:58–71. https://doi.org/10.1007/978-0-387-37754-4_4

Sutovsky P (2015) New Approaches to Boar Semen Evaluation, Processing and Improvement. Reprod Domest Anim 50(S2):11–19. https://doi.org/10.1111/rda.12554

Sutovsky P, Oko R (2011) Spermatozoa: the good, the bad and the ugly. Mol Reprod Dev 78(2):67–67. https://doi.org/10.1002/mrd.21270

Swanson EW, Boatman JP (1953) The Effect of Thiouracil Feeding upon the Seminal Characteristics of Dairy Bulls1. J Dairy Sci 36(3):246–252. https://doi.org/10.3168/jds.S0022-0302(53)91486-6

Swanson EW, Boyd LJ (1962) Factors affecting coiled-tail spermatozoa in the bull. Am J Vet Res 23:300–309

Thilander G, Settergren I, Plöen L (1985) Abnormalities of testicular origin in the neck region of bull spermatozoa. Anim Reprod Sci 8(1):151–157. https://doi.org/10.1016/0378-4320(85)90082-X

Thundathil J, Barth AD, Palasz AT, Mapletoft RJ (1998a) Fertilization characteristics of bovine sperm with pyriform heads. Theriogenology 49(1). https://doi.org/10.1016/s0093-691x(98)90727-0

Thundathil J, Meyer R, Palasz AT, Barth AD, Mapletoft RJ (2000) Effect of the knobbed acrosome defect in bovine sperm on IVF and embryo production. Theriogenology 54(6):921–934. https://doi.org/10.1016/S0093-691X(00)00402-7

Thundathil J, Palasz A, Mapletoft R, Barth A (1999) An investigation of the fertilizing characteristics of pyriform-shaped bovine spermatozoa. Anim Reprod Sci 57(1–2):35–50

Thundathil J, Palasz AT, Barth AD, Mapletoft RJ (1998b) Fertilization characteristics and in vitro embryo production with bovine sperm containing multiple nuclear vacuoles. Mol Reprod Dev 50(3):328–333. https://doi.org/10.1002/(sici)1098-2795(199807)50:3<328::aid-mrd9>3.0.co;2-l

Thundathil J, Palasz AT, Barth AD, Mapletoft RJ (2001a) The use of in vitro fertilization techniques to investigate the fertilizing ability of bovine sperm with proximal cytoplasmic droplets. Anim Reprod Sci 65(3):181–192. https://doi.org/10.1016/S0378-4320(00)00231-1

Thundathil J, Palomino J, Barth A, Mapletoft R, Barros C (2001b) Fertilizing characteristics of bovine sperm with flattened or indented acrosomes. Anim Reprod Sci 67(3):231–243. https://doi.org/10.1016/S0378-4320(01)00127-0

Truitt-Gibert AJ, Johnson LA (1980) The crater defect in boar spermatozoa: a correlative study with transmission electron microscopy, scanning electron microscopy, and light microscopy. Gamete Res 3(3):259–266. https://doi.org/10.1002/mrd.1120030308

van Leyen K (1998) Un-LOXing the door to organelle degradation. Trends Cell Biol 8(11):436. https://doi.org/10.1016/S0962-8924(98)01386-5

Vogt PH, Edelmann A, Kirsch S, Henegariu O, Hirschmann P, Kiesewetter F, Köhn FM, Schill WB, Farah S, Ramos C, Hartmann M, Hartschuh W, Meschede D, Behre HM, Castel A, Nieschlag E, Weidner W, Gröne HJ, Jung A, Engel W, Haidl G (1996) Human Y chromosome azoospermia factors (AZF) mapped to different subregions in Yq11. Hum Mol Genet 5(7):933–943. https://doi.org/10.1093/hmg/5.7.933

Waberski D, Meding S, Dirksen G, Weitze KF, Leiding C, Hahn R (1994) Fertility of long-term-stored boar semen: influence of extender (Androhep and Kiev), storage time and plasma droplets in the semen. Anim Reprod Sci 36(1):145–151. https://doi.org/10.1016/0378-4320(94)90061-2

Ward WS, Coffey DS (1991) DNA packaging and organization in mammalian spermatozoa: comparison with somatic cells. Biol Reprod 44(4):569–574. https://doi.org/10.1095/biolreprod44.4.569

Welsh TH, McCraw RL, Johnson BH (1979) Influence of corticosteroids on testosterone production in the bull. Biol Reprod 21(3):755–763. https://doi.org/10.1095/biolreprod21.3.755

Wenkoff MS (1978) A sperm mid-piece defect of epididymal origin in two Hereford bulls. Theriogenology 10(4):275–281. https://doi.org/10.1016/0093-691X(78)90105-X

Williams G (1965) An abnormality of the spermatozoa of some Hereford bulls. Vet Rec 77(41):1204–1206. https://doi.org/10.1136/vr.77.41.1204

Williams W, Savage A (1925) Observations on seminal micropathology of bulls. Cornell Veterinarian:353–375

Yanagimachi R (1994) Mammalian fertilization. In: Knobil E, Neil JD (eds) The physiology of reproduction, 2nd edn. Raven Press, New York, pp 189–317

Yogev L, Segal S, Zeharia E, Gamzu R, Maymon BB, Paz G, Botchan A, Hauser R, Yavetz H, Kleiman SE (2004) Sex chromosome alignment at meiosis of azoospermic men with azoospermia factor microdeletion. J Androl 25(1):110–116. https://doi.org/10.1002/j.1939-4640.2004.tb02765.x

Zhang J, Ricketts SW, Tanner SJ (1990) Antisperm antibodies in the semen of a stallion following testicular trauma. Equine Vet J 22(2):138–141. https://doi.org/10.1111/j.2042-3306.1990.tb04228.x

Zralý Z, Bendova J, Diblíková I, Svecova D, Kummer V, Mašková J, Veznik Z (2002) Antisperm Antibodies in Blood Sera of Bulls and Correlations with Age, Breed and Ejaculate Quality. Acta Vet Brno 71:303–308

Chapter 6
Mechanisms of Development of Sperm Defects

Contents

Abstract The intention of this chapter is to provide insights on the possible causes and disruptive mechanisms in play of some commonly occurring bull sperm abnormalities. Consideration is given to mutations of genes, inhibition of enzymes, deficiencies of trace elements, toxins, simulated stress, hormonal changes, hypoosmolarity and cold shock that cause similar disruptive phenotypes in development and structure seen in commonly occurring bull sperm abnormalities. The possible causes and disruptive mechanisms of the following bull sperm defects, arising from the above research, are evaluated: Tail stump, Decapitations, Dag, Dag-like, Short tail, Pseudodoplet, Segmental aplasia of the mitochondrial sheath, Coiled tails, Knobbed and Nuclear vacuolation. In addition, the idea arising from murid research, that mutations affecting sperm head shape most always affect motility, while mutations affecting sperm tail formation rarely affect sperm head shape is considered. Examples of mutations in genes or inhibition of enzymes involved with the early stages of acrosome formation are given that lead to a variety of 'globozoospermic-like' sperm head phenotypes all of which are associated with various degrees of aberrant sperm tail morphologies.

Keywords Sperm abnormalities: Causes · Disruptive mechanisms · Tail stump defect · Decapitations · Dag defect · Dag-like defects · Short tail · Pseudodoplet · Segmental aplasia of the mitochondrial sheath · Coiled tails · Knobbed defect · Nuclear vacuolation · Globozoospermia

© The Author(s), under exclusive license to Springer Nature Switzerland AG 2025
A. Barth et al., *Abnormal Morphology of Bovine Spermatozoa*, Advances in Anatomy, Embryology and Cell Biology 240, https://doi.org/10.1007/978-3-031-70126-9_6

6.1 Sperm Tail Defects

As documented in preceding Chapters there are several research directions available for obtaining insights on the mechanisms and ultimately the causes behind common occurring bull sperm defects listed in Chap. 4. Insights on the disruption of normal mechanisms leading to abnormal development of spermatozoa derive from morphological/developmental, biochemical, proteomic, genomic and environmental research. Correlation of the results between these methodologies can ultimately point to causes behind commonly occurring bull sperm defects. The advent of the gene knockout has made it possible to correlate morphological consequences of eliminating the function of respective male germ cell proteins in mice with commonly occurring bull sperm defects.

A few gene knockout phenotypes appear to match common bull sperm abnormalities. As pointed out in the Sperm Tail Composition section in Chap. 1, sperm head–tail separations or decapitations occur in gene knockouts of most characterized Odf genes, as they encode for proteins that are prominent constituents of the ODF, connecting piece (CP) and basal plate. However, at what level of the CP the separation occurs is dependent on type of ODF protein and whether the separation even occurs is dependent upon the region that is deleted in a particular Odf gene. Heterozygous Odf2$^{+/-}$ male mice with deletions of exons 6 and 7 led to sperm heads that retained their necks or CP and conversely tails without necks (Ito et al. 2019), ***resembling the Tail Stump bull sperm defect*** *(Fig. 9.22* in *Chap. 4)*. On the other hand, heterozygous mice carrying a gene trap insertion in exon 9 of Odf2 had normal sperm counts and motility with no sign of sperm-head separations (Salmon et al. 2006). Similarly, chimaeric male mice, with a gene trap between exons 4 and 5 of Odf2, were infertile but without obvious head–tail separations (Tarnasky et al. 2010). It should be pointed out that the sperm Tail Stump defect most commonly found in bulls results from a dysfunctional distal centriole that fails to form the axoneme during spermiogenesis (Blom and Birch-Andersen 1980). In this case free tails are rarely seen in the ejaculate.

It appears that a genetic disturbance unrelated to the Odf genome can give rise to a defect resembling the Tail Stump bull sperm defect. Whole-genome sequencing data revealed a recessively inherited frameshift mutation in bovine ARMC3 that is ***associated with a Tail Stump-like sperm defect in Swedish Red (Bos taurus)*** cattle (Pausch et al. 2016). Spermatids from mice with Armc3 deletion showed blocked ribophagy, low energy levels of mitochondria and motionless flagella, which caused male infertility (Lei et al. 2021a, b). These findings point to a germ tissue-specific autophagic function of ARMC3. However, in this case an ultrastructural exploration of the developmental steps during spermatogenesis leading to the Tail Stump-like sperm defect is required to better understand the underlying mechanism.

Homozygous Odf1-knockout mice lacking the promoter region and exon 1 were infertile and one of the major phenotypes reported was head-neck separation, although it was difficult to discern from this article at exactly what level of the connecting piece this separation occurred (Yang et al. 2012). The third most prominent ODF protein after ODF1 and ODF2 is OAZ3. Homozygous Oaz3 knock-out male mice were found to be infertile due to the separation of sperm heads and tails between the basal plate and capitulum of the CP (Tokuhiro et al. 2009). Petersen et al. (2002) first characterized another ODF protein, PMFBP. The disruption of the gene encoding this protein led to the separation of the head and tail between the nuclear envelope and basal plate (Zhu et al. 2018). The phenotypes of the latter three Odf-gene knockouts appear to have a ***resemblance to the Decapitated bull sperm defect*** *(Fig. 4.12* in *Chap. 4)*. Since ODF proteins are a prominent structural part of the outer shell (i. e., capitulum, striated collar and basal belt) of the CP one would expect that defects in the composition of the ODFs could lead to abnormalities in the CP such as, head–tail detachments.

Separation of heads and tails or decapitations deserve the distinction of being classified as either head or tail defects, depending upon which component of the head-neck junction is predominantly affected or where the fracture line occurs. An example of a decapitation, of a specific morphogenetic nature, that appears to originate from the sperm head has been reported in Guernsey, Swedish, and Hereford bulls (Blom and Birch-Andersen 1970; Blom 1977; Hancock and Rollinson 1949; Settergren and Nicander 1968). The characteristic seminal features are that 80–100% of the sperm heads and tails are separated; a large percentage of the tails remain alive, motile, and bent in the midpiece around the cytoplasmic droplet; the sperm concentration is normal; and no other defects are discernible. Ultrastructural examination of spermiogenesis reveals an undersized and irregular implantation groove and corresponding basal plate, resulting in a weak anchorage to the capitulum (Blom and Birch-Andersen 1970; Settergren and Nicander 1968). During sperm passage in the caput epididymis, the fragile connection between the basal plate of the head and the capitulum of the tail is severed. Because the predisposition for head–tail separation occurs at the level of the implantation fossa, which is misshapen and structurally undersized, it most likely is a head rather than tail defect. Although the basal plate shares proteins with the ODF and CP, its assembly in this case would be undersized and irregular in shape. A similar sterilizing decapitated sperm defect has been investigated in man (Perotti et al. 1981).

Head–tail detachments are not the only consequences of targeted ODF protein deficiencies. Because of the weakness in ODF structure, often represented by missing or thinned ODFs, secondary effects occur concomitant with the acquisition of motility in the epididymis (Tarnasky et al. 2010; Yang et al. 2012; Zhu et al. 2018). These are bending, coiling and breaking of the tail, leading to membrane fusions of bent or broken regions of the tail as well as extrusion and disarrangement of ODFs and microtubular doublets from the axial fiber bundle. ***Many of these characteristics are evident in the Dag bull sperm defect*** *(Fig. 4.14* in *Chap. 4)*. However, the Dag defect also shows a derangement of mitochondria in the mitochondrial sheath, which originates in the testis. This seemingly unrelated characteristic is also found in Odf1-knockout mice (Yang et al. 2012) and may be due to the inability of mitochondrial bound KLC3 to bind $ODF1^{-/-}$ during mitochondrial sheath assembly in the testis (Zhang et al. 2012). As will be discussed, ***mitochondrial sheath defects*** alone can induce bending, coiling and breaking of the tail, concomitant with the acquisition of motility in the epididymis, and therefore ***are contenders for causing 'Daglike' sperm defects***.

AKAP4 makes up close to 50% of FS structure and thus it is not surprising that the ribs and longitudinal columns of the FS do not assemble in targeted disruption of mouse Akap4 (Miki et al. 2002; Brown et al. 2003). Consequently, proteins normally anchored to the FS, that are essential for motility, capacitation and redox regulation, are missing and the principal piece of the tail is shortened substantially, ***resembling the Short tail bull sperm defect*** *(Fig. 4.23* in *Chap. 4)*. Unlike the stunted assembly of the FS, Miki et al. (Miki et al. 2002) observed that the axoneme, ODF and mitochondrial sheath of the sperm tail appear normal, ***as also found in the Short tail bull sperm defect*** *(Fig. 4.23* in *Chap. 4)*. A similar phenotype is associated with the Akap3 gene null mouse, in which the ribs of the FS fail to develop but the longitudinal columns of the FS appear to assemble normally (Xu et al. 2020). In this case, the principal piece of the tail is shorter than normal but not as short as in the Akap4 gene knockout.

Sperm tail defects rarely fall into easily distinguishable categories such as the dynein-less arm and axoneme-less defects seen in immotile human spermatozoa (Gibbons and Rowe 1965; Afzelius et al. 1975; Pedersen and Rebbe 1975; Eliasson et al. 1977; Baccetti et al. 1979, 1980; Inaba 2011) In fact, many tail anomalies are complicated by associated structural changes that accompany and divert attention from the primary or predominant structural defect. A case in point is the aplastic mitochondrial sheath sperm defect induced in rats by gossypol, a cottonseed extract (Oko and Hrudka 1982a, b, 1984a, b; Oko 1984). This defect is easily recognized in the late maturation phase of spermiogenesis and in the caput epididymis as a missing segment (gap) of the mitochondrial sheath and as a dispersal of a granular substance at this level of the tail that normally binds the mitochondrial sheath and outer dense fibers (Figs. 1.31, 1.32a in Chap. 1). However, as the affected sperm tail travels through the epididymis, concomitant with the acquisition of motility, an interplay of the structural weakness in the sheath and motility predispose the axial fibers to split and erupt at the level of the missing segment (Figs. 1.32b, 1.33, Chap. 1). The protrusion of axial fibers at this level is accompanied by extraction of fibers (specifically 4, 5, 6, and 7) from the parts of the tail below the gap. Consequently, the tails are prone to bend, reflect over the midpiece and fuse with it, or break and separate at the gap level. The development of the associated or secondary lesions is reconstructed diagrammatically in Chap. 1 in Fig. 1.34. This particular research reinforces the importance of studying the pathogenicity of a defect rather than resting the diagnosis (as so often is the case) solely on the structural impairment that the sperm displays in the ejaculate or distal regions of the epididymis. It would be so easy in this case to overemphasize the associated (secondary) changes and overlook the primary or major structural change in the mitochondrial sheath. A misinterpretation arising from such an analysis may suggest a malformation in the axoneme or axial fiber bundle. Similar sequelae of secondary sperm tail defects as described above for gossypol treatment occur when the normal supply of selenium to the testis is depleted (Olson et al. 2005, 2007) or when GPx4/PHGPx, the selenium retaining protein of the mitochondrial capsule, is eliminated (Imai et al. 2009; Schneider et al. 2009).

The above examples reinforce the idea that the outer structural components of the sperm tail, namely the mitochondrial and fibrous sheaths, play a supporting role during motility. These structural collars must be elastic and strong enough to sustain and contain the excessive bending of the axial fiber bundle during motility. A weakness or malformation in these structures at any given point in the tail could lead to an eruption of axial fibers at this level and/or to a focal folding and breaking of the tail, ***resembling the Dag bull sperm defect*** *(Fig. 4.14* in *Chap. 4)*. This hypothesis is further supported by the investigations of Olson and Linck (Olson and Linck 1977), in which they removed the mitochondrial sheath of the rat sperm tail but left all other flagellar components intact. On the addition of ATP and resumption of motility, the flagellar geometry disrupted in a manner reminiscent of gossypol-affected spermatozoa. Outer dense fibers 4, 5, 6, and 7 (and their corresponding microtubule doublets) were ejected from the proximal end of the fibrous sheath but remained attached at their proximal end to the striated columns of the connecting piece. It is important to gage the extent of damage to the mitochondrial sheath. If mitochondrial damage is severe, the sperm will be less motile with less disruption of axonemal and outer dense fibers, ***resembling the Pseudodroplet bull sperm defect*** *(Fig. 4.16* in *Chap. 4)*. This interplay between the extent of mitochondrial damage and axial fiber disruption is reinforced when gossypol treatment is extended resulting in a less motile sperm phenotype and correspondingly less disruption of axial fibers.

Why the dislocation of axial fibers most often concerns the larger half of the axial fiber bundle (namely fibers 4, 5, 6, and 7) is enigmatic. Many tail defects, whether induced or naturally occurring, display a similar propensity. Examples are the Dag defect in the bull (Blom 1966; Koefoed-Johnsen and Pedersen 1971), a Daglike midpiece defect in the bull (Hellmén et al. 1980), gossypol-induced sperm defect in the rat (Oko and Hrudka 1982b), the coiled tail defect in the boar (Holt 1982), selenium deficiency-induced sperm tail defect in mice and rats (Wu et al. 1979; Olson et al. 2005, 2007; Imai et al. 2009; Schneider et al. 2009), the wobbler mutant gene-induced sperm tail defect in mice (Leestma and Sepsenwol 1980), the T-locus-induced sperm tail defect in mice (Yanagisawa 1965; Olds 1971) and the ODF gene knockout induced tail defects in mice (Tarnasky et al. 2010; Yang et al. 2012; Zhu et al. 2018). Cooper and Hamilton (Cooper and Hamilton 1977) also found a small percentage of dissolving (probably older) spermatozoa displaying the same symptoms in the epididymis of normal rodents. Although the structural mechanisms that allow the preferential extrusion of axial fibers 4–7 remains to be elucidated, it should be kept in mind that the doublet microtubule-dense fibers 3 and 8 are attached to the longitudinal columns of the fibrous sheath, effectively partitioning the flagellum into two unequal compartments, the smaller one containing microtubule doublets-dense fibers 9, 1, and 2 (Figs. 1.22, 1.33D in Chap. 1). A more common denominator for this pattern of fiber dislocation may very well represent an initial disintegration or malformation of the outer structural sheaths rather than of the axial fibers. It should kept in mind that extrusion of these fibers is usually associated with sperm tails that are bent, broken, folded or coiled, which leads to membrane fusions between different parts of the tail and head. Thus, there is a strong possibility that the dislocation of axial fibers and accompanying changes are symptomatic of a primary change in the outer investments of the sperm tail.

The function of the outer dense fibers, in sperm motility remains poorly understood. It was proposed that the outer dense fibers, which are larger than their doublet counterparts, are nothing more than passive stiffening elements (Hellmén et al. 1980; Pedersen 1974; Phillips 1972; Price 1973). This was based on the observation of their 'keratin-like' composition in mammalian sperm and that the thicker the dense fibers, the stiffer the flagellar beat (Baccetti et al. 1973; Baccetti et al. 1976a, b; Calvin et al. 1975; Olson and Sammons 1980). However, protein compositional studies, covered in Chap. 1, show that ODFs are made up of unique proteins unrelated to the keratin family. Furthermore, evidence that the ODFs and corresponding microtubule doublets function in unison during motility is based on their structural linkage and on the fact that the ODFs and microtubule doublets are often extracted from defective mitochondrial- or fibrous-sheaths together, usually retaining their linkage. From a consideration of this phenomenon it is possible that the mechanical energy generated by the microtubule doublets sliding on each other is transferred to the elastic ODFs, causing them to bend and consequently recoil, thus assisting in the propulsive thrust. Presumably, structural weaknesses in the outer sheath would fail to contain the bending of the dense fibers. Tail abnormalities associated with obvious developmental changes in the outer dense fibers are few. Dinsdale and Williams (Dinsdale and Williams 1980) reported a zinc deficiency-induced change in the sperm tail of rats associated with atrophy of the seminiferous epithelium. The most common sperm defects were an absence of the central and subfibril B microtubules of the axoneme and pronounced shape changes in the outer dense fibers. Dense fibers often assumed elongate crescentic forms in transverse sections. Zinc deficiency as a direct factor in the malformation of dense fibers is supported by the finding that during development they are the principal components of the tail that bind zinc (Baccetti et al. 1976b; Calvin et al. 1975). Zinc excesses have also been implicated as a causal factor in the bull Dag defect (Blom and Wolstrup 1976) and in the coiling of human sperm tails in suspension (Eliasson and Lindholm 1974). However, in these cases, if zinc indeed had an effect it was most probably at the functional level because the structural development of the dense fibers did not appear impaired as was the case in zinc deficiency.

Several studies have shown that during epididymal maturation, sperm components with a high cysteine or sulfhydryl (-SH) content, especially the ODFs, become stabilized by disulfide (-S–S-) cross-links formed by the oxidation of -SH groups (Calvin et al. 1975, 1973). Zinc is incorporated into the structure of the dense fibers during spermiogenesis, complexes and stabilizes the -SH-rich proteins and hence regulates the extent and arrangement of -S–S- cross-linking that occurs during epididymal maturation. During maturation both the zinc and -SH groups decrease as the latter are oxidized to form -S–S- cross-links, suggesting that zinc depletion is necessary for oxidation to occur (Calvin et al. 1975). That the addition of zinc to spermatozoa has the effect of inhibiting further oxidation of –SH (Calvin et al. 1973) is confirmatory evidence. Thus the possibility exists that the folding and coiling of the tail associated with zinc excesses in the bull (Blom and Wolstrup 1976) and human (Eliasson and Lindholm 1974) could be due to a weakness in ODF caused by lack of disulfide cross-linking.

In the context of zinc's regulatory role in fertility, four distinct types of sperm zinc ion distribution patterns (coined zinc signature) occur during in vitro capacitation (IVC) (Kerns et al. 2018). This zinc signature can be manipulated experimentally by altering the levels of zinc during IVC (by the addition of zinc chelators or $ZnCl_2$). It was shown that zinc serves as a de-capacitating factor, by decreasing the proteolytic activities of the sperm 26S proteasome and MMP2 (Kerns et al. 2020), by preventing sperm-release from glycans of the oviductal reservoir (Kerns et al. 2020) and by inhibiting hydrogen voltage channel HVCN1 (Lishko and Kirichok 2010). There are also an abundance of zinc proteins that differ in quantity between non-capacitated (ejaculated) and IVC spermatozoa, three more closely studied ones being, carbonic anhydrase 2 (CA2), calicin (CCIN), and coiled-coil domain containing 39 (CCDC39) (Zigo et al. 2022). The findings above highlight the importance of zinc in fertilization and pave the way for improved semen analysis using image-based flow cytometry to detect the state of the sperm-zinc signature.

Folding or coiling of the sperm tail, probably the most common symptom of tail defects, appears to be predisposed by many unrelated factors and often creates ambiguities in classification of tail defects. In the preceding discussion, evidence was presented that folding and coiling may result from malformation or weakness in the outer supporting sheaths or outer dense fibers. However, as so often occurs, interpretation of the structural mechanism is not clear-cut. A primary example is the Dag defect, which develops early in epididymal transit. There are two, possibly three, interpretations as to the causes of folding and coiling, both of which are symptomatic of this defect. One theory, as discussed above, suggests a weakness in the dense fibers caused by a zinc excess. Another theory, proposed by Koefoed-Johnsen and Pedersen (Koefoed-Johnsen and Pedersen 1971) and based upon the close relation between the bending and breaking of the flagellum at the distal half of the midpiece and the remnant of the distal cytoplasmic droplet, implicates a premature release of lysosomal enzymes by the distal droplet and a resultant decomposition of structural components of the tail. More than likely the outer structural sheaths would be the first component disrupted by the release of hydrolytic enzymes. The latter theory may appear more plausible because of the focal nature of the disruption and because a variety of experimental manipulations of the sperm medium known to induce bending and coiling, such as hypoosmolarity (Drevius 1963; Lindahl and Drevius 1964) and cold shock (Pedersen and Lebech 1971), can be associated with a disintegration of the cytoplasmic droplet and release of its hydrolases (Dott and Dingle 1968). Bending, folding, or coiling of the sperm tail has also been induced by surgical cryptorchidism (Ploen and Hakansson 1978) and is associated with morphological changes in the cauda epididymal epithelium of estrogen-treated bulls (Cupps and Briggs 1965; Cupps et al. 1960). Presumably, both these treatments produce changes in epididymal secretions that affect the environment and hence maturation of the spermatozoa.

The predisposition for folding or coiling is not necessarily restricted to changes in the epididymis but can arise at the level of the testes. As discussed earlier, segmental aplasia of the spermatid mitochondrial sheath in the testes caused by gossypol ingestion (Oko and Hrudka 1982b) or selenium deficiency (Olson et al. 2005, 2007) leads to sperm tail folding in the epididymis. Consideration of a testicular origin should be made when tail defects are concurrent with degenerative changes in the testes or accompanied by acrosomal and nuclear sperm anomalies. Examples of such occurrences are found in DICA- (an aromatic compound) treated monkeys (Buthala and Lobl 1979; Lobl and Mathews 1978), ethylene dibromide-treated rams (ElJack and Hrudka 1979; Hrudka and ElJack 1979), and cryptorchid rats and rabbits (Nagano 1963; Ploen and Hakansson 1978). However, as cautioned by Blom (Blom 1966) in reference to the Dag and other defects, the lack of morphological changes in the testes does not necessarily indicate that the spermatozoa are normal from a functional point of view.

6.2 Sperm Head Defects

As mentioned in Chap. 2, in a developmental survey of targeted mouse mutants known to affect sperm fertility it was concluded that mutations that affect sperm head shape most always affect motility, while mutations that affect sperm tail formation rarely affect sperm head shape (de Boer et al. 2015). The authors state, 'the striking dependency of motility on proper sperm head development is not easy to understand, but likely is of evolutionary benefit'. From our point of view, part of the explanation can be attributed to the fact that a large part of head formation occurs before tail formation during spermiogenesis. Thus if the framework or assembly of the spermatid head is incomplete it could affect the subsequent implantation of the axoneme and/or downstream assembly of accessory components around it. Furthermore, physiological pathways disrupted during assembly of the head may also adversely affect tail assembly. Cases in point are that most mutations of genes or inhibitions of enzymes involved with the early stages of acrosome formation, which prevent fusion of proacrosomic vesicles to form the acrosomic vesicle and hence the acrosomic cap, lead to a variety of 'globozoospermic-like' sperm head phenotypes associated with various degrees of aberrant sperm tail morphologies and/or immotility (Koscinski et al. 2011; Harbuz et al. 2011; Dam et al. 2007; Pierre et al. 2012; Sotomayor and Handel 1986; Kang-Decker et al. 2001; Yildiz et al. 2006; Oko et al. 2011; van der Spoel et al. 2002; Guidi et al. 2018). These globozoospermic defects and their development are discussed in detail in Chap. 2 under the heading ***'Defects associated with the formation of the acrosome and the manchette'***. However, it is important to point out that whether globozoospermia is 'complete' (no acrosome, no postacrosomal sheath, nucleus is round) or 'incomplete' (no acrosome, postacrosomal sheath is present, nucleus is oblong) there is usually low motility and sometimes even a low sperm count evident on examination of the spermiogram. If all three factors are involved, i. e. abnormal sperm shape, poor sperm movement and low sperm count, then the condition is defined as oligoasthenoteratozoospermia.

Oligoasthenoteratozoospermia (OATS) is a common symptom of male subfertility in humans and indicates a major disturbance (genetic or environmental) in spermatogenesis, which does not allow the post-meiotic phase or spermiogenesis to progress normally. It can be broken down into 3 parts: Oligozoospermia—low number of sperm, Asthenozoospermia—poor sperm movement and Teratozoospermia—abnormal sperm shape. In bull, Kopp et al. (2007) described a oligoasthenoteratozoospermic and sterilizing sperm defect, which based on its morphological characteristic was named the multinuclear-multiflagellar sperm (MNMFS) defect. All spermatozoa were abnormal, lacking an acrosome and having rounded and condensed nuclei. Many of the spermatozoa had multiple nuclei and multiple sperm tails. The tails seldom displayed evidence of a mitochondrial sheath. The bull sperm defects resembled many of the phenotypic characteristics of a targeted gene knockout $Hrb^{-/-}$ mutant mouse strain (Juneja and van Deursen 2005; Kang-Decker et al. 2001), displaying globozoospermic-like symptoms.

There are of course bull sperm head defects that show up in various bovine breeds that are not associated with tail impairment and/or immotility. One such inherited sperm defect affecting the acrosome is the Knobbed Sperm defect of the beaded variety (Hancock 1953), associated with sterility (Fig 4.2d-i in Chap.4), and another is the Knobbed Sperm defect of the indented variety (Barth 1986), associated with subfertility (Fig. 4.2j-o in Chap. 4). These two defects are characterized in Chap. 5. Most recently, a rare, non-synonymous and homozygous point mutation located within a WD40 signaling domain of Eml5 ($Eml5^{R1654W}$) was identified in an Angus bull used extensively in artificial insemination for its outstanding progeny production traits (Nogueira et al. 2022). Phenotypically, ***spermatozoa from this subfertile bull resembled the Knobbed Sperm defect*** of the beaded variety as well as presenting some tapered/pyriform sperm heads. A search of 1,000 Bull Genomes Run9 database revealed an additional 74 heterozygous animals and 8 homozygous animals harboring this exact mutation across several different breeds (0.7% frequency within the 6,191 sequenced animals). This frequency approximates the incidence of bulls severely affected with the knobbed acrosome sperm defect (0.57%) in 1758 beef bulls examined (Barth 1986). EML5 encodes a microtubule-associated protein with high expression in the testis and brain and appears to be involved in vesicular trafficking and protein complex scaffolding. The nonsynonymous nucleotide substitution alters the amino acid sequence of the protein and hence its normal, yet unknown function, as can be surmised by the increased retention of EML5 in the head of heterozygous and homozygous $Eml5R^{1654W}$ bulls (Nogueira et al. 2022). The authors rightly point out, that although the resultant compromised fertility of these bulls is tolerated by the AI industry for their superior progeny production traits, the extensive use of these animals could lead to decreased reproductive fitness within herds.

Another sperm head defect causing subfertility that is usually not associated with immotility is nuclear vacuolation (Fig. 4.7 in Chap. 4). Varieties of causes for nuclear vacuolation are stress, inadequate thermoregulation of the testis, toxins, inheritance and an interplay between stress and inheritance, meaning that some bulls are more susceptible to stress than others are. As shown in Chap. 5 the vacuoles develop from invaginations of the nuclear membrane into the spermatid nucleoplasm, usually at the end of round spermatid formation (Step 7 of spermiogenesis) just before the beginning of spermatid elongation (Oko 1977; Barth and Oko 1989). Under simulated stress (dexamethasone) the incidence of nuclear vacuolation (termed crater or diadem defect) in ejaculated spermatozoa of susceptible bulls was seen to increase an average of 25 days after initiation of the stress. This duration of 25 days corresponds with the timing of the origin or appearance of the nuclear vacuolation in step 7 spermatids found in stage VII of the cycle of the seminiferous epithelium. After the dexamethasone treatment, the incidence of the nuclear defect remained elevated for as long as 35 days and as short as 5 days, depending upon the bull. One hypothesis for the cause of nuclear vacuolation is that its results from a testicular hormonal imbalance. Favoring this hypothesis is that the injection of dexamethasone into bulls suppresses luteinizing hormone (LH) and testosterone levels (Thibier and Rolland 1976) and that stage VII of the cycle of the seminiferous epithelium is the most testosterone sensitive stage (Russell and Clermont 1977; Russell et al. 1981; Oko and Hrudka 1984a). In androgen deficiency studies step 7 spermatids and stage VII pachytene spermatocytes are the first cells of the testis to show degenerative changes.

Compliance with Ethical Standards
All applicable international, national and institutional guidelines for the care and use of animals were followed. This current work was supported by the Natural Science and Engineering Research Council of Canada (Grant number RGPIN/05305) to Richard Oko at Queen's University, along with the Agriculture and Food Research Initiative Competitive Grant number 2023–67015-39262 from the USDA National Institute of Food and Agriculture, Animal Breeding, Genetics & Genomics Program and seed funding from the College of Agriculture, Food and Natural Resources of the University of Missouri to Peter Sutovsky. Work for this book was also supported by a 2021–2022 Lalor Foundation Postdoctoral Research Fellowship to Lauren Hamilton. The research support to Viv Perry was funded by Australian Research Council (Grant number LP110100649), English Beef and Lamb Executive a division of the Agriculture and Horticulture Development Board (Grant number RC3815), Queensland Sperm Morphology Laboratories, Ridley AgriProducts Pty Ltd and the Australian cattle industry. The Canadian Association of Animal Breeders provided the foundation for research by Albert Barth. They supported the Development of the Diagnostic Spermatology Laboratory at the Western College of Veterinary Medicine, Saskatoon, SK., along with related support for numerous other research directives. Many grants were also received

from Saskatchewan Agriculture, including the Cattle Marketing Deductions Fund, the Horned Cattle Trust Fund, the Cattle Breeders Research Council, the Beef Development Fund and the Agriculture Development Fund. Animal research approval for research or clinical studies summarized in these chapters were given by respective University Animal Care Committees. The authors declare no conflict of interest as there is no perceived interest in prejudicing the impartiality of the research reported.

References

Afzelius BA, Eliasson R, Johnsen O, Lindholmer C (1975) Lack of dynein arms in immotile human spermatozoa. J Cell Biol 66(2):225–232. https://doi.org/10.1083/jcb.66.2.225

Baccetti B, Burrini AG, Maver A, Pallini V, Renieri T (1979) 9 + 0" immotile spermatozoa in an infertile man. Andrologia 11(6):437–443

Baccetti B, Burrini AG, Pallini V (1980) Spermatozoa and cilia lacking axoneme in an infertile man. Andrologia 12(6):525–532. https://doi.org/10.1111/j.1439-0272.1980.tb01344.x

Baccetti B, Pallini V, Burrini AG (1973) The accessory fibers of the sperm tail. 1. Structure and chemical composition of the bull coarse fibers. J Submicrosc Cytol 5:237–256

Baccetti B, Pallini V, Burrini AG (1976a) The accessory fibers of the sperm tail. High-sulfur and low-sulfur components in mammals and cephalopods. J Ultrastruct Res 57(3):289–308. https://doi.org/10.1016/s0022-5320(76)80118-9

Baccetti B, Pallini V, Burrini AG (1976b) The accessory fibers of the sperm tail. II. Their role in binding zinc in mammals and cephalopods. J Ultrastruct Res 54(2):261–275. https://doi.org/10.1016/s0022-5320(76)80155-4

Barth AD (1986) The knobbed acrosome defect in beef bulls. Can Vet J 27(10):379–384

Barth AD, Oko R (1989) Abnormal morphology of bovine spermatozoa. Abnormal morphology of bovine spermatozoa.

Blom E (1966) A new sterilizing and hereditary defect (the "Dag defect") located in the bull sperm tail. Nature 209(5024):739–740. https://doi.org/10.1038/209739a0

Blom E (1977) A decapitated sperm defect in two sterile Hereford bulls. Nord Vet Med 29(3):119–123

Blom E, Birch-Andersen A (1970) Ultrastructure of the "decapitated sperm defect" in Guernsey bulls. J Reprod Fertil 23(1):67–72. https://doi.org/10.1530/jrf.0.0230067

Blom E, Birch-Andersen A (1980) Ultrastructure of the tail stump sperm defect in the bull. Acta Pathol Microbiol Scand A 88(6):397–405. https://doi.org/10.1111/j.1699-0463.1980.tb02512.x

Blom E, Wolstrup C (1976) Zinc as a possible causal factor in the sterilizing sperm tail defect, the 'Dag-defect.' Jersey Bulls. Nord Vet Med 28(10):515–518

Brown PR, Miki K, Harper DB, Eddy EM (2003) A-kinase anchoring protein 4 binding proteins in the fibrous sheath of the sperm flagellum. Biol Reprod 68(6):2241–2248. https://doi.org/10.1095/biolreprod.102.013466

Buthala DA, Lobl TJ (1979) Electron microscope study of 1-(2,4-dichlorobenzyl)-1H-indazole-3-carboxylic acid, an exfoliative antispermatogenic agent, in the rat testis. Cytobios 25(97):23–28

Calvin HI, Hwang FH, Wohlrab H (1975) Localization of zinc in a dense fiber-connecting piece fraction of rat sperm tails analogous chemically to hair keratin. Biol Reprod 13(2):228–239. https://doi.org/10.1095/biolreprod13.2.228

Calvin HI, Yu CC, Bedford JM (1973) Effects of epididymal maturation, zinc (II) and copper (II) on the reactive sulfhydryl content of structural elements in rat spermatozoa. Exp Cell Res 81(2):333–341. https://doi.org/10.1016/0014-4827(73)90523-5

Cooper TG, Hamilton DW (1977) Observations on destruction of spermatozoa in the cauda epididymidis and proximal vas deferens of non-seasonal male mammals. Am J Anat 149(1):93–110. https://doi.org/10.1002/aja.1001490107

Cupps PT, Briggs JR (1965) Changes in the epididymis associated with morphological changes in the spermatozoa. J Dairy Sci 48(9):1241–1244. https://doi.org/10.3168/jds.S0022-0302(65)88433-8

Cupps PT, Laben RC, Rahlmann DF (1960) Effects of estradiol benzoate injections on the characteristics of bovine semen. J Dairy Sci 43:1135–1139

Dam AH, Koscinski I, Kremer JA, Moutou C, Jaeger AS, Oudakker AR, Tournaye H, Charlet N, Lagier-Tourenne C, van Bokhoven H, Viville S (2007) Homozygous mutation in SPATA16 is associated with male infertility in human globozoospermia. Am J Hum Genet 81(4):813–820. https://doi.org/10.1086/521314

de Boer P, de Vries M, Ramos L (2015) A mutation study of sperm head shape and motility in the mouse: lessons for the clinic. Andrology 3(2):174–202. https://doi.org/10.1111/andr.300

Dinsdale D, Williams RB (1980) Ultrastructural changes in the sperm-tail of zinc-deficient rats. J Comp Pathol 90(4):559–566. https://doi.org/10.1016/0021-9975(80)90104-8

Dott HM, Dingle JT (1968) Distribution of lysosomal enzymes in the spermatozoa and cytoplasmic droplets of bull and ram. Exp Cell Res 52(2):523–540. https://doi.org/10.1016/0014-4827(68)90493-x

Drevius LO (1963) Spiralization in tails of mammalian spermatozoa in hypotonic media. Nature 197:1123–1124

Eliasson R, Lindholm C Relationship between cellular zinc concentration and morphology of human sperm tails. In: Proc. 8th World Congr. Fertil. Steril., Buenos Aires, 1974.

Eliasson R, Mossberg B, Camner P, Afzelius BA (1977) The immotile-cilia syndrome. A congenital ciliary abnormality as an etiologic factor in chronic airway infections and male sterility. N Engl J Med 297 (1):1–6. https://doi.org/10.1056/nejm197707072970101

ElJack AH, Hrudka F (1979) Pattern and dynamics of teratospermia induced in rams by parenteral treatment with ethylene dibromide. J Ultrastruct Res 67(2):124–134. https://doi.org/10.1016/s0022-5320(79)80002-7

Gibbons IR, Rowe AJ (1965) Dynein: a Protein with Adenosine Triphosphatase Activity from Cilia. Science 149(3682):424–426. https://doi.org/10.1126/science.149.3682.424

Guidi LG, Holloway ZG, Arnoult C, Ray PF, Monaco AP, Molnár Z, Velayos-Baeza A (2018) AU040320 deficiency leads to disruption of acrosome biogenesis and infertility in homozygous mutant mice. Sci Rep 8(1):10379. https://doi.org/10.1038/s41598-018-28666-6

Hancock J (1953) The spermatozoa of sterile bulls. J Exp Biol 30(1):50–56

Hancock JL, Rollinson DHL (1949) A seminal defect associated with sterility of Guernsey bulls. Vet Rec 61:742

Harbuz R, Zouari R, Pierre V, Ben Khelifa M, Kharouf M, Coutton C, Merdassi G, Abada F, Escoffier J, Nikas Y, Vialard F, Koscinski I, Triki C, Sermondade N, Schweitzer T, Zhioua A, Zhioua F, Latrous H, Halouani L, Ouafi M, Makni M, Jouk PS, Sèle B, Hennebicq S, Satre V, Viville S, Arnoult C, Lunardi J, Ray PF (2011) A recurrent deletion of DPY19L2 causes infertility in man by blocking sperm head elongation and acrosome formation. Am J Hum Genet 88(3):351–361. https://doi.org/10.1016/j.ajhg.2011.02.007

Hellmén E, Plöen L, Settergren I, Nicander L (1980) Middle piece defects of testicular origin in bull sperm. Nord Vet Med 32(10):423–426

Holt WV (1982) Epididymal origin of a coiled-tail sperm defect in a boar. J Reprod Fertil 64(2):485–489. https://doi.org/10.1530/jrf.0.0640485

Hrudka F, ElJack AH (1979) The effect of ethylene dibromide on differentiation of the acrosome, nucleus, and transient nuclear appendages in ram spermatids. J Ultrastruct Res 67(2):135–151. https://doi.org/10.1016/s0022-5320(79)80003-9

Imai H, Hakkaku N, Iwamoto R, Suzuki J, Suzuki T, Tajima Y, Konishi K, Minami S, Ichinose S, Ishizaka K, Shioda S, Arata S, Nishimura M, Naito S, Nakagawa Y (2009) Depletion of selenoprotein GPx4 in spermatocytes causes male infertility in mice. J Biol Chem 284(47):32522–32532. https://doi.org/10.1074/jbc.M109.016139

Inaba K (2011) Sperm flagella: comparative and phylogenetic perspectives of protein components. Mol Hum Reprod 17(8):524–538. https://doi.org/10.1093/molehr/gar034

Ito C, Akutsu H, Yao R, Yoshida K, Yamatoya K, Mutoh T, Makino T, Aoyama K, Ishikawa H, Kunimoto K, Tsukita S, Noda T, Kikkawa M, Toshimori K (2019) Odf2 haploinsufficiency causes a new type of decapitated and decaudated spermatozoa, Odf2-DDS, in mice. Sci Rep 9(1):14249. https://doi.org/10.1038/s41598-019-50516-2

Juneja SC, van Deursen JM (2005) A mouse model of familial oligoasthenoteratozoospermia. Hum Reprod 20(4):881–893. https://doi.org/10.1093/humrep/deh719

Kang-Decker N, Mantchev GT, Juneja SC, McNiven MA, van Deursen JM (2001) Lack of acrosome formation in Hrb-deficient mice. Science 294(5546):1531–1533. https://doi.org/10.1126/science.1063665

Kerns K, Sharif M, Zigo M, Xu W, Hamilton LE, Sutovsky M, Ellersieck M, Drobnis EZ, Bovin N, Oko R, Miller D, Sutovsky P (2020) Sperm Cohort-Specific Zinc Signature Acquisition and Capacitation-Induced Zinc Flux Regulate Sperm-Oviduct and Sperm-Zona Pellucida Interactions. Int J Mol Sci 21(6). https://doi.org/10.3390/ijms21062121

Kerns K, Zigo M, Drobnis EZ, Sutovsky M, Sutovsky P (2018) Zinc ion flux during mammalian sperm capacitation. Nat Commun 9(1):2061. https://doi.org/10.1038/s41467-018-04523-y

Koefoed-Johnsen HH, Pedersen H (1971) Further observations on the Dag-defect of the tail of the bull spermatozoon. J Reprod Fertil 26(1):77–83. https://doi.org/10.1530/jrf.0.0260077

Kopp C, Sukura A, Tuunainen E, Gustavsson I, Parvinen M, Andersson M (2007) Multinuclear-multiflagellar sperm defect in a bull–a new sterilizing sperm defect. Reprod Domest Anim 42(2):208–213. https://doi.org/10.1111/j.1439-0531.2006.00754.x

Koscinski I, Elinati E, Fossard C, Redin C, Muller J, Velez de la Calle J, Schmitt F, Ben Khelifa M, Ray PF, Kilani Z, Barratt CL, Viville S (2011) DPY19L2 deletion as a major cause of globozoospermia. Am J Hum Genet 88(3):344–350. https://doi.org/10.1016/j.ajhg.2011.01.018

Leestma JE, Sepsenwol S (1980) Sperm tail axoneme alterations in the Wobbler mouse. J Reprod Fertil 58(1):267–270. https://doi.org/10.1530/jrf.0.0580267

Lei Y, Zhang X, Xu Q, Liu S, Li C, Jiang H, Lin H, Kong E, Liu J, Qi S, Li H, Xu W, Lu K (2021a) Autophagic elimination of ribosomes during spermiogenesis provides energy for flagellar motility. Dev Cell 56(16):2313-2328.e2317. https://doi.org/10.1016/j.devcel.2021.07.015

Lei Y, Zhang X, Xu Q, Liu S, Li C, Jiang H, Lin H, Kong E, Liu J, Qi S, Li H, Xu W, Lu K (2021b) A conserved Vac8/ARMC3-PtdIns3K-CI cascade regulates autophagy initiation and functions in spermiogenesis by promoting ribophagy. Autophagy 17(12):4512–4514. https://doi.org/10.1080/15548627.2021.1988813

Lindahl PE, Drevius LO (1964) OBSERVATIONS ON BULL SPERMATOZOA IN A HYPOTONIC MEDIUM RELATED TO SPERM MOBILITY MECHANISMS. Exp Cell Res 36:632–646. https://doi.org/10.1016/0014-4827(64)90319-2

Lishko PV, Kirichok Y (2010) The role of Hv1 and CatSper channels in sperm activation. J Physiol 588(Pt 23):4667–4672. https://doi.org/10.1113/jphysiol.2010.194142

Lobl TJ, Mathews J (1978) Effect of 1-(2,4-dichlorobenzyl)-indazole-3-carboxylic acid on sperm tails in rhesus monkeys. J Reprod Fertil 52(2):275–278. https://doi.org/10.1530/jrf.0.0520275

Miki K, Willis WD, Brown PR, Goulding EH, Fulcher KD, Eddy EM (2002) Targeted disruption of the Akap4 gene causes defects in sperm flagellum and motility. Dev Biol 248(2):331–342. https://doi.org/10.1006/dbio.2002.0728

Nagano T (1963) FINE STRUCTURAL CHANGES IN THE FLAGELLUM OF THE SPERMATID IN EXPERIMENTAL CRYPTORCHIDISM OF THE RAT. J Cell Biol 18(2):337–344. https://doi.org/10.1083/jcb.18.2.337

Nogueira E, Tirpák F, Hamilton LE, Zigo M, Kerns K, Sutovsky M, Kim J, Volkmann D, Jovine L, Taylor JF, Schnabel RD, Sutovsky P (2022) A Non-Synonymous Point Mutation in a

WD-40 Domain Repeat of EML5 Leads to Decreased Bovine Sperm Quality and Fertility. Front Cell Dev Biol 10:872740. https://doi.org/10.3389/fcell.2022.872740
Oko R (1977) Normal and defective bovine spermatogenesis. University of Calgary, Canada
Oko R (1984) The effects and possible mechanisms of gossypol action on sperm organelles, spermatogenesis and accessory sex glands of the adult rat. University of Saskatchewan,
Oko R, Donald A, Xu W, van der Spoel AC (2011) Fusion failure of dense-cored proacrosomal vesicles in an inducible mouse model of male infertility. Cell Tissue Res 346(1):119–134. https://doi.org/10.1007/s00441-011-1248-9
Oko R, Hrudka F (1982a) Effect of gossypol on spermatozoa. Arch Androl 9:39–43
Oko R, Hrudka F (1982b) Segmental aplasia of the mitochondrial sheath and sequelae induced by gossypol in rat spermatozoa. Biol Reprod 26(1):183–195. https://doi.org/10.1095/biolreprod26.1.183
Oko R, Hrudka F (1984a) Comparison of the effects of gossypol, estradiol-17 beta and testosterone compensation on male rat reproductive organs. Biol Reprod 30(5):1198–1207. https://doi.org/10.1095/biolreprod30.5.1198
Oko R, Hrudka F (1984b) Gossypol-induced early and delayed effects in the seminiferous epithelium of the adult rat. Contraception Delivery Syst 5:335–356
Olds PJ (1971) Effect of the T locus on sperm ultrastructure in the house mouse. J Anat 109(Pt 1):31–37
Olson GE, Linck RW (1977) Observations of the structural components of flagellar axonemes and central pair microtubules from rat sperm. J Ultrastruct Res 61(1):21–43. https://doi.org/10.1016/s0022-5320(77)90004-1
Olson GE, Sammons DW (1980) Structural chemistry of outer dense fibers of rat sperm. Biol Reprod 22(2):319–332. https://doi.org/10.1095/biolreprod22.2.319
Olson GE, Winfrey VP, Nagdas SK, Hill KE, Burk RF (2005) Selenoprotein P is required for mouse sperm development. Biol Reprod 73(1):201–211. https://doi.org/10.1095/biolreprod.105.040360
Olson GE, Winfrey VP, Nagdas SK, Hill KE, Burk RF (2007) Apolipoprotein E receptor-2 (ApoER2) mediates selenium uptake from selenoprotein P by the mouse testis. J Biol Chem 282(16):12290–12297. https://doi.org/10.1074/jbc.M611403200
Pausch H, Venhoranta H, Wurmser C, Hakala K, Iso-Touru T, Sironen A, Vingborg RK, Lohi H, Söderquist L, Fries R, Andersson M (2016) A frameshift mutation in ARMC3 is associated with a tail stump sperm defect in Swedish Red (Bos taurus) cattle. BMC Genet 17:49. https://doi.org/10.1186/s12863-016-0356-7
Pedersen H (1974) The human spermatozoan. Dan Med Bull 21:1–36
Pedersen H, Lebech PE (1971) Ultrastructural changes in the human spermatozoon after freezing for artificial insemination. Fertil Steril 22(2):125–133. https://doi.org/10.1016/s0015-0282(16)38048-7
Pedersen H, Rebbe H (1975) Absence of arms in the axoneme of immobile human spermatozoa. Biol Reprod 12(5):541–544. https://doi.org/10.1095/biolreprod12.5.541
Perotti ME, Giarola A, Gioria M (1981) Ultrastructural study of the decapitated sperm defect in an infertile man. J Reprod Fertil 63(2):543–549. https://doi.org/10.1530/jrf.0.0630543
Petersen C, Aumüller G, Bahrami M, Hoyer-Fender S (2002) Molecular cloning of Odf3 encoding a novel coiled-coil protein of sperm tail outer dense fibers. Mol Reprod Dev 61(1):102–112. https://doi.org/10.1002/mrd.1136
Phillips DM (1972) Comparative analysis of mammalian sperm motility. J Cell Biol 53(2):561–573. https://doi.org/10.1083/jcb.53.2.561
Pierre V, Martinez G, Coutton C, Delaroche J, Yassine S, Novella C, Pernet-Gallay K, Hennebicq S, Ray PF, Arnoult C (2012) Absence of Dpy19l2, a new inner nuclear membrane protein, causes globozoospermia in mice by preventing the anchoring of the acrosome to the nucleus. Development 139(16):2955–2965. https://doi.org/10.1242/dev.077982
Ploen L, Hakansson N (1978) Abnormal epididymal spermatozoa two to thirty-five days after a brief experimental cryptorchidism in the rabbit. Int J Androl 1:250–261

Price JM (1973) Biochemical and morphological studies of outer dense fibers of rat spermatozoa. J Cell Biol 59:272A

Russell LD, Clermont Y (1977) Degeneration of germ cells in normal, hypophysectomized and hormone treated hypophysectomized rats. Anat Rec 187(3):347–366. https://doi.org/10.1002/ar.1091870307

Russell LD, Malone JP, Karpas SL (1981) Morphological pattern elicited by agents affecting spermatogenesis by stimulation. Tissue Cell 13(2):369–380. https://doi.org/10.1016/0040-8166(81)90011-2

Salmon NA, Reijo Pera RA, Xu EY (2006) A gene trap knockout of the abundant sperm tail protein, outer dense fiber 2, results in preimplantation lethality. Genesis 44(11):515–522. https://doi.org/10.1002/dvg.20241

Schneider M, Förster H, Boersma A, Seiler A, Wehnes H, Sinowatz F, Neumüller C, Deutsch MJ, Walch A, Hrabé de Angelis M, Wurst W, Ursini F, Roveri A, Maleszewski M, Maiorino M, Conrad M (2009) Mitochondrial glutathione peroxidase 4 disruption causes male infertility. Faseb j 23(9):3233–3242. https://doi.org/10.1096/fj.09-132795

Settergren I, Nicander L Ultrastructure of disintegrated bull sperm. In: Proc. 6th Int. Congr. Anim. Reprod. A.I., Paris, 1968. pp 191–194

Sotomayor RE, Handel MA (1986) Failure of acrosome assembly in a male sterile mouse mutant. Biol Reprod 34(1):171–182. https://doi.org/10.1095/biolreprod34.1.171

Tarnasky H, Cheng M, Ou Y, Thundathil JC, Oko R, van der Hoorn FA (2010) Gene trap mutation of murine outer dense fiber protein-2 gene can result in sperm tail abnormalities in mice with high percentage chimaerism. BMC Dev Biol 10:67. https://doi.org/10.1186/1471-213x-10-67

Thibier M, Rolland O (1976) The effect of dexamethasone (DXM) on circulating testosterone (T) and luteinizing hormone (LH) in young postpubertal bulls. Theriogenology 5(2):53–60. https://doi.org/10.1016/0093-691x(76)90168-0

Tokuhiro K, Isotani A, Yokota S, Yano Y, Oshio S, Hirose M, Wada M, Fujita K, Ogawa Y, Okabe M, Nishimune Y, Tanaka H (2009) OAZ-t/OAZ3 is essential for rigid connection of sperm tails to heads in mouse. PLoS Genet 5(11):e1000712. https://doi.org/10.1371/journal.pgen.1000712

van der Spoel AC, Jeyakumar M, Butters TD, Charlton HM, Moore HD, Dwek RA, Platt FM (2002) Reversible infertility in male mice after oral administration of alkylated imino sugars: a nonhormonal approach to male contraception. Proc Natl Acad Sci U S A 99(26):17173–17178. https://doi.org/10.1073/pnas.262586099

Wu AS, Oldfield JE, Shull LR, Cheeke PR (1979) Specific effect of selenium deficiency on rat sperm. Biol Reprod 20(4):793–798. https://doi.org/10.1095/biolreprod20.4.793

Xu K, Yang L, Zhang L, Qi H (2020) Lack of AKAP3 disrupts integrity of the subcellular structure and proteome of mouse sperm and causes male sterility. Development 147(2). https://doi.org/10.1242/dev.181057

Yanagisawa K (1965) Studies on the mechanism of abnormal transmission ratios at the T-locus in the house mouse. IV. Some morphological studies on the mature sperm in males heterozygous for T-alleles. Jpn J Genet 40:793–798

Yang K, Meinhardt A, Zhang B, Grzmil P, Adham IM, Hoyer-Fender S (2012) The small heat shock protein ODF1/HSPB10 is essential for tight linkage of sperm head to tail and male fertility in mice. Mol Cell Biol 32(1):216–225. https://doi.org/10.1128/mcb.06158-11

Yildiz Y, Matern H, Thompson B, Allegood JC, Warren RL, Ramirez DM, Hammer RE, Hamra FK, Matern S, Russell DW (2006) Mutation of beta-glucosidase 2 causes glycolipid storage disease and impaired male fertility. J Clin Invest 116(11):2985–2994. https://doi.org/10.1172/jci29224

Zhang Y, Ou Y, Cheng M, Saadi HS, Thundathil JC, van der Hoorn FA (2012) KLC3 is involved in sperm tail midpiece formation and sperm function. Dev Biol 366(2):101–110. https://doi.org/10.1016/j.ydbio.2012.04.026

Zhu F, Liu C, Wang F, Yang X, Zhang J, Wu H, Zhang Z, He X, Zhang Z, Zhou P, Wei Z, Shang Y, Wang L, Zhang R, Ouyang YC, Sun QY, Cao Y, Li W (2018) Mutations in PMFBP1

Cause Acephalic Spermatozoa Syndrome. Am J Hum Genet 103(2):188–199. https://doi.org/10.1016/j.ajhg.2018.06.010

Zigo M, Kerns K, Sen S, Essien C, Oko R, Xu D, Sutovsky P (2022) Zinc is a master-regulator of sperm function associated with binding, motility, and metabolic modulation during porcine sperm capacitation. Commun Biol 5(1):538. https://doi.org/10.1038/s42003-022-03485-8

Chapter 7
Assessing Bovine Male Fertility in a Technological Age

Contents

Abstract New and emerging technologies allow for a deeper and more comprehensive understanding of sperm physiology that can be harnessed to improve bull fertility selection. This chapter focuses on (1) the use of conventional and emerging flow cytometry techniques to further enhance functional sperm assessments; (2) new developments in proteomic and metabolomic biomarkers of bull fertility and how they can better inform fertility evaluations; and (3) the use of sperm selection technologies to optimize the fertility outcomes of bulls in artificial insemination service. As our knowledge of sperm physiology continues to expand, technology will allow for a faster translational capacity and continuous development of techniques. The technologies and techniques presented are current tools that can be used to enhance the efficiency, precision and accuracy of bull fertility assessments and better inform herd management.

Keywords Flow Cytometry · Image-Based Flow Cytometry · Biomarkers · DNA Integrity · Acrosome Integrity · Mitochondrial Function · Oxidative Stress · Viability · Membrane Integrity · Sperm Selection · Proteomics · Metabolomics · Swim-up · Density gradient centrifugation · Glass Wool Filtration · Nanopurification · Sperm sex selection

© The Author(s), under exclusive license to Springer Nature Switzerland AG 2025

A. Barth et al., *Abnormal Morphology of Bovine Spermatozoa*, Advances in Anatomy, Embryology and Cell Biology 240, https://doi.org/10.1007/978-3-031-70126-9_7

7.1 Introduction

The evolution of techniques used for bovine sperm quality evaluations and male fertility predictions has advanced concurrently with both technological innovation and new scientific discoveries in fundamental areas of sperm biology. While traditional techniques, such as breeding soundness evaluations (BSEs) still form the bedrock of assessments in the field, new methodologies have emerged that can enhance our understanding of bovine male fertility and allow for a more effective management of herd health and fitness. In this chapter we will provide an overview of the different assessment methods, both currently in use and emerging, with a focus on biomarker discovery, sperm purification and flow cytometry techniques.

7.2 Physiological Biomarkers and the Use of Flow Cytometry for Sperm Analysis

7.2.1 Conventional Flow Cytometry

Flow cytometry (Fig. 7.1) has become an increasingly popular tool for sperm analysis, providing a high throughput alternative to traditional methods, while also increasing the objectivity of the assessment. Its ability to measure thousands of cells within seconds also gives a more comprehensive understanding compared to the conventional 200–300 cells investigated by microscopy. Moreover, the advancement of dyes and probes and the refinement of staining procedures allow for the investigation of multiple cellular features simultaneously, resulting in a greater overall understanding of a samples' functional potential. Briefly outlined below are physiological biomarker assays routinely used to assess functional parameters of sperm fitness through conventional flow cytometry.

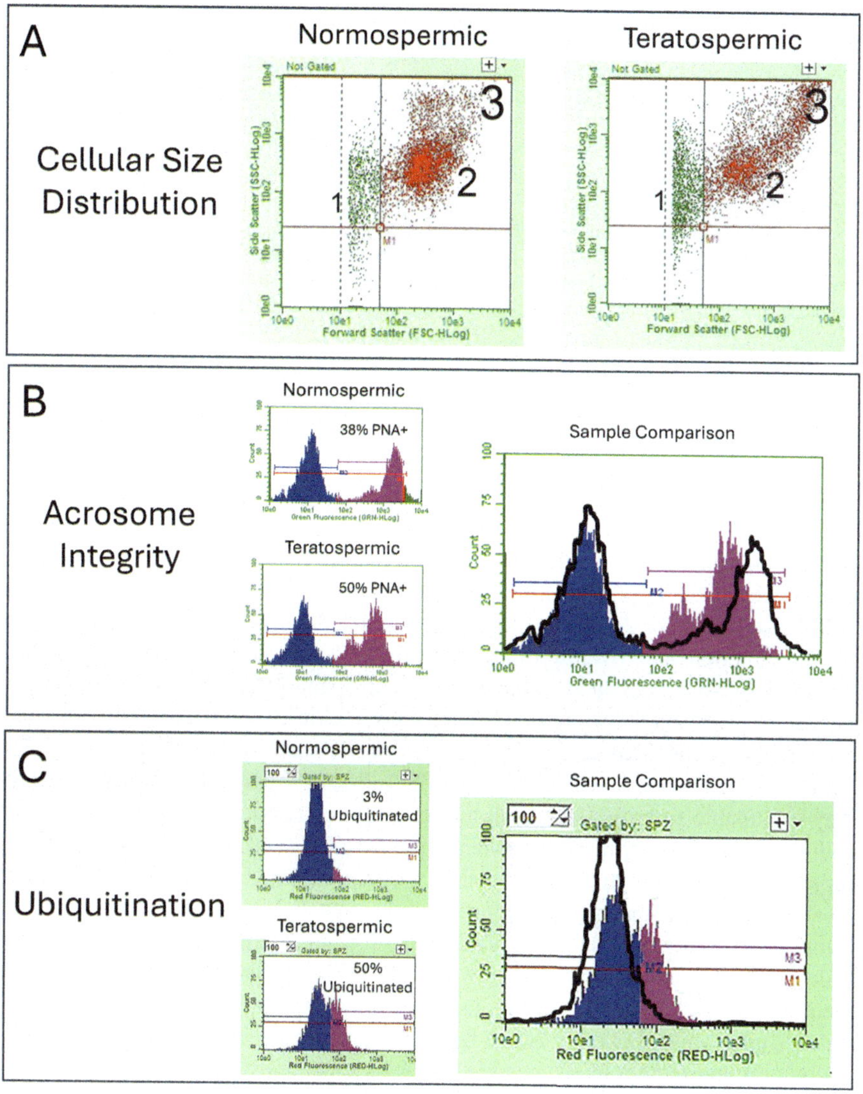

Fig. 7.1 Conventional Flow Cytometry Assessment Using PNA and Ubiquitin as Biomarkers of Sperm Fitness. Graphic outputs of a conventional flow cytometer include brightfield scatter diagrams of forward and side photomultiplier tubes (hence forward and side scatter) capturing the footprints of individual cells/events passing in front of them; and histograms of fluorescence induced by the excitation (with laser beam) of a probe with which the cells are labeled, again captured by cytometer's photomultiplier tubes. **(A)** Two comparative dot plots showing side scatter on the y-axis and forward scatter on the x-axis show the differences in cellular size between normospermic and teratospermic samples. On each plot, (1) refers to the population in green that represents small cells, cellular fragments as well as debris; (2) represents the cellular size range for normal spermatozoa; and (3) represents large cells (immature forms of spermatogenic as well as somatic cells contaminating semen), multi-nuclear and/or multi-flagellar spermatozoa or spermatozoa that may be stuck together. (**B**) A comparison of the acrosome damage found within

normospermic and teratospermic samples, measured by lectin PNA labelling. The top histogram on the left represents the normospermic sample, whereas the histogram on the bottom left represents the teratospermic sample. The percentage of PNA labelled cells is represented by the pink gate and the percentage of labelled cells has been added above the gate. On the right is a histogram overlay of both samples. The normospermic sample is represented by the black line with the teratospermic sample histogram shown behind. (**C**) A comparison of the level of ubiquitination found within normospermic and teratospermic samples. The top histogram on the left represents the normospermic sample, whereas the histogram on the bottom left represents the teratospermic sample. The percentage of ubiquitinated spermatozoa is represented by the pink gate on each histogram and the percentage of ubiquitinated spermatozoa has been added above the gate. On the right is a histogram overlay of both samples. The normospermic sample is represented by the black line with the teratospermic samples histogram shown behind

7.2.2 Cellular Viability and Membrane Integrity

Two essential metrics measured for sperm analysis are cellular viability and membrane integrity. The percentage of viable cells within an ejaculate at any one time is an important parameter that can influence many downstream semen processing events. The percentage of live cells can dictate the sperm concentration per dose that is used both for AI or freezing and dead cells can have a detrimental impact on the function and overall fitness of surrounding spermatozoa, through their metabolic by-products such as reactive oxygen species (ROS). Viability measurements are often made through the indirect evaluation of the sperm plasmalemma. Fluorescent probes that target dead or dying cells such as Propidium iodide (PI), SYTOX™, Zombie™, TO-PRO™ and YO-PRO™ dyes, are effective because they are unable to cross the membrane of intact cells but are able to accumulate and/or bind to intracellular components like DNA in spermatozoa that have a disrupted plasmalemma. In contrast, viability dyes that target live cells are often permeable to cell membranes but require cellular metabolic activity to activate their fluorescent conjugates such as SYBR-14™, fluorescein diacetate (FDA), and Calcein™. For many viability assays both sets of dyes are used allowing for the identification of three subpopulations: live – metabolically active and membrane intact, dying – metabolically active but membrane damaged, and dead – metabolically inert and membrane damaged (Fig. 7.2A). The most commonly used viability assay combines SYBR-14™ and PI, and was developed by Garner and Johnson (1995). Lastly, membrane integrity can also be investigated by probes that target changes in the membrane architecture such as Merocyanine 540 and Annexin-V.

7.2.3 DNA Integrity

Sperm DNA integrity is routinely investigated as part of a fertility assessment as it is often a functional dysregulation endpoint that can be indicative of a decreased fertilization potential. Increased DNA damage has also been associated with an increased amount of immotile, moribund and dead spermatozoa within a sample.

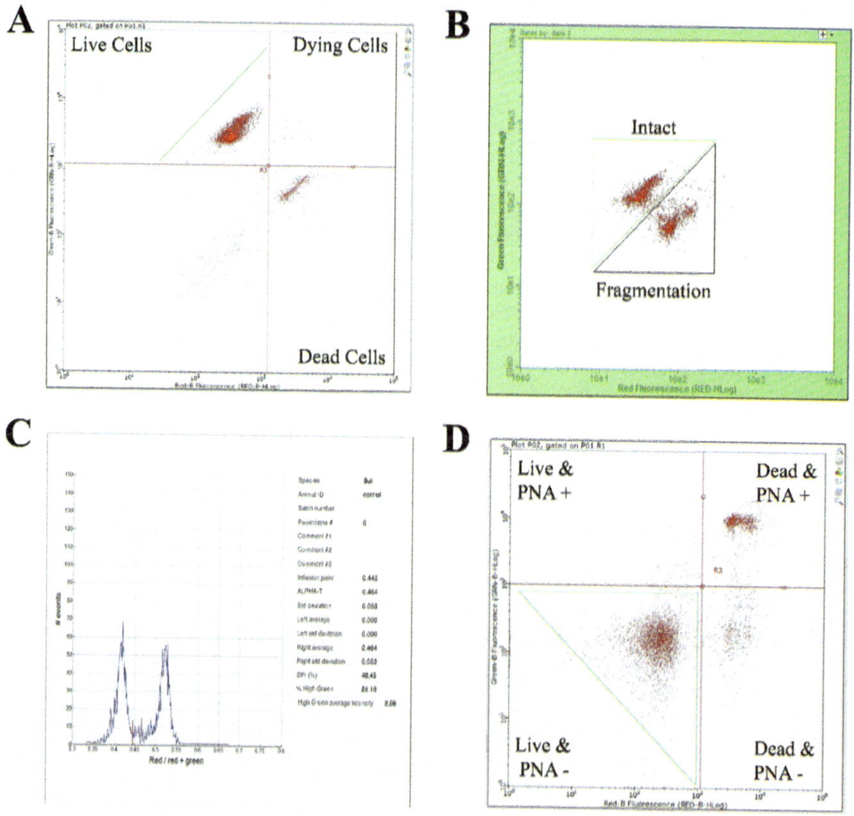

Fig. 7.2 Representative Images of Flow Cytometric Outputs from Sperm Physiological Biomarker Assessments and Computer-Assisted Sperm Analysis (CASA). (**A**)A sperm viability scatter/dot plot with SYBR14 ™ intensity (live cell dye) on the y-axis and Propidium Iodide (PI; dead and moribund cell dye) intensity on the x-axis. The population of live spermatozoa, staining with only SYBR14™ are highlighted with a green triangle in the upper left quadrant, dying spermatozoa that are dually stained with SYBR14 ™ and PI and are in the upper right quadrant and dead spermatozoa stained with only PI are within the bottom right quadrant. The bottom left quadrant shows cellular debris. (**B**) A dot plot of a bovine semen sample processed for DNA compaction assessment by using acridine orange. The y-axis shows green fluorescence intensity, and the x-axis shows red fluorescence intensity. Intact DNA has a strong green fluorescence intensity and is shown in the upper green triangle, whereas fragmented DNA fluoresces red and is highlighted by the black triangle. (**C**) A histogram output from the EasySoft software by IMV Technologies of the same DNA compaction assay shown in plot (**B**). (**D**) A dot plot that is representative of the acrosome integrity of spermatozoa within the sample. The y-axis represents PNA-FITC fluorescence intensity, and the x-axis represents PI fluorescence intensity. The bottom left quadrant shows live spermatozoa with intact acrosomes, highlighted by a green triangle, the upper left quadrant shows live spermatozoa with disrupted acrosomes, the bottom right quadrant represents dead spermatozoa with intact acrosomes and the upper right quadrant represents dead spermatozoa with disrupted acrosomes. (**E**) A dot plot showing mitochondrial membrane potential measured by the dual emission probe JC-1. The y-axis represents red fluorescence intensity, and the x-axis represents green fluorescence intensity. When mitochondrial membranes become polarized, they accrue red fluorescent JC-1-aggregates which are represented by cells highlighted by the green triangle as polarized cells. Conversely, spermatozoa with mitochondria that are depolarized emit green fluorescence and those spermatozoa are highlighted by the black

triangle. **(F)** A dot plot of the oxidative potential assay using dihydrorhodamine123 (DHR123) that includes an oxidative challenge with hydrogen peroxide. The y-axis represents the fluorescence intensity of DHR123, and the x-axis shows the fluorescence intensity of the viability dye PI. The lower left quadrant represents the live spermatozoa that did not respond to the oxidative challenge, the upper left quadrant represents live spermatozoa that effectively respond to the oxidative challenge, highlighted by the green triangle, the lower right quadrant represents the dead spermatozoa that did not respond to the oxidative challenge and the upper right quadrant are the dead spermatozoa that did respond to the oxidative challenge. **(G)** A sperm motility assessment done using a portable Androscope and analyzed by the AndroVision® software developed by MiniTube. The representative image is marked with dark green lines for fast motile spermatozoa, light green lines for slow motile spermatozoa, orange lines for spermatozoa demonstrating circular motility, yellow lines for spermatozoa with only static motility and red dots for immotile spermatozoa. Beside the image there is a corresponding output that states the number of spermatozoa counted, the number of fields of view analyzed as well as the percentages of each type of motility identified. Panels A, D, E, F & G are from assessments on freshly collected bovine spermatozoa and panels B & C are representative of a frozen-thawed sample that was mixed 1:1 with an UV irradiated sample of bovine spermatozoa to use as a positive control for the sperm chromatin structure/DNA compaction assay. The spermatozoa that are highlighted by the green triangle in each plot represent physiologically superior spermatozoa for fertilization. Data for plots A, D, E, & F were acquired using GuavaSoft InCyte 4.5 with a Guava easyCyte HT from Cytek. Data for plots B & C were acquired using GuavaSoft 1.0 from IMV technologies on a Guava easyCyte 5HT. Plots were provided courtesy of Abigail Zezeski and Dr. Tom Geary, Fort Keogh Livestock and Range Laboratory, Miles City, Montana USA

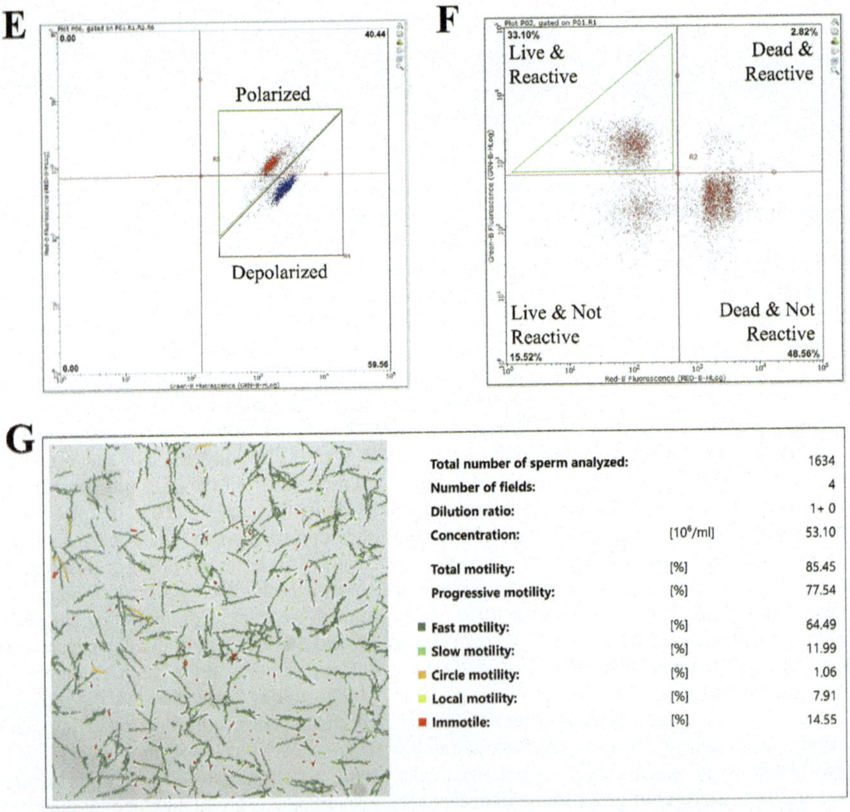

Fig. 2 (Continued)

Furthermore, DNA fragmentation within spermatozoa that successfully fertize an oocyte has been suggested to contribute to defects in embryo and/or fetal development, leading to early pregnancy loss (Evenson 1999; Carrell et al. 2003; Lewis and Simon 2010). The two main flow cytometry-based DNA integrity assays are the Sperm Chromatin Structure Assay (SCSA®) (Evenson et al. 1980a, b) and the terminal deoxynucleotidyl transferase mediated dUTP nick end-labelling assay (TUNEL) (Gorczyca et al. 1993).

Many DNA compaction assays, such as the SCSA®, harness the properties of acridine orange (AO), a dual emission dye that emits a green fluorescence when bound to double stranded DNA and emits a red fluorescence when bound to single stranded DNA. In most assays, an initial denaturation step is required, and staining must be done at an acidic pH. The quantification of the ratio of green to red fluorescent intensity evaluates the potential for DNA damage and the susceptibility to denaturation, resulting in a metric that has been shown to correlate with fertility (Fig. 7.2B,C) (Evenson et al. 1980a, b; Larson-Cook et al. 2003; Bungum et al. 2004; Zini et al. 2005; Evenson 2016). Furthermore, the SCSA® is also able to evaluate the proportion of immature sperm cells within a sample based on the high DNA stainability (HDS) metric (Evenson et al. 1999).

The TUNEL assay measures DNA fragmentation using a TdT transferase that labels free 3' ends of DNA (Gorczyca et al. 1993). The DNA breaks get fluorescently labelled within the nucleus, and increased intensity positively correlates with increased fragmentation. The TUNEL assay can be visualized by either flow cytometry or epifluorescent microscopy and shows the current DNA damage within the cell. The amount of DNA damage demonstrated by the TUNEL assay has been shown to correlate with fertilization and pregnancy rates (Høst et al. 2000; Sutovsky et al. 2002; Henkel et al. 2004).

It is of note that while both the SCSA® and TUNEL assays are used to investigate DNA integrity, they do target different aspects of DNA damage, with the SCSA evaluating the susceptibility for DNA damage and denaturation, whereas TUNEL reports current DNA breaks (Henkel et al. 2010). Furthermore, while this section does focus on flow cytometry techniques for sperm assessment, other techniques have also been shown to effectively assess DNA integrity such as Feulgen staining (see appendix for methodology) (Dobrinski et al. 1994), the sperm chromatin dispersion test (Fernández et al. 2003), and the gel-electrophoresis based COMET assay (Singh et al. 1988; Hughes et al. 1996).

7.2.4 Acrosome Integrity

Acrosome integrity is most commonly assessed through the use of plant-based lectins such as *Arachis hypogaea* (peanut) agglutinin (PNA), or *Pisum sativum* agglutinin (PSA) that bind to acrosomal components and are only able to be incorporated into live cells when the acrosomal outer membrane becomes compromised (Fig. 7.2D)(Cross et al. 1986; Mortimer et al. 1987; Nagy et al. 2003). While both PNA and PSA can effectively stain the acrosome, PNA is currently more favoured as it shows a decreased level of non-specific binding (Dalimata and Graham 1997). These lectins are commercially available, easy to use and can be purchased conjugated to a variety of fluorophores allowing them to be incorporated into multiplex assays or used for the independent assessment of the acrosome.

7.2.5 Mitochondrial Function

Mitochondrial function is of pivotal importance for energy production, which helps power both sperm motility and redox equilibrium. Several approaches can be taken to assess mitochondrial function such has the measurement of mitochondrial membrane potential (MMP), the assessment of the oxidative status of the mitochondria or the measurement of levels of antioxidants present within the organelle. One of the most common probes used to measure MMP is 5,5,6,6-tetrachloro-1,1,3,3-tetraethylbenzimidazolylcarbocyanine iodide (JC-1), or its more water-soluble alternative JC-10. Both variations are dual emission probes that emit a green fluorescence when the probe is in its monomer state, signifying low membrane potential but shift to produce orange fluorescing J-aggregates when mitochondrial membrane potential is high. The ratio of fluorescent intensities is used to estimate the overall mitochondrial function (Fig. 7.2E) (Garner and Thomas 1999). The MMP can also be measured using MitoProbe $DiOC_2(3)$, which functions in a similar capacity to JC-1 or by probes such as tetramethylrhodamine ethyl ester (TmRE) and tetramethylrhodamine methyl ester (TmRM), which enter live and active mitochondrial and emit a fluorescent intensity that correlates with changes in MMP. Mitochondrial respiration can be measured by using probes such as Rhodamine123 or certain MitoTracker probes; however, while the presence of fluorescence can indicate function it is not able to distinguish between different respiration rates (Graham 2001). Moreover, one benefit of the MitoTracker dyes is that many of them are compatible with fixation for multi-probe immunocytochemistry. Superoxide production by the mitochondria can also be assessed by using probes such as MitoSOX, that do not cross react with other reactive oxygen species, whereas cellular levels of ROS scavenging glutathione (GSH) can be probed with monobromobimane (mBBr) to gain a estimation of the overall oxidative state of mitochondria.

7.2.6 Reactive Oxygen Species and Oxidative Stress

Reactive Oxygen Species (ROS) and Oxidative Stress (OS) levels and their implications on sperm function have been widely documented (review- Aitken and Drevet 2020). It has become clear that low levels of ROS are required for normal sperm function, but when levels exceed the redox capacity of the cell, damage can occur. This leads to a dual assessment opportunity when investigating sperm function. Specific probes can investigate oxidative potential; a cell's ability to react to an oxidation challenge, whereas others can detect the current levels of different ROS. Furthermore, the extent of oxidative damage accrued can also be readily assessed through investigations of different endpoints such as protein oxidation or membrane lipid peroxidation. One of the main groups of probes used to assess the overall oxidation state of spermatozoa is the range of DCFDA (dichlorofluorescein diacetate) dyes that are cell permeable and require the action of a cellular esterase to cleave the diacetate group to form the non-fluorescent DCFH. In the presence of ROS such as hydrogen peroxide, DCFH is oxidized to dichlorofluorescein (DCF) which then emits a green fluorescence (LeBel et al. 1992; Carter et al. 1994). Similar probes that are also used to measure ROS in bovine spermatozoa are dihydroethidium (DHE) and dihydrorhodamine 123 (DHR123), which can be used to measure both basal levels of ROS and how spermatozoa react to an oxidative challenge, such has the introduction of H_2O_2 (Fig. 7.2F). Lastly, indirect implications of ROS levels can also be assessed through endpoints such as lipid peroxidation of the plasmalemma using probes such as the lipophilic dual emission dye BODIPY™, that exhibits a fluorescent shift when in contact with peroxyl radicals.

7.2.7 Sperm Motility

Whilst not a physiological assay done using flow cytometry, the evaluation of spermatozoan motility is a metric routinely investigated when completing a bull fertility assessment (Fig. 7.2G). Sperm kinetics and how sperm motility can inform on the underlying sperm physiology has long been of interest (Rikmenspoel and Van Herpen 1957; Gray 1958). Back in the 1970's photo- and cine-micrography were used to manually track sperm movement and quantify sperm motility (Katz and Dott 1975; David et al. 1981; Katz et al. 1981), however, starting in the late 1980's and commensurate with the increased use of computers, an automated assessment of sperm kinetics was developed and coined computer assisted sperm analysis (CASA)(Mortimer 1990; Mortimer et al. 2015).

Despite the use of the universal term, CASA technologies and software may employ different algorithms when calculating the final percent progressively motile value. Similarly, sperm kinematics employed for sperm tracking usually include: Curvilinear Velocity (VCL, μm/sec), Straight Line Velocity (VSL, μm/sec), Average Path Velocity (VAP, μm/s), Straightness (STR = VSL/VAP × 100, %), Beat Cross Frequency (BCF, Hz), and Amplitude of Lateral Head Displacement (ALH, μm) but there is no universally accepted grouping. As discussed in Chap. 9 significant differences (16%–37%) can occur between laboratories (Brito 2016). It is therefore vital that technical standards in parameters, particularly, pipetting techniques, the type of media or extender used, software settings and sample chamber used be standardized and reported if results are to be meaningful (Farrell et al. 1998).

Similarly, whether a CASA system or microscope assessment of motility is employed, there is inherent difficulty in relating in vitro assessment of sperm motility with fertility: Some studies investigating frozen thawed semen found a good correlation between the assessment of progressive sperm motility and fertility (Broekhuijse et al. 2012; Waberski et al. 2022), while others did not. Furthermore, in fresh semen there is limited evidence to show a correlation between progressive motility observed during the standard BBSE and bull fertility (Perry et al. 1990; Fitzpatrick et al. 2002).

This poor relationship is not surprising as it is now realised that the random forward movement of spermatozoa we observe, and measure in vitro is not what occurs in the female reproductive tract. In the oviduct the spermatozoa move using positive rheotaxis against the downstream fluid flow whereby the head of the sperm orientates toward the wall of the tract (Tung et al. 2017). The wall offers some protection for the sperm from the midstream fluid flow in the centre of the lumen and, during this wall passage, sperm swim in clusters, not separately as we observe in vitro, enabling faster passage through the visco-elastic fluids of the tract. Once sperm reach the oviductal isthmus they form a sperm reservoir, bound to the epithelium, with motility and capacitation suppressed. Sperm released from these epithelial cells swim to the oocyte in the ampulla in a different hyperactivated motility state, again dissimilar to that observed in vitro.

Despite these difficulties of, a) relating observed in vitro motility behaviour to that in the tract and b) divergence in methods of assessment, sperm motility remains a mainstay of sperm assessment. Progress in motility assessment is ongoing and aims to increase our understanding of the physiology that regulates sperm motility in the tract. In the last few years new portable CASA-based systems have been introduced commercially such as the mCASA by iSperm (Aidmics, Taiwan) and Androscope by Miniatube that only require a connection to a suitable laptop, or tablet. The iSperm device has been reported as a viable option for the field examination of motility in stallions and dogs (Dini et al. 2019; Moraes et al. 2019; Domain et al. 2022) with some caveats advanced on the required improvements in software (Domain et al. 2022). Similarly, author (LH) found the Androscope device satisfactory for use in BBSE although there are no published reports on the use of these devices in bulls. While these new technologies and their corresponding software do alleviate the accessibility issues of the traditional CASA systems and are more affordable alternatives, comparison studies are needed to ensure they can be reliable and accurate tools to use in the field.

7.2.8 *Image-based Flow Cytometry*

Image-based flow cytometry (IBFC) combines the high throughput capabilities of conventional flow cytometry with the advantages of microscopy allowing for the acquisition of large image galleries (Fig. 7.3) that correlate with the quantitative flow cytometry measurements (George et al. 2004; Rees et al. 2022). The added visual and

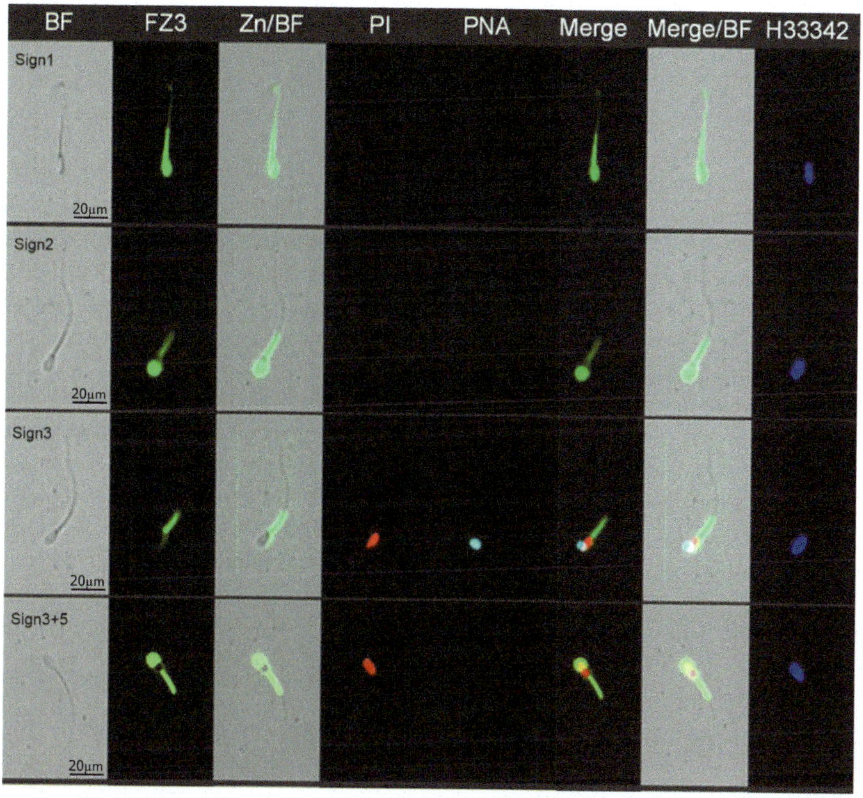

Fig. 7.3 An Image Gallery Captured by Using Image-Based Flow Cytometry (IBFC). Image gallery of multiplex sperm labeling generated by an image-based flow cytometer (Cytek Amnis FlowSight imaging cytometer). Frozen-thawed bull spermatozoa were labeled with FluoZin-3, AM (zinc ion reporter and capacitation status indicator; green); propidium iodide (nuclear envelope integrity stain; red); Hoechst 33,342 (cell-permeant DNA stain; blue); peanut agglutinin lectin conjugated to AlexaFluor 647 (outer acrosomal membrane integrity stain; cyan). The image gallery also contains bright field and composite channels. Zinc signature Sign1 – non-capacitated status; Sign2 – early capacitation state; Sign3 – late capacitation state; Sign3+5 – transitional state between signatures 2 and 3 (observed mainly in frozen-thawed spermatozoa); not shown: zinc signature 4 – moribund/death spermatozoa with no zinc signal. Zinc has been suggested as a master regulator of capacitation processes, and the zinc ion redistributions shown in each zinc signature are a novel way to visualize sperm capacitation progression in vitro. Images courtesy of Prof. Michal Zigo, University of Missouri

spatial data unlocks a powerful tool that allows for a deeper understanding of the data as well as novel applications that can focus not only on signal intensity changes but signal translocations in varied cell types, including bovine spermatozoa (Kennedy et al. 2014). In relation to sperm evaluation, image-based flow cytometry allows for increasingly complex analysis of cellular features and a more dynamic visualization of functional processes. One example of how IBFC has enhanced our knowledge of sperm processes is the implication of zinc as a regulator of capacitation, first reported as transient zinc signatures in boars, but further validated for humans and bulls (Kerns et al. 2018a, b, 2020). The high throughput visualization of single cells at increasingly high magnifications and speeds can allow for augmented automation and sensitivity to changes that can be combined with multiplex assays to greatly enhance assessment efficiency. The large data sets generated through IBFC also lend themselves to the use of artificial intelligence. The integration of artificial intelligence into the IBFC analysis pipeline is currently being developed for a label free sperm morphology assessment system, that trains on brightfield images of known morphological defects, limiting the need for specialized probes, additional staff, and processing time (Keller and Kerns 2022). A deep learning approach could also be used to harness fluorescent biomarker data in combination with sperm morphometric parameters, resulting in a powerful and streamlined approach for sire selection.

The development of flow cytometry-based assays to assess sperm function and performance has greatly increased the efficiency and accuracy of semen analysis. The use of advanced and specialized flow cytometry assays continue to enhance our understanding of a bulls' sperm function and could prove highly valuable when investigating bulls that have an uncertain fertility status based on traditional breeding soundness evaluations. Overall, technological innovation within the field of cytometry and the integration of deep learning should only enhance our knowledge, accuracy and efficiency moving forward.

7.3 The use of "Omics" in Bovine Fertility Screening

7.3.1 Introduction to Omics

Bull fertility is a complex trait that can be impacted by age, nutrition, environment, and management as well as the bulls' underlying genetics. To unravel its complexity, many have turned to bioinformatics, using genomic, proteomic, transcriptomic and metabolomic approaches, summed up as sperm phenomics (Sutovsky et al. 2024), in combination with large repositories of fertility data available for breeding sires to uncover predictive biomarkers of fertility outcomes.

The emergence and increased accessibility to "omic" technologies have greatly enhanced our ability to investigate the innate differences between bulls that have been characterized as "high" and "low" fertility. However, it is of note that a large proportion of fertility correlation studies have been performed in dairy breeds, leaving gaps of knowledge with respect to beef breeds. Additionally, recent findings have suggested that sire conception rate (SCR) classification and field fertility do not always correlate with an animals' embryo production potential in vitro (Fallon et al. 2023; Lockhart et al. 2023), highlighting that caution must be taken when comparing findings on bull fertility across studies. Moreover, the vast amount of data that can be retrieved using "omics" technologies, has the potential to reveal an endless number of possible correlations and causational links between fertility metrics and different genes, proteins, transcripts, and small molecules.

Therefore, due diligence must be taken to also ensure that new biomarkers are comprehensively vetted in the quest to unravel the intricacies of bovine male fertility.

Reproductive fitness is under polygenic control with many male and female reproductive traits showing low heritability (Kuhn and Hutchison 2008; Taylor et al. 2018a, b). Despite the challenge, great advances have been made investigating how genetic variants impact male reproductive success in bulls. While the breath and complexity of work done in this area is out of the scope of this section, research based on investigating runs of homozygosity as markers of inbreeding depression and areas of low heterozygosity in bulls of varied fertility (Nani and Peñagaricano 2020; Pacheco et al. 2023), as well as studies looking at quantitative trait loci (QTL) (Han and Peñagaricano 2016; Puglisi et al. 2016; Butler et al. 2022), single nucleotide polymorphisms (SNPs) (Dai et al. 2009; Sang et al. 2011; Yang et al. 2011) and copy number variants (Hamilton et al. 2009; Liu and Bickhart 2012; Mukherjee et al. 2013; Yue et al. 2013) have provided invaluable information in this area. Similarly work done on sperm transcriptome differences between bulls with differing fertility status has provided insight into how both coding and non-coding transcripts from spermatozoa as well as epididymal and accessory sex gland secretions can impact spermatozoon fitness and early embryo development (Card 2014; Donnellan et al. 2022; Indriastuti et al. 2022; Selvaraju et al. 2022).

The subsections below will highlight work done investigating proteomic and metabolomic biomarkers of bull fertility, which can be thought of as the functional end products of the upstream genetic variability and transcriptomic regulation within the male reproductive system. This section will also briefly highlight how validated biomarkers can be effectively incorporated in the current sperm assessment pipelines to further inform our fertility assessments.

7.3.2 Proteomic Biomarkers Associated with Sperm Quality

At ejaculation, spermatozoa mix with secretions from the epididymis and accessory sex glands, both of which have been shown to modulate spermatozoon function and influence fertilization potential (Amann et al. 1963; Voglmayr et al. 1972; Yanagimachi 1994). In bulls, the fluid from accessory sex glands has been shown to enhance the oocyte-penetrating capacity of spermatozoa from the caudal epididymis (Henault et al. 1995; Moura et al. 2006; Killian 2018) as well as mediate other fertility associated events (Killian et al. 1993; Bellin et al. 1998; Moura et al. 2006). Thus, the proteomic composition and differential expression of proteins within bovine accessory sex gland secretions may hold insights into sperm quality and function. Key seminal plasma and epididymal proteins that have been identified as candidate biomarkers for fertility include osteopontin (Killian et al. 1993; Cancel et al. 1997), albumin (Go and Wolf 1985; Kasimanickam et al. 2019), clusterin (Ibrahim et al. 2000), constituents of the binder of sperm (BSP) protein family (Odhiambo et al. 2014; Diaz-Miranda et al. 2024) and ubiquitin (Sutovsky et al. 2002; Kumar et al. 2012).

Osteopontin (OPN) is an extracellular matrix protein expressed as multiple isoforms within the bovine male reproductive tract. Increased expression of OPN has been found to correlate with high fertility in bulls; however, it should be noted that only the presence of the 55 kDa isoform found within seminal plasma has been positively correlation with fertility in Holsteins (Cancel et al. 1997). Functionally, OPN present in epididymal fluid has been suggested to have a protective effect against calcium crystallization (Erikson et al. 2007), whereas after ejaculation it may function as a regulator of intracellular calcium changes during capacitation (Handrow et al. 1989; Boccia et al. 2013). Differential localizations of OPN have also been reported on the head and midpiece of ejaculated bovine spermatozoa where it is suggested to function in mediating sperm-egg interactions (Gonçalves et al. 2008; Monaco et al. 2009). Therefore, an increase in OPN concentration within bovine seminal plasma may function in multiple ways to improved fertilization potential in bulls.

Albumin has been highlighted as a positive biomarker for bovine fertility, showing increased expression within the seminal plasma of high fertility bulls (Kasimanickam et al. 2019). Albumin demonstrates a protective effect on sperm membrane integrity, decreasing the adverse impacts of free radicals and limiting membrane lipid peroxidation (Kasimanickam et al. 2019). Furthermore, albumin has also been shown to facilitate cholesterol outflow from the plasmalemma and therefore may act as a temporal mediating factor of capacitation progression (Go and Wolf 1985).

Clusterin, abundant in epididymal fluid and seminal plasma, is a negative biomarker of bull fertility (Ibrahim et al. 2000) that functions as an extracellular chaperone (Wilson and Easterbrook-Smith 2000) through ATP dependent interactions with hydrophobic regions of misfolded or stressed proteins (Bailey et al. 2001; Wyatt et al. 2009a, b). Investigations in human semen have implicated its chaperone activity in both extracellular and intracellular proteostasis of toxic protein aggregates (Wyatt et al. 2013; Merlotti et al. 2015; Roan et al. 2017); however, it has not been shown to independently function in refolding proteins (Poon et al. 2000). In bulls, clusterin has been associated with spermatozoa exhibiting head, midpiece and principal piece defects (Ibrahim et al. 2000). Moreover, it has also been shown to agglutinate abnormal spermatozoa, preventing apoptotic or necrotic cells from increasing oxidative stress within their immediate environment. In humans, seminal clusterin was found to have a differential glycosylation pattern than that which is found in serum, as well as a 20-fold higher expression level (Choi et al. 1990; Kapron et al. 1997). Merlotti et al. (2015) suggests that unlike serum clusterin, the role of seminal clusterin may not only be to clear misfolded protein complexes, but to also promote immunological tolerance to sperm antigens in both the male and female reproductive tracts. Therefore, increased clusterin on abnormal spermatozoa may be indicative of dysregulation during sperm development or maturation, that was not able to be cleared due to a testicular or epididymal immune privilege (Merlotti et al. 2015).

The family of binder of sperm proteins (BSPs) are the most abundant proteins found within bovine seminal plasma and have been shown to function in a multifaceted manner (Manjunath et al. 1994; Gwathmey et al. 2006). Functionally, within the seminal plasma, BSPs are involved in regulating sperm motility before capacitation and facilitating the establishment of the oviductal reservoir (Thérien et al. 1999; Gwathmey et al. 2006). The BSPs have binding specificity to phosphorylcholine and have been shown to modulate cholesterol and lipid efflux from the bovine plasmalemma, a function that may lead to detrimental effects to the sperm plasma membrane of frozen-thawed samples causing increased cryo-capacitation like changes (Thérien et al. 1999; Medeiros et al. 2002; D'Amours et al. 2012; Diaz-Miranda et al. 2024). Furthermore, due to their ability to change the membrane lipid concentration, studies have shown that enrichment of BSPs on the sperm surface can be an indicator the cell is defective or of low quality (Moura et al. 2006; Roncoletta et al. 2006; D'Amours et al. 2010, 2012; Lessard et al. 2011; Odhiambo et al. 2014; Diaz-Miranda et al. 2024).

Like clusterin and BSP proteins, ubiquitin (UBB) has been proposed as a marker of negative sperm quality. Ubiquitin is a small molecule that is secreted through apocrine pathways in the epididymis and has been shown to binds preferentially to the surface of defective spermatozoa (Sutovsky et al. 2001a, b). Ubiquitination of defective spermatozoa within the epididymis is proposed to signal for their degradation during epididymal transit (Sutovsky et al. 2002). Therefore, the presence of UBB labelled cells within the ejaculate may be suggestive of the incomplete or overloaded degradation pathways within the epididymis and could be indicative of increased dysregulation within spermatogenesis or sperm maturation (Sutovsky et al. 2002).

7.3.3 Proteomic Biomarkers Reflecting Sperm Capacitation Status and Acquisition of Fertilizing Ability

Once ejaculated, spermatozoa must undergo physiological changes associated with the sperm capacitation process to functionally prepare for interactions with the oocyte (Chang 1951; Austin 1952). Capacitation is an energy intensive, apoptotic-like process that creates large amounts of oxidative stress that must be regulated effectively for fertilization to be successful (Aitken 2017). Therefore, proteins that function in energy metabolism, modulation of sperm motility and oxidative-reductive capacities are of great interest. Enolase 1/α (ENO1) has been put forth as a positive fertility biomarker in bulls based on proteomic comparisons between high and low fertility animals (Aslam et al. 2018). Functionally, ENO1 is a glycolytic enzyme that catalyzes the conversion of 2-phosphoglycerate to phosphoenolpyruvate during glycolysis. However, ENO1 has also been shown to have the ability to interact with microtubules in the tail of rat spermatozoa, which may allow it to act as a mechanism for controlling the activity of the glycolytic pathway and thus sperm motility (Srere and Knull 1998; Gitlits et al. 2000). The relative abundance of PEBP4 was demonstrated to show a significant positive correlation with high fertility in bulls, a correlation that may be attributed to its regulatory role in the maintenance and remodeling of the sperm plasmalemma (An et al. 2012; Somashekar et al. 2017), an essential process during sperm capacitation. ProAKAP4, the precursor protein to AKAP4, a protein kinase A anchoring protein of the sperm tail, has also been shown to have a positive correlation with bovine male fertility in both proteomic (Aslam et al. 2018; Sergeant et al. 2019; Dordas-Perpinyà et al. 2022) and transcriptomic studies (Singh et al. 2019). The AKAP4 protein is a highly abundant scaffold protein of the sperm tail fibrous sheath, which functionally regulates the flagellar motion (Miki et al. 2002). Concomitantly, the expression of outer dense fiber 2 (ODF2), another structural protein of the sperm tail has also been positively correlated with bovine fertility (Kaya et al. 2022), highlighting how good structure and proper sperm development facilitates increased sperm function. Proteins that function in an antioxidant capacity have also recently been shown to correlate positively with high fertility such as peroxiredoxin 6 (PDRX6) (Rabaglino et al. 2022), apoptosis-stimulating of p53 (ASPP2), and ubiquinol-cytochrome c reductase complex core protein 2 (UQCRC2) suggesting that an increased antioxidant capacity has a positive influence on fertility (Özbek et al. 2021).

Fertilization and the initiation of zygotic development require coordinated interactions between spermatozoon and oocyte factors, as well as successful oocyte activation and paternal DNA incorporation. While hormones, metabolites and extracellular factors within the female reproductive tract are known to modulate fertilization events, sperm proteins that have direct functional participation in sperm-egg interactions are of keen interest as fertility biomarkers. One such candidate is testis specific serine kinase 6 (TSSK6) which has a positive correlation with bull fertility and has been suggested to function in the relocation of IZUMO1 from the anterior sperm head to overlaying the postacrosomal sheath region, a requirement for sperm-egg fusion (Sosnik et al. 2009). Moreover, once the sperm and egg have successfully fused, sperm-borne oocyte activating factors (SOAFs) are required to initiate calcium oscillations and the polyspermy block, the resumption and completion of meiosis II by the female gamete, the extrusion of the second polar body, as well as downstream signalling events. Currently, two candidate SOAFs have been identified, WW-domain binding protein (WBP2NL/PAWP) and phospholipase C zeta 1 (PLCZ1) (Swann et al. 2004; Aarabi et al. 2014). Investigations into the predictive value of PAWP have shown that abnormal PAWP expression correlated with aberrant sperm phenotypes and that PAWP expression has predictive value for conception rate when analyzed using a stepwise regression analysis (Kennedy et al. 2014; Sutovsky et al. 2015; Kaya et al. 2022). Phospholipase C zeta 1 has also been proposed as a fertility biomarker in cattle; however the activating capacity of bovine PLCZ1 has been shown to differ from that of other species such as mice and humans (Cooney et al. 2010). Finally, once incorporated successfully, embryonic development is at least partially dependent on the quality of the paternal DNA. While some histones are retained within the haploid spermatid nucleus, studies have shown that the ratio of histones to protamines within the mature sperm nucleus correlates with fertility status (Evenson 1999; Zini et al. 2007). Furthermore, the expression of protamine 1 (PRM1), the most abundant nuclear protein within the mature bovine sperm nucleus, has also been identified as a candidate fertility biomarker, with PRM1 expression being shown to positively correlate with bull fertility status (Feugang et al. 2010; Dogan et al. 2015). Conversely, the expression of proteasomal subunit 6 (PSMA6) has been found to negatively correlate with sperm DNA fragmentation (D'Amours et al. 2010).

Proteomic investigations of bovine spermatozoa and seminal plasma highlight the complexity of sperm fitness and the numerous physiological changes that must be coordinated and functionally regulated for spermatozoa to reach the site of fertilization. While many markers of functional regulation, energy metabolism, proteostasis, antioxidant capacity and sperm-egg interaction proteins have been discovered as fertility biomarkers, at the current time, no one proteomic biomarker can effectively predict bull fertility outcomes. Differential protein expression studies between high and low fertility bulls have provided great insight into potential proteins and pathways that regulate sperm function; however, one must take care when comparing animals or studies as differences in collection methods, environment and management have been shown to alter protein composition (Rego et al. 2015). These compositional changes may influence sperm processing and can further complicate our ability to accurately compare and categorize bulls. Therefore, as we look to the future, a more holistic and standardized approach may be required to provide better predictive accuracy.

7.3.4 Metabolomic Approach for Investigating Sperm Fitness

Metabolomics is the study and quantification of low molecular weight molecules and represents both the biochemical environment and downstream end-products from cellular and regulatory pathways within a cell, tissue, or overall system (Idle and Gonzalez 2007; Dunn et al. 2011). Metabolites can include carbohydrates, organic compounds, amino acids, peptides, steroids, polyamines, and fatty acids. Through the analysis of small molecules in bovine seminal plasma and spermatozoa, insights from the field of metabolomics can offer a comprehensive understanding of the underlying biochemical pathways involved in sperm physiology and their impact on fertility (Kovac et al. 2013; Bieniek et al. 2016; Menezes et al. 2019; Saraf et al. 2020; Park et al. 2023).

By using comparative metabolomic approaches several studies have identified specific metabolites that show differential expression between high and low fertility sires, shedding light on potential metabolic signatures that could be used as predictive biomarkers (Menezes et al. 2019; Saraf et al. 2020). In a study by

Saraf et al. (2020) that focused on the endogenous metabolites in crossbred bulls with varied fertility status, metabolites uniquely found in high fertility animals were associated with β-alanine metabolism, glutathione metabolism, fatty acid elongation in mitochondria, and taurine and hypotaurine metabolism, whereas metabolites uniquely identified in low fertility bulls associated with fatty acid metabolism, fatty acid elongation in mitochondria, and inositol phosphate metabolism. Furthermore, two metabolites, spermine and L-cysteine were only present in high fertility bulls. Functionally, the spermine concentration may act as a decapacitating factor and temporal regulator of acrosomal exocytosis (Rubinstein et al. 1995) limiting spontaneous or early exocytosis from occurring, and L-cysteine may function in oxidative and thermal stress protection of the plasmalemma (Iqbal et al. 2016). Conversely dihydrolipoamide, leukotriene LTC4, and inositol 1,4,5-trisphosphate were exclusively found in metabolomic profiles of low fertility bulls and may be by-products of metabolic pathways responsible for hyperactivation during sperm capacitation (Saad and Burka 1983; Ho and Suarez 2001; Mitra and Shivaji 2004; Mitra et al. 2005; Kumar et al. 2006), suggesting premature or increased cryo-capacitation may be occurring in the low fertility sires after freeze-thawing. Studies on the exogenous metabolites from fresh bovine spermatozoa revealed the predominant metabolites were oleic acid, phosphoric acid, phosphine, carbamate, and glycerol, whereas benzoic acid, acetic acid, L-serine, carbonate, and 2-ketobutyric acid were of low abundance (Menezes et al. 2019). Furthermore, a higher abundance of GABA, benzoic acid, Carbamate, and lactic acid was shown to be predictive of high fertility, whereas higher palmitic acid abundance was predictive of low fertility (Menezes et al. 2019).

Association studies investigating the links between metabolomic molecules and sperm quality parameters have also been performed (Narud et al. 2020). Findings revealed that expression levels of several amino acids and trace elements correlated to functional sperm parameters. Cysteine and glutamic acid showed negative correlations with chromatin integrity in spermatozoa, which was proposed to be facilitated by a decreased capacity to mitigate oxidative stress. Furthermore, a decreased abundance of aspartic acid was shown to correlate with increased non-return rates at 56 days as well as a decreased presence of immature spermatozoa, suggesting the abundance of aspartic acid may be marker for proper chromatin compaction and maturation of spermatozoa (Narud et al. 2020). Additionally, trace elements such as iron (Fe) and zinc (Zn) exhibited negative associations with sperm motility and chromosome integrity, highlighting their roles in sperm function and fertility.

Overall, metabolomic profiling provides a more systemic view of bovine spermatozoa and seminal plasma function and its interplay with the extracellular environment. By elucidating the metabolic pathways involved in fertility regulation, metabolomics hold promise for identifying novel biomarkers and therapeutic targets for improving bovine reproductive success.

7.3.5 *The Integration of "Omics" Biomarkers in Bovine Fertility Assessments*

The ability to translate research from biomarker discovery to high throughput sperm phenotype assessment pipelines can at times be a slow process. However, with increasingly sophisticated technologies, higher throughput multiplexing capabilities and the identification of known and validated biomarkers, greater predictive efficiency can be achieved. Several studies have shown the effective implementation of integrated approaches using a combination of conventional physiological and newly discovered proteomic biomarkers with promising success [e.g., (Kennedy et al. 2014; Nogueira et al. 2022; Diaz-Miranda et al. 2024)]. Furthermore, the added information that comes from integrating metabolomic investigations into studies that also use traditional physiological biomarkers, such as combining the mitochondrial membrane potential assay JC-1, with sperm motility assessments and metabolomic information showing oxidative-phosphorylation efficiencies of the sperm mitochondria elevate the level of knowledge about the functional capacities of spermatozoa from animals with differing fertility potential (de Agostini Losano et al. 2023). Our ability to harness the knowledge provided from "omics" technologies and combine it with current sperm analysis techniques shows great potential for increasing our predictive potential when assessing male fertility.

7.4 Sperm Purification Techniques

7.4.1 *Introduction to Semen Purification*

The integration of advanced reproductive technologies within agriculture has reshaped traditional approaches to bovine semen distribution and preservation, as the industry strives to continuously increase operational efficiencies. With regards to male reproductive management, this has led to enhanced precision when selecting for both genetics and physical traits and more widespread implementation of semen sorting and purification techniques. Some of the fundamental work investigating semen sorting techniques was done in the 1970s when methodologies were developed to sort based on differing mass, motility, swimming patterns, volumetric differences as well as immunological properties with varied success (Barlow and Vosa 1970; Ericsson et al. 1973; Prasad et al. 2010; Jain et al. 2011).

7.4.2 Sex Selection in Semen Processing

It was the discovery that sperm DNA quantification could differentiate between X- and Y- bearing spermatozoa using flow cytometry that greatly advanced the field (Otto et al. 1979; Gledhill et al. 1982; Pinkel et al. 1982; Garner et al. 1983). These findings demonstrated a reliable technique to differentiate between X- and Y-bearing spermatozoa, the only caveat to the technique was that the spermatozoa needed to be killed and permeabilized to allow for dye penetration and thus DNA quantification. This hurdle was overcome by Johnson et al. (1987a, b), who modified the staining procedure by using vital DNA stain Hoechst 33342, thus eliminating the need for permeabilization of cells, resulting in the first live offspring to be produced by sexed semen, through work done in rabbits (Johnson et al. 1989). The first commercial licence for the use of sexing technology in AI was granted in 2002 (Seidel 2014) and the rights to the technology were sold to Sexing Technologies in 2007, through its procurement of XY Inc. Whether aiming to increase the number of female replacements in a dairy herd or produce superior breeding stock for meat production, this technological advancement created a lucrative tool for both dairy and beef industries, offering producers a level of control over the genetic outcomes of offspring that was once unimaginable.

7.4.3 Swim-up Sperm Purification

Concurrent with work being done to select for spermatozoa based on sex chromosomes, researchers have also been actively working on innovative techniques to isolate the best quality spermatozoa from an ejaculate, in a fashion that limits cell damage, and reduces harmful biologics and cellular debris, all in a cost and time effective manner. Presently, the swim-up (Mahadevan and Baker 1984), and density gradient centrifugation (DGC) are the most widely used techniques in both livestock and human assisted reproductive technology (ART) procedures. The swim-up is one of the oldest methods of sperm purification, originally developed for human sperm selection, that relies on the self-propelled movement of spermatozoa from either neat semen, extended semen, or a washed pellet into an overlying medium (Mahadevan and Baker 1984). The technique effectively removes dead and immotile cells along with cellular debris resulting in a sample that is highly enriched for motile spermatozoa, however, recovery rates can be low. The swim-up technique was quickly adopted for use in sperm preparation for bovine in vitro fertilization (IVF), with selected spermatozoa showing increased in vitro fertilization rates (Keefer et al. 1985; Halangk and Bohnensack 1986; Parrish et al. 1986; Parrish and Foote 1987); however, it is not easily scalable for large volume use in semen processing for artificial insemination (AI). Furthermore, there has been concern that post-selection wash steps required to remove any remaining decapacitating factors from accessory sex gland secretions of fresh semen or components of the freezing solutions of frozen-thawed semen may be harmful to spermatozoa. To address the post-selection centrifugation concerns, a modified swim-up that incorporated a hyaluronidase wash media interface between the sample and the overlying media was developed (Shamsuddin and Rodriguez-Martinez 1994); however, it is unclear how widely used it has become.

7.4.4 Density Gradient Centrifugation

Density gradient centrifugation (DGC) is an alternative sperm purification technique that is commonly used for bovine IVF and embryo production, especially when using cryopreserved semen samples. It utilizes centrifugation in combination with density gradient media to fractionate different populations of spermatozoa as well as cellular debris, and non-cellular semen and extender components. The DGC technique was not developed specifically for sperm purification, but was shown to be well suited, allowing for the recovery of motile and mostly morphologically normal cells, while limiting osmotic and mechanical stress. This may be attributed to the fact that mature spermatozoa that have good nuclear morphology are denser than their immature counterparts and therefore sediment further down within the density gradient (Le Lannou and Blanchard 1988; Oshio 1988). Furthermore, motile cells may be able to align themselves with the centrifugal force resulting in their enrichment in the denser layers (Rhemrev et al. 1989). Early studies by Beatty (1964) demonstrated that uncoated colloidal silica could effectively separate spermatozoa; however, the uncoated colloidal silica had a detrimental effect on fertilization ability. Since that time several media have been put forth for the use in DGC including Sephadex (Steeno et al. 1975), albumin (Ericsson 1977), Percoll® (Pertoft et al. 1978; Gorus and Pipeleers 1981), Ficoll (Bongso et al. 1989), polysaccharide beads (Ohashi et al. 1992) and Accudenz (Sbracia et al. 1996). To date, Percoll®, the density gradient medium composed of colloidal silica particles coated with polyvinylpyrrolidone, is the most widely used medium within bovine IVF operations, used in both single layer and discontinuous gradient separation, with the polyvinylpyrrolidone coating alleviating the toxicity of the uncoated colloidal silica seen by Beatty (1964) (Gorus and Pipeleers 1981; Töpfer-Petersen and Schill 1981; Vierula and Rajaniemi 1981; Lessley and Garner 1983). Nevertheless, some practitioners are moving to alternative density media, to stay concurrent with the halted use of Percoll® in human ART by the manufacturer (Biotech 1996), for which the exact reason for its removal from medical use was never explicitly revealed. A few of the alternative density gradients that are currently being used are BoviPure®, Isolate™, Optiprep® and *PureSperm*®. The gradient media used, the speed of centrifugation as well as the precision in layering both the gradients and the sample can all impact the effectiveness of the technique. However, when prepared carefully and done at low enough centrifugal force not to damage cells, DGC leads to the effective isolation and high recovery rate of motile spermatozoa that are largely morphologically normal, while simultaneously removing unwanted semen and extender components, as well as certain bacteria.

7.4.5 *Glass-Wool Filtration*

Glass-wool filtration (GWF) is a lesser used technique for livestock sperm purification, that dually selects cells based on their differential adherence to the glass and their ability to pass through the glass wool fibers. First developed by Paulson and Polakoski (1977) to decrease sample viscosity and improve motility of human samples, and later refined by Jeyendran and Perez-Pelaez (1987), GWF has been shown to enrich for increased viability, progressive motility, nuclear maturity, and DNA integrity compared to the raw sample when the appropriate types of glass wool are used (Grunewald et al. 2007). Whilst more commonly used in human sperm preparations, GWF has proven to be as effective as other traditional sperm preparation techniques in increasing progressive motility, DNA integrity and fertilization potential in livestock (Larson et al. 1999). Glass-wool filtration is especially effective in cases where samples have increased viscosity or suffer from oligozoospermia or asthenozoospermia, where it outperforms both DGC and swim-up techniques (Larson et al. 1999). Studies have confirmed that GWF is an effective technique for bovine semen samples, and should be considered a good sperm preparation alternative to the more traditional procedures (Arzondo et al. 2012).

7.4.6 *Nanopurification*

Procedures that function on negative selection techniques are highly desirable for sperm selection, as they limit the impact of processing on the high-quality spermatozoa while removing the detrimental cells from the population. One negative selection procedure that is gaining popularity within both human clinics and livestock production is that of nanopurification. Most nanopurification procedures implement magnetic nanoparticles that are coated with specific lectins, proteins, antibodies, or compounds. The nanoparticles are mixed within the samples to allow adherence of the particles to the spermatozoa of interest before being placed on a magnet to draw out all particles and target cells. This procedure limits harsh processing steps such as centrifugation while enriching for high motility, increased viability, and improved acrosome integrity. Nanopurification with lectin peanut agglutinin (PNA) particles has been shown to improve field fertility in cattle (Odhiambo et al. 2014), while use of PNA particles as well as Annexin V coated particles have been shown to improve motility in liquefied boar semen (Durfey et al. 2019). Nanopurification has also been used in both fresh and frozen semen samples from donkey (Yousef et al. 2020), camel (Rateb 2021), and buffalo (Bisla et al. 2020) with promising results. Nanopurification is an encouraging new technology that can adapt as new surface biomarkers emerge, allowing for increased precision in the selection process.

7.4.7 Microfluidics

Microfluidic devices are an additional technology that has great promise in sperm selection as they can better mimic the selection process that occurs in vivo, through manipulation of the microgeometry of the devices and the use of thermotaxis, rheotaxis and chemotaxis. However, with a few exceptions (Nagata et al. 2018), the use of microfluidic devices is mostly confined to clinical settings and not widely used in livestock sperm processing.

The integration of advanced reproductive technologies into agricultural practices has allowed for great advancement in the precision and efficiency of genetic and physical trait selection as well as the ability to refine the quality of semen being used for both agricultural IVF and AI. Innovations in both semen sexing and semen purification technologies have allowed for improved reproductive success; however, trade-offs exist for each technique. The pros and cons must be weighed for both large AI companies and producers to decide if the potential benefits of a technique are worth the risks and financial investment.

7.5 Conclusion

The need for greater reproductive efficiency in livestock breeding programs is a driving force for the improved understanding of the spermatozoon contributions to bovine fertilization and pregnancy success. The continuous advancements in technology allow us to garner a better understanding of sperm function that can inform the development of superior ways to harness and manipulate sperm fitness for the betterment of the agriculture industry. This chapter highlighted how technology is currently being used to augment bull fertility, and the innovative ways technology could be employed in the future to improve bull reproductive efficiency in this new and emerging age of precision agriculture.

Compliance with Ethical Standards
All applicable international, national and institutional guidelines for the care and use of animals were followed. This current work was supported by the Natural Science and Engineering Research Council of Canada (Grant number RGPIN/05305) to Richard Oko at Queen's University, along with the Agriculture and Food Research Initiative Competitive Grant number 2023–67015-39262 from the USDA National Institute of Food and Agriculture, Animal Breeding, Genetics & Genomics Program and seed funding from the College of Agriculture, Food

and Natural Resources of the University of Missouri to Peter Sutovsky. Work for this book was also supported by a 2021–2022 Lalor Foundation Postdoctoral Research Fellowship to Lauren Hamilton. The research support to Viv Perry was funded by Australian Research Council (Grant number LP110100649), English Beef and Lamb Executive a division of the Agriculture and Horticulture Development Board (Grant number RC3815), Queensland Sperm Morphology Laboratories, Ridley AgriProducts Pty Ltd and the Australian cattle industry. The Canadian Association of Animal Breeders provided the foundation for research by Albert Barth. They supported the Development of the Diagnostic Spermatology Laboratory at the Western College of Veterinary Medicine, Saskatoon, SK, along with related support for numerous other research directives. Many grants were also received from Saskatchewan Agriculture, including the Cattle Marketing Deductions Fund, the Horned Cattle Trust Fund, the Cattle Breeders Research Council, the Beef Development Fund and the Agriculture Development Fund. Animal research approval for research or clinical studies summarized in these chapters were given by respective University Animal Care Committees. The authors declare no conflict of interest as there is no perceived interest in prejudicing the impartiality of the research reported.

References

Aarabi M, Balakier H, Bashar S, Moskovtsev SI, Sutovsky P, Librach CL, Oko R (2014) Sperm-derived WW domain-binding protein, PAWP, elicits calcium oscillations and oocyte activation in humans and mice. FASEB J 28(10):4434–4440. https://doi.org/10.1096/fj.14-256495

Aitken RJ (2017) Reactive oxygen species as mediators of sperm capacitation and pathological damage. Mol Reprod Dev 84(10):1039–1052. https://doi.org/10.1002/mrd.22871

Aitken RJ, Drevet JR (2020) The importance of oxidative stress in determining the functionality of mammalian spermatozoa: a two-edged sword. Antioxidants 9(2):111

Amann RP, Hokanson JF, Almquist JO (1963) Cannulation of the bovine ductus deferens for quantitative recovery of epididymal spermatozoa. Reproduction 6(1):65–69

An L-P, Maeda T, Sakaue T, Takeuchi K, Yamane T, Du P-G, Ohkubo I, Ogita H (2012) Purification, molecular cloning and functional characterization of swine phosphatidylethanolamine-binding protein 4 from seminal plasma. Biochem Biophys Res Commun 423(4):690–696

Arzondo MM, Caballero JN, Marín-Briggiler CI, Dalvit G, Cetica PD, Vazquez-Levin MH (2012) Glass wool filtration of bull cryopreserved semen: a rapid and effective method to obtain a high percentage of functional sperm. Theriogenology 78(1):201–209

Aslam MKM, Sharma VK, Pandey S, Kumaresan A, Srinivasan A, Datta TK, Mohanty TK, Yadav S (2018) Identification of biomarker candidates for fertility in spermatozoa of crossbred bulls through comparative proteomics. Theriogenology 119:43–51

Austin CR (1952) The 'Capacitation' of the Mammalian Sperm. Nature 170:326

Bailey RW, Dunker AK, Brown CJ, Garner EC, Griswold MD (2001) Clusterin, a binding protein with a molten globule-like region. Biochemistry 40(39):11828–11840

Barlow P, Vosa CG (1970) The Y chromosome in human spermatozoa. Nature 226(5249):961–962

Beatty RA (1964) Density gradient media for mammalian spermatozoa.

Bellin ME, Oyarzo JN, Hawkins HE, Zhang H, Smith RG, Forrest DW, Sprott LR, Ax RL (1998) Fertility-associated antigen on bull sperm indicates fertility potential. J Anim Sci 76(8):2032–2039

Bieniek JM, Drabovich AP, Lo KC (2016) Seminal biomarkers for the evaluation of male infertility. Asian J Androl 18(3):426–433

Biotech P (1996) Important notice: Percoll® NOT to be used in Assisted Reproduction Technologies in Humans. Pharm Biotech Inc

Bisla A, Rautela R, Yadav V, Singh P, Kumar A, Ghosh S, Kumar A, Bag S, Kumar B, Srivastava N (2020) Nano-purification of raw semen minimises oxidative stress with improvement in post-thaw quality of buffalo spermatozoa. Andrologia 52(9):e13709

Boccia L, Di Francesco S, Neglia G, De Blasi M, Longobardi V, Campanile G, Gasparrini B (2013) Osteopontin improves sperm capacitation and in vitro fertilization efficiency in buffalo (Bubalus bubalis). Theriogenology 80(3):212–217

Bongso A, Ng S-C, Mok H, Lim MN, Teo HL, Wong PC, Ratnam S (1989) Improved sperm concentration, motility, and fertilization rates following Ficoll treatment of sperm in a human in vitro fertilization program. Fertil Steril 51(5):850–854

Brito LFC (2016) A multilaboratory study on the variability of bovine semen analysis. Theriogenology 85(2):254–266

Broekhuijse M, Šoštarić E, Feitsma H, Gadella BM (2012) Application of computer-assisted semen analysis to explain variations in pig fertility. J Anim Sci 90(3):779–789

Bungum M, Humaidan P, Spano M, Jepson K, Bungum L, Giwercman A (2004) The predictive value of sperm chromatin structure assay (SCSA) parameters for the outcome of intrauterine insemination. IVF and ICSI. Hum Reprod 19(6):1401–1408

Butler ML, Hartman AR, Bormann JM, Weaber RL, Grieger DM, Rolf MM (2022) Genome-wide association study of beef bull semen attributes. BMC Genomics 23(1):74

Cancel AM, Chapman DA, Killian GJ (1997) Osteopontin is the 55-kilodalton fertility-associated protein in Holstein bull seminal plasma. Biol Reprod 57(6):1293–1301

Card CJ (2014) The bovine spermatozoal transcriptome and sire fertility. University of Rhode Island

Carrell DT, Liu L, Peterson CM, Jones KP, Hatasaka HH, Erickson L, Campbell B (2003) Sperm DNA fragmentation is increased in couples with unexplained recurrent pregnancy loss. Arch Androl 49(1):49–55

Carter WO, Naryanan PK, Robinson JP (1994) Intracellular hydrogen peroxide and superoxide anion detection in endothelial cells. J Leucoc Biol 55(2):253–258

Chang MC (1951) Fertilizing capacity of spermatozoa deposited into the Fallopian tubes. Nature 168:697

Choi N-H, Tobe T, Hara K, Yoshida H, Tomita M (1990) Sandwich ELISA assay for quantitative measurement of SP-40, 40 in seminal plasma and serum. J Immunol Methods 131(2):159–163

Cooney MA, Malcuit C, Cheon B, Holland MK, Fissore RA, D'Cruz NT (2010) Species-Specific Differences in the Activity and Nuclear Localization of Murine and Bovine Phospholipase C Zeta. Biol Reprod 83(1):92–101

Cross NL, Morales P, Overstreet JW, Hanson FW (1986) Two simple methods for detecting acrosome-reacted human sperm. Gamete Res 15(3):213–226

D'Amours O, Bordeleau L-J, Frenette G, Blondin P, Leclerc P, Sullivan R (2012) Binder of sperm 1 and epididymal sperm binding protein 1 are associated with different bull sperm subpopulations. Reproduction 143(6):759

D'Amours O, Frenette G, Fortier M, Leclerc P, Sullivan R (2010) Proteomic comparison of detergent-extracted sperm proteins from bulls with different fertility indexes. Reproduction 139(3):545–556

Dai L, Zhao Z, Zhao R, Xiao S, Jiang H, Yue X, Li X, Gao Y, Liu J, Zhang J (2009) Effects of novel single nucleotide polymorphisms of the FSH beta-subunit gene on semen quality and fertility in bulls. Anim Reprod Sci 114(1–3):14–22

Dalimata AM, Graham JK (1997) Cryopreservation of rabbit spermatozoa using acetamide in combination with trehalose and methyl cellulose. Theriogenology 48(5):831–841
David G, Serres C, Jouannet P (1981) Kinematics of human spermatozoa. Gamete Res 4(2):83–95
de Agostini Losano JD, Souders CL, Martyniuk CJ, Daigneault BW (2023) Characterization of bioenergetic and kinematic plasticity of bovine sperm reveals novel and dynamic temporal traits. Reproduction 166(2):135–147
Diaz-Miranda EA, Hamilton LE, Zigo M, Fallon L, Ortega MS, Assumpção MEOD, Guimarães JD, Sutovsky P (2024) Regional abundances of binder of sperm (BSP) proteins are negatively associated with the quality of frozen-thawed bovine spermatozoa. Reproduction 167(1)
Dini P, Troch L, Lemahieu I, Deblende P, Daels P (2019) Validation of a portable device (iSperm®) for the assessment of stallion sperm motility and concentration. Reprod Domest Anim 54(8):1113–1120
Dobrinski I, Hughes HPA, Barth AD (1994) Flow cytometric and microscopic evaluation and effect on fertility of abnormal chromatin condensation in bovine sperm nuclei. Reproduction 101(3):531–538
Dogan S, Vargovic P, Oliveira R, Belser LE, Kaya A, Moura A, Sutovsky P, Parrish J, Topper E, Memili E (2015) Sperm protamine-status correlates to the fertility of breeding bulls. Biol Reprod 92(4):91–92
Domain G, Banchi P, Ali Hassan H, Eilers A, Lannoo J, Wydooghe E, Niżański W, Van Soom A (2022) Sperm gone smart: a portable device (iSperm®) to assess semen concentration and motility in dogs. Animals 12(5):652
Donnellan EM, Perrier J-P, Keogh K, Štiavnická M, Collins CM, Dunleavy EM, Sellem E, Bernecic NC, Lonergan P, Kenny DA (2022) Identification of differentially expressed mRNAs and miRNAs in spermatozoa of bulls of varying fertility. Front Vet Sci 9:993561
Dordas-Perpinyà M, Sergeant N, Ruelle I, Bruyas J-F, Charreaux F, Michaud S, Carracedo S, Catalán J, Miró J, Delehedde M (2022) ProAKAP4 semen concentrations as a valuable marker protein of post-thawed semen quality and bull fertility: a retrospective study. Vet Sci 9(5):224
Dunn WB, Broadhurst DI, Atherton HJ, Goodacre R, Griffin JL (2011) Systems level studies of mammalian metabolomes: the roles of mass spectrometry and nuclear magnetic resonance spectroscopy. Chem Soc Rev 40(1):387–426
Durfey CL, Swistek SE, Liao SF, Crenshaw MA, Clemente HJ, Thirumalai RVKG, Steadman CS, Ryan PL, Willard ST, Feugang JM (2019) Nanotechnology-based approach for safer enrichment of semen with best spermatozoa. J Anim Sci Biotechnol 10:1–12
Ericsson RJ (1977) Isolation and storage of progressively motile human sperm. Andrologia 9(1):111–114
Ericsson RJ, Langevin CN, Nishino M (1973) Isolation of fractions rich in human Y sperm. Nature 246(5433):421–424
Erikson DW, Way AL, Chapman DA, Killian GJ (2007) Detection of osteopontin on Holstein bull spermatozoa, in cauda epididymal fluid and testis homogenates, and its potential role in bovine fertilization. Reproduction 133(5):909–917
Evenson DP (1999) Loss of livestock breeding efficiency due to uncompensable sperm nuclear defects. Reprod Fertil Dev 11(1):1–16
Evenson DP (2016) The Sperm Chromatin Structure Assay (SCSA®) and other sperm DNA fragmentation tests for evaluation of sperm nuclear DNA integrity as related to fertility. Anim Reprod Sci 169:56–75
Evenson DP, Darzynkiewicz Z, Melamed MR (1980a) Relation of mammalian sperm chromatin heterogeneity to fertility. Science (80-) 210(4474):1131–1133
Evenson DP, Darzynkiewicz Z, Melamed MR (1980b) Comparison of human and mouse sperm chromatin structure by flow cytometry. Chromosoma 78(2):225–238
Evenson DP, Jost LK, Marshall D, Zinaman MJ, Clegg E, Purvis K, De Angelis P, Claussen OP (1999) Utility of the sperm chromatin structure assay as a diagnostic and prognostic tool in the human fertility clinic. Hum Reprod 14(4):1039–1049

Fallon L, Diaz-Miranda E, Hamilton L, Sutovsky P, Zigo M, Spencer TE, Ortega MS (2023) The development of new biomarkers of spermatozoa quality in cattle. Front Vet Sci 10

Farrell PB, Presicce GA, Brockett CC, Foote RH (1998) Quantification of bull sperm characteristics measured by computer-assisted sperm analysis (CASA) and the relationship to fertility. Theriogenology 49(4):871–879

Fernández JL, Muriel L, Rivero MT, Goyanes V, Vazquez R, Alvarez JG (2003) The sperm chromatin dispersion test: a simple method for the determination of sperm DNA fragmentation. J Androl 24(1):59–66

Feugang JM, Rodriguez-Osorio N, Kaya A, Wang H, Page G, Ostermeier GC, Topper EK, Memili E (2010) Transcriptome analysis of bull spermatozoa: implications for male fertility. Reprod Biomed Online 21(3):312–324

Fitzpatrick LA, Fordyce G, McGowan MR, Bertram JD, Doogan VJ, De Faveri J, Miller RG, Holroyd RG (2002) Bull selection and use in northern Australia: Part 2. Semen Traits. Anim Reprod Sci 71(1–2):39–49

Garner DL, Gledhill BL, Pinkel D, Lake S, Stephenson D, Van Dilla MA, Johnson LA (1983) Quantification of the X-and Y-chromosome-bearing spermatozoa of domestic animals by flow cytometry. Biol Reprod 28(2):312–321

Garner DL, Johnson LA (1995) Viability assessment of mammalian sperm using SYBR-14 and propidium iodide. Biol Reprod 53(2):276–284

Garner DL, Thomas CA (1999) Organelle-specific probe JC-1 identifies membrane potential differences in the mitochondrial function of bovine sperm. Mol Reprod Dev Inc Gamete Res 53(2):222–229

George TC, Basiji DA, Hall BE, Lynch DH, Ortyn WE, Perry DJ, Seo MJ, Zimmerman CA, Morrissey PJ (2004) Distinguishing modes of cell death using the ImageStream® multispectral imaging flow cytometer. Cytom Part A J Int Soc Anal Cytol 59(2):237–245

Gitlits VM, Toh BH, Loveland KL, Sentry JW (2000) The glycolytic enzyme enolase is present in sperm tail and displays nucleotide-dependent association with microtubules. Eur J Cell Biol 79(2):104–111

Gledhill BL, Pinkel D, Garner DL (1982) Identifying X-and Y-chromosome-bearing sperm by DNA content: retrospective perspectives and prospective opinions

Go KJ, Wolf DP (1985) Albumin-mediated changes in sperm sterol content during capacitation. Biol Reprod 32(1):145–153. https://doi.org/10.1095/biolreprod32.1.145

Gonçalves RF, Chapman DA, Bertolla RP, Eder I, Killian GJ (2008) Pre-treatment of cattle semen or oocytes with purified milk osteopontin affects in vitro fertilization and embryo development. Anim Reprod Sci 108(3–4):375–383

Gorczyca W, Gong J, Darzynkiewicz Z (1993) Detection of DNA strand breaks in individual apoptotic cells by the in situ terminal deoxynucleotidyl transferase and nick translation assays. Cancer Res 53(8):1945–1951

Gorus FK, Pipeleers DG (1981) A rapid method for the fractionation of human spermatozoa according to their progressive motility. Fertil Steril 35(6):662–665

Graham JK (2001) Assessment of sperm quality: a flow cytometric approach. Anim Reprod Sci 68(3–4):239–247

Gray J (1958) The Movement of the Spermatozoa of the Bull. J Exp Biol 35(1):96–108. https://doi.org/10.1242/jeb.35.1.96

Grunewald S, Miska W, Miska G, Rasch M, Reinhardt M, Glander H-J, Paasch U (2007) Molecular glass wool filtration as a new tool for sperm preparation. Hum Reprod 22(5):1405–1412

Gwathmey TM, Ignotz GG, Mueller JL, Manjunath P, Suarez SS (2006) Bovine Seminal Plasma Proteins PDC-109, BSP-A3, and BSP-30-kDa Share Functional Roles in Storing Sperm in the Oviduct1. Biol Reprod 75(4):501–507. https://doi.org/10.1095/biolreprod.106.053306

Halangk W, Bohnensack R (1986) Quantification of sperm motility by a turbidimetric assay. Correlation to cellular respiration. Biomed Biochim Acta 45(3):331–341

Hamilton CK, Favetta LA, Di Meo GP, Floriot S, Perucatti A, Peippo J, Kantanen J, Eggen A, Iannuzzi L, King WA (2009) Copy number variation of testis-specific protein, Y-encoded (TSPY) in 14 different breeds of cattle (Bos taurus). Sex Dev 3(4):205–213

Han Y, Peñagaricano F (2016) Unravelling the genomic architecture of bull fertility in Holstein cattle. BMC Genet 17:1–11

Handrow RR, First NL, Parrish JJ (1989) Calcium requirement and increased association with bovine sperm during capacitation by heparin. J Exp Zool 252(2):174–182

Henault MA, Killian GJ, Kavanaugh JF, Griel LC Jr (1995) Effect of accessory sex gland fluid from bulls of differing fertilities on the ability of cauda epididymal sperm to penetrate zona-free bovine oocytes. Biol Reprod 52(2):390–397

Henkel R, Hajimohammad M, Stalf T, Hoogendijk C, Mehnert C, Menkveld R, Gips H, Schill W-B, Kruger TF (2004) Influence of deoxyribonucleic acid damage on fertilization and pregnancy. Fertil Steril 81(4):965–972

Henkel R, Hoogendijk CF, Bouic PJD, Kruger TF (2010) TUNEL assay and SCSA determine different aspects of sperm DNA damage. Andrologia 42(5):305–313

Ho H-C, Suarez SS (2001) An inositol 1, 4, 5-trisphosphate receptor-gated intracellular Ca2+ store is involved in regulating sperm hyperactivated motility. Biol Reprod 65(5):1606–1615

Høst E, Lindenberg S, Smidt-Jensen S (2000) The role of DNA strand breaks in human spermatozoa used for IVF and ICSI. Acta Obstet Gynecol Scand 79(7):559–563

Hughes CM, Lewis SEM, McKelvey-Martin VJ, Thompson W (1996) A comparison of baseline and induced DNA damage in human spermatozoa from fertile and infertile men, using a modified comet assay. MHR Basic Sci Reprod Med 2(8):613–619

Ibrahim NM, Gilbert GR, Loseth KJ, Crabo BOG (2000) Correlation between clusterin-positive spermatozoa determined by flow cytometry in bull semen and fertility. J Androl 21(6):887–894

Idle JR, Gonzalez FJ (2007) Metabolomics. Cell Metab 6(5):348–351

Indriastuti R, Pardede BP, Gunawan A, Ulum MF, Arifiantini RI, Purwantara B (2022) Sperm transcriptome analysis accurately reveals male fertility potential in livestock. Animals 12(21):2955

Iqbal S, Riaz A, Andrabi SMH, Shahzad Q, Durrani AZ, Ahmad N (2016) L-Cysteine improves antioxidant enzyme activity, post-thaw quality and fertility of Nili-Ravi buffalo (Bubalus bubalis) bull spermatozoa. Andrologia 48(9):943–949

Jain A, Yathish HM, Jain T, Sharma A (2011) Efficient production of sexed semen by Flow Cytometry: a Review. Agric Rev 32(1):36–45

Jeyendran RS, Van der Ven HH, Perez-Pelaez M, Al-Hasani S, Diedrich K (1987) Separation of viable spermatozoa by standardized glass wool column. UK Parthenon Publ, Vitr Fertil Carnforth, p 49

Johnson LA, Flook JP, Hawk HW (1989) Sex preselection in rabbits: live births from X and Y sperm separated by DNA and cell sorting. Biol Reprod 41(2):199–203

Johnson LA, Flook JP, Look MV (1987a) Flow cytometry of X and Y chromosome-bearing sperm for DNA using an improved preparation method and staining with Hoechst 33342. Gamete Res 17(3):203–212

Johnson LA, Flook JP, Look MV, Pinkel D (1987b) Flow sorting of X and Y chromosome-bearing spermatozoa into two populations. Gamete Res 16(1):1–9

Kapron JT, Hilliard GM, Lakins JN, Tenniswood MPR, West KA, Carr SA, Crabb JW (1997) Identification and characterization of glycosylation sites in human serum clusterin. Protein Sci 6(10):2120–2133

Kasimanickam RK, Kasimanickam VR, Arangasamy A, Kastelic JP (2019) Sperm and seminal plasma proteomics of high-versus low-fertility Holstein bulls. Theriogenology 126:41–48

Katz DF, Bloom TD, Bondurant RH (1981) Movement of bull spermatozoa in cervical mucus. Biol Reprod 25(5):931–937

Katz DF, Dott HM (1975) Methods of measuring swimming speed of spermatozoa. Reproduction 45(2):263–272

Kaya A, Dogan S, Vargovic P, Kutchy NA, Ross P, Topper E, Oko R, van der Hoorn F, Sutovsky P, Memili E (2022) Sperm proteins ODF2 and PAWP as markers of fertility in breeding bulls. Cell Tissue Res 387(1):159–171

Keefer CL, Brackett BG, Troop CG (1985) Bovine fertilization after in vitro insemination with motile sperm fractions. Theriogenology 23(1):198

Keller A, Kerns K (2022) Deep learning, artificial intelligence methods to predict boar sperm acrosome health. Anim Reprod Sci 247:107110

Kennedy CE, Krieger KB, Sutovsky M, Xu W, Vargovič P, Didion BA, Ellersieck MR, Hennessy ME, Verstegen J, Oko R, Sutovsky P (2014) Protein expression pattern of PAWP in bull spermatozoa is associated with sperm quality and fertility following artificial insemination. Mol Reprod Dev 81(5):436–449. https://doi.org/10.1002/mrd.22309

Kerns K, Sharif M, Zigo M, Xu W, Hamilton LE, Sutovsky M, Ellersieck M, Drobnis EZ, Bovin N, Oko R (2020) Sperm cohort-specific zinc signature acquisition and capacitation-induced zinc flux regulate sperm-oviduct and sperm-zona pellucida interactions. Int J Mol Sci 21(6):2121

Kerns K, Zigo M, Drobnis EZ, Sutovsky M, Sutovsky P (2018a) Zinc ion flux during mammalian sperm capacitation. Nat Commun 9(1):2061

Kerns K, Zigo M, Sutovsky P (2018b) Zinc: a necessary ion for mammalian sperm fertilization competency. Int J Mol Sci 19(12):4097

Killian G (2018) Fertility-associated proteins in male and female reproductive fluids of cattle. Anim Reprod 9(4):703–712

Killian GJ, Chapman DA, Rogowski LA (1993) Fertility-associated proteins in Holstein bull seminal plasma. Biol Reprod 49(6):1202–1207

Kovac JR, Pastuszak AW, Lamb DJ (2013) The use of genomics, proteomics, and metabolomics in identifying biomarkers of male infertility. Fertil Steril 99(4):998–1007

Kuhn MT, Hutchison JL (2008) Prediction of dairy bull fertility from field data: use of multiple services and identification and utilization of factors affecting bull fertility. J Dairy Sci 91(6):2481–2492

Kumar P, Kumar D, Singh I, Yadav PS (2012) Seminal plasma proteome: promising biomarkers for bull fertility. Agric Res 1:78–86

Kumar V, Rangaraj N, Shivaji S (2006) Activity of pyruvate dehydrogenase A (PDHA) in hamster spermatozoa correlates positively with hyperactivation and is associated with sperm capacitation. Biol Reprod 75(5):767–777

Larson-Cook KL, Brannian JD, Hansen KA, Kasperson KM, Aamold ET, Evenson DP (2003) Relationship between the outcomes of assisted reproductive techniques and sperm DNA fragmentation as measured by the sperm chromatin structure assay. Fertil Steril 80(4):895–902

Larson KL, Brannian JD, Timm BK, Jost LK, Evenson DP (1999) Density gradient centrifugation and glass wool filtration of semen remove spermatozoa with damaged chromatin structure. Hum Reprod 14(8):2015–2019

Le Lannou D, Blanchard Y (1988) Nuclear maturity and morphology of human spermatozoa selected by Percoll density gradient centrifugation or swim-up procedure. Reproduction 84(2):551–556

LeBel CP, Ischiropoulos H, Bondy SC (1992) Evaluation of the probe 2', 7'-dichlorofluorescin as an indicator of reactive oxygen species formation and oxidative stress. Chem Res Toxicol 5(2):227–231

Lessard C, Siqueira LG, D'Amours O, Sullivan R, Leclerc P, Palmer C (2011) Infertility in a beef bull due to a failure in the capacitation process. Theriogenology 76(5):891–899

Lessley BA, Garner DL (1983) Isolation of motile spermatozoa by density gradient centrifugation in Percoll®. Gamete Res 7(1):49–61

Lewis SEM, Simon L (2010) Clinical implications of sperm DNA damage. Hum Fertil 13(4):201–207

Liu GE, Bickhart DM (2012) Copy number variation in the cattle genome. Funct Integr Genomics 12:609–624

Lockhart KN, Drum JN, Balboula AZ, Spinka CM, Spencer TE, Ortega MS (2023) Sire modulates developmental kinetics and transcriptome of the bovine embryo. Reproduction 166(5):337–348

Mahadevan M, Baker G (1984) Assessment and preparation of semen for in vitro fertilization. In: Clinical in vitro fertilization. Springer, pp 83–97

Manjunath P, Chandonnet L, Leblond E, Desnoyers L (1994) Major proteins of bovine seminal vesicles bind to spermatozoa. Biol Reprod 50(1):27–37

Medeiros CMO, Forell F, Oliveira ATD, Rodrigues JL (2002) Current status of sperm cryopreservation: why isn't it better? Theriogenology 57(1):327–344. https://doi.org/10.1016/S0093-691X(01)00674-4

Menezes EB, Velho ALC, Santos F, Dinh T, Kaya A, Topper E, Moura AA, Memili E (2019) Uncovering sperm metabolome to discover biomarkers for bull fertility. BMC Genomics 20:1–16

Merlotti A, Dantas E, Remes Lenicov F, Ceballos A, Jancic C, Varese A, Rubione J, Stover S, Geffner J, Sabatté J (2015) Fucosylated clusterin in semen promotes the uptake of stress-damaged proteins by dendritic cells via DC-SIGN. Hum Reprod 30(7):1545–1556

Miki K, Willis WD, Brown PR, Goulding EH, Fulcher KD, Eddy EM (2002) Targeted disruption of the Akap4 gene causes defects in sperm flagellum and motility. Dev Biol 248(2):331–342

Mitra K, Rangaraj N, Shivaji S (2005) Novelty of the Pyruvate Metabolic Enzyme Dihydrolipoamide Dehydrogenasein Spermatozoa: CORRELATION OF ITS LOCALIZATION, TYROSINE PHOSPHORYLATION, AND ACTIVITY DURING SPERMCAPACITATION. J Biol Chem 280(27):25743–25753

Mitra K, Shivaji S (2004) Novel tyrosine-phosphorylated post-pyruvate metabolic enzyme, dihydrolipoamide dehydrogenase, involved in capacitation of hamster spermatozoa. Biol Reprod 70(4):887–899

Monaco E, Gasparrini B, Boccia L, De Rosa A, Attanasio L, Zicarelli L, Killian G (2009) Effect of osteopontin (OPN) on in vitro embryo development in cattle. Theriogenology 71(3):450–457

Moraes CR, Runcan EE, Blawut B, Coutinho da Silva MA (2019) The use of iSperm technology for on-farm measurement of equine sperm motility and concentration. Transl Anim Sci 3(4):1513–1520

Mortimer D (1990) Objective analysis of sperm motility and kinematics. Handb Lab diagnosis Treat Infertil :97–133

Mortimer D, Curtis EF, Miller RG (1987) Specific labelling by peanut agglutinin of the outer acrosomal membrane of the human spermatozoon. Reproduction 81(1):127–135

Mortimer ST, Van der Horst G, Mortimer D (2015) The future of computer-aided sperm analysis. Asian J Androl 17(4):545–553

Moura AA, Chapman DA, Koc H, Killian GJ (2006) Proteins of the cauda epididymal fluid associated with fertility of mature dairy bulls. J Androl 27(4):534–541

Mukherjee A, Dass G, Gohain M, Brahma B, Datta TK, De S (2013) Absolute copy number differences of Y chromosomal genes between crossbred (Bos taurus× Bos indicus) and Indicine bulls. J Anim Sci Biotechnol 4:1–7

Nagata MPB, Endo K, Ogata K, Yamanaka K, Egashira J, Katafuchi N, Yamanouchi T, Matsuda H, Goto Y, Sakatani M (2018) Live births from artificial insemination of microfluidic-sorted bovine spermatozoa characterized by trajectories correlated with fertility. Proc Natl Acad Sci 115(14):E3087–E3096

Nagy S, Jansen J, Topper EK, Gadella BM (2003) A triple-stain flow cytometric method to assess plasma-and acrosome-membrane integrity of cryopreserved bovine sperm immediately after thawing in presence of egg-yolk particles. Biol Reprod 68(5):1828–1835

Nani JP, Peñagaricano F (2020) Whole-genome homozygosity mapping reveals candidate regions affecting bull fertility in US Holstein cattle. BMC Genomics 21:1–9

Narud B, Klinkenberg G, Khezri A, Zeremichael TT, Stenseth E-B, Nordborg A, Haukaas TH, Morrell JM, Heringstad B, Myromslien FD (2020) Differences in sperm functionality and intracellular metabolites in Norwegian Red bulls of contrasting fertility. Theriogenology 157:24–32

Nogueira E, Tirpák F, Hamilton LE, Zigo M, Kerns K, Sutovsky M, Kim J, Volkmann D, Jovine L, Taylor JF, Schnabel RD, Sutovsky P (2022) A Non-Synonymous Point Mutation in a WD-40 Domain Repeat of EML5 Leads to Decreased Bovine Sperm Quality and Fertility. Front cell Dev Biol 10

Odhiambo JF, DeJarnette JM, Geary TW, Kennedy CE, Suarez SS, Sutovsky M, Sutovsky P (2014) Increased conception rates in beef cattle inseminated with nanopurified bull semen. Biol Reprod 91(4):91–97

Ohashi K, Saji F, Wakimoto A, Kato M, Tsutsui T, Tanizawa O (1992) Preparation of oligozoospermic and/or asthenozoospermic semen for intrauterine insemination using the SpermPrep semen filtration column. Fertil Steril 57(4):866–870

Oshio S (1988) Apparent densities of spermatozoa of various mammalian species. Gamete Res 20(2):159–164

Otto FJ, Hacker U, Zante J, Schumann J, Göhde W, Meistrich ML (1979) Flow cytometry of human spermatozoa. Histochemistry 61:249–254

Özbek M, Hitit M, Kaya A, Jousan FD, Memili E (2021) Sperm functional genome associated with bull fertility. Front Vet Sci 8:610888

Pacheco HA, Rossoni A, Cecchinato A, Peñagaricano F (2023) Identification of runs of homozygosity associated with male fertility in Italian Brown Swiss cattle. Front Genet 14:1227310

Park Y-J, Pang W-K, Pang M-G (2023) Integration of omics studies indicates that species-dependent molecular mechanisms govern male fertility. J Anim Sci Biotechnol 14(1):28

Parrish JJ, Foote RH (1987) Quantification of bovine sperm separation by a Swim-up method relationship to sperm motility, integrity of acrosomes, sperm migration in polyacrylamide gel and fertility. J Androl 8(4):259–266

Parrish JJ, Susko-Parrish JL, Leibfried-Rutledge ML, Critser ES, Eyestone WH, First NL (1986) Bovine in vitro fertilization with frozen-thawed semen. Theriogenology 25(4):591–600

Paulson JD, Polakoski KL (1977) A glass wool column procedure for removing extraneous material from the human ejaculate. Fertil Steril 28(2):178–181

Perry VEA, Chenoweth PJ, Post TB, Munro RK (1990) Fertility indices for beef bulls. Aust Vet J 67(1):13–16

Pertoft H, Laurent TC, Låås T, Kågedal L (1978) Density gradients prepared from colloidal silica particles coated by polyvinylpyrrolidone (Percoll). Anal Biochem 88(1):271–282

Pinkel D, Lake S, Gledhill BL, Van Dilla MA, Stephenson D, Watchmaker G (1982) High resolution DNA content measurements of mammalian sperm. Cytom J Int Soc Anal Cytol 3(1):1–9

Poon S, Easterbrook-Smith SB, Rybchyn MS, Carver JA, Wilson MR (2000) Clusterin is an ATP– independent chaperone with very broad substrate specificity that stabilizes stressed proteins in a folding-competent state. Biochemistry 39(51):15953–15960

Prasad CS, Rangasamy S, Satheshkumar S (2010) Sex preselection in domestic animals-current status and future prospects.

Puglisi R, Gaspa G, Balduzzi D, Severgnini A, Vanni R, Macciotta NPP, Galli A (2016) Genomewide analysis of bull sperm quality and fertility traits. Reprod Domest Anim 51(5):840–843

Rabaglino MB, Le Danvic C, Schibler L, Kupisiewicz K, Perrier JP, O'Meara CM, Kenny DA, Fair S, Lonergan P (2022) Identification of sperm proteins as biomarkers of field fertility in Holstein-Friesian bulls used for artificial insemination. J Dairy Sci 105(12):10033–10046

Rateb SA (2021) Purification of cryopreserved camel spermatozoa following protease-based semen liquefaction by lectin-functionalized DNA-defrag magnetic nanoparticles. Reprod Domest Anim 56(1):183–192

Rees P, Summers HD, Filby A, Carpenter AE, Doan M (2022) Imaging flow cytometry. Nat Rev Methods Prim 2(1):86

Rego JPA, Moura AA, Nouwens AS, McGowan MR, Boe-Hansen GB (2015) Seminal plasma protein profiles of ejaculates obtained by internal artificial vagina and electroejaculation in Brahman bulls. Anim Reprod Sci 160:126–137

Rhemrev J, Jeyendran RS, Vermeiden JPW, Zaneveld LJD (1989) Human sperm selection by glass wool filtration and two-layer, discontinuous Percoll gradient centrifugation. Fertil Steril 51(4):685–690

Rikmenspoel R, Van Herpen G (1957) Photoelectric and cinematographic measurements of the motility of bull sperm cells. Phys Med Biol 2(1):54

Roan NR, Sandi-Monroy N, Kohgadai N, Usmani SM, Hamil KG, Neidleman J, Montano M, Ständker L, Röcker A, Cavrois M (2017) Semen amyloids participate in spermatozoa selection and clearance. Elife 6:e24888

Roncoletta M, da Morani E, SC, Esper CR, Barnabe VH, Franceschini PH, (2006) Fertility-associated proteins in Nelore bull sperm membranes. Anim Reprod Sci 91(1–2):77–87

Rubinstein S, Lax Y, Shalev Y, Breitbart H (1995) Dual effect of spermine on acrosomal exocytosis in capacitated bovine spermatozoa. Biochim Biophys Acta (BBA)-Molecular Cell Res 1266(2):196–200

Saad MH, Burka JF (1983) Isolation of leukotriene C4 from human seminal fluid. Prostaglandins 26(6):943–954

Sang L, Du Q-Z, Yang W-C, Tang K-Q, Yu J-N, Hua G, Zhang X-X, Yang L-G (2011) Polymorphisms in follicle stimulation hormone receptor, inhibin alpha, inhibin bata A, and prolactin genes, and their association with sperm quality in Chinese Holstein bulls. Anim Reprod Sci 126(3–4):151–156

Saraf KK, Kumaresan A, Dasgupta M, Karthikkeyan G, Prasad TSK, Modi PK, Ramesha K, Jeyakumar S, Manimaran A (2020) Metabolomic fingerprinting of bull spermatozoa for identification of fertility signature metabolites. Mol Reprod Dev 87(6):692–703

Sbracia M, Sayme N, Grasso J, Vigue L, Huszar G (1996) Sperm function and choice of preparation media: comparison of Percoll and Accudenz discontinuous density gradients. J Androl 17(1):61–67

Seidel GE (2014) Update on sexed semen technology in cattle. Animal 8(s1):160–164

Selvaraju S, Ramya L, Swathi D, Parthipan S, Archana SS, Binsila BK, Krishnappa B, Arangasamy A (2022) Sperm Transcriptome Sequencing for Predicting Bull Fertility: concepts, Facts and Future Directions. In: Frontier Technologies in Bovine Reproduction. Springer, pp 133–146

Sergeant N, Briand-Amirat L, Bencharif D, Delehedde M (2019) The sperm specific protein proAKAP4 as an innovative marker to evaluate sperm quality and fertility. J Dairy Vet Sci 11:555803

Shamsuddin M, Rodriguez-Martinez H (1994) A simple, non-traumatic swim-up method for the selection of spermatozoa for in vitro fertilization in the bovine. Anim Reprod Sci 36(1–2):61–75

Singh NP, McCoy MT, Tice RR, Schneider EL (1988) A simple technique for quantitation of low levels of DNA damage in individual cells. Exp Cell Res 175(1):184–191

Singh R, Junghare V, Hazra S, Singh U, Sengar GS, Raja TV, Kumar S, Tyagi S, Das AK, Kumar A (2019) Database on spermatozoa transcriptogram of catagorised Frieswal crossbred (Holstein Friesian X Sahiwal) bulls. Theriogenology 129:130–145

Somashekar L, Selvaraju S, Parthipan S, Patil SK, Binsila BK, Venkataswamy MM, Karthik Bhat S, Ravindra JP (2017) Comparative sperm protein profiling in bulls differing in fertility and identification of phosphatidylethanolamine-binding protein 4, a potential fertility marker. Andrology 5(5):1032–1051

Sosnik J, Miranda PV, Spiridonov NA, Yoon S-Y, Fissore RA, Johnson GR, Visconti PE (2009) Tssk6 is required for Izumo relocalization and gamete fusion in the mouse. J Cell Sci 122(15):2741–2749

Srere PA, Knull HR (1998) Location—location—location. Trends Biochem Sci 23(9):319–320

Steeno O, Adimoelja A, Steeno J (1975) Separation of X-and Y-bearing human spermatozoa with the sephadex gel-filtration method. Andrologia 7(2):95–97

Sutovsky P, Aarabi M, Miranda-Vizuete A, Oko R (2015) Negative biomarker-based male fertility evaluation: sperm phenotypes associated with molecular-level anomalies. Asian J Androl 17(4):554

Sutovsky P, Hamilton LE, Zigo M, Ortiz D'Avila Assumpção ME, Jones A, Tirpak F, Agca Y, Kerns K, Sutovsky M (2024) Biomarker-based human and animal sperm phenotyping: the good, the bad and the ugly. Biol Reprod :ioae061

Sutovsky P, Moreno R, Ramalho-Santos J, Dominko T, Thompson WE, Schatten G (2001a) A putative, ubiquitin-dependent mechanism for the recognition and elimination of defective spermatozoa in the mammalian epididymis. J Cell Sci 114(9):1665–1675

Sutovsky P, Neuber E, Schatten G (2002) Ubiquitin-dependent sperm quality control mechanism recognizes spermatozoa with DNA defects as revealed by dual ubiquitin-TUNEL assay. Mol Reprod Dev 61(3):406–413

Sutovsky P, Terada Y, Schatten G (2001b) Ubiquitin-based sperm assay for the diagnosis of male factor infertility. Hum Reprod 16(2):250–258

Swann K, Larman MG, Saunders CM, Lai FA (2004) The cytosolic sperm factor that triggers Ca2+ oscillations and egg activation in mammals is a novel phospholipase C: PLCζ. Reproduction 127(4):431–439

Taylor JF, Schnabel RD, Sutovsky P (2018a) Genomics of Bull Fertility. Animal 12(s1):s172–s183

Taylor JF, Schnabel RD, Sutovsky P (2018b) Identification of genomic variants causing sperm abnormalities and reduced male fertility. Anim Reprod Sci 194:57–62

Thérien I, Moreau R, Manjunath P (1999) Bovine seminal plasma phospholipid-binding proteins stimulate phospholipid efflux from epididymal sperm. Biol Reprod 61(3):590–598

Töpfer-Petersen E, Schill WB (1981) A new separation method of subcellular fractions of boar spermatozoa. Andrologia 13(2):174–176

Tung C, Lin C, Harvey B, Fiore AG, Ardon F, Wu M, Suarez SS (2017) Fluid viscoelasticity promotes collective swimming of sperm. Sci Rep 7(1):3152

Vierula M, Rajaniemi H (1981) Changes in surface protein structure of bull spermatozoa during epididymal maturation. Int J Androl 4(1–6):314–320

Voglmayr JK, Kavanaugh JF, Griel LC, Amann RP (1972) A modified technique for cannulating the rete testis of the bull. Reproduction 31(2):291–294

Waberski D, Suarez SS, Henning H (2022) Assessment of sperm motility in livestock: perspectives based on sperm swimming conditions in vivo. Anim Reprod Sci 246:106849

Wilson MR, Easterbrook-Smith SB (2000) Clusterin is a secreted mammalian chaperone. Trends Biochem Sci 25(3):95–98

Wyatt A, Yerbury J, Poon S, Dabbs R, Wilson M (2009a) The chaperone action of Clusterin and its putative role in quality control of extracellular protein folding. Adv Cancer Res 104:89–114

Wyatt AR, Yerbury JJ, Ecroyd H, Wilson MR (2013) Extracellular chaperones and proteostasis. Annu Rev Biochem 82:295–322

Wyatt AR, Yerbury JJ, Wilson MR (2009b) Structural characterization of clusterin-chaperone client protein complexes. J Biol Chem 284(33):21920–21927

Yanagimachi R (1994) Fertility of mammalian spermatozoa: its development and relativity. Zygote 2(4):371–372

Yang W-C, Li S-J, Xie Y-H, Tang K-Q, Hua G-H, Zhang C-Y, Yang L-G (2011) Two novel SNPs of the type I gonadotropin releasing hormone receptor gene and their associations with superovulation traits in Chinese Holstein cows. Livest Sci 136(2–3):164–168

Yousef MS, López-Lorente AI, Diaz-Jimenez M, Consuegra C, Dorado J, Pereira B, Ortiz I, Cardenas S, Hidalgo M (2020) Nano-depletion of acrosome-damaged donkey sperm by using lectin peanut agglutinin (PNA)-magnetic nanoparticles. Theriogenology 151:103–111

Yue XP, Chang TC, DeJarnette JM, Marshall CE, Lei CZ, Liu W-S (2013) Copy number variation of PRAMEY across breeds and its association with male fertility in Holstein sires. J Dairy Sci 96(12):8024–8034

Zini A, Meriano J, Kader K, Jarvi K, Laskin CA, Cadesky K (2005) Potential adverse effect of sperm DNA damage on embryo quality after ICSI. Hum Reprod 20(12):3476–3480

Zini A, San Gabriel M, Zhang X (2007) The histone to protamine ratio in human spermatozoa: comparative study of whole and processed semen. Fertil Steril 87(1):217–219

Chapter 8
Prenatal and Preweaning Environmental Effects upon Pubertal Development and Sperm Production

Contents

Abstract This chapter examines the role of early life perturbations upon the developing bull calf. Surprisingly, we observe effects upon sexual development and sperm quality, including sperm morphology, as early as the periconception period (−60 to 23 days post conception). Similarly, during postnatal life (prior to 6mths of age) dietary perturbations retard sexual development and sperm quality. Herein we discuss the ontology of development and examine why these periods of development are fundamental to fertility in the bull. Ultimately, this leads to recommendations for changes to current husbandry protocols.

Keywords Fetal programming · Testis development · Early calfhood · Puberty in the bull

8.1 Prenatal Development: Embryonic and Fetal Programming in the Male Calf

Feto placental growth is responsive to available nutrients from the earliest stage of life during periconception (Copping et al. 2020a). Reduced periconception nutrition affects fetal organ development and postnatal development including the reproductive development of male progeny (Copping et al. 2018) (Fig. 8.1).

© The Author(s), under exclusive license to Springer Nature Switzerland AG 2025

A. Barth et al., *Abnormal Morphology of Bovine Spermatozoa*, Advances in Anatomy, Embryology and Cell Biology 240, https://doi.org/10.1007/978-3-031-70126-9_8

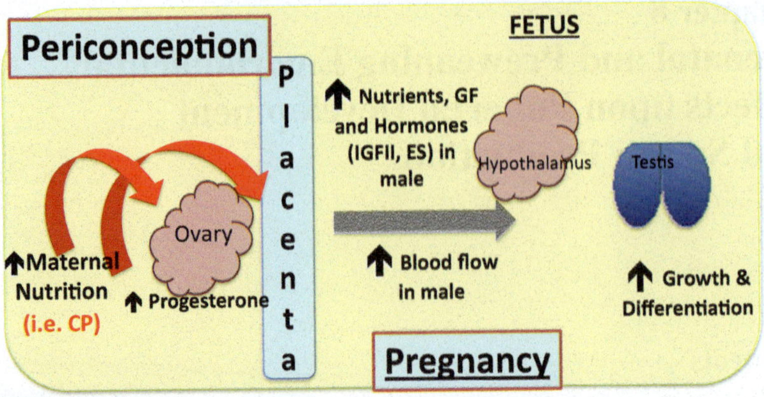

Fig. 8.1 The effects of Periconception diet in the heifer upon the developing male fetus (CP – crude protein, GF – growth factors, IGF – insulin like growth factor II, ES – Estrone Sulphate)

Concurrently, this reduced nutrition decreases conception rates, reduces progesterone levels (Copping et al. 2020b) (Fig. 8.2) and may alter the number of male calves conceived, as in other species (Rosenfeld and Roberts 2004).

Embryonic development requires extensive epigenetic remodelling of the gamete nuclei in preparation for the establishment of transcriptional networks which underpin pluripotency (Pant and Keefer 2009; van Leeuwen et al. 2015). The

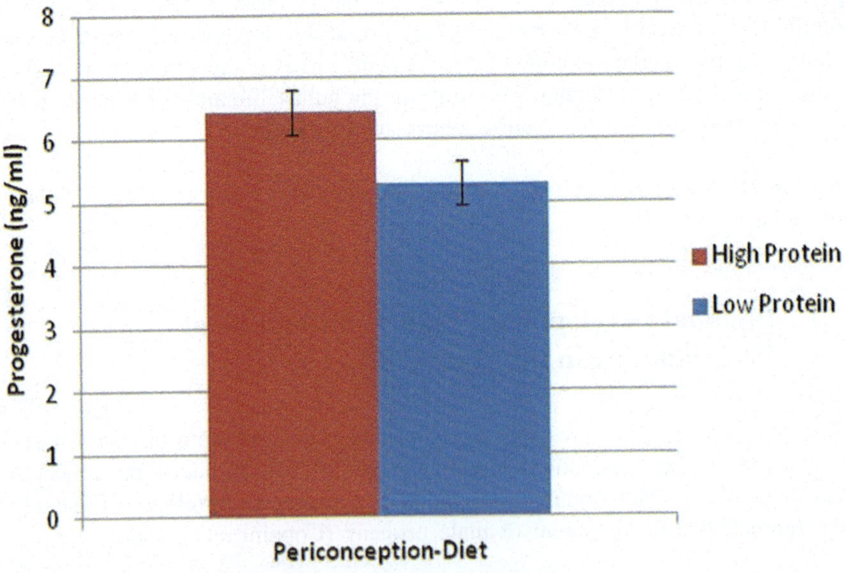

Fig. 8.2 Pregnancy rate 8% higher in high protein periconception group with progesterone higher at 23 and 36 days post conception. Adapted from (Copping et al., 2020)

bovine embryo remains transcriptionally silent until the 8 to 16 cell stage when differences in genes expressed in male and female embryos can be detected (Heras et al. 2016). It is suggested that this difference in gene expression may be associated with X linked genes and the differential time points at which these are silenced (Planells et al. 2020). Functional dimorphism between male and female bovine embryos is driven by this differential gene expression with both the blastocyst and embryo responding in a sex specific manner to fluctuating maternal nutrient status (Heras et al. 2016; Bermejo-Alvarez et al. 2010) (Micke et al. 2010). The male embryo responds to growth promoting influences by increasing cell divisions more rapidly and increasing embryonic metabolism (Bermejo-Alvarez et al. 2011a). This earlier rate of development of the male fetoplacental unit, shown in Figs. 8.3 and 8.4, may be linked to its increased vulnerability to nutritional perturbations during periconception (Copping et al. 2014). We consider the consequences of this sexual dimorphism.

The consequence of this sexual dimorphism is an epigenome that is specific to both the prevailing nutritional environment in utero and the sex of the embryo. Furthermore, the resultant offspring are programmed to cope more effectively with the same nutritional conditions postnatally as those that prevailed prenatally (Barker 1992). For example, if nutritional restriction is experienced in utero the fetus expects these same restrictions *ex utero* during both neonatal (Perry et al. 2019) and adult life (Sullivan et al. 2009b; Micke et al. 2015). When this does not occur the health and survival of the offspring may be compromised (Bermejo-Alvarez et al. 2011a).

8.2 The Sex Specific Placenta and Male Development

Concomitant with the development of the embryo, the formation of the placenta also requires a coordinated series of epigenetic modifications that precisely regulate gene expression at key developmental time points (Hemberger 2007). As the outer cell mass forms the fetal portion of the placenta, the placenta is sex specific and responds to nutritional perturbations in a sex specific manner (O'Connell et al. 2013; Copping et al. 2020a; Hernandez-Medrano et al. 2015) (Fig. 8.4). Placental hormones are secreted from the binucleate giant cells which originate in the fetal portion of the placenta, migrate across the maternal portion and secrete their contents into the maternal circulation (Wooding 1992). They modulate fetal growth by stimulating repartitioning of maternal nutrients to the fetus (Bertolini et al. 2006), their concentrations in maternal circulation being indicators of placental development (Breukelman et al. 2005; Langley-Evans and McMullen 2010). The steroid hormones, progesterone and estrogen, also contribute to placental growth, differentiation and function (Hoffmann and Schuler 2002; Reik 2007). The placenta is the major source of both hormones with estrogen being metabolised to estrone sulphate by the cotyledons in the bovine.

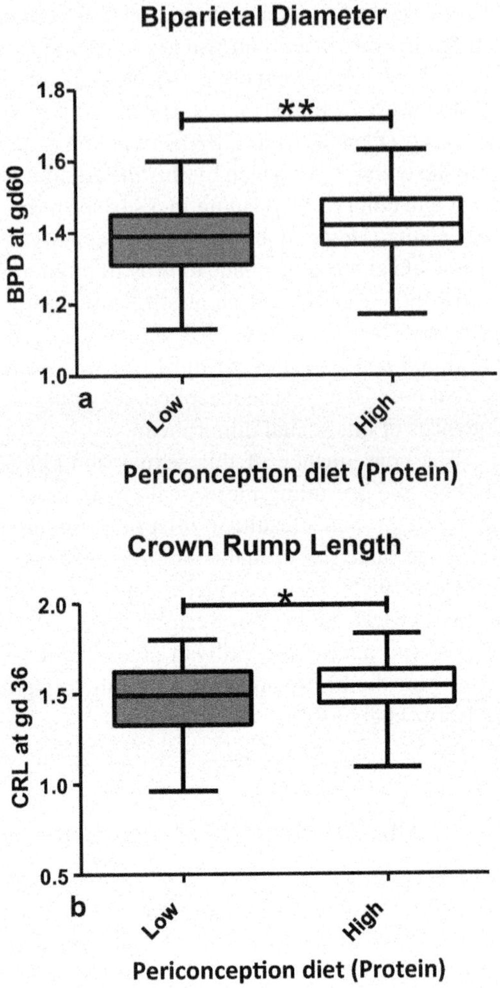

Fig. 8.3 High protein peri conception diet increased BPD at gestation day 60 (gd60) and CRL at gd36 particularly in males. Adapted from (Copping et al. 2014)

The production of placental hormones is affected by fetal gender with estrone sulphate and bovine placental lactogen production increased by a deficient maternal diet in the male fetoplacental unit only (Sullivan et al. 2009a). Similarly, sex specific effects of maternal dietary restriction upon blood flow to the placenta via the middle uterine artery were reported (Hernandez-Medrano et al. 2015) with the male fetoplacental unit being more responsive to these restrictions (Fig. 8.5 (a,b).

Consequently, studies on the gender-specific placenta which concentrate on the term (Clifton 2010; Ursula 2004) or mid gestation (Rasby et al. 1990), consider the female placenta to be responsive to changes in glucocorticoid concentration, with increases in cortisol correlating with changes in female cortisol metabolism,

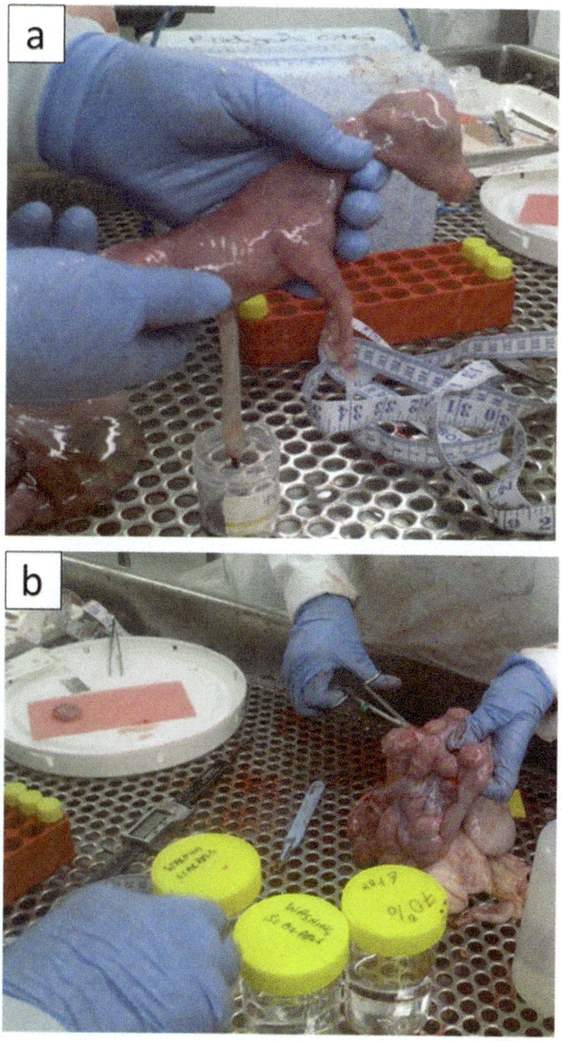

Fig. 8.4 (a) Excised fetus 98d and (b) excised 98d placenta. Reduced protein in the periconception period reduced the size of the male feto-placental unit more than the female

cytokine mRNA expression, the IGF axis, adrenal function and growth. In contrast, the male placenta appears glucocorticoid-resistant. This suggests that the male strategy for responding to an adverse maternal environment is minimalist with few gene, protein or functional changes in the placenta (Clifton 2010).

However, more recent research (Estrella et al. 2024), including our own, suggest the male displays the reverse of this phenomenon; cell divisions among male embryos occur more rapidly, and embryonic metabolism is increased and is more responsive to growth-promoting influences (Bermejo-Alvarez et al. 2011b) than in female embryos. Protein restriction affects middle uterine artery blood flow

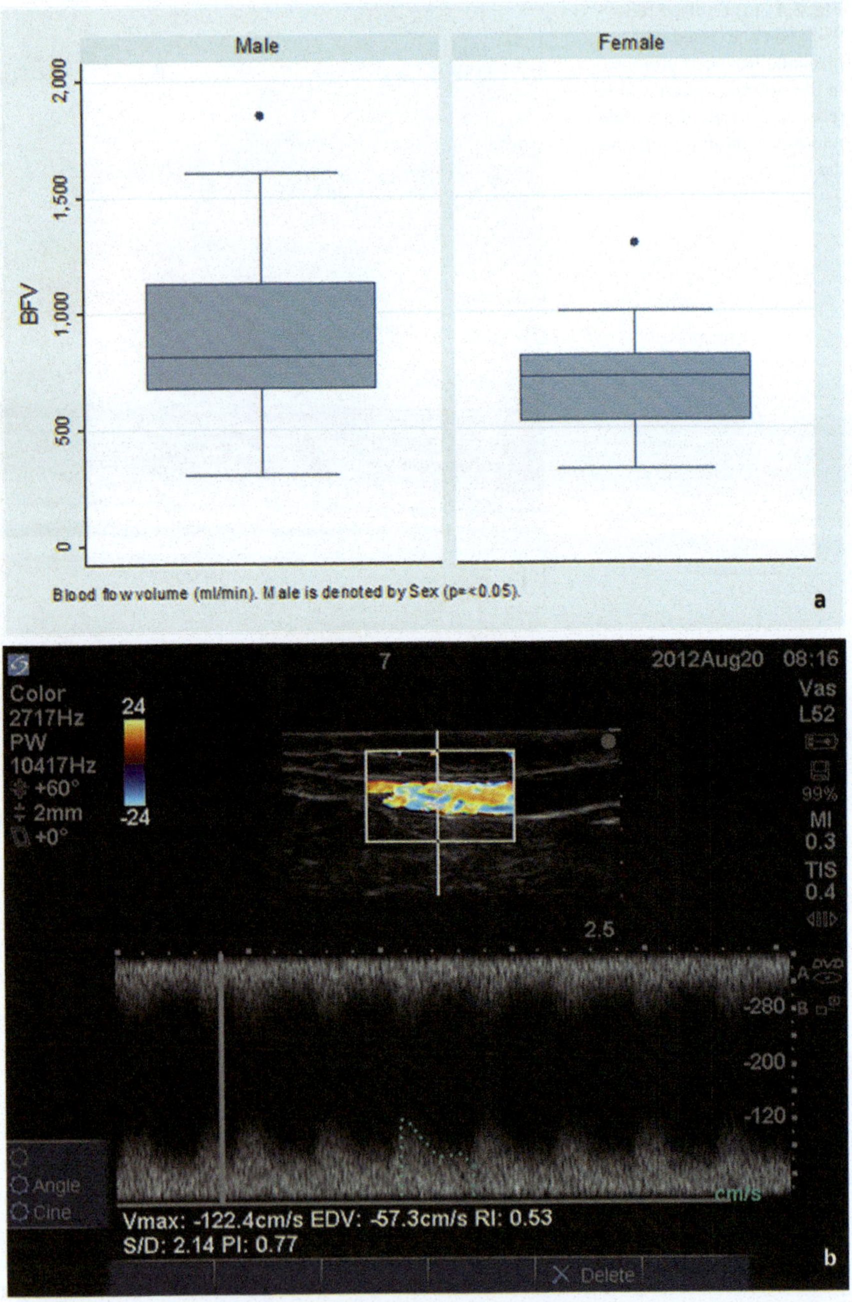

Fig. 8.5 (a) Low periconception protein increased blood flow to fetus by reducing PI (pulsatility index) and increasing TAMV (time averaged mean velocity). Interestingly, blood flow volume was shown to be significantly greater to the male fetus (b) Blood flow to the fetus: Periconception high protein increased the Pulsatility index at the middle uterine artery at 4 months and Vmax (maximum velocity), TAMV at 5 months of gestation ($p = 0.01$). Adapted from (Hernandez-Medrano et al. 2015)

to the male fetus more signicantly than in the female (Hernandez-Medrano et al. 2015), associated with more significant effects in the male on placental growth, fetal organ growth and gene expression (Copping et al. 2020b) (Fig. 8.5). We have proposed that the earlier rate of development of the male fetoplacental unit makes the male embryo more vulnerable to nutritional perturbations periconception and in the first trimester than the slower developing female embryo which seems unable to exert an influence on production of placental hormones or fetal growth until the second trimester (Sullivan et al. 2009a; Copping et al. 2020a, b). As the male reproductive tract is developing during this period of maximal response to nutritional epigenetic effects it is not surprising that we observe significant effects upon postnatal sexual development from maternal nutritional perturbations at this time (Copping et al. 2018).

Reduced protein in the second trimester increases the production of bovine placental hormones in the female fetoplacental unit (Sullivan et al. 2009a). In models which consider the mid to late term placenta in other species, the female placenta responds to an adverse maternal environment with multiple placental gene/protein changes, resulting in an adjustment in growth ensuring the survival of the female foetus in the presence of any additional adverse events which may further compromise nutrient or oxygen supply. We have reported that the female fetus, as well as appearing less affected by early restriction, was able to counteract the perturbation by compensatory growth during the second and third trimesters (Copping et al. 2020b; Hernandez-Medrano et al. 2015; Micke et al. 2015). The male was unable to do this (Copping et al. 2020a) which may partly explain the increased neonatal loss observed in male compared to female offspring.

These studies (Micke et al. 2015; Hernandez-Medrano et al. 2015; Copping et al. 2018) suggest, a critical early time period may exist where the male feto-placental unit is responsive to an adverse maternal environment. This is defined by the early ability of the male placenta to alter hormonal production to enable either cessation or instigation of fetal growth as well as the ability of the male embryo and early fetus to respond to maternal dietary perturbation.

8.3 The Embryonic Testis

The primordial germ cells arise from the yolk sac entoderm of the early embryo. These cells are amoeboid in nature and migrate to the genital ridges located on the inner surface of the dorsal body wall of the embryo. Their numbers increase in number by mitotic division while en route and after reaching the genital ridge by dpc 25 there are 1–2000 primordial germ cells on the genital ridge (McGowan et al. 2018). Several mutations are known to severely reduce the number of primordial germ cells reaching and populating the genital ridges. Steel factor and cytokine leukemia inhibitory factor (Gu et al. 2009) may be responsible for the high mitotic rate of division. The undifferentiated gonad lies adjacent to the kidney. It is first observed at dpc 28–29 and remains morphologically undifferentiated for several weeks. Three cell types occur in the genital ridge that invade the gonadal tissue under the influence of Wilms tumor-1 (WT-1). This gene is also responsible for the development of Leydig cells and with Lim1 is responsible for early gonad development (McGowan et al. 2018).

In the bull testicular hypoplasia is congenital and possibly hereditary in origin and is caused by a marked lack of spermatogonia in the gonads during fetal life. Hypoplasia may be partial, or total, due to a failure of germinal cells to develop in the yolk sac, failure to migrate to the gonad, failure to multiply in the gonad, or they undergo extensive degeneration after they have entered the gonad.

As the bull embryo develops it has three distinct renal stages: The pronephros (a non functional evolutionary remnant), the mesonephros; used during the first trimester, and the metanephros which is the final kidney. The metanephros later buds off to form the mesonephric duct. The ureter for the mesonephros is the mesonephric (Wolfian) duct which will become the epididymis. The efferent ductules (ductuli efferentes, vasa efferentia) develop from the mesonephric tubules and join the mesonephric duct (epididymis) to the sex cords (seminiferous tubules) via the rete tubules during fetal development.

Failure of the ductuli efferentes to join the rete results in blind ending ductules leading to sperm impaction in the adult. Rupture of these impacted ductuli lead to sperm granulomas. If all of the efferent ductules fail to connect to the rete, seminiferous fluid and sperm are unable to leave the testes, resulting enlarged, edematous structures which then degenerate.

The SRY gene (sex determining region on the short arm of the Y chromosome) is required for the differentiation of the fetal testis including the regression of the Mullerian ducts. Expression of SRY begins at dpc 37 and peaks at dpc 39 (Ross et al. 2009). Other sex determining genes have also been identified (Windley and Wilhelm 2015). Sox9 is the direct target gene for SRY in the mouse but in the bovine Sox10 has recently been reported as more important (Planells et al. 2020). This gene leads to the differentiation of superficial epithelium of the cords to differentiate into Sertoli cells. Sox9 controls most of the other sex determining genes in the mouse with Sox10 proposed to act similarly in the bovine (Planells et al. 2020). Sertoli cells secrete anti-Müllerian hormone, which suppresses the development of the female reproductive tract. SRY is expressed at low levels throughout life in the bovine unlike the mouse where it occurs for a short critical period. Sox9 (and perhaps in cattle Sox10) continues to orchestrate gonadal function via the Sertoli cell throughout life (Ungewitter and Yao 2013).

In the bull the the differentiation of the reproductive tract, apart from the epididymis is complete prior to the end of the first trimester. Commencement of differentiation of the male bovine gonad occurs at dpc 41 to 42 of gestation following the peak expression of SRY, with masculinization of the external genitalia occurring around dpc 47. This is driven by testosterone and androstenedione secretion from the newly differentiated fetal Leydig cells. By dpc 60 the scrotum is well differentiated with regression of the paramesonephric ducts complete by dpc 80 (Rota et al.

2002). The masculinization of the internal genitalia occurs in two distinct phases. In the first phase (dpc 56 to 58) the early buds of the seminal vesicles and prostate appear, and the bulbourethral gland begins to develop. In the second phase (from dpc70) branching of the seminal vesicles begins, differentiation of the epididymis begins, and stabilization of the mesonephric ducts occurs (Ross et al. 2009).

8.4 In Utero Environmental Perturbation of Fetal Development

This highly complex cascade of events is dependent upon a program of regulated gene expression with epigenetic marks known to alter critical components of this system. For this reason environmental perturbations prior to the end of the first trimester may have long term effects upon male progeny (Copping et al. 2018). We observed that the spermiogram of such bulls showed that percent normal sperm (PNS) was achieved at a later date in these bulls concomitant with reduced SC and increased age at sexual maturity (Copping et al. 2018) Figs. 8.6 and 8.7).

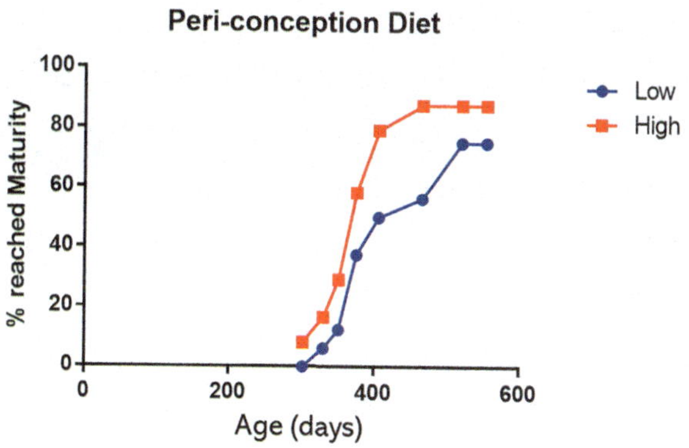

Fig. 8.6 Periconception low protein delays % of bulls reaching sexual maturity. Adapted from (Copping et al., 2018)

These effects of delayed sperm production were associated with lower levels of circulating anti-Müllerian hormone (AMH) and higher follicle stimulating hormone (FSH) concentrations (Fig. 8.8) (at 300 days of age) known to be essential for initiation and maintainance of reproductive development including the production of sperm. We propose that a high protein periconception diet may initiate this earlier elevation of FSH, in accord with the earlier downregulation of AMH, thereby enhancing the development of the hypothalamic–pituitary–gonadal axis (essential to sexual development) and increased receptivity to circulating hormones. In practice protein restriction, such as that observed in the rangelands of Australia during the breeding season, may be sufficient to initiate thes effects upon the male bovine fetus (Copping et al. 2014, 2020a, b).

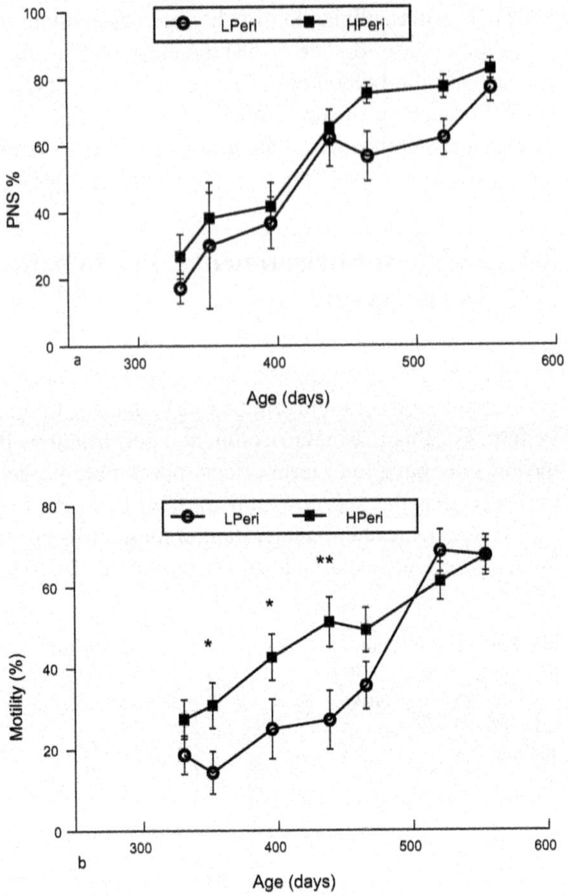

Fig. 8.7 Periconception low protein significantly delays bulls reaching 70%PNS (percent normal sperm) and progressive motility. Reproduced with permission from (Copping et al., 2018) https://doi.org/10.1071/RD17102

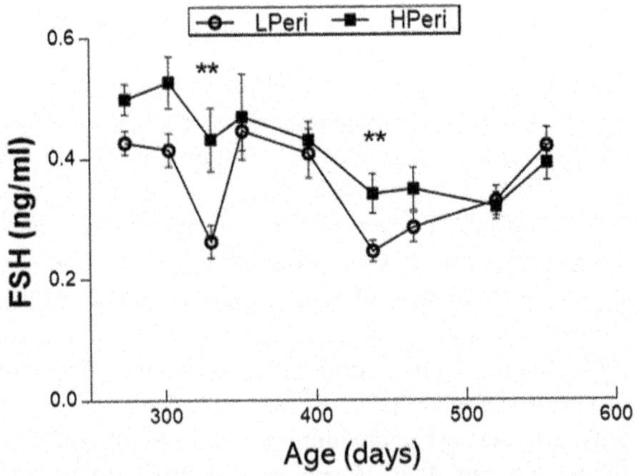

Fig. 8.8 Periconception low protein reduces FSH in the developing yearling bull. Reproduced with permission from (Copping et al. 2018)

8.5 The Effects of Nutrition During Calfhood Prior to Weaning upon Bull Development and Sperm Output

In the bovine, within the first 4 months after birth the male germ cells (gonocytes) give rise to smaller "primitive type-A spermatogonia" which sit on the basement membrane and are characterized by an elongated nucleus with fine chromatin. These primitive type-A spermatogonia remain throughout life as a source of regeneration for functional A1 spermatogonia.

In the bull calf transitory LH rise that occurs between 12–16 weeks of age which is critical to the functional development of the Sertoli cell (Rawlings et al. 2005). Poor nutrition during this calfhood period results in a reduced or delayed LH rise and is associated with delayed puberty (Brito et al. 2007a). In recent years, following Canadian and Irish studies, the dairy industry has benefited from improved feeding of housed bull calves (Byrne et al. 2018b, a). These studies showed that improved preweaning feeding hastened sexual maturity, whereas, post weaning improved nutrition did not (Barth et al. 2008; Brito et al. 2007a, b, c; Byrne et al. 2018a). This effect may be reproduced in the male progeny of range beef heifers by varying maternal protein intake during lactation (Callaghan et al. 2020a, b) (Fig. 8.9). In these protein supplemented heifers milk intake was increased in their male progeny with associated increased weight gains preweaning (Callaghan et al. 2020a). This was concomitant with increased eye muscle area in this cohort of bulls presale and importantly earlier sexual maturity; with both scrotal circumference and PNS (percent normal sperm) being increased (Callaghan et al. 2020b) (Fig. 8.10). Similarly bulls reared by primiparous dams supplemented during lactation were 1.7 times more likely to reach puberty earlier than L bulls ($P = 0.076$) when puberty was defined as attaining $SC \geq 28$ cm (Fig. 8.11).

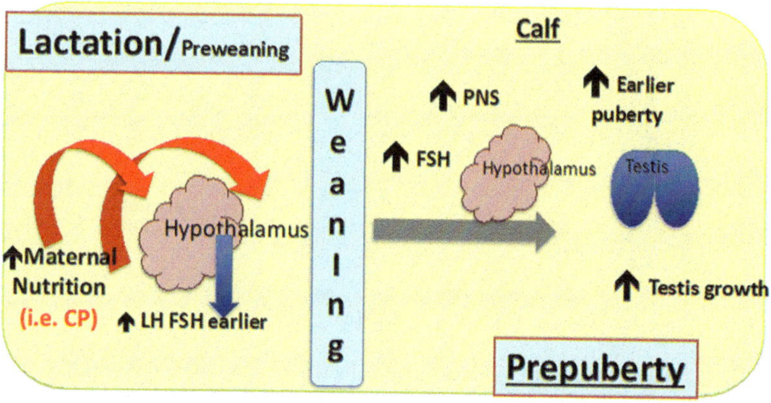

Fig. 8.9 Diet during lactation of the breeder cow affects the sexual development of the male calf. (CP- crude protein, FSH- follicle stimulating hormone, PNS- percent normal sperm, LH- luteinizing hormone)

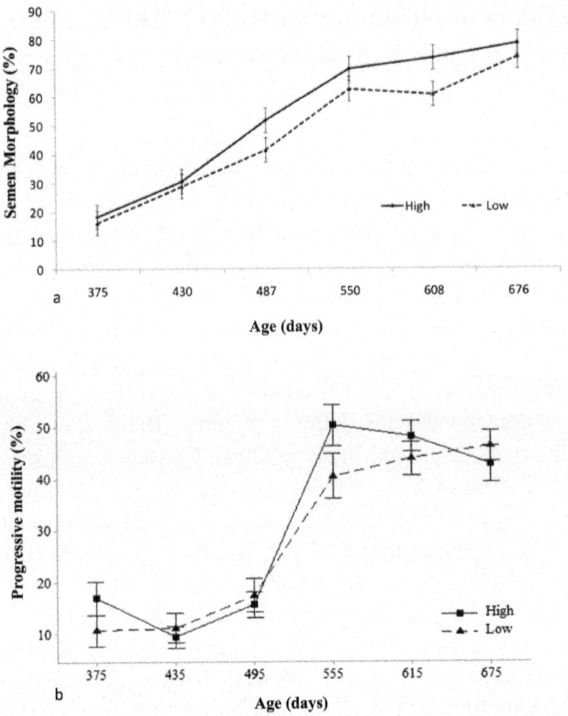

Fig. 8.10 Supplementation of the heifer during lactation increased the rate at which satisfactory semen quality was achieved in male offspring. Adapted from (Callaghan et al., 2020a, b)

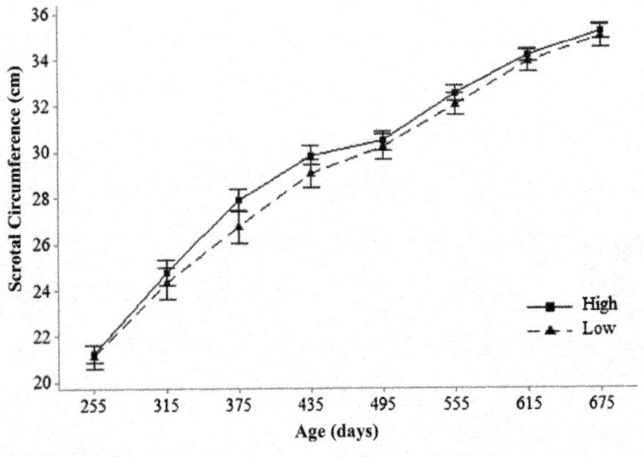

Fig. 8.11 Scrotal circumference of developing bulls whose dams were supplemented with protein (high) during lactation was significantly greater than those bulls whose dams were unsupplemented (low) at 375 days of age. Adapted from (Callaghan, 2020)

8.6 Puberty

With the onset of puberty the sex cords hollow out and form the seminiferous tubules which lengthen producing pronounced waviness and folding. At the beginning of spermatogenesis, every first generation of cells is prone to die, mainly via apoptosis, but the survivors give rise to the next higher generation. For example, A1 spermatogonia have a wave of proliferation and then most die off, survivors give rise to A2 spermatogonia. The A2 cells then make several attempts to become established, the survivors giving rise to A3's. In bulls it takes 4 to 5 months after the initial spermatogonial wave of proliferation until complete spermatozoa are produced. During this period the bull may be particularly susceptible to environmental (including dietary) perturbations upon sperm production (Cusack and Perry 1995; Callaghan et al. 2016). High energy diets are often fed to young, peripubertal bulls, to increase weight, scrotal circumference and semen output. This practice may have negative effects upon FSH and testosterone concentrations sufficient to cause disruption to sperm development (Callaghan et al. 2016). The reduction in PNS (percent normal sperm) and particularly the effects upon vacuoles observed in this study may be sufficient to preclude these bulls from passing a bull breeding soundness exam for a period of at least 90d. It would appear logical for producers of both dairy and beef bulls to initiate supplementation preweaning either to the dams in a range situation or directly to the calf in intensive rearing systems. This should enable the expression of puberty earlier without the negative effects of high energy diets in the peripubertal period.

8.7 Conclusion

We conclude that early gestation and preweaning are two periods of life in the bull which are critical to the sexual development of a bull. Both during the prenatal and preweaning stages of life dietary intervention to enhance development may lead to an economically and physiologically sound result.

Current practices, particularly in the beef industry, however, continue to try to improve bull maturity by giving high energy diets post weaning (Callaghan et al. 2016; Persson and Soderquist 2007). This practice may be deleterious to sperm output and and the long term productivity of the bull (Callaghan et al. 2016).

Compliance with Ethical Standards
All applicable international, national and institutional guidelines for the care and use of animals were followed. This current work was supported by the Natural Science and Engineering Research Council of Canada (Grant number RGPIN/05305) to Richard Oko at Queen's University, along with the Agriculture and Food Research Initiative Competitive Grant number 2023-67015-39262 from the USDA National Institute of Food and Agriculture, Animal Breeding, Genetics & Genomics Program and seed funding from the College of Agriculture, Food

and Natural Resources of the University of Missouri to Peter Sutovsky. Work for this book was also supported by a 2021–2022 Lalor Foundation Postdoctoral Research Fellowship to Lauren Hamilton. The research support to Viv Perry was funded by Australian Research Council (Grant number LP110100649), English Beef and Lamb Executive a division of the Agriculture and Horticulture Development Board (Grant number RC3815), Queensland Sperm Morphology Laboratories, Ridley AgriProducts Pty Ltd and the Australian cattle industry. The Canadian Association of Animal Breeders provided the foundation for research by Albert Barth. They supported the Development of the Diagnostic Spermatology Laboratory at the Western College of Veterinary Medicine, Saskatoon, SK., along with related support for numerous other research directives. Many grants were also received from Saskatchewan Agriculture, including the Cattle Marketing Deductions Fund, the Horned Cattle Trust Fund, the Cattle Breeders Research Council, the Beef Development Fund and the Agriculture Development Fund. Animal research approval for research or clinical studies summarized in these chapters were given by respective University Animal Care Committees. The authors declare no conflict of interest as there is no perceived interest in prejudicing the impartiality of the research reported.

References

Barker DJP (1992) The effect of nutrition of the fetus and neonate on cardiovascular disease in adult life. Proceedings of the Nutrition Society 51:135–144

Barth AD, Brito LFC, Kastelic JP (2008) The effect of nutrition on sexual development of bulls. Theriogenology 70(3):485–494

Bermejo-Alvarez P, Rizos D, Lonergan P, Gutierrez-Adan A (2011a) Transcriptional sexual dimorphism during preimplantation embryo development and its consequences for developmental competence and adult health and disease. Reproduction 141(5):563–570. https://doi.org/10.1530/rep-10-0482

Bermejo-Alvarez P, Rizos D, Lonergan P, Gutierrez-Adan A (2011b) Transcriptional sexual dimorphism in elongating bovine embryos: implications for XCI and sex determination genes. Reproduction 141(6):801–808. https://doi.org/10.1530/rep-11-0006

Bermejo-Alvarez P, Rizos D, Rath D, Lonergan P, Gutierrez-Adan A (2010) Sex determines the expression level of one third of the actively expressed genes in bovine blastocysts. Proc Natl Acad Sci 107(8):3394–3399. https://doi.org/10.1073/pnas.0913843107

Bertolini M, Wallace CR, Anderson GB (2006) Expression profile and protein levels of placental products as indirect measures of placental function in in vitro-derived bovine pregnancies. Reproduction 131(1):163–173

Breukelman SP, Perenyi Z, Ruigh Ld, Leeuw AMvW-d, Jonker FH, Vernooij JCM, Beckers JF, Weijden GCvd, Vos PLAM, Dieleman SJ, Taverne MAM (2005) Plasma concentrations of bovine pregnancy-associated glycoprotein (bPAG) do not differ during the first 119 days between ongoing pregnancies derived by transfer of in vivo and in vitro produced embryos. Theriogenology 63(5):1378–1389

Brito LFC, Barth AD, Rawlings NC, Wilde RE, Crews DHJ, Boisclair YR, Kastelic JP (2007a) Effect of feed restriction during calfhood on serum concentrations of metabolic hormones, gonadotropins, testosterone, and on sexual development in bulls. Reproduction 134:171–181

Brito LFC, Barth AD, Rawlings NC, Wilde RE, Crews DH Jr, Mir PS, Kastelic JP (2007b) Effect of improved nutrition during calfhood on serum metabolic hormones, gonadotropins, and testosterone concentrations, and on testicular development in bulls. Domest Anim Endocrinol 33(4):460–469

Brito LFC, Barth AD, Rawlings NC, Wilde RE, Crews DH Jr, Mir PS, Kastelic JP (2007c) Effect of nutrition during calfhood and peripubertal period on serum metabolic hormones, gonadotropins and testosterone concentrations, and on sexual development in bulls. Domest Anim Endocrinol 33(1):1–18

Byrne CJ, Fair S, English AM, Cirot M, Staub C, Lonergan P, Kenny DA (2018a) Plane of nutrition before and after 6 months of age in Holstein-Friesian bulls: I. Effects on performance, body composition, age at puberty, and postpubertal semen production. Journal of Dairy Science 101(4):3447–3459. https://doi.org/10.3168/jds.2017-13719

Byrne CJ, Fair S, English AM, Urh C, Sauerwein H, Crowe MA, Lonergan P, Kenny DA (2018b) Plane of nutrition before and after 6 months of age in Holstein-Friesian bulls: II. Effects on metabolic and reproductive endocrinology and identification of physiological markers of puberty and sexual maturation. J Dairy Sci 101(4):3460–3475. https://doi.org/10.3168/jds.2017-13720

Callaghan MJ, McAuliffe P, Rodgers RJ, Hernandez-Medrano J, Perry VEA (2016) Subacute ruminal acidosis reduces sperm quality in beef bulls1. J Anim Sci 94(8):3215–3228. https://doi.org/10.2527/jas.2015-0235

Callaghan MJ, Rodgers RJ, Perry VEA (2020a) Supplementation of rangeland primiparous Bos indicus x Bos taurus beef heifers during lactation. 1. Effects on dam milk production and liveweight, bull calf growth, live carcass characteristics and metabolic hormone concentrations. Theriogenology 152:69–82. https://doi.org/10.1016/j.theriogenology.2020.04.030

Callaghan MJ, Rodgers RJ, Perry VEA (2020b) Supplementation of rangeland primiparous Bos indicus x Bos taurus beef heifers during lactation. 2. Effects upon the reproductive development of bull calf progeny. Theriogenology 152:83–93. https://doi.org/10.1016/j.theriogenology.2020.04.032

Clifton VL (2010) Review: sex and the human placenta: mediating differential strategies of fetal growth and survival. Placenta 31(Suppl):S33-39. https://doi.org/10.1016/j.placenta.2009.11.010

Copping KJ, Hernandez-Medrano J, Hoare A, Hummitzsch K, McMillen IC, Morrison JL, Rodgers RJ, Perry VEA (2020a) Maternal periconceptional and first trimester protein restriction in beef heifers: effects on placental parameters and fetal and neonatal calf development. Reprod Fertil Dev 32(5):495–507. https://doi.org/10.1071/RD19017

Copping KJ, Hoare A, Callaghan M, McMillen IC, Rodgers RJ, Perry VEA (2014) Fetal programming in 2-year-old calving heifers: peri-conception and first trimester protein restriction alters fetal growth in a gender-specific manner. Animal Production Science 54(9):1333–1337. https://doi.org/10.1071/AN14278

Copping KJ, Hoare A, McMillen IC, Rodgers RJ, Wallace CR, Perry VEA (2020b) Maternal periconceptional and first trimester protein restriction in beef heifers: effects on maternal performance and early fetal growth. Reprod Fertil Dev 32:835–850. https://doi.org/10.1071/RD19149

Copping KJ, Ruiz-Diaz MD, Rutland CS, Mongan NP, Callaghan MJ, McMillen IC, Rodgers RJ, Perry VEA (2018) Peri-conception and first trimester diet modifies reproductive development in bulls. Reprod Fertil Dev 30(5):703–720. https://doi.org/10.1071/RD17102

Cusack PMV, Perry V (1995) The effect of feeding whole cottonseed on the fertility of bulls. Aust Vet J 72(12):463–466

Estrella CAS, Gatford KL, Xiang R, Javadmanesh A, Ghanipoor-Samami M, Nattrass GS, Shuaib E, McAllister MM, Beckman I, Thomsen DA, Clifton VL, Owens JA, Roberts CT, Hiendleder S, Kind KL (2024) Asymmetric growth-limiting development of the female conceptus. Frontiers in Endocrinology 14. https://doi.org/10.3389/fendo.2023.1306513

Gu Y, Runyan C, Shoemaker A, Surani A, Wylie C (2009) Steel factor controls primordial germ cell survival and motility from the time of their specification in the allantois, and provides a continuous niche throughout their migration. Development 136(8):1295–1303. https://doi.org/10.1242/dev.030619

Hemberger M (2007) Epigenetic landscape required for placental development. Cellular and Molecular Life Sciences: CMLS 64(18):2422–2436. https://doi.org/10.1007/s00018-007-7113-z

Heras S, De Coninck DIM, Van Poucke M, Goossens K, Bogado Pascottini O, Van Nieuwerburgh F, Deforce D, De Sutter P, Leroy JLMR, Gutierrez-Adan A, Peelman L, Van Soom A (2016) Suboptimal culture conditions induce more deviations in gene expression in male than female bovine blastocysts. BMC genomics 17(1):72. https://doi.org/10.1186/s12864-016-2393-z

Hernandez-Medrano JH, Copping KJ, Hoare A, Wapanaar W, Grivell R, Kuchel T, Miguel-Pacheco G, McMillen IC, Rodgers RJ, Perry VEA (2015) Gestational Dietary Protein Is Associated with Sex Specific Decrease in Blood Flow, Fetal Heart Growth and Post-Natal Blood Pressure of Progeny. PLoS ONE 10(4):e0125694. https://doi.org/10.1371/journal.pone.0125694

Hoffmann B, Schuler G (2002) The bovine placenta; a source and target of steroid hormones: observations during the second half of gestation. Domestic Animal Endocrinology- Fourth International Conference on Farm Animal Endocrinology 23(1–2):309–320

Langley-Evans SC, McMullen S (2010) Developmental origins of adult disease. Med Princ Pract 19(2):87–98. https://doi.org/10.1159/000273066

McGowan M, Holland MK, Boe-Hansen G (2018) Review: ontology and endocrinology of the reproductive system of bulls from fetus to maturity. Animal 12:s19–s26. https://doi.org/10.1017/S1751731118000460

Micke GC, Sullivan TM, Gatford KL, Owens JA, Perry VE (2010) Nutrient intake in the bovine during early and mid-gestation causes sex-specific changes in progeny plasma IGF-I, liveweight, height and carcass traits. Anim Reprod Sci 121(3–4):208–217. https://doi.org/10.1016/j.anireprosci.2010.05.017

Micke GC, Sullivan TM, Kennaway DJ, Hernandez-Medrano J, Perry VEA (2015) Maternal endocrine adaptation throughout pregnancy to nutrient manipulation: consequences for sexually dimorphic programming of thyroid hormones and development of their progeny. Theriogenology 83(4):604–615. https://doi.org/10.1016/j.theriogenology.2014.10.022

O'Connell BA, Moritz KM, Walker DW, Dickinson H (2013) Sexually dimorphic placental development throughout gestation in the spiny mouse (Acomys cahirinus). Placenta 34(2):119–126. https://doi.org/10.1016/j.placenta.2012.11.009

Pant D, Keefer CL (2009) Expression of pluripotency-related genes during bovine inner cell mass explant culture. Cloning Stem Cells 11(3):355–365. https://doi.org/10.1089/clo.2008.0078

Perry VEA, Copping KJ, Miguel-Pacheco G, Hernandez- Medrano J (2019) The Effects of Developmental Programming upon Neonatal Mortality. Veterinary Clinics: Food Animal Practice 35(2):289–302. https://doi.org/10.1016/j.cvfa.2019.02.002

Persson Y, Soderquist LE, Stina (2007) Joint Disorder; a Contributory Cause to Reproductive Failure in Beef Bulls? Acta veterinaria scandinavica 49.:31

Planells B, Gómez-Redondo I, Sánchez JM, McDonald M, Cánovas Á, Lonergan P, Gutiérrez-Adán A (2020) Gene expression profiles of bovine genital ridges during sex determination and early differentiation of the gonads†. Biol Reprod 102(1):38–52. https://doi.org/10.1093/biolre/ioz170

Rasby RJ, Wettemann RP, Geisert RD, Rice LE, Wallace CR (1990) Nutrition, body condition and reproduction in beef cows: fetal and placental development, and estrogens and progesterone in plasma. J Anim Sci 68(12):4267–4276

Rawlings NC, Bartlewski PM, Aravindakshan J, Cook SJ (2005) The relationship between secretory patterns of gonadotrophic hormones and the attainment of puberty in bull and heifer calves born early or late during the spring calving season. Anim Reprod Sci 86(3–4):175–186

Reik W (2007) Stability and flexibility of epigenetic gene regulation in mammalian development. Nature 447(7143):425–432

Rosenfeld CS, Roberts RM (2004) Maternal Diet and Other Factors Affecting Offspring Sex Ratio: a Review. Biol Reprod 71(4):1063–1070. https://doi.org/10.1095/biolreprod.104.030890

Ross DGF, Bowles J, Hope M, Lehnert S, Koopman P (2009) Profiles of Gonadal Gene Expression in the Developing Bovine Embryo. Sexual Development 3(5):273–283

Rota A, Ballarin C, Vigier B, Cozzi B, Rey R (2002) Age dependent changes in plasma anti-Müllerian hormone concentrations in the bovine male, female, and freemartin from birth to puberty: relationship between testosterone production and influence on sex differentiation. Gen Comp Endocrinol 129(1):39–44. https://doi.org/10.1016/S0016-6480(02)00514-2

Sullivan TM, Micke GC, Magalhaes RS, Martin GB, Wallace CR, Green JA, Perry VEA (2009a) Dietary protein during gestation affects circulating indicators of placental function and fetal development in heifers. Placenta 30(4):348–354

Sullivan TM, Micke GC, Perry VE (2009b) Influences of diet during gestation on potential postpartum reproductive performance and milk production of beef heifers. Theriogenology 72(9):1202–1214. https://doi.org/10.1016/j.theriogenology.2009.07.016

Ungewitter EK, Yao HHC (2013) How to Make a Gonad: cellular Mechanisms Governing Formation of the Testes and Ovaries. Sexual Development 7(1–3):7–20. https://doi.org/10.1159/000338612

Ursula M (2004) The elusive action of sex-determining genes: mitochondria to the rescue? J Theor Biol 228(3):359–365. https://doi.org/10.1016/j.jtbi.2004.02.001

van Leeuwen J, Berg DK, Pfeffer PL (2015) Morphological and Gene Expression Changes in Cattle Embryos from Hatched Blastocyst to Early Gastrulation Stages after Transfer of In Vitro Produced Embryos. PLoS ONE 10(6):e0129787. https://doi.org/10.1371/journal.pone.0129787

Windley SP, Wilhelm D (2015) Signaling Pathways Involved in Mammalian Sex Determination and Gonad Development. Sexual Development 9(6):297–315. https://doi.org/10.1159/000444065

Wooding FBP (1992) Current topic: the synepitheliochorial placenta of ruminants: binucleate cell fusions and hormone production. Placenta 13:101–113

Chapter 9
Standardization of the Assessment of Bovine Spermatozoal Abnormalities, in Relation to Fertility, as Part of the Bull Breeding Soundness Examination

Contents

Abstract Research has determined that the percentage morphologically normal sperm in bulls is highly repeatable and strongly correlated with days to conception and calf output in both dairy and beef herds. Currently, however, standardization of the assessment of morphology between laboratories is lacking, leading to inability of the practitioner to gain meaningful evaluations that relate to bull fertility.

Here we outline protocols that permit a standardized approach among laboratories that has been successfully used for 20 years. The scheme includes skill updates and training of laboratory personnel along with annual examination of their work. This results in a reliable analysis and report which is easily interpreted by practitioners and used as part of the standard BBSE.

This model has enabled increased accuracy of prognosis for practitioners and is well regarded both in Australia and overseas.

Keywords Standardization sperm morphology · Bull breeding soundness examination (BBSE) · University of Queensland Sperm Morphology Standardization Program (UQSMSP)

The original version of the chapter has been revised. A correction to this chapter can be found at https://doi.org/10.1007/978-3-031-70126-9_10

© The Author(s), under exclusive license to Springer Nature Switzerland AG 2025, Corrected Publication 2025

A. Barth et al., *Abnormal Morphology of Bovine Spermatozoa*, Advances in Anatomy, Embryology and Cell Biology 240, https://doi.org/10.1007/978-3-031-70126-9_9

9.1 Introduction

In the examination of bull sperm the PNS (percentage morphologically normal sperm) is highly repeatable and correlated with both days to conception and calf output in dairy (Al-Makhzoomi et al. 2008; Nagy et al. 2013; Attia et al. 2016)} and beef herds (Holroyd et al. 2002). This measure of fertility is, however, often lacking with ejaculates appraised only for sperm concentration and motility (Attia et al. 2016). The speed and efficiency of flow cytometry also means that it is replacing microscopy as a measure of ejaculate quality (Hossain et al. 2011).

The interpretation of morphology (and other sperm quality) assessments from laboratories is considered unreliable due to the variability of results and classification systems (Brito 2016). This study reported significant differences between 13 laboratories. This variation was facilitated by the difference in magnification of samples, staining procedures, and technician protocols. Similarly differences arise under automated computer assisted sperm morphology analysis (ASMA). These systems permit only the assessment of head measures and surface images allowing many abnormalities to remain unobserved (Auger 2010) leading to low correlations with fertility (Gravance et al. 2009).

A standard approach to PNS assessment was reported in 2006 (Fordyce et al. 2006), as part of the ACV BBSE (bull breeding soundness examination). This protocol promotes the use of central laboratories to complete PNS (Perry et al. 2002; McAuliffe et al. 2010; Perry 2021). This chapter discusses this advances in this system under the UQSMSP (University of Queensland Sperm Morphology Standardisation Program).

9.2 Discussion

The assessments of PNS using differential interference contrast microscopy (DIC), or phase contrast microscopy under × 1000 magnification is used in the Australian scheme as this has been found to yield a more accurate assessment of PNS, particularly of head abnormalities (Sekoni et al. 1981; Sprecher and Coe 1996; Al-Makhzoomi et al. 2008; Freneau et al. 2010; Palmer et al. 2013), in comparison with bright field microscopy.

Samples are collected during the BBSE crush (chute) side into isotonic neutral buffered formal saline, and sent to specialized laboratories where the wet preparations are examined by trained morphologists. Focussing on each sperm individually is vital to the accurate assessment of abnormalities at the limit of resolution. Some morphologists reduce the requirement for the refocusing by drying down the wet mount on a warm box or stage prior to rehydrating by using the same fixative under the coverslip. The Australian scheme also considers that the use of DIC, compared to phase contrast (Perry 2021) is preferable. These techniques under DIC microscopy permit abnormalities of the spermatozoon nucleus, e.g., pale centres, to be viewed without the aid of Feulgen staining, however, Feulgen is always used if a bull is to fail PNS on the pale centre abnormality.

To pass the BBSE a bull must have < 30% abnormal sperm (Wiltbank and Parish 1986; Fitzpatrick et al. 2002) in his spermiogram. This PNS threshold level is generally accepted throughout the world, however, in Australia individual thresholds for each abnormality (Perry et al. 2002) built upon the reported effect upon fertility are recorded. All defects upon individual sperm are recorded which is important in a standard program as one technician may count a different abrnormality to another. The online morphology counting system developed by the ACV (Australian Cattle Veterinarians Association) allows a single sperm to be counted with many abnormalities. Each abnormality has a number (Fig. 9.1) which are entered into the software which collates the data entered and generates a morphology report which directly feeds into the BBSE report.

Andrologists who are members of the UQSMSP Australian scheme are required to attend at an annual workshop and execute microscopy techniques which maximise both the number and randomisation of fields of view, e.g., only sperm in the middle of the field of view are counted. Normally a minimum of 100 sperm are counted per bull (McAuliffe et al. 2010), however, 200 are counted in borderline samples with (62–77) counts (Kuster et al. 2004). These procedures all reduce variability.

This system is directed by the research of Evenson (Evenson 1999) who developed a system of nomenclature whereby sperm abnormalities are divided into compensable or uncompensable traits. The female tract (Mitchell et al. 1985) and the vestments of the oocyte act as barriers to motile sperm (Saacke et al. 1998) precluding many from access to the oolemma of the oocyte. Abnormalities that prevent the sperm from either reaching or penetrating the secondary oocyte are considered compensable traits. These traits when in the presence of sufficient numbers of normal sperm in an inseminate do not reduce fertility. Compensable traits would include those that may cause abnormal or nil motility (filtered out in the female tract), and abnormal acrosomes (filtered out by the oviduct, cumulus oophorous or zona pellucida). The threshold for such abnormalities is set at 30% (Johnson 1997; Perry et al. 2002; McAuliffe et al. 2010). Uncompensable traits on the other hand are those that cause a reduction in fertility despite sufficient numbers of normal sperm in the inseminate. This is because sperm with these abnormalities are able to reach the ovum, initiate fertilization and block polyspermy but do not go on to form a viable embryo. These traits include such things as diadem defects and pyriform head. The threshold of these abnormalities is therefore higher at 20% (Perry et al. 2002; McAuliffe et al. 2010).

Members of UQSMSP follow guidelines on laboratory protocols including equipment required. The main sperm morphology categories (Fordyce et al., 2006) with threshold levels are, in order:

a. normal sperm, including aberrations considered not to affect conception rates
b. proximal droplets (PD) (20%)
c. midpiece abnormalities (MP) (30%)
d. loose heads and principal piece abnormalities (HT) (30%)
e. pyriform heads (PY) (20%)
f. knobbed acrosomes (KA) (30%)
g. vacuoles and teratoids (VT) including abnormalities of DNA condensation (20%)

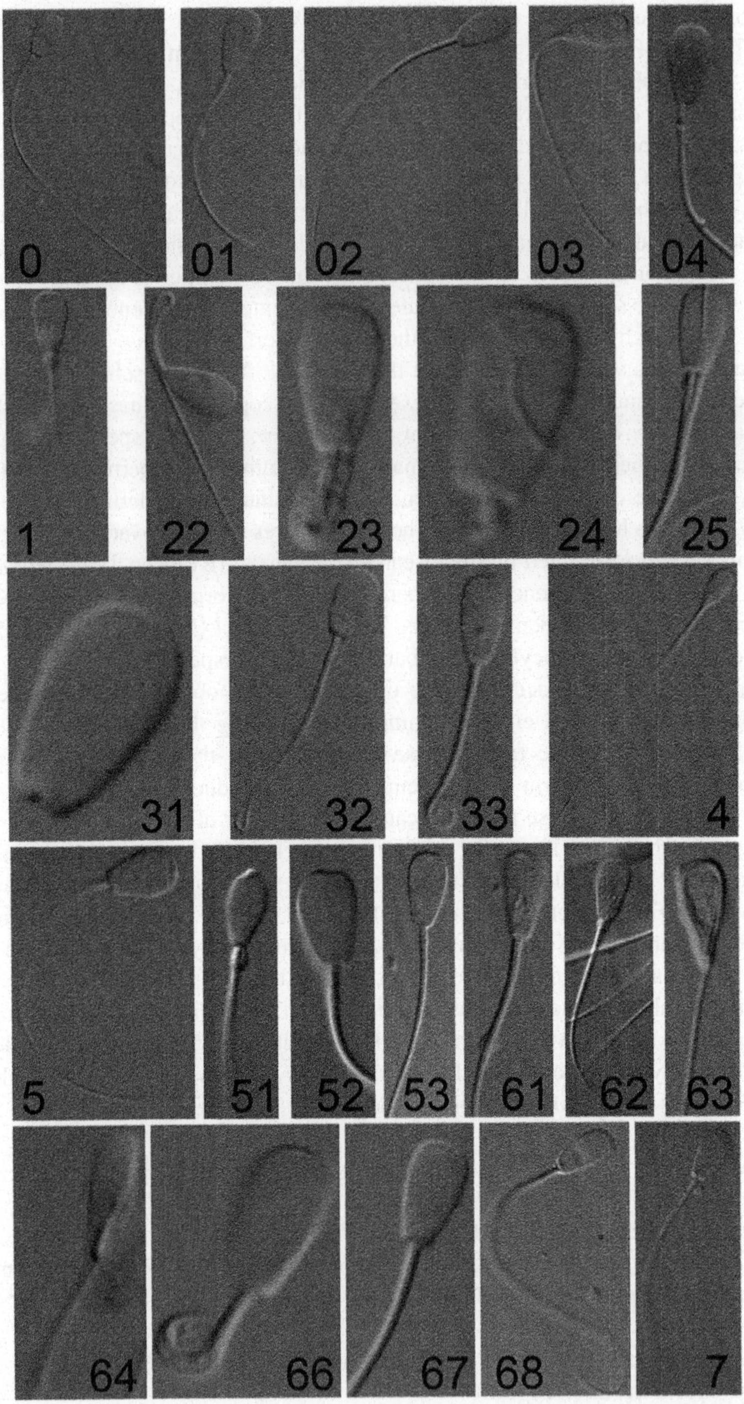
0
01
02
03
04
1
22
23
24
25
31
32
33
4
5
51
52
53
61
62
63
64
66
67
68
7

Fig. 9.1 Categorization of Sperm morphology where each abnormality is allocated a numerical value. These numbers are entered into the Bull breeding soundness morphology software which collates the number of abnormalities and creates the morphology report (Perry, 2021). The numbers applied to each category and subcategory under UQSMSP guidelines are:
0 Normal sperm category. Subcategories of Normal: 01 Distal Cytoplasmic Droplets, 02 Abaxial Tails, 03 Bent Midpieces, 04 Segmental Aplasia, 05 Slightly Pyriform Heads, 06 Narrow Heads (see Chapter 4 in *Figs. 4.5, 4.6* for clarification of slightly pyriform, narrow, tapered)
1 Proximal Cytoplasmic Droplets category
2 Midpieces Abnormalities category. Subcategories: 22 Distal Midpiece Reflex 23 Dag Defect 24 Stumped Tails 25 Multiple Tails
3 Loose Heads and Abnormal Tails category. Subcategories: 31 Detached Heads 32 Reflex Tails 33 Coiled Tails
4 Pyriform Heads category (see Chap. 4 in *Figs. 4.4, 4.5* for differentiation from milder forms which do not affect fertility).
5 Knobbed Acrosome category. Subcategories: beaded 51, indented 52, flattened 53
6 Vacuoles and Teratoids category. Subcategories: 61 Nuclear Vacuoles 62 Diadem Defect 63 Teratoid Heads 64 Rolled Crest Heads 65 Microcephalic Heads (see Chap. 4 in *Fig. 4.6*) 66 Macrocephalic Heads, 67 Single apical vacuoles, 68 Pale centres
7 Swollen Acrosome Category (These Are not Morphologically Abnormal but Are Noted in the Report as Aged Non-Functional Sperm)
N.B. Abnormalities for which there are no current number are counted and listed in comments, for example, abnormal condensation of the DNA, and considered in their appropriate category. In this case this would be the vacuole and teratoid as this is a problem of the nuclear material.

The current system in Australia includes swollen acrosome (SA) in the count as a morphological abnormality which is physiologically incorrect but is there currently to ensure these are not missed. SA is an issue of viability. The authors herein have therefore deleted this classification from the morphology abnormality categories. The authors suggest that the number of SA be counted separately and reported in the comments section of the BBSE report. This is already practiced in the assessment of frozen thawed semen evaluation (as a count of percent intact acrosomes) as it has been established that percent intact acrosome is an important measure of semen quality in frozen thawed sperm (Berndtson et al. 1981) and affects the dosage of functional sperm.

Swollen and missing acrosomes are common in aged deteriorating sperm. They do not necessarily indicate a disturbance in spermatogenesis. Whether they indicate a lack of critical proteins or osmotic imbalance in the seminal fluid is open to discussion. Swollen acrosomes can occasionally occur in high numbers and on such occasions the authors consider that they should be counted in a separate section as they are not functional sperm. They may not be picked up as immotile and therefore may be missed in the current Australian system between the practitioner's assessment crush side and referral laboratory.

These main sperm morphology categories are further divided into subcategories in the advanced ACV sheet view, which has been further supplemented by UQSMSP. For example, there is a differentiation between flat acrosomes, beaded acrosomes and indented knobbed acrosomes, and the various subcategories of nuclear vacuolation (Perry 2021). All morphologists must complete their counts by using this advanced sheet with all subcategories to enable comparisons between morphologists if any queries arise on the counts completed. UQSMSP laboratories must keep samples for 3 years and enable these samples to be made available to UQSMSP examiners when a disagreement arises. Each morphologist is required to perform competency checks on 5 samples per annum which are sent out from UQSMSP and results must be submitted back within 2 weeks. Unacceptable variation from the median results in the morphologists being sent a second round of test samples. These are to be completed with the option of additional tuition, if requested, supplied by UQSMSP.

The assessment of PNS must be completed by the andrologist with full knowledge of the stages of spermatogenesis. Similarly they must also be aware of studies that show the effects of environmental and developmental perturbations upon spermatogenesis and subsequent spermatozoon morphology. Environmental perturbations that cause elevation in circulating cortisol are enough to affect sperm morphology (Barth and Bowman 1994; Holroyd et al. 2002). The developmental process of spermiogenesis is testosterone dependent. Studies have shown that the formation of the acrosome from the Golgi apparatus, the compaction of sperm chromatin, alignment of the mitochondrial sheath, the organization of the centriole to form the flagellum and initiation of resorption of residual sperm cytoplasm are all testosterone dependent. This is discretely shown in the study by Callaghan (Callaghan et al. 2016). In this study a single subacute ruminal acidosis event was associated with elevated cortisol, reduced FSH and testosterone followed by the sequential appearance of sperm abnormalities in the subsequent weeks. A similar sequential appearance of abnormalities have been reported following testicular thermoregulation studies whether due to obesity, scrotal abnormality, climate, or fever (Barth 2013). Sequential deterioration of the spermiogram has also been reported after transport and relocation (Perry 2017) or dexamethasone treatment (Barth and Bowman 1994).

Homeothermy of the testis is therefore essential to sperm production and maintained mechanically by the cremaster and dartos muscles and the testicular vascular cone (Hees et al. 1984). The vasculate cone consists of the pampiniform plexus and the incoming testicular artery. The functioning of this structure requires a distinct scrotal neck to enable heat exchange between the venous and arterial blood. In obese animals fat in the neck of the scrotum disables efficient heat exchange. As the testes operate in a marginally hypoxic environment an increase in temperature may increase metabolic rate without a corresponding increase in blood flow. The testis tissues are therefore susceptible to hypoxia (Waites and Moule 1960). Heat itself, rather than hypoxia, has been reported to affect testicular fumction (Rizzoto and Kastelic 2020). Heat may also contribute (Felton-Taylor et al. 2020) to the reduced number of bulls passing the sperm morphology test at 70% in Far Northern tropical Australia. As may season either via temperature or nutritional intake (Entwistle and Fordyce 2003). Increased numbers of bulls with knobbed acrosome and nuclear vacuole defects have also been reported in the summer months (Felton-Taylor et al. 2020).

Prenatal (Copping et al. 2018), preweaning (Brito et al. 2007; Callaghan et al. 2020), or peripubertally (Callaghan et al. 2016, 2020) nutritional deficiencies have been shown to affect spermatogenesis. Toxic substances contained within feedstuffs may affect sperm morphology such as the gossypol in cottonseed (Chenoweth et al. 2000). Althought a supply of metallic cations may alleviate this effect by binding the gossypol in the rumen (Cusack and Perry 1995).

Environmental effects upon sperm morphology may overlap inherited conditions such as the knobbed acrosome, (Hancock 1953) and the Dag defect (Blom 1966; Barth 2013). Similarly, breed influences upon sperm morphology, have been reported in some studies (Hoflack et al. 2006) (Felton-Taylor et al. 2020) but not others (Menon et al. 2011).

Standardisation under UQSMSP: Images of each category are given in **Fig.** 9.1 with references to instructive images in Chapter 4 included.

Normal Sperm: This category includes both normal forms as well as slightly abnormal versions which have been shown not to affect fertility such as: abaxial tails, minor segmental aplasia (<3 micron), distal droplets, slightly bent midpiece, and slightly pyriform. These variations are discussed below.

Proximal Droplets (PD): These are more prevalent in the pubertal bull with frequency reducing with age (Amann et al. 2000) (Fig. 9.2). In the mature they indicate abnormal spermiogenesis and/or epididymal function. They may be observed in the ejaculate 7–10 days following a temperature or stress event

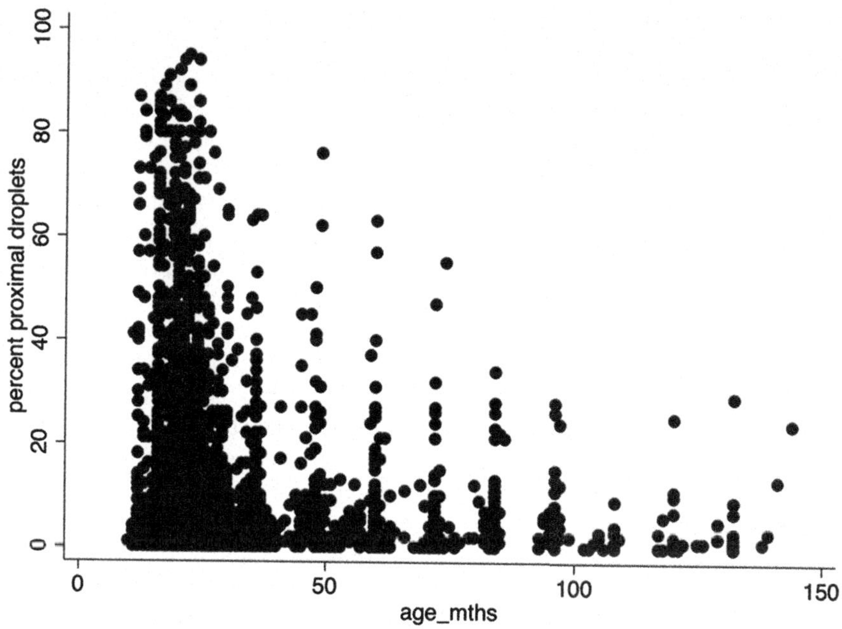

Fig. 9.2 Regression of percent of proximal droplets observed in 11,387 bulls in Australia. Reprinted with permission from (Felton-Taylor et al. 2020) https://doi.org/10.1016/j.theriogenology.2019.09.001 published Open Access CC BY 4.0. As bulls age proximal droplet numbers are reduced with the highest numbers occurring peripubertally

(Barth and Bowman 1994) and 15 days subsequent to a subacute ruminal acidosis (Callaghan et al. 2016).

The type of abnormalities associated with the proximal droplets in the ejaculate may alter the prognosis for the bull. This trait is considered uncompensable as affected sperm fail to bind to ova and furthermore morphologically normal sperm associated with high numbers of proximal droplet sperm also have impaired ability to bind with the ova (Thundathil et al. 2001a). Similarly, counts of 10–15% proximal droplets (Barth et al. 2008) have been associated with decreased fertility. (Amann et al. 2000) found that in bulls with > 30% proximal droplets, the associated, apparently normal looking sperm had a reduced ability to fertilize ova. Furthermore, after freeze thawing PDs present in semen are associated with decreased membrane integrity and increased chromatin damage post-thaw (Carreira et al. 2012). This uncompensable defect therefore has a threshold of 20% as it is associated with poor conception rates (Söderquist et al. 1991).

***Distal Droplets (DD)*:** The percentage of sperm in ejaculates with distal droplets may vary widely between sequential ejaculates (McAuliffe et al. 2010) and bulls with high numbers of distal droplets, but otherwise normal spermiograms, achieve normal pregnancy rates (Barth 2013). Distal droplets are, therefore, not considered to be a sperm defect by the author or by other researchers (Johnson 1997) and are placed in the normal category. This is dissimilar to reports in the boar where distal droplets in frozen sperm have been associated with infertility.

Development. The Sertoli cell assists in the elongation of the round spermatid from step 8 to 12 along with the exertions of the manchette (Meistrich et al. 1990). The surplus cytoplasm and organelles from this process and from the addition of the accessory components of the tail (i.e., ODF, FS and MS) form the residual body which is removed by the Sertoli cell during spermiation. In this selective removal process a stalk forms between the residual body and the connecting piece of the spermatozoon tail. When the stalk ruptures, a proximal cytoplasmic droplet is left with the spermatozoon at its head–tail interface. All sperm entering the caput epididymis have a proximal droplet. During passage through the epididymis the proximal droplet normally moves to the distal midpiece position.

During spermatozoon maturation the membrane proteins are remodeled (Jones 1989). In the case of proximal droplets, membrane proteins that should have been acquired during epididymal passage and which are needed for sperm to bind to the zona pellucida of secondary oocyte may be missing (Jones 1989). Sperm membrane proteins that are acquired from seminal plasma are needed both for distal droplet shedding and to enable the binding of sperm to the oviductal epithelium, thus establishing the sperm reservoir within the isthmus of the oviduct (Suarez 2015). This oviductal sperm reservoir is essential for fertility. Frozen boar semen with normal spermiograms but containing a high percentage of distal-droplet-sperm had impaired fertility (Waberski et al. 1994) confirming the theory that membrane-binding proteins might be deficient in such sperm. However, there are no reports of distal droplets in bulls causing reduced fertility in natural service. The use of frozen bull semen with a high percentage of distal droplets has not been reported.

As seminal fluid is critical to the availability of these sperm membrane proteins, it is important that massage of the ampullae and seminal vesicles is sufficient to illicit a quantity of seminal fluid during the semen collection process (McAuliffe et al. 2010).

***Pyriform Heads (PY)*:** Narrow in the postacrosomal region. It is very important to note, that there is variation in the degree of pyriformity in the bull: In a series of experiments (Barth et al. 1992) reported that fertility was related to the severity of pyriformity of the head. A moderate degree of pyriformity, in the absence of other signs of disturbed spermatogenesis, is not detrimental to fertility. Such sperm are considered in the normal category. However, extreme tapering in the postacrosomal region results in significant reductions in fertility (see Chap. 4 in Figs. 4.4, 4.5 for clarification). Young bulls in good body condition have a greater likelihood of recovery from this condition than older bulls. In peripubertal bulls (Felton-Taylor et al. 2020) (Fig. 9.2) this condition is prevalent. Pyriform heads are partially compensable as some pyriform head sperm have been reported to fertilize oocytes but these oocytes displayed a reduced ability to cleave (Thundathil et al. 1998a) it therefore has a threshold of no more than 20%.

Development. Sperm with pyriform heads appear about 20d after an environmental perturbation such as dexamethasone treatment or scrotal insulation (Barth and Bowman 1994), with some bulls showing a predisposition to this abnormality (Barth 2013). Pyriform heads are differentially excluded from advancement in the female reproductive tract, specifically at the cervix, uterus, utero-tubal junction (Mitchell et al. 1985).

***Knobbed Acrosomes (KA)*:** *This defect may be inherited or arise following a stress event* (Larsen and Chenoweth 1990; Callaghan et al. 2016) (Chap. 5 in Fig. 5.3)**.** It is often observed in the peripubertal bull (Copping et al. 2018) and to rise 30d after a single ruminal acidotic event (Callaghan et al. 2016) (Fig. 3).

Two forms of this defect are regularly observed; beaded and indented, however a third form; flattened, a subcategory of the indented form (Thundathil et al. 2000) is essential to differentiate as the fertility prognosis of each differ. *The beaded form* is considered inherited by an autosomal recessive gene (Hancock 1953; Barth 1986; Nogueira et al. 2022) and is often associated with sterility and occurs in a high percentage of sperm in the ejaculate. *The indented form* is described as an enlargement of the apical ridge that folds back on the apex of the sperm head resulting in a flattened appearance of the sperm apex. It is much more common than the beaded form. In the pig this form is associated with gene aberrations on chromosome 15 (Sironen et al. 2010).

In non-competitive matings bulls affected with indented or flattened KA may achieve near normal fertility. This may reflect that normal sperm coexisting with these sperm are present in sufficient numbers to achieve conception as sperm with the flattened or indented form have been reported as unable to penetrate the zona pellucida (Thundathil et al. 2000). As a compensable abnormality the threshold is therefore 30%. This threshold is confirmed by the work of Andersson (Andersson et al. 1990) who found that, when present in less than 25% of sperm there was no decrease in fertility. This effect differs from that observed in

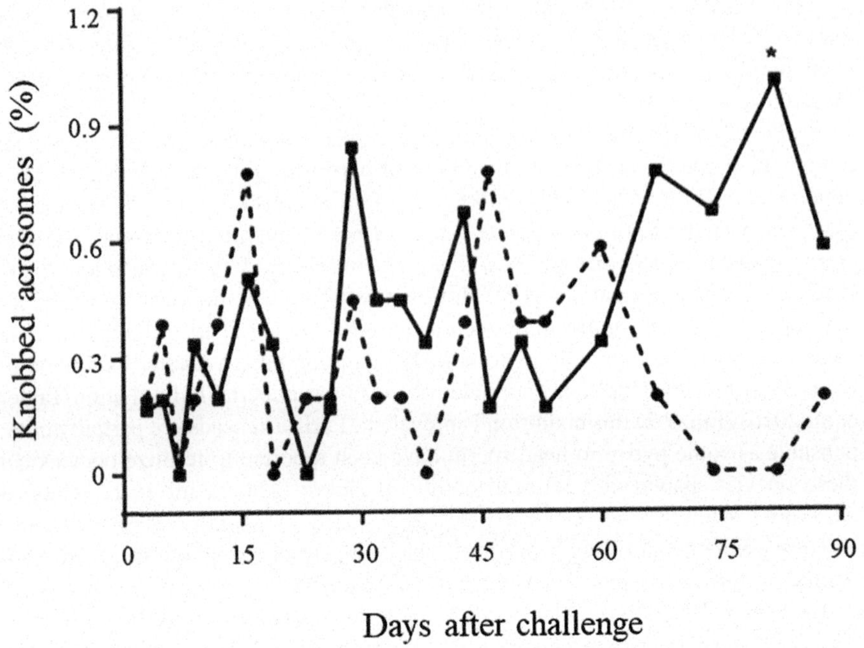

Fig. 9.3 Knobbed acrosome appearance in the ejaculate following a single induction of mild ruminal acidosis (solid line; ■) or water (dashed line; •) until 88 d after challenge. Differences between treatments at time points are shown (*$P<0.05$) Reprinted from (Callaghan et al., 2016), originally published under an Open Access CC BY 4.0 license

bulls with > 80% indented or flattened acrosomes as despite the presence of apparently normal sperm in the ejaculate which bound to the zona the resultant embryos had a reduced ability to cleave (Thundathil et al. 2001b).

Development. The presence of an intact acrosome provides assurance that the cytoplasmic membrane covering it is unimpaired and will bind to the oviductal sperm reservoir. The knobbed acrosome defect is observed from the stage 7 spermatid onward. In the pig the KA defect has been linked to genes associated with ubiquitination, a prerequisite for both chromatin remodeling, acrosome formation (Sironen et al. 2010) and embryo development (Mao et al. 2020). Interestingly, normal sperm within an ejaculate containing KA defective sperm will attain the secondary oocyte preferentially compared with sperm with the KA (Mitchell et al. 1985).

***Swollen Acrosomes (SA)*:** As swelling and sloughing of the acrosome is a normal progression during spermatozoon aging, due to delayed spermatozoon transport through the ampullae and possibly the epididymis, SA is not a morphological abnormality of the spermatozoon and the authors consider should not be included in the percent normal sperm (PNS) count. Similarly, this defect may also be caused by poor handling or osmotic damage and should be recorded when observed separately to the PNS count.

In the assessment of frozen thawed sperm, swollen acrosomes and lost acrosomes are classified under a separate percent intact acrosome (PIA) category as the PIA is recognized (Berndtson et al. 1981) as reducing the fertilizing ability of frozen sperm. Similar studies into the effects of SA, observed in the ejaculate following BBSE, upon bull fertility have not been completed. Studies have shown that increased numbers of SA sperm are associated with sperm accumulation in the bull (McAuliffe et al. 2010; Barth 2007) (see the section below on accumulator bulls under loose head). Ageing of the sperm causes the acrosome to undergo a similar reaction to capacitation resulting in the lifting of the acrosome; such sperm will be unable to attach to the secondary oocyte.

It is important to consider that swollen acrosomes are often seen in conjunction with other head abnormalities such as knobbed acrosomes. This is because the KA abnormality causes premature initiation of the acrosome reaction (Thundathil et al. 2001b). The swollen acrosome in these cases may hide the knobbed defect in initial observations. In these cases, spermiogenesis has obviously been disrupted. This defect seldom occurs in very high numbers except in accumulator bulls or when examining frozen thawed sperm that have undergone iatrogenic stress.

***Vacuoles and Teratoids (VT)*:** Nuclear vacuolation may be observed following environmental stress with evidence also of heritability of the defect, particularly in some breeds of bull (Callaghan et al. 2016). In *Bos indicus* cross bulls (Felton-Taylor et al. 2020) nuclear vacuoles are the most common defect recorded. Predisposition to this abnormality exists in some bulls, perhaps due to a hormone imbalance in the testis, after environmental perturbation (Callaghan et al. 2016). Vacuoles appeared 30d after ruminal acidosis (Callaghan et al. 2016) (Fig. 9.4) or 20d after dexamethasone treatment (Barth and Bowman 1994). Three forms of vacuolation occur: large confluent vacuoles, diadem (crater) defect, and small apical vacuoles.

Large confluent vacuoles (LCV), or craters, look under DIC as if large pieces are missing from the side of the head (see Fig. 9.1) and may appear after a stress event (Callaghan et al. 2016). Alternately this abnormality may be observed as an inherited defect (Olley 2001) in combination with smaller craters in the same ejaculate. These bulls were infertile which concurs with similar Canadian studies, which found that, at levels of>20%, pregnancy rates were reduced (Barth 2013).

Apical vacuoles are termed SAV under the Australian system (small apical vacuoles as they were initially observed in the apical position, however, they may be found throughout the head of the sperm. They are often associated with the diadem defect and may exist as multiple small vacuoles throughout the nucleus. Ejaculates with>80% SAV have reduced conception rates and in an IVF study no sperm with these vacuoles were observed inside the zona pellucida (Barth 2013). Upon evaluation of a series of ejaculates this defect unusually appears more transient than the other forms of vacuolation.

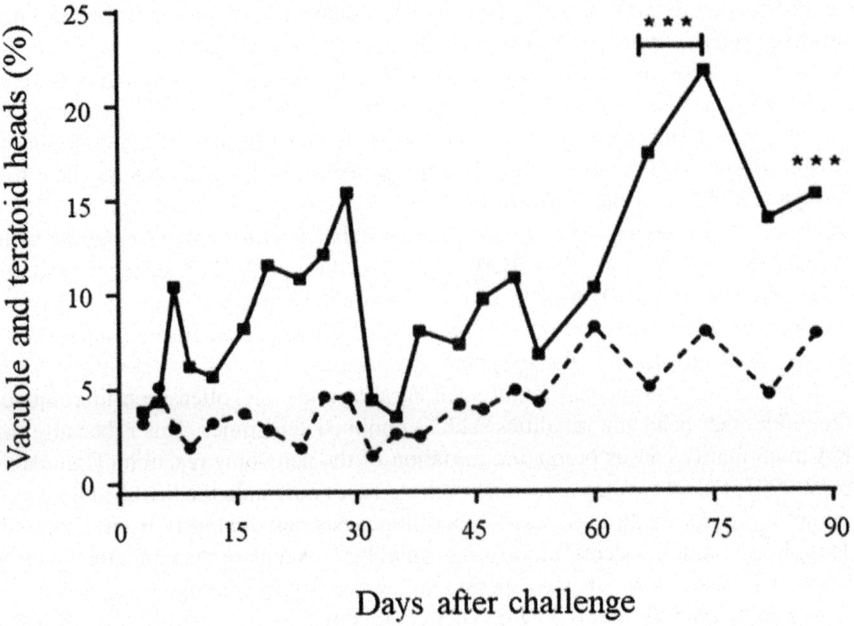

Fig. 9.4 The appearance of vacuoles after a single induction of mild ruminal acidosis in bulls (solid line; ■) or water (dashed line; •). Differences between treatments at time points are shown (*$P<0.05$; (***$P<0.001$) Reprinted from (Callaghan et al., 2016), originally published under Open Access CC BY 4.0 license

There have been many studies on the effect of the diadem defect as this is associated with severe decrease in fertility in the bull. It appears as a string of small vacuoles along the equatorial segment of the sperm, The numbers of this defect may differ between ejaculates following a stress event or remain stable (Larsen and Chenoweth 1990). A severe reduction in fertility occurs in bulls with > 60% of this abnormality in the ejaculate (Pilip et al. 1996; Thundathil et al. 1998b). Some sperm with this defect may bind to the zona and initiate fertilization, but studies have been unable to determine if this fertilization produced viable zygotes (Thundathil et al. 1998a). Further, most of the sperm with this abnormality did not reach the zona pellucida. Similarly, (Pilip et al. 1996), reported that sperm with multiple nuclear vacuoles had a reduced ability to fertilize secondary oocytes. Nuclear vacuoles are given a threshold level of not more than 20% (Thundathil et al. 1998b) reflecting the results of the studies completed to date.

***Abnormal DNA Condensation*:** Feulgen staining is a convenient method to evaluate Sperm DNA condensation by light microscopy using × 1000 phase contrast or DIC microscopy. Similarly this abnormality may be can be seen using a flow cytometric assay called the sperm chromatin structure assay (SCSA). This uses the metachromatic properties of acridine orange to measure the susceptibility of chromatin to denaturation (Barth 2013). The Feulgen and SCSA methods are highly correlated (Dobrinski et al. 1994). Advances in DIC microscopy permit detection of abnormal DNA condensation to a limited extent although Feulgen staining is completed if bulls are to fail on this abnormality alone.

***Pale Centres*:** As with abnormal DNA condensation this abnormality may be observed with Feulgen staining as the gold standard, by using phase, or DIC at × 1000. It may also be observed under excellent DIC alone at × 1000. On electron microscopy, the abnormality is seen as a localized narrowing of the sperm nucleus in the post acrosomal region (Chap. 5 in Fig. 5.14) with vesiculation of the nuclear envelope. This abnormality has been reported to cause decreased fertility in a case study in Canada (Barth pers com) and in two bulls in Australia (Perry). The abnormality may appear in conjunction with vacuole abnormalities or as the sole nuclear abnormality and is currently under study (Perry).

***Rolled Head- Nuclear Crest- Giant Head Syndrome*:** This abnormality is thought to be an inherited condition (Barth and Oko 1989). The threshold for this defect is at 20% as such sperm may penetrate the zona pellucida but do not go on to produce a viable embryo (Barth 2013). The latter study reported conception rates between 27–74% when the defect appeared at 20–30% of the ejaculate.

***Teratoid Sperm*:** The determining identification of these sperm is that they are scarcely unrecognizable as sperm. The sperm nucleus varies from normal to grossly misshapen, may be vacuolated and the tail is often coiled up completely and lies superimposed on the head. Although they are often present in 0.5 to 2 percent of sperm in normal ejaculates, their percentage may increase to 5 to 10 percent in severe disturbances of spermatogenesis. An inherited condition, somewhat similar to azoospermia in men, has been encountered in closely related Charolais bulls as well as other breeds of bulls (Barth 2013). In all these bulls there appeared few normal sperm. Teratoid sperm are tolerated at 15% in association with 70% normal sperm.

Midpiece Defects

***Distal Midpiece*:** This is the most common defect reported in *Bos taurus* bull ejaculates (Perry et al. 2002; Menon et al. 2011). The cytoplasmic droplet at the bend of the midpiece is indicative of where the abnormality occurs, i.e. the distal half of the cauda epididymis. It must involve a bending of the midpiece and not just the principal piece. A similar bending may be produced iatrogenically by hypotonic or high pH solutions but such bending does not usually contain a cytoplasmic droplet within the bend as this droplet is shed during passage though the cauda epididymis in normal sperm. This defect may occur soon after a stress event as it is found in the cauda epididymis 16 days after an acidotic event (Callaghan et al. 2016) and 4–11 days after dexamethasone treatment (Barth and Bowman 1994). This defect may improve within 16 days following initial observation.

The prognosis for recovery is dependent upon the condition of the bull and the abnormalities concurrently observed. Disturbed spermatogenesis is to be suspected if these occur in conjunction with segmental aplasia, dag-like defects or even a fracture at the tail bend. There may also be a predisposition for this defect which may change during the year. This abnormality causes the sperm to swim backwards. It is therefore a compensable abnormality with a threshold of 30%.

***Dag Defect*:** The term 'Dag defect' should be used only when the actual inherited Dag defect condition is observed. This is an inherited condition with a serious effect upon fertility (Koefoed-Johnsen et al. 1980; Barth and Oko 1989). Bulls without the inherited condition, however, may produce a small percentage of Dag-like defects during a disturbance of spermatogenesis. This displays as a few sperm resembling those seen in the inherited Dag defect are present, usually along with many other defects. This can occur in a variety of scenarios and is not normally present at>4%. That is, unlike the true Dag defect, the Dag-like defects are found only in small percentages and are mixed with a variety of different defects depending on the severity and duration of a stress for example.

The Dag and Dag-like defects must be considered a compensable trait as the sperm are not forwardly motile (Perry et al. 2002) and therefore accepted at<30% in the ejaculate with less than 70% normal sperm. However, this defect usually occurs either at extremely high levels as in the inherited condition, or in the single digits where it is present as part of a general disturbance of spermatogenesis. The presence of fractured axonemal elements, with filaments protruding from the mitochondrial sheath are characteristic of this abnormality. The most common form of Dag observed is an extrusion of one or more microtubular doublets and corresponding ODFs from the mitochondrial and fibrous sheaths, often affecting only one side of the axoneme. The half of the axoneme containing doublets 4, 5, 6 and 7 is characteristically the first and the more seriously affected portion. Sometimes, there is an extrusion of microtubular doublets from both halves of the axoneme, and generally the last microtubular doublets to disappear are 3 and 8, which maintain their connection with the fibrous sheath (FS) through the two normally occurring FS laminae. Both the fibrous sheath bundle, made up of the microtubular doublets and ODFs, and the mitochondrial sheath are most impaired in the Dag defect.

***Segmental Aplasia of the Mitochondrial Sheath and the Pseudodroplet Defect*:** Segmental aplasia even at levels of 90% may not reduce fertility (Barth 2013). This defect is therefore considered under the normal category unless the gap in the midpiece is larger than 3 microns in which case fracture of the midpiece may occur. These large gaps are recorded as a midpiece abnormality (McAuliffe et al. 2010). This defect may be transient or permanently observed in the ejaculate of a bull. Gossypol in the diet (Chenoweth et al. 2000) and a viral disease (Bovine Ephemeral fever) have been reported to result in this midpiece defect occurring in the ejaculate (Chenoweth and Burgess 1972).

The pseudodroplet abnormality was first reported more than 50 years ago (Blom 1968; Barth and Oko 1989) with a similar defect more recently reported by Rocha (Rocha et al. 2006) The defect in the Rocha report occurred more apically than that previously reported by Blom and Barth compared (Barth 2013). Furthermore, the defect reported by Rocha did not affect fertility unlike that recorded in the Blom study. The abnormality presents as a thickening of the midpiece often attendant to a bend or fracture. TEM shows an accumulation of dense granules within these thickened areas. The light microscope images of the apical defect are similar to a broken connecting piece appearance.

Other Tail Defects

***Abaxial Tails*:** Ejaculates containing 60–100% sperm with abaxial tails alone (Barth 1989) do not display lower fertility, that is, abaxial tails are counted as normal. However, if abaxial tails are seen in an ejaculate in combination with other sperm with accessory tails (Aughey and Renton 1968) this scenario can cause a significant drop in fertility. As such these are then moved to the abnormal midpiece category. Therefore, the prognosis differs dependent upon the presence of accessory tails.

This differentiation rests in the formation of the tail which is initiated by the migration of the proximal and central centriole to the nucleus base. The axoneme of the tail arises from the distal centriole. The proximal centriole forms part of the connecting piece, awaiting its role as a precursor of the zygotic centrosome after fertilization. Replication of the centrioles within spermatids is necessarily suppressed giving rise to one flagellum after the completion of second meiosis. Additional tails and/or implantation fossa(e) may, therefore, imply additional centrioles are present. The sperm centriole-contributed zygotic centrosome is critical to the separation of chromosomes during the first cleavage of the ovum. This knowledge permits the distinction between abaxial tails, even when present at low levels of 12–20%, when in combination with > 17% accessory tails, to be considered an abnormality within the ejaculate (Barth 2013). To summate: abaxial tails, with accessory tails, are counted within the midpiece defect category, however, within the normal category if present on their own.

***Tail Stump Defect*:** This abnormality is easily missed as the cytoplasmic droplet often covers the tail stump giving the impression of merely a loose head. Bulls with a high proportion of this defect are usually sterile. It is considered compensable as the sperm are immotile.

***Loose Heads /Tail abnormalities (HT)*:** If motility is low in the initial crush side assessment of the semen, then further ejaculates (up to 3) should be taken so that sperm that may have become senescent in ampullary storage can be eliminated, and a representative sample collected. When this technique is not employed a large number of loose heads may be observed. This combination of abnormalities can however also occur due to testicular degeneration or hypoplasia, inflamed ampullae or epididymitis, heat stress, etc. Discussion exists on whether detached heads should be included in the percentage of abnormal sperm, when they have normal shape and appear to be due to senescence, rather than an abnormality of testis function. However, whether they occur due to an abnormal head–tail junction should also be considered. These are compensable abnormalities due to the poor ability to swim.

Sperm accumulators: It is important to discern here the presence of a sperm accumulator as 30–40% loose heads is characteristic of this widespread condition occurring in 1% of bulls. Such bulls must be ejaculated (by EEJ or AV) to depletion (to watery semen) every second day for 7 to 10 days in order to get back to normal viability. Sperm accumulators do not recover and fail to impregnate the first 10 females they breed. They may then re-accumulate when breeding pressure becomes low. These bulls should therefore considered subfertile, rather than fertile.

This may be a place where the Australian standardization system may fail—unless counts of membrane intact sperm are carefully completed to determine sperm viability. It is important to note whether acrosomes are present in such bulls which is why swollen acrosomes are counted in the Australian system. Sperm accumulators occur at any age of bull at about 1% of the bull population. Such bulls usually have normal sperm morphology except that the sperm are dead. Their use will affect the bottom line on total weaning weight so, it is important to detect them in the semen submissions. This could be overcome by an additional crush side EN smear for ease of viability analysis although few practitioners complete both a formal saline vial and a EN smear due to time constraints.

Decapitated head: This defect is difficult to separate from loose heads unless a fresh specimen has been also observed where vigorously swimming tails with proximal droplet attached may be seen. This defect has been found to be heritable in both Guernsey and Hereford bulls and may affect 80–100% of sperm in the ejaculate (Barth 2013).

***Principal Piece/Tail Defects*:** These defects, including coiled principal pieces, and bent tails, are usually observed in low numbers. They may be caused by temperature shock or stress event and occur during passage through the epididymis. These defects are the easiest to create iatrogenically by mismanagement of temperature during collection or pH of the diluent. Levels of 30% are acceptable with 70% normal sperm as this is a compensable abnormality.

9.3 Conclusion

The sperm morphologist or andrologist should have the ability to assist in prognosis based upon their familiarity with the process of spermatogenesis together with information such as that provided by a full bull breeding soundness examination where environmental stressors are recorded (e.g., vaccination history, puberty, age, body condition, etc.). This combination of knowledge enables accuracy.

The standard of>70% normal sperm in the ejaculate is well recognized throughout the world but determining what abnormalities actually reduce fertility is less well appreciated. Recording all of the abnormalities present in the sample does not produce the best advice to the practitioner when a prognosis for a bull is needed. A system whereby only those abnormalities that are associated with infertility, and at what threshold level is required.

When the standardised Australian model (Fordyce et al. 2006) is considered, it has enabled increased accuracy of prognosis for practitioners. It is well regarded both in Australia and overseas (Irons et al. 2007; Penny 2010). The updated UQSMSP continues the standardization ethic involving skill updates and training of personnel in morphology laboratories along with annual examination of their work which should result in easily interpreted, reliable analysis.

Compliance with Ethical Standards
All applicable international, national and institutional guidelines for the care and use of animals were followed. This current work was supported by the Natural Science and Engineering Research Council of Canada (Grant number RGPIN/05305) to Richard Oko at Queen's University, along with the Agriculture and Food Research Initiative Competitive Grant number 2023-67015-39262 from the USDA National Institute of Food and Agriculture, Animal Breeding, Genetics & Genomics Program and seed funding from the College of Agriculture, Food and Natural Resources of the University of Missouri to Peter Sutovsky. Work for this book was also supported by a 2021–2022 Lalor Foundation Postdoctoral Research Fellowship to Lauren Hamilton. The research support to Viv Perry was funded by Australian Research Council (Grant number LP110100649), English Beef and Lamb Executive a division of the Agriculture and Horticulture Development Board (Grant number RC3815), Queensland Sperm Morphology Laboratories, Ridley AgriProducts Pty Ltd and the Australian cattle industry. The Canadian Association of Animal Breeders provided the foundation for research by Albert Barth. They supported the Development of the Diagnostic Spermatology Laboratory at the Western College of Veterinary Medicine, Saskatoon, SK., along with related support for numerous other research directives. Many grants were also received from Saskatchewan Agriculture, including the Cattle Marketing Deductions Fund, the Horned Cattle Trust Fund, the Cattle Breeders Research Council, the Beef Development Fund and the Agriculture Development Fund. Animal research approval for research or clinical studies summarized in these chapters were given by respective University Animal Care Committees. The authors declare no conflict of interest as there is no perceived interest in prejudicing the impartiality of the research reported.

References

Al-Makhzoomi A, Lundeheim N, Haard M, Rodriguez-Martinez H (2008) Sperm morphology and fertility of progeny-tested AI dairy bulls in Sweden. Theriogenology 70(4):682–691. https://doi.org/10.1016/j.theriogenology.2008.04.049

Amann R, Seidel G, Mortimer R (2000) Fertilizing potential in vitro of semen from young beef bulls containing a high or low percentage of sperm with a proximal droplet. Theriogenology 54(9):1499–1515

Andersson M, Vierula M, Alanko M (1990) Three types of acrosomal aberrations of bull spermatozoa and their relation to fertility. Acta Vet Scand 31(2):175–179

Attia S, Katila T, Andersson M (2016) The Effect of Sperm Morphology and Sire Fertility on Calving Rate of Finnish Ayrshire AI Bulls. Reprod Domest Anim 51(1):54–58. https://doi.org/10.1111/rda.12645

Auger J (2010) Assessing human sperm morphology: top models, underdogs or biometrics? Asian J Androl 12(1):36–46. https://doi.org/10.1038/aja.2009.8

Aughey E, Renton JP (1968) Abnormal spermatozoa in an Ayrshire bull. Veterinary Record 82:129–131

Barth A (2013) Bull Breeding Soundness. Third Edition edn. WCABP, Western Canadian Association of Bovine Practitioners, Saskatoon, Canada

Barth AD (1986) The Knobbed Acrosome Defect in Beef Bulls. Can Vet J 27(10):379–384
Barth AD (1989) Abaxial tail attachment of bovine spermatozoa and its effect on fertility. The Canadian veterinary journal = La revue veterinaire canadienne 30 (8):656–662
Barth AD (2007) Sperm accumulation in the ampullae and cauda epididymides of bulls. Anim Reprod Sci 102(3–4):238–246. https://doi.org/10.1016/j.anireprosci.2006.11.005
Barth AD, Bowman PA (1994) The sequential appearance of sperm abnormalities after scrotal insulation or dexamethasone treatment in bulls. Can Vet J 35(2):93–102
Barth AD, Bowman PA, Bo GA, Mapletoft RJ (1992) Effect of narrow sperm head shape on fertility in cattle. The Canadian veterinary journal = La revue veterinaire canadienne 33(1):31–39
Barth AD, Brito LFC, Kastelic JP (2008) The effect of nutrition on sexual development of bulls. Theriogenology 70(3):485–494
Barth AD, Oko RJ (1989) Abnormal morphology of bovine spermatozoa, 1st edn. Iowa State University Press, Iowa, US
Berndtson WE, Olar TT, Pickett BW (1981) Correlation between post-thaw motility and acrosomal integrity of bovine sperm. J Dairy Sci 64(2):346–349. https://doi.org/10.3168/jds.S0022-0302(81)82572-6
Blom E (1966) A New Sterilizing and Hereditary Defect (the 'Dag Defect') located in the Bull Sperm Tail. Nature 209:739. https://doi.org/10.1038/209739a0
Blom E (1968) A new sperm defect—"pseudo-droplets"—in the middle piece of the bull sperm. [State Veterinary Serum Laboratory], [Copenhagen]
Brito LFC (2016) A multilaboratory study on the variability of bovine semen analysis. Theriogenology 85(2):254–266. https://doi.org/10.1016/j.theriogenology.2015.05.027
Brito LFC, Barth AD, Rawlings NC, Wilde RE, Crews DH Jr, Mir PS, Kastelic JP (2007) Effect of nutrition during calfhood and peripubertal period on serum metabolic hormones, gonadotropins and testosterone concentrations, and on sexual development in bulls. Domest Anim Endocrinol 33(1):1–18
Callaghan MJ, McAuliffe P, Rodgers RJ, Hernandez-Medrano J, Perry VEA (2016) Subacute ruminal acidosis reduces sperm quality in beef bulls1. J Anim Sci 94(8):3215–3228. https://doi.org/10.2527/jas.2015-0235
Callaghan MJ, Rodgers RJ, Perry VEA (2020) Supplementation of rangeland primiparous Bos indicus x Bos taurus beef heifers during lactation. 2. Effects upon the reproductive development of bull calf progeny. Theriogenology 152:83–93. https://doi.org/10.1016/j.theriogenology.2020.04.032
Carreira JT, Mingoti GZ, Rodrigues LH, Silva C, Perri SH, Koivisto MB (2012) Impact of proximal cytoplasmic droplets on quality traits and in-vitro embryo production efficiency of cryopreserved bull spermatozoa. Acta Vet Scand 54(1):1. https://doi.org/10.1186/1751-0147-54-1
Chenoweth PJ, Burgess GW (1972) Mid-piece abnormalities in bovine semen following ephemeral fever. Aust Vet J 48(1):37–38
Chenoweth PJ, Chase CC Jr, Risco CA, Larsen RE (2000) Characterization of gossypol-induced sperm abnormalities in bulls. Theriogenology 53:11193–12003
Copping KJ, Ruiz-Diaz MD, Rutland CS, Mongan NP, Callaghan MJ, McMillen IC, Rodgers RJ, Perry VEA (2018) Peri-conception and first trimester diet modifies reproductive development in bulls. Reprod Fertil Dev 30(5):703–720. https://doi.org/10.1071/RD17102
Cusack PMV, Perry V (1995) The effect of feeding whole cottonseed on the fertility of bulls. Aust Vet J 72(12):463–466
Dobrinski I, Hughes HPA, Barth AD (1994) Flow cytometric and microscopic evaluation and effect on fertility of abnormal chromatin condensation in bovine sperm nuclei. J Reprod Fertil 101(3):531–538. https://doi.org/10.1530/jrf.0.1010531
Entwistle KW, Fordyce G (2003) Evaluating and Reporting Bull Fertility. Australian Cattle Veterinarians, Eight Mile Plains Australia
Evenson DP (1999) Loss of livestock breeding efficiency due to uncompensable sperm nuclear defects. Reprod Fertil Dev 11(1):1–16. https://doi.org/10.1071/RD98023

Felton-Taylor J, Prosser KA, Hernandez-Medrano JH, Gentili S, Copping KJ, Macrossan PE, Perry VEA (2020) Effect of breed, age, season and region on sperm morphology in 11,387 bulls submitted to breeding soundness evaluation in Australia. Theriogenology 142:1–7. https://doi.org/10.1016/j.theriogenology.2019.09.001

Fitzpatrick LA, Fordyce G, McGowan MR, Bertram JD, Doogan VJ, De Faveri J, Miller RG, Holroyd RG (2002) Bull selection and use in northern Australia. Part 2. Semen traits. Animal Reproduction Science 71(1–2):39–49. https://doi.org/10.1016/s0378-4320(02)00024-6

Fordyce G, Entwistle K, Norman S, Perry V, Gardiner B, Fordyce P (2006) Standardising bull breeding soundness evaluations and reporting in Australia. Theriogenology 66(5):1140–1148. https://doi.org/10.1016/j.theriogenology.2006.03.009

Freneau GE, Chenoweth PJ, Ellis R, Rupp G (2010) Sperm morphology of beef bulls evaluated by two different methods. Anim Reprod Sci 118(2–4):176–181. https://doi.org/10.1016/j.anireprosci.2009.08.015

Gravance CG, Casey ME, Casey PJ (2009) Pre-freeze bull sperm head morphometry related to post-thaw fertility. Anim Reprod Sci 114(1):81–88. https://doi.org/10.1016/j.anireprosci.2008.09.014

Hancock JL (1953) The spermatozoa of sterile bulls. J Exp Biol 30:50–58

Hees H, Leiser R, Kohler T, Wrobel KH (1984) Vascular morphology of the bovine spermatic cord and testis - I. Light- and scanning electron-microscopic studies on the testicular artery and pampiniform plexus. Cell and Tissue Research 237(1):31–38. https://doi.org/10.1007/BF00229196

Hoflack G, Opsomer G, Van Soom A, Maes D, de Kruif A, Duchateau L (2006) Comparison of sperm quality of Belgian Blue and Holstein Friesian bulls. Theriogenology 66(8):1834–1846. https://doi.org/10.1016/j.theriogenology.2006.05.007

Holroyd RG, Doogan W, De Faveri J, Fordyce G, McGowan MR, Bertram JD, Vankan DM, Fitzpatrick LA, Jayawardhana GA, Miller RG (2002) Bull selection and use in northern Australia. 4. Calf output and predictors of fertility of bulls in multiple-sire herds. Animal Reproduction Science 71(1–2):67–79. https://doi.org/10.1016/s0378-4320(02)00026-x

Hossain MS, Johannisson A, Wallgren M, Nagy S, Siqueira AP, Rodriguez-Martinez H (2011) Flow cytometry for the assessment of animal sperm integrity and functionality: state of the art. Asian J Androl 13(3):406–419. https://doi.org/10.1038/aja.2011.15

Irons PC, Nöthling JO, Bertschinger HJ (2007) Bull breeding soundness evaluation in Southern Africa. Theriogenology 68(6):842–847. https://doi.org/10.1016/j.theriogenology.2007.06.013

Johnson WH (1997) The Significance to Bull Fertility of Morphologically Abnormal Sperm. Veterinary Clinics: Food Animal Practice 13(2):255–270. https://doi.org/10.1016/S0749-0720(15)30339-X

Jones R (1989) Membrane remodelling during sperm maturation in the epididymis. Oxf Rev Reprod Biol 11:285–337

Koefoed-Johnsen H, Andersen J, Andresen E, Blom E, Philipsen H (1980) The Dag defect of the tail of the bull sperm. Studies on the inheritance and pathogenesis. Theriogenology 14(6):471–475

Kuster CE, Singer RS, Althouse GC (2004) Determining sample size for the morphological assessment of sperm. Theriogenology 61(4):691–703

Larsen RE, Chenoweth PJ (1990) Diadem/crater defects in spermatozoa from two related angus bulls. Mol Reprod Dev 25(1):87–96. https://doi.org/10.1002/mrd.1080250115

Mao J, Zigo M, Zuidema D, Sutovsky M, Sutovsky P (2020) NEDD4-like ubiquitin ligase 2 protein (NEDL2) in porcine spermatozoa, oocytes, and preimplantation embryos and its role in oocyte fertilization†. Biol Reprod 104(1):117–129. https://doi.org/10.1093/biolre/ioaa186

McAuliffe P, Johnston P, Johnston H, Perry VEA (2010) Ejaculators, Morphology and Microscopes. Australian Cattle Veterinarian, Barton ACT

Meistrich ML, Trostle-Weige PK, Russell LD (1990) Abnormal manchette development in spermatids of azh/azh mutant mice. Am J Anat 188(1):74–86. https://doi.org/10.1002/aja.1001880109

Menon AG, Barkema HW, Wilde R, Kastelic JP, Thundathil JC (2011) Associations between sperm abnormalities, breed, age, and scrotal circumference in beef bulls. Can J Vet Res 75(4):241–247

Mitchell JR, Senger PL, Rosenberger JL (1985) Distribution and Retention of Spermatozoa with Acrosomal and Nuclear Abnormalities in the Cow Genital Tract2. J Anim Sci 61(4):956–967. https://doi.org/10.2527/jas1985.614956x

Nagy S, Johannisson A, Wahlsten T, Ijäs R, Andersson M, Rodriguez-Martinez H (2013) Sperm chromatin structure and sperm morphology: their association with fertility in AI-dairy Ayrshire sires. Theriogenology 79(8):1153–1161. https://doi.org/10.1016/j.theriogenology.2013.02.011

Nogueira E, Tirpák F, Hamilton LE, Zigo M, Kerns K, Sutovsky M, Kim J, Volkmann D, Jovine L, Taylor JF, Schnabel RD, Sutovsky P (2022) A Non-Synonymous Point Mutation in a WD-40 Domain Repeat of EML5 Leads to Decreased Bovine Sperm Quality and Fertility. Front Cell Dev Biol 10:872740. https://doi.org/10.3389/fcell.2022.872740

Olley T (2001) Nuclear Vacuole Defects in Bovine Spermatozoa. University of Queensland, Brisbane, Honours

Palmer CW, Persson Y, Söderquist L (2013) Classification of the potential breeding ability of range beef bulls based on semen quality parameters in samples collected by transrectal massage—A comparison of the Swedish and Canadian systems. Anim Reprod Sci 140(3):124–130. https://doi.org/10.1016/j.anireprosci.2013.06.001

Penny CD (2010) Case Reports: bull pre-breeding examination. Livestock 15(7):15–18. https://doi.org/10.1111/j.2044-3870.2010.tb00321.x

Perry VE, Phillips N, Fordyce G, Gardiner B, Entwistle K, Chenoweth P, Doogan VJ Semen collection and evaluation. In: Bull Fertility: selection & Management in Australia, 2002. Australian Assoc. of Cattle Veterinarians, pp 1–19

Perry VEA (2017) Increasing standardisation in sperm morphology. Paper presented at the Australian Cattle Veterinarians, Sydney, Australia

Perry VEA (2021) The Role of Sperm Morphology Standards in the Laboratory Assessment of Bull Fertility in Australia. Frontiers in Veterinary Science 8(509). https://doi.org/10.3389/fvets.2021.672058

Pilip R, Del Campo MR, Barth AD, Mapletoft RJ (1996) In vitro fertilizing characteristics of bovine spermatozoa with multiple nuclear vacuoles: a case study. Theriogenology 46(1):1–12. https://doi.org/10.1016/0093-691X(96)00136-7

Rizzoto G, Kastelic JP (2020) A new paradigm regarding testicular thermoregulation in ruminants? Theriogenology 147:166–175. https://doi.org/10.1016/j.theriogenology.2019.11.019

Rocha A, Oliveira E, Vilhena MJ, Diaz J, Sousa M (2006) A novel apical midpiece defect in the spermatozoa of a bull without an apparent decrease in motility and fertility. A Case Study. Theriogenology 66(4):913–922. https://doi.org/10.1016/j.theriogenology.2006.02.032

Saacke RG, Dejarnette JM, Bame JH, Karabinus DS, Whitman SS (1998) Can spermatozoa with abnormal heads gain access to the ovum in artificially inseminated super- and single-ovulating cattle? Theriogenology 50(1):117–128. https://doi.org/10.1016/S0093-691X(98)00119-8

Sekoni VO, Gustafsson BK, Mather EC (1981) Influence of wet fixation, staining techniques, and storage time on bull sperm morphology. Nord Vet Med 33(4–5):161–166

Sironen A, Uimari P, Nagy S, Paku S, Andersson M, Vilkki J (2010) Knobbed acrosome defect is associated with a region containing the genes STK17b and HECW2 on porcine chromosome 15. BMC Genomics 11:699–699. https://doi.org/10.1186/1471-2164-11-699

Söderquist L, Janson L, Larsson K, Einarsson S (1991) Sperm Morphology and Fertility in A. I. Bulls. Journal of Veterinary Medicine Series A 38(1–10):534–543. https://doi.org/10.1111/j.1439-0442.1991.tb01045.x

Sprecher DJ, Coe PH (1996) Differences in bull spermiograms using eosin-nigrosin stain, feulgen stain, and phase contrast microscopy methods. Theriogenology 45(4):757–764

Suarez SS (2015) Gamete and zygote transport. Knobil and Neill's Physiology of Reproduction 1:113–145

Thundathil J, Barth AD, Palasz AT, Mapletoft RJ (1998a) Fertilization characteristics of bovine sperm with pyriform heads. Theriogenology 49(1). https://doi.org/10.1016/s0093-691x(98)90727-0

Thundathil J, Palasz AT, Barth AD, Mapletoft RJ (1998b) Fertilization characteristics and in vitro embryo production with bovine sperm containing multiple nuclear vacuoles. Mol Reprod Dev 50(3):328–333. https://doi.org/10.1002/(sici)1098-2795(199807)50:3<328::aid-mrd9>3.0.co;2-l

Thundathil J, Meyer R, Palasz AT, Barth AD, Mapletoft RJ (2000) Effect of the knobbed acrosome defect in bovine sperm on IVF and embryo production. Theriogenology 54(6):921–934. https://doi.org/10.1016/S0093-691X(00)00402-7

Thundathil J, Palasz AT, Barth AD, Mapletoft RJ (2001a) The use of in vitro fertilization techniques to investigate the fertilizing ability of bovine sperm with proximal cytoplasmic droplets. Anim Reprod Sci 65(3):181–192. https://doi.org/10.1016/S0378-4320(00)00231-1

Thundathil J, Palomino J, Barth A, Mapletoft R, Barros C (2001b) Fertilizing characteristics of bovine sperm with flattened or indented acrosomes. Anim Reprod Sci 67(3):231–243. https://doi.org/10.1016/S0378-4320(01)00127-0

Waberski D, Meding S, Dirksen G, Weitze KF, Leiding C, Hahn R (1994) Fertility of long-term-stored boar semen: influence of extender (Androhep and Kiev), storage time and plasma droplets in the semen. Anim Reprod Sci 36(1):145–151. https://doi.org/10.1016/0378-4320(94)90061-2

Waites GMH, Moule GR (1960) Blood pressure in the internal spermaticarrtery of the ram. J Reprod Fertil Suppl 1:223–229

Wiltbank JN, Parish NR (1986) Pregnancy rate in cows and heifers bred to bulls selected for semen quality. Theriogenology 25(6):779–783. https://doi.org/10.1016/0093-691X(86)90093-2

Correction to: Standardization of the Assessment of Bovine Spermatozoal Abnormalities, in Relation to Fertility, as Part of the Bull Breeding Soundness Examination

Correction to: Chapter 9 in: A. Barth et al., Abnormal Morphology of Bovine Spermatozoa, https://doi.org/10.1007/978-3-031-70126-9_9

This Chapter 9 was unintentionally published before all corrections had been made. It has therefore been updated after publication. The fundamental content was not affected.

The updated version of this chapter can be found at https://doi.org/10.1007/978-3-031-70126-9_9

© The Author(s), under exclusive license to Springer Nature Switzerland AG 2025

A. Barth et al., *Abnormal Morphology of Bovine Spermatozoa*,
Advances in Anatomy, Embryology and Cell Biology 240,
https://doi.org/10.1007/978-3-031-70126-9_10

Appendix

Formulas and Staining Methods in Light Microscopy for Sperm Examination

Isotonic Neutral Buffered Formal Saline

As mentioned in Chapter 9, ***the use of isotonic neutral buffered formal saline to examine sperm in wet mounts has become the preferred method of light microscopic examination in many laboratories. This is due to the significantly greater ability to detect head abnormalities, such as nuclear defects and deleterious effects upon the acrosome when viewed by Differential Interference (DIC) and phase-contrast Microscopy.***

To make 1L of ***isotonic neutral buffered saline***:

DiSodium Hypophosphate*[1]	Na_2HPO_4	6.1936 g
Potassium Dihydrogen*[2]	KH_2PO_4	2.542 g
Sodium Chloride	NaCl	5.406 g

Dissolve each of the three chemicals in @100 mLs of distilled water and then add to a 1L volumetric flask. Make up to 1L with distilled H_2O.

To make ***isotonic neutral buffered formal saline (BFS),*** to stop motility and for long-term sperm preservation, add 125 mL formaldehyde to the above before making up to 1 L volume (Campbell et al. 1960).

Formaldehyde (AR)	HCHO	125 mL
(Univar Formaldehyde Solution HCHO 30.03 UN No 2209 or equivalent)		

© The Editor(s) (if applicable) and The Author(s), under exclusive license to Springer Nature Switzerland AG 2025
A. Barth et al., *Abnormal Morphology of Bovine Spermatozoa*, Advances in Anatomy, Embryology and Cell Biology 240, https://doi.org/10.1007/978-3-031-70126-9

*Other Names [1] DiSodium Hydrogen Phosphate (anhydrous) / Sodium phosphate dibasic
[2] Potassium di Hydrogen orthophosphate

Phosphate Buffered Glutaraldehyde (Johnson et al. 1976)

There has been a shift away from the use of glutaraldehyde based buffered formal saline following research that showed this caused increased clumping of sperm. This method of sperm preservation may still have some useful applications.

To prepare ***phosphate-buffered saline*** (PBS), pH 7.2:

Stock	*Amount Required*
NaH_2PO_4 –8 g (add dH_2O up to 1000 mL)	30 mL
Na_2HPO_4 –9.47 g (add dH2O up to 1000 mL)	70 mL
NaCl	0.45 g
	100 mL PBS

To prepare ***0.2% glutaraldehyde in PBS***: weigh 0.4 g glutaraldehyde 50% w/w solution (add PBS up to 100 mL).

Formal Citrate (Barth 2013)

Formal citrate has similar qualities to BFS listed above and is used to stop motility and for sperm preservation. Most veterinarians have ready access to 37% formalin.

To prepare ***sodium citrate*** add 29 g of sodium citrate dehydrate to 1 L of distilled H_2O.

To prepare ***formal citrate*** add 40 mL of 37% formalin to 960 mL of sodium citrate.

Making Wet Mounts

During the BBSE examination fresh semen is added to a 1 mL vial of BFS (see formula above). The amount of semen added to the vial is dependent upon the concentration of the fresh semen at ejaculation. Sufficient semen to give the vial liquid a slightly opaque image is sufficient. At the laboratory a small drop of the diluted preserved semen is added to a clean slide. A coverslip is placed over and a drop of oil applied to the coverslip.

Frozen Thawed Semen Examination

Transfer a 2- to 3-mm diameter drop of thawed semen to the center of a clean microscope slide. A 2- to 3-mm drop of BFS is then placed beside the semen on the slide and the two drops are mixed. A coverslip is then placed over the mixture; 30–60s are needed for the drop to spread completely and for streaming motion to stop. A drop of oil is applied to the coverslip and the semen is examined at $\times$1000 DIC or phase-contrast. If the preparation is too thick, examination will be difficult because many of the sperm heads will be lying on their edges rather than flat. Each cell, even in thin preparations, is usually not totally in one focal plane and one must therefore focus up and down slightly on each cell. When examining the morphology or acrosome integrity of spermatozoa in a milk extender, it is critical for the wet mount to be very thin. Slight finger pressure on the coverslip to make the drop as thin as possible does not injure the cells and will reduce the obscurity caused by milk globules and particles.

Alternatively, add a very small drop of thawed semen to a 1 ml vial of BFS, agitate gently, and immediately add a very small drop of this mixture to a clean slide and coverslip. The fluid should not quite reach the edges of the coverslip. Add immersion oil as above and view as soon as possible after the sperm have settled. Focusing on each cell will be required as above. *For making a very thin preparation,* add a very small drop of thawed semen to a clean slide and coverslip as above. Using a fresh pipette tip gradually add a very small amount of BFS, slowly, to the edge of the coverslip slowly. You may observe the BFS migrating across under the coverslip and stop prior to reaching the far edge. Add immersion oil and examine under 1000 DIC as above.

When frozen thawed semen is examined the Percent Intact Acrosomes (PIA) is assessed separately to the morphology of the sperm. Determination of the PIA is a morphologic method of measuring post-thaw viability which is also correlated to fertility (Saacke and White 1972). It acts as a valuable addition to motility estimates in determining the viability and potential fertilizing capacity of spermatozoa since all immotile cells are not necessarily dead, whereas sperm without acrosomes are irreversibly damaged.

To examine the acrosome, cell motion must be stopped. This is best accomplished by mixing the semen with BFS which fixes the membranes and prevents further deterioration. The examination of PIA employs similarly methodology to the examination of morphology whereby: A small drop of semen is placed onto slide and cover slip applied. The amount of fluid on the slide must be sufficient so that it barely reaches the edges of the coverslip to minimize the depth of field, otherwise, examination of the acrosomes will be difficult due to constant focusing adjustments on each spermatozoon. It is particularly important to make the coverslip preparation very thin when semen is extended in milk as the milk fat globules tend to obscure the spermatozoa. A small drop of BFS is added to the edge of the coverslip, which creeps gradually across the slide immobilizing the sperm.

Acrosomes are examined at 1000 × magnification with DIC. Two hundred cells are examined, the first and second counts of 100 are compared. If these counts are within 5% of each other no further counts are necessary. Acrosome counts are done immediately after thawing the straw and after a two-hour stress test in which semen is maintained at 37C in a water bath. Minimum standards for percent intact acrosomes are 0 h = 60% intact acrosomes, 2 h = 40% intact acrosomes.

A select few staining methods, that are reliable for routine use:

Eosin-Nigrosin Stain

Hancock's stain (Hancock 1951)		***Blom's stain*** (Blom 1968)	
Eosin Y	5 g	Eosin B	5 g
Nigrosin	30 g	Nigrosin	40 g
Distilled water	300 mL	Distilled water	500 mL

To prepare both stains, add the nigrosin to the water while stirring and heating until all is dissolved. Then add the eosin to the nigrosin solution in the same fashion. Boiling should be avoided throughout the procedure.

Since these stains are quite hypotonic, artifactual formation of bent principal pieces may occur, especially if smears are not dried very rapidly. In addition, these stains should be refrigerated to prevent bacterial growth. The ***modified formula*** *below* based on a paper by (Bamba 1988) overcomes hypotonic shock and the stain may be kept at room temperature without bacterial growth.

Modified Eosin-Nigrosin Stain (Bamba 1988)

Eosin Y	6.7 g
Nigrosin	50.0 g
0.3 M Tris	100 mL
Thimol	0.1 g
dH20	600 mL

(0.3 M Tris hydroxymethyl aminomethane = 3.63 g in 100 mL)
(300 mL of 5% Glucose could be used in place of 100 mL of 0.3 M Tris)

Use a laboratory magnetic stirrer if at all possible. Gradually add the Nigrosin to 600 mL of hot water while stirring continuously. Then add in the Eosin Y and stir until all is dissolved. Continue stirring for another 30 min or more. Stir in the remaining ingredients. Filter with Whatman #1 or #2 filter paper in a Buchner Filtration apparatus over a 1L Erlenmeyer flask.

Eosin-Aniline Blue Stain (Shaffer and Almquist 1948)

Eosin B	1 g
Aniline blue	4 g
M/8 phosphate buffer	100 mL

Dissolve the stains in the buffer, which has been heated to 85 °C. Prepare the M/8 phosphate buffer in the following way:

1. Dissolve 1.702 g KH_2PO_4 in 100 mL dH_2O
2. Dissolve 1.776 g anhydrous Na_2HPO_4 in 100 mL dH_2O
3. Mix 28.5 mL KH_2PO_4 solution and 71.5 mL Na_2HPO_4 solution. The pH of this solution should be about 7.2

Stock solutions of eosin-nigrosin and eosin-aniline blue stains will keep well for several years if refrigerated. These stains may be left in dropper bottles at room temperature without harm to the staining properties for up to 6 months; however, bacteria and fungi usually will multiply in the stains and soon appear in large numbers on semen smears.

Both eosin-nigrosin and eosin-aniline blue stains give good results for routine morphology evaluations. The eosin component of these stains will penetrate the injured or broken cell membrane, causing the cell to become pink. Sperm cells with intact membranes look white against the dark background provided by the nigrosin or aniline blue components of the stain. The stain cracks in areas of the smear where it is too thick. The cracks often involve the sperm cells and to those inexperienced in examining smears, may be mistaken for an abnormality.

Smears

There are several methods for making smears. The following is recommended:

1. Place a 4–5 mm drop of warm stain near the frosted end of a warm microscope slide.
2. Place a drop of semen near the stain and mix the two on the slide. The size of the drop of semen varies with the density of the semen sample. For very concentrated semen use a drop 2 mm in diameter. For dilute and frozen semen use a larger drop-up to 6 mm in diameter.

3. To make the smear, a second slide held at a 30–40° angle is pushed against the drop of stained semen and pulled back slowly as shown below.

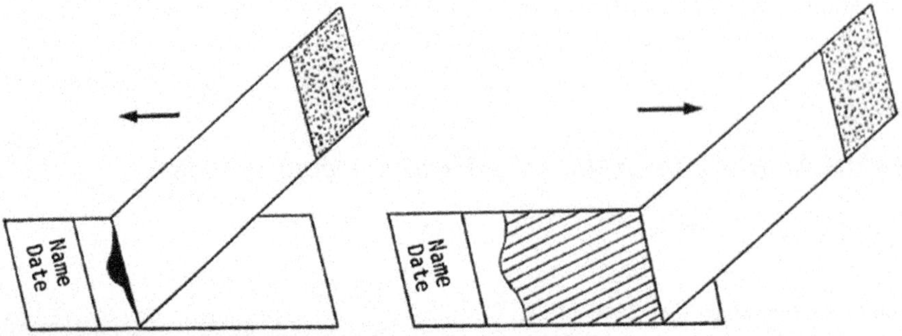

4. Dry the smear quickly by blowing air across it.

Nuclear Chromatin Condensation and the Use of Feulgen Staining

Abnormal DNA condensation cannot be detected by light microscopic examination of unstained semen preparations, or semen smears prepared with routine morphology stains. Feulgen staining overcomes this problem and shows similar results to the Sperm Chromatin Structure Assay (SCSA) indicating a decreased structural integrity of the abnormally condensed nuclei.

Other methods that have been designed to determine DNA damage include the single-cell gel electrophoresis assay (Comet), transferase-mediated dUTP nick end-labeling (TUNEL), tritium-labelled ^{3}H-actinomycin D incorporation assay, the in situ nick translation (ISNT) assay, and DNA breakage detection-fluorescence in situ hybridisation (DBD-FISH). These are basically microscopy methods, but TUNEL and Sperm Chromatin Structure Assay (SCSA) can be used in flow cytometry. Although all methods indicate DNA strand breaks, the SCSA is most commonly used.

Feulgen Stain (Feulgen and Rossenbeck 1924; Kasten 1964).

The Feulgen procedure results in a magenta staining of DNA only, revealing the sperm nucleus clearly. The acrosome and tail are not stained. The procedure involves acid hydrolysis, which unmasks aldehyde groups in the deoxyribose moiety of DNA but not the ribose moiety of RNA. The aldehyde groups are then exposed to Schiff's reagent, which links to them through a bi-sulfite bond. This results in the formation of a colored complex. The unbound reagent must then be washed away, since it oxidizes readily to produce a colored complex that is not a Schiff complex and is spurious. Schiff's reagent loses SO_2, when kept for any length of time, and the pH becomes altered. The intensity of the stain depends quite critically on the pH at which it is used. Schiff's reagent should be stored in a dark-colored bottle in a refrigerator; if crystals form or it starts to turn pink, the reagent is no longer useful.

Procedure

1. Allow semen smears to dry thoroughly to prevent loss of cells during processing
2. Place smears into 5 N hydrochloric acid for 30 min
3. Wash slides by running water slowly into the corner of a staining dish for about 30 s
4. Place smears into Schiff's reagent for 45 min
5. Wash as in step 3 and allow to dry
6. Examine at 1000 × magnification with phase contrast.

All steps are carried out in staining jars or dishes. Washing in running water can be done by allowing a steady gentle stream of water to run from a tap directly into a corner of a staining dish. Sometimes the loss of too many cells from the smear is a problem. This can be overcome by prolonging the air-drying time immediately after making the smear. When frozen semen is extended in milk, making the smear too thick may result in a greater loss of sperm cells from the slide than when a thinner smear is made.

Smears can be examined at × 1000 with bright-field, DIC or phase-contrast microscopy. Phase-contrast gives the nuclei a deeper color with greater contrast. All the nuclei will be in the same focal plane. This makes examination easier than with wet mounts using phase-contrast or DIC which may require up and down focusing on each nucleus examined.

Giemsa Stain *(Saacke 1970; Watson 1975).*

This method stains the acrosome dark purple.

Stock Solution

1. Using a pestle and mortar, grind 3.8 g of solid Giemsa stain with 375 mL of absolute methanol. Add 125 mL of glycerol and store at 37 °C for 1 week. During this time shake it for a few minutes each day.
2. A second method of preparation has been described as follows. Dissolve 1 g of Giemsa powder in 66 mL of glycerin. Mix thoroughly and store for 12 h at room temperature. Add 66 mL of absolute methanol and store at 5 °C.

Working Solution

Stock Giemsa solution	3 mL
Sorensen's 0.1 M phosphate buffer at pH 7	2 mL
Distilled water	35 mL
0.1 M Sorensen's buffer:	
KH_2PO_4 (13.609 g/L dH_2O)	17 mL
Na_2HPO_4 (14.198 g/L dH_2O)	33 mL
	50 mL

A second method of preparation has been described as follows:

Giemsa stock solution	10 drops
Double distilled water	10 ml

Working solutions must always be freshly prepared immediately before use.

Staining Method

1. Dilute fresh ejaculate in 2.9% sodium citrate 1:5–1:10 in a test tube.
2. Smear semen and air-dry quickly.
3. Fix in 5 or 10% neutral formal saline for 15 min.
4. Wash smears in running water for 20 min.
5. Stain for 90 min or longer in Giemsa working solution.
6. Rinse by dipping in 2 changes of distilled water and air-dry.

Acid-Fast Stain

Acid-fast stain colors mature nucleoplasm light blue. Examination of acid-fast-stained smears with phase-contrast microscopy improves resolution.

Method

1. Dilute fresh ejaculate in 2.9% sodium citrate 1:5 to 1:10. Thawed frozen
2. semen does not need dilution.
3. Make semen smears and allow them to dry thoroughly (at least 1 h).
4. Fix the dried smear in 10% neutral buffered formal saline for 15 min.
5. Wash smears in running water for 10 min.
6. Stain in a solution of crystal violet for 15 min.
7. Allow excess stain to drain, then dip in absolute methanol for 30 s.
8. Dip in a second bath of absolute ethanol for 60 s.
9. Differentiate in 5% acetic acid in absolute ethanol for 5 min.
10. Dip in absolute ethanol for 1 min and air-dry.

Materials

1. Crystal violet stem solution: Saturated crystal violet (approximately 14% crystal violet in absolute ethanol) should be prepared and stored in darkness. New stem solution does not work well; use older solutions.
2. Crystal violet staining solution: This working solution is prepared by mixing 1 part stem solution with 4 parts Michaelis buffer.
3. Michaelis buffer, pH 7.5:
4. 16.1 mL 0.1 N HCI
   23.9 mL 0.1 M barbital sodium (0.1 M barbital sodium is 825 mg barbital sodium + dH_2O up to 40 mL)
5. Differentiator:
   100 mL absolute ethanol
   5 mL Acetic Acid

Fluorescent Dyes for Observing Spermatozoa in Opaque Media

Van Demark (Van Demark et al. 1959) evaluated 14 fluorochromes for their usefulness in studying spermatozoa in opaque extenders by fluorescent microscopy. They concluded that coriphosphin HK, euchrysin 2 GNX, and acridine orange in that order were the most useful. We have found that acridine orange is readily available from most chemical companies and gives very satisfactory results. We have not studied the effect of these dyes on sperm viability; however, it was reported that they caused little or no reduction in the motility of bovine spermatozoa when used at low concentrations of 1:20,000–1:25,000. The motility of spermatozoa in milk extenders is very easily studied and causes less eyestrain than when motility is studied by phase-contrast or DIC. Morphology of spermatozoa is clearly visible with fluorescent microscopy, permitting study of the integrity of acrosomes. However, we found that with properly prepared wet mounts, DIC microscopy is superior for studying acrosomal integrity in milk extenders. The method for stain preparation and wet-mount preparation for fluorescent microscopy is as follows:

1. Dissolve 1 part dye in 2000 parts warm 0.9% saline. Filter and adjust to pH 6.8 with 0.1 N NaOH.
2. Add dye to semen to give a 1:20,000–1:25,000 concentration. This usually amounts to 1 drop of dye solution to 0.5 mL of semen.
3. Prepare a wet mount in routine fashion and observe with fluorescent microscopy at × 1000 in a darkened room.

Spermatozoa take up the dye very rapidly and incubation with the stain is not usually necessary, although fluorescence may improve if the semen is incubated with the dye for 15 min at 37 °C.

Acrosomal Integrity Staining Using Lectin Peanut Agglutinin (PNA)

Peanut Agglutinin (PNA) is a highly specific plant lectin that has an affinity for β-galactose-containing carbohydrates that are located within the outer acrosomal membrane.

PNA can be purchased conjugated to a wide range of fluorophores and can be used to assess acrosome integrity in live spermatozoa, through its preferential binding to acrosomes with compromised integrity. Furthermore, PNA staining is often combined with the viability dye propidium iodide (PI) to allow for detection of both viability and acrosome status simultaneously.

Materials Required

Lectin PNA (1 mg/mL)
Propidium Iodide (PI) (1 mg/mL) (Optional)
Hoechst 33,342 (1 mg/mL) (Optional)
600 uL Eppendorf tube
Glass Slide
Coverslip

Methods

1. Dilute an aliquot of live spermatozoa to a concentration between $5–20 \times 10^6$cells/mL. Spermatozoa can be diluted in warm PBS (no azide) or in a HEPES buffered media such as Sperm TL.
2. In a small Eppendorf tube add 200 μL of diluted sample.
3. Subsequently add 1uL PNA (1 mg/mL), and 1μL PI (1 mg/mL) and let incubate for approximately 10–15 min shielded from light at 38° C. Additionally, 1 μL of Hoechst 33,342 (1 mg/mL) can also be added as a nuclear counterstain if desired.
4. After time has elapsed gently mix the sample by pipetting slowly up and down before adding approximately 10 μL of sample to a clean glass slide
5. Gently add a coverslip by first lowering one side onto the sample and then slowly lowering down the rest of the coverslip to let out any air bubbles that may be present.
6. Visualize immediately.

When assessing for acrosome integrity in live cells specifically, the best accuracy is achieved through immediate assessment after staining. Fixation of spermatozoa before staining is not recommended as it may increase acrosomal membrane permeability, resulting in higher PNA fluorescence. Furthermore, wet mount staining is preferred as doing a smear and drying slides will disrupt the acrosomes, and PNA will essentially become a way of detecting the absence of outer acrosomal matrix layers rather than an indicator of acrosomes with disrupted outer membranes and intact matrices.

References

Bamba K (1988) Evaluation of acrosomal integrity of boar spermatozoa by bright field microscopy using an eosin-nigrosin stain. Theriogenology 29(6):1245–1251. https://doi.org/10.1016/0093-691X(88)90004-0

Barth AD (2013) Bull Breeding Soundness. Third edn. Western Canadian Association of Bovine Practitioners (WCABP), Saskatoon, Canada

Blom E (1968) A new sperm defect—"pseudodroplets"—in the middle piece of the bull sperm. Nord Vet Med 20:279–283

Campbell RC, Hancock JL, Shaw IG (1960) Cytological characteristics and fertilizing capacity of bull spermatozoa. J Agric Sci 55(1):91–99. https://doi.org/10.1017/S002185960002164X

Feulgen R, Rossenbeck H (1924) Mikroskopisch-chemischer Nachweis einer Nucleinsäure vom Typus der Thymonucleinsäure und die- darauf beruhende elektive Färbung von Zellkernen in mikroskopischen Präparaten. Biol Chem 135:203–248

Hancock JL (1951) A staining technique for the study of temperature-shock in semen. Nature 167(4243):323–324. https://doi.org/10.1038/167323b0

Johnson L, Berndtson WE, Pickett BW (1976) An improved method for evaluating acrosomes of bovine spermatozoa. J Anim Sci 42(4):951–954. https://doi.org/10.2527/jas1976.424951x

Kasten FH (1964) The Feulgen reaction—an enigma in cytochemistry. Acta Histochem 17:88–99

Saacke RG (1970) Morphology of the sperm and its relationship to fertility. In: Proc 3rd Tech Conf AI Reprod, Nat Assoc Anim. Breeders Chicago. p 28

Saacke RG (1972) White JM Semen quality tests and their relationship to fertility. In: Proc 4th Tech. Conf, Anim. Reprod & AI, Nat Assoc Anim. Breeders. pp 22–27

Shaffer HE, Almquist JO (1948) Vital staining of bovine spermatozoa with an eosin-aniline blue staining mixture. J Dairy Sci 31:677

Van Demark NL, Estergreen VL, Schorr R, Kuhlman DF (1959) Use of fluorescent dyes for observing bovine spermatozoa in opaque media. J Dairy Sci 42:1314–1319

Watson PF (1975) Use of a Giemsa stain to detect changes in acrosomes of frozen ram spermatozoa. Vet Rec 97(1):12–15. https://doi.org/10.1136/vr.97.1.12

© The Editor(s) (if applicable) and The Author(s), under exclusive license to Springer Nature Switzerland AG 2025

A. Barth et al., *Abnormal Morphology of Bovine Spermatozoa*, Advances in Anatomy, Embryology and Cell Biology 240, https://doi.org/10.1007/978-3-031-70126-9